Circuit Process Mechanism

Circuit Process Mechanism

Sridhar Chitta
Associate Professor
Electronics and Communication Engineering
Mahaveer Institute of Science and Technology
Hyderabad, India

I.K. International Publishing House Pvt. Ltd.
NEW DELHI • BANGALORE

Published by
I.K. International Publishing House Pvt. Ltd.
S-25, Green Park Extension
Uphaar Cinema Market
New Delhi – 110 016 (India)
E-mail: info@ikinternational.com
Website: www.ikbooks.com

ISBN: 978-93-81141-43-4

1st reprint (revised and updated) 2013

Published by Krishan Makhijani for I.K. International Publishing House Pvt. Ltd., S-25, Green Park Extension, Uphaar Cinema Market, New Delhi – 110 016 and Printed by Rekha Printers Pvt. Ltd., Okhla Industrial Area, Phase-II, New Delhi – 110 020.

Dedicated to
my father
Late Sri C.S. Suryanarayana
and
my mother
Smt. C.N. Sita

Foreword

I feel it a great privilege to introduce the book "Circuit Process Mechanism", a special book written in a style of its own covering the fundamentals of electric and magnetic fields, circuits and electronic processes in a single unit.

There are several books in Physics which deal with the principles and physics of basic electricity and magnetism. Further, at a university level, many books deal with the subject in a detailed manner with a mathematical approach. However, not many books present the real physical processes and implied aspects with suitable examples and engineering applications.

At this juncture, Professor Sridhar Chitta has come out with a lucid presentation of the underlying physical phenomenon and application of electricity and electromagnetic field theory.

Professor Sridhar has vast experience as a practicing instrumentation and controls engineer and has also wide experience in electrical engineering in general. His rich experience, wide knowledge-base and practical touch have been thoroughly used in the writing and development of this book.

He has combined textual descriptions with several illustrative figures full of details and included animated power point slides covering every pertinent principle and phenomenon that will aid readers in visualizing circuit processes. Emphasis has been laid on dealing with a few fundamental topics like "Conduction process" (Chapter 1), electromagnetic fields and its application to electrical power (Chapters 5 and 7), electronic processes of capacitors and semiconductors (Chapters 2, 3 and 9).

The author has given an in-depth and lucid presentation for an applied physicist or engineer to get an accurate and correct viewpoint of these very fundamental electric and magnetic processes. Every case has been dealt so that a present-day electrical and electronic engineer understands these physical and applied aspects at the first level before he/she studies topics of electromagnetic field theory and electrical circuits or networks.

A unified treatment of electrostatics and electric circuits is comprehensively covered. The conventional and basic electrical or electronic engineering textbooks do not touch upon these details and this book fulfils this need.

While "Electric and Magnetic Interactions" by Ruth W. Chabay and Bruce A. Sherwood is a primary book that deals with these principles and processes (which is of an advanced level) this book by Professor Sridhar Chitta forms an excellent fundamental level book on the unified approach especially with respect to ac circuits.

The author has spent more than four years conducting a survey and in-depth search for related literature on the contents. With critical inputs, suggestions and encouragement from renowned Professors Hermann Haertel, Bruce A. Sherwood, David J. Griffiths, Norris W. Preyer, Satyabrata Jit and Dipankar and several other experts in this important field, this book becomes unique by itself.

I feel that anyone who wishes to have a clear and basic idea of the conceptual and physical basis of electricity, magnetism and electric and magnetic fields should read the book. The book can be used for basic electrical and electronic courses, both as a text and a reference.

I congratulate the author in taking a bold step for presenting a tough and difficult part of electrical science in a natural and simple manner to the reader.

I wish all success for the endeavour.

Professor (Dr.) V. Kamaraju
Principal and Professor (Retd.),
Dept. of Electrical Engineering,
JNTU College of Engineering,
Kakinada, A.P., India

Preface

Most text books presently available neither describe accurately nor precisely the mechanism of the processes involved in the operation of electrical and electronic circuits. Students are deprived of an opportunity to learn the processes at a critical juncture in their University education. In present day University syllabi, in the topics of electrical and electronics subjects, the electric field plays little or no role in circuit analysis. Students therefore, lose an important opportunity for additional practice with electric field, and by the end of the discussion of circuits, students' understanding of the field concept may have faded.

It is with an intention to remedy this situation that I have prepared this book that could serve as a textbook on the physical and applied aspects of the principles of electricity and magnetism. I began with the preparation of power point slides with animations and then, based on them, ventured to prepare this book that could have a more universal reach to students of engineering and science. The concepts are presented combining text, diagrams depicting circuit processes with states evolving in time, and mathematics to create a correct concept.

The contents of the chapters are aimed to develop a strong sense of feel for the *"mechanism"* of electric and electromagnetic phenomenon and focus on the role of *surface charges* in conduction, the role of *electric and magnetic fields* in circuit processes and the *transport of energy* in circuits.

The electric and magnetic processes are described based on the concepts in "Electric and Magnetic Interactions" by Ruth Chabay and Bruce Sherwood (John Wiley & Sons, 1995)(there now exists a later version, the second volume of their textbook "Matter and Interactions", John Wiley & Sons, 3rd edition, 2011) and several other sources.

Keeping in view the requirements of undergraduate Electronics, Electrical, Electronics and Communication, and Instrumentation engineering students and the common misconceptions they have about electrical and electronic circuit processes, the extent and nature of the contents of the book have been designed to provide a preliminary exposure before they learn more advanced topics in electromagnetic field theory and electrical circuits or networks.

Some of the common student misconceptions I have addressed are:

- *the electric field inside a metal is always zero (even when the system is not in static equilibrium);*
- *drifting electrons push each other through a wire just as water molecules push each other through a pipe (despite charge neutrality inside the metal);*
- *there cannot be any potential difference across an open switch because $V = IR$, and there is no I;*
- *electrons travel at the speed of light in wires; and*
- *a circuit requires two separate wires for operation.*

Numbered references in the text are listed after the last Appendix.

The CD included with this textbook contains powerpoint presentations with animations covering visualizations of circuit processes described in the text and is organized in chapters conveniently. The animations cover several circuit processes and those related to capacitive and inductive circuits subject to sinusoidal voltages are treated in vivid detail.

Sridhar Chitta,
Hyderabad, A.P. India

Acknowledgements

This book would not have been possible without the textbook of Professors Ruth W. Chabay and Bruce A. Sherwood titled *Electric and Magnetic Interactions*. Even as I began reading their monumental work, I realized that I had found the book that would guide me through the fundamentals of electricity and magnetism from an atomic viewpoint in a very comprehensive manner. When I found the going difficult, as in the concept of the electric field of charged particles and their interactions, I contacted Professor Sherwood and he guided me through detailed emails to my queries through very sticky patches with patience and understanding, often late in the night. Words will fail to express the gratitude I owe him for the encouragement he gave me to continue my writing during this very difficult phase in the preparation of the content.

The transport of energy in circuits is tangled heavily with the presence of closely linked electric and magnetic fields and the source charges and currents and here I have borrowed from Professor David Griffiths' *Introduction to Electrodynamics*. The lucid and humour laced descriptions in his text with a simplified treatment of the complex mathematics involved are, in my opinion, unmatched in most textbooks that treat this topic. I owe him gratitude for replying to my queries on the energy stored in charged particles and permanent magnets.

William Beaty's website amasci.com was one among the first internet resources I browsed and it led me to believe that there was a missing link in the dry textbook treatments of the topic of electricity. After reading "Misconceptions spread by K-6 textbooks", I mailed him a query about how there is a voltage on a single wire connected to a battery and, the explanation he gave in an email made me realize that all the concepts of electricity I had at that time needed to be reviewed. I followed up his suggestion to refer to Chabay and Sherwood's articles and textbook for which I express my sincere gratitude. Thus, I began my voyage of discovery of electricity and magnetism beginning right from the atomic level and the field level impact on the behaviour and operation of simple circuits.

I gratefully acknowledge Professor Norris W. Preyer for permitting me to use images of computer simulation results from his own files showing charge movement and potential variations during the very initial transient periods of a capacitor when it is discharged and for editing the textual descriptions.

I also express my sincere thanks to Professor (Dr.) Dipankar of IIT, Mumbai who gave me valuable insights on the variation of the potentials of objects when brought near charged objects from an energy point of view.

Atomic level descriptions of what happens when an object is touched is a very difficult topic and I express thanks to John M. Clement and John Denker of Physics Forums hosted by Buffalo University for the detailed inputs on this topic. I have included these descriptions to provide a continuum to the descriptions of the average field inside matter and to enable students gain an intuitive insight into what happens when voltmeter probes are touched on conductors when making voltage measurements.

Professor Satyabrata Jit, co-author of *Electronic Devices and Circuits*, gave me valuable comments and suggestions on the descriptions of electric field variations within diodes and suggested modifications to the figures in the chapter on diodes and I am deeply indebted to him for this.

Professor (Dr.) V. Kamaraju who has co-authored *High Voltage Engineering* along with Professor (Dr.) M.S. Naidu, helped me edit content related to the potentials of objects and the energy of electrons and was a pillar of strength while I was preparing the manuscript and I am deeply indebted to him for his patience and understanding in answering my queries.

I express sincere thanks to Professor (Dr.) R.L. Narasimham, author of *Analysis of Linear Control Systems*, and who is my mentor and guide towards my research in the area of large-scale control systems, for his continued support and encouragement in the preparation of the manuscript.

I express my sincere thanks to Ms. K. Sreelatha who read the manuscript and assisted me in editing the text at several places thereby enriching it.

I am thankful to my wife Mayura and son Sudeendra who were patient and understanding while I was preparing the manuscript and the slides over the past few years, even as I could spend very few evenings and holidays with them.

Sridhar Chitta

Contents

1

Surface Charges and Conduction Processes

Chapter 1 introduces *surface charges* broadly and their role in enabling the conduction process. Surface charges maintain the potential around conductors and provide the electric field according to Hermann Haertel, a scientist at the Institute for Science Education, University of Kiel.

The chapter begins by introducing the need for electric fields to be present in conductors to enable the conduction process and the manner in which this is achieved by the natural phenomenon of surface charge development and their distribution. The idea of charge separation that produces voltages and emfs is also presented followed by a description of the conduction mechanism in conductors and resistors.

The next section is devoted to developing a visualization of current flow in resistors and continues with sample calculations of the potential profiles inside the wires and resistors. The chapter concludes with a description of the energy processes involved in conduction inside wires and resistors.

1.1 SURFACE CHARGES

Excerpt from a "A Unified treatment of electrostatics and circuits" by Ruth W. Chabay and Bruce A. Sherwood[1].

Ohm's (German physicist, 1789-1854) steady-state stationary charge distribution consisted of a gradient of volume charge *inside the conductor*, whereas on the basis of what was known about electrostatics there could *not* be a volume charge *inside a* conductor.

"Kirchhoff resolved the remaining fundamental issues, not just with his loop and node rules but more importantly by *unifying electrostatics and circuits*."

Kirchhoff (German physicist, 1824-1887) put the driving charge gradient on the *"surface"* of the conductors and showed that electrostatic and circuit phenomena belonged to one science, not two.

Unfortunately, this unification was later lost sight of as the role of potential came to dominate the analysis of circuits and, the ***surface charges.........***

.....disappeared from view.

The above short introduction to the history of developments in the study of surface charge phenomena was made to put into perspective some of the reasons why the topic is not treated in modern-day textbooks notwithstanding its importance in enabling the fundamental process of electrical conduction.

Charges will move when there is a force acting on them as in the presence of an electric field, and they will continue to move unless acted upon by opposing forces such as:

(i) those that impede their motion due to collisions with the vibrating lattice of atomic cores in metals or;

(ii) electric fields produced by charges that accumulate on the surfaces of capacitor plates or;

(iii) Non-coulomb electric fields (Faraday's law) or fields due to the motion of conductors in steady magnetic fields.

We will learn more about these opposing *fields* (ii and iii) and how these are produced in subsequent chapters.

1.2 CURRENT

Moving charges constitute *a current.*

Current strength in conductors is determined by the charge (electron) collisions with vibrating atomic cores (or lattice ions, that we shall describe later) that physically resist them and by other electric fields that oppose the driving electric fields of the moving charges (electrons).

Such opposing electric fields are produced by "accumulated" charges on the surfaces of capacitor plates (*fringe fields*), *by Faraday's law* (*induced emf*) and, *by* the motion of conductors in steady magnetic fields.

The current strength in a conductor of low resistance (large number of free electrons per volume or with a high electron mobility) and a field applied to it, will be more than .....

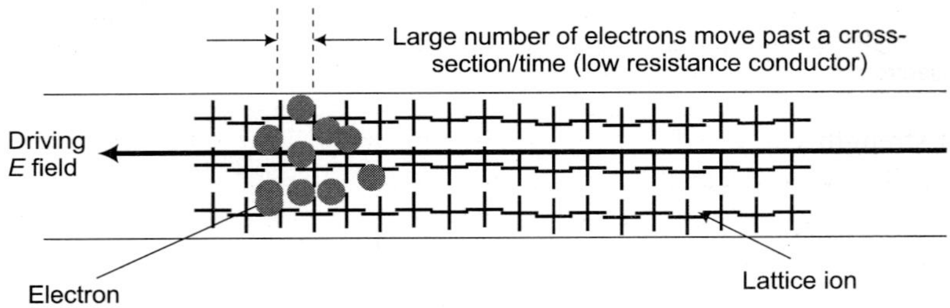

...the current strength in a conductor of high resistance (small number of free electrons per volume or with a low electron mobility) given the same strength of the applied field.

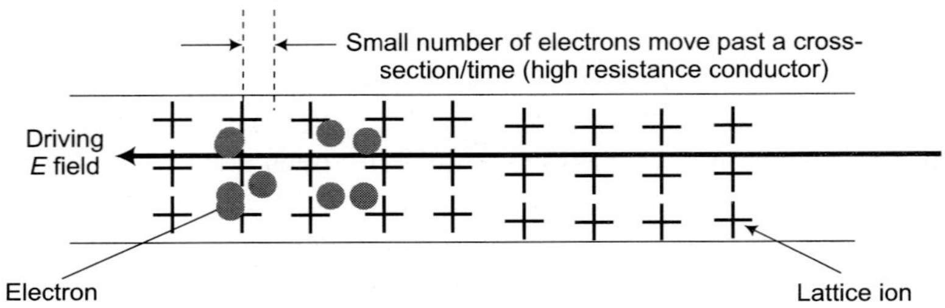

In a conductor with high resistance, only a small number of electrons would have moved past a cross-section/time given the same strength of the driving field. Such conductors are called resistors. Resistance will be discussed in section 1.19 in more detail.

Current Strength in a Conductor with Opposing Fields

The figure below shows an opposing field in the same conductor to which a driving electric field is applied.

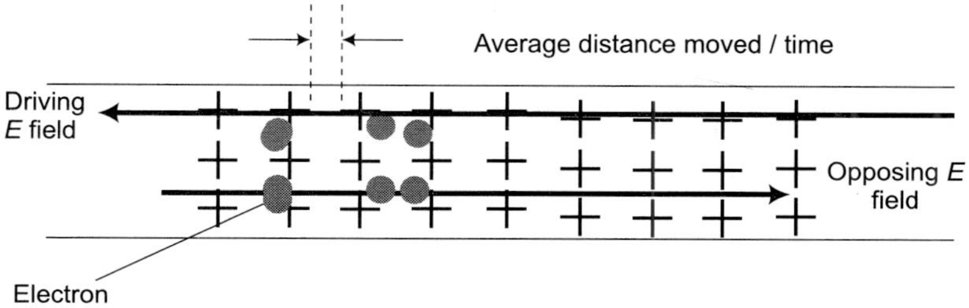

The driving field from a voltage source and the opposing field, say, from a capacitor (fringe) or from an inductor (Faraday's law) or in a conductor moving in a steady magnetic field, will combine to produce a resultant or net field inside the conductor.

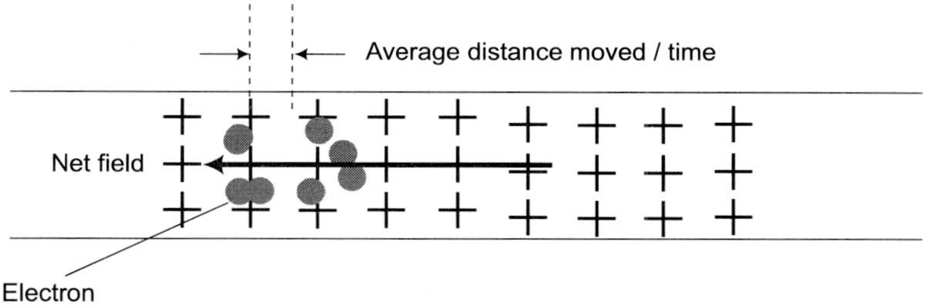

The current strength will be reduced by such opposing fields in the conductor. However, do not forget that while we consider the 'net' electric field.....in the conductor.......the 'net' field is a result of two opposing fields and that both these fields are alive and existing like two teams straining in a tug of war competition.

Sometimes, other fields could so act on an applied driving field as to *increase* the resultant field strength in conductors as with the use of precharged capacitors. We will deal with this aspect of the electric field in Chapter 3.

1.3 ELECTRIC FIELD

In this section, we will examine the nature of an electric field, its sources and action.

Electric Charges are "Sources" of Electric Fields.

It is the very same electric field that causes a charged comb (or a charged plastic pen) to attract tiny bits of paper and, that is present in current carrying conductors. I urge the reader to perform this activity several times (a plastic comb or a plastic pen can be charged by rubbing it in one's hair) and make oneself fully *aware* of the *presence* and force that this field exerted on the paper; and, then, to visualize the force such a field will exert on conduction band electrons inside a conductor. Also observe how the charged plastic comb can make the hair on your hand, stand.

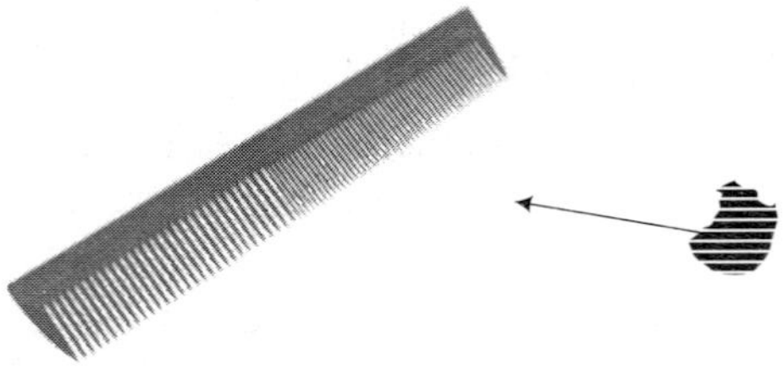

Charges and electric field

Charge creates ***something invisible*** throughout the space surrounding it, and if another charge happens to be present, it will feel the effect of this "something".

We call this "something" ***electric field***. The electric field created by a charge is present throughout the space at all times, ***whether or not there is another charge around to feel its effects***.[2]

A charge "*alters invisibly*" the space surrounding it and we call this altered space *electric field*. The *space* between the charged comb and the tiny bit of paper was *altered* resulting in the attraction.

The electric field created by a charge "penetrates" through matter.

When unlike charges are bunched together then we can imagine that these are connected together by invisible compression and tension springs.

Just as we know that stretched springs and compressed springs can store potential energy, we can visualize that when we pull unlike charges apart and compress like charges they store electric potential energy.

1.4 CHARGES AND ELECTROSTATIC POTENTIAL ENERGY

A single positive charge is surrounded by an *electric field*.

If you *bring another* positive charge near this, then you have to do work to bring it near (because like charges repel each other), and this *work done* energy is stored in the *electric field* that extends to the entire universe in this *system* of charges—comprising the 2 charges....

...like compressing a spring.

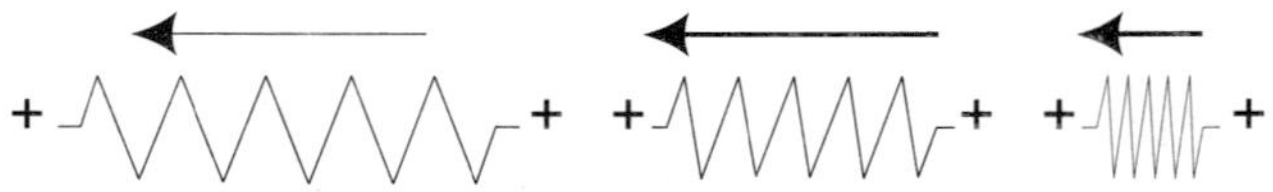

When you *release* the *positive charge* the *electrostatic energy stored in the system of 2 charges is released* and is converted into *kinetic energy* in the *positive charge*

...like releasing a coiled spring.

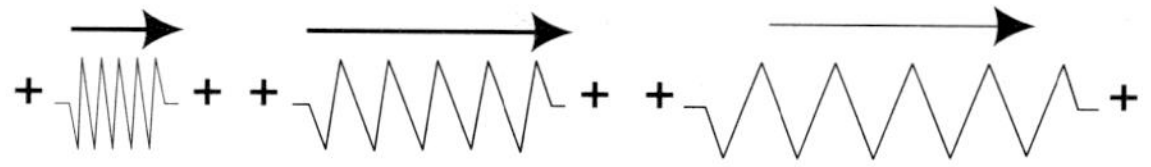

The potential energy is stored in the electric field of the system of charges.

For separating the charges of a neutral atom (electrons and positive atomic cores), you have to do work to separate them (because unlike charges attract each other), and this *work done* energy is stored in the *electric field* of this *system* of charges comprising the 2 charges...

...***like stretching a spring.***

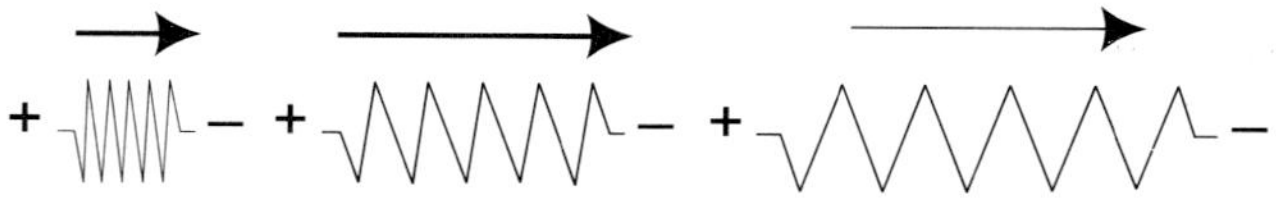

When you *release* the *negative charge* the *electrostatic energy stored in the system of 2 charges is released* and is converted to *kinetic energy* in the *negative and positive charges...*

...***like releasing a stretched spring.***

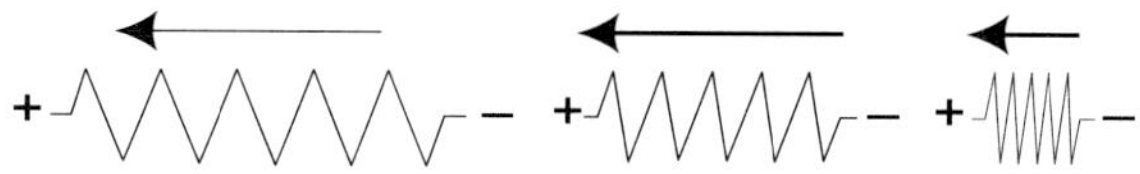

Compressing charges together to make potential energy.

You may imagine that in the palms of your hands you are holding positive charges and by bringing your palms near one another (compress the charges) you can store energy.... you may release them to make kinetic energy....ever and ever....

You may repeat this as many times as you please.

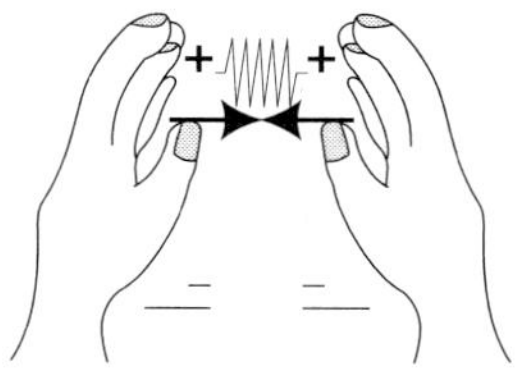

The same can be accomplished with *"opposite"* charges that are attracted and near each other.

Separate them to make "potential energy" as in batteries, release them to make "kinetic energy".

Do this as many times as you please....over and over....separate opposite charges and make potential energy....*release* and *convert* into kinetic energy.

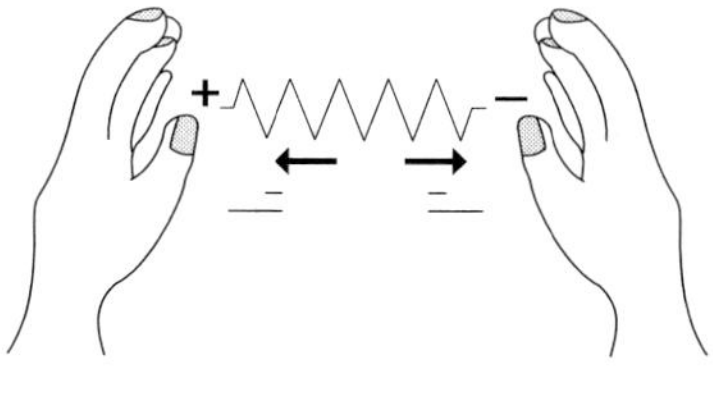

Visit http://ocw.mit.edu/ans7870/8/8.027/fo4/visulizations/electrostatics/38-videogame/38-EMViideoGame320.html and play the videogame (click "Help" for instruction) to get a feel of charge interactions.

1.5 POTENTIAL ENERGY

In this section we will study the different types of potential energy and then understand the nature of electric potential energy. Potential energy is the *energy* which results from position or configuration.

An object may have the capacity for doing *work* as a result of its position in a gravitational field (gravitational potential energy), an electric field (electric potential energy), or a magnetic field (magnetic potential energy).

We are familiar with the potential energy stored in an object raised above the ground.

You would not like this to happen to you! Do you?

Stored energy, which can be used to do work, is called potential energy. *Stored energy* can be made portable as in batteries and is capable of doing work *on demand*.

We consider next the types of potential energy.

Gravitational Potential Energy

When a rock is raised above the earth and released, the gravitational attraction between the rock and the earth causes the rock to fall to the ground.

A raised object has *gravitational potential energy* stored in the system of the object and the earth. Potential energy should be associated with PAIRS of interacting objects, not with single objects. When raised, the object does not change in mass, volume, color or shape. It gained potential energy due to its position in the earth's gravitational field. This is the type of potential energy we are most familiar with. Take away the earth from a raised stone and gravitational potential energy loses all meaning.

Other Types of Potential Energy

Among the less familiar ones though important types are:

Strain (or Elastic) Potential Energy

If we *stretch* or *compress* a spring and then release it, the spring moves back to its original length. The stretched or compressed spring has *strain potential energy* because it has the *potential to move.* It stores potential energy by straining the bonds between the atoms of an elastic material.

Examples of strain potential energy

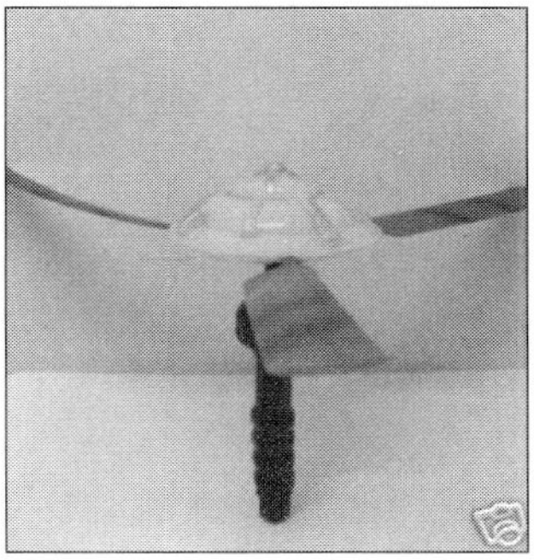

As a child, most of us must have played with this toy!

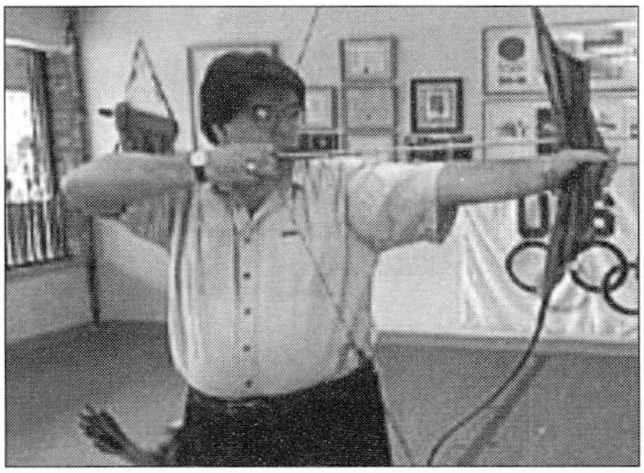

When the bowstring is released the arrow gains kinetic energy

One more type which is of importance to students of electricity and magnetism is:

Potential Energy of Chemical Reactions

Chemical Potential Energy: Chemical potential energy exists because atoms and molecules can *take in* or *give off energy* when their chemical bonds are formed or broken.

Batteries store chemical potential energy, which is converted into electrical energy when the battery *operates*.

Potential Energy of Electric Fields

"Electrical potential energy" is stored in electric fields when positive and negative electric charges *are separated* by an external agent against their force of attraction. The *amount* of stored energy depends on the number of separated charges *and* on the distance they are separated.

Potential Energy of Magnetic Moments

Potential energy is stored in magnetic moments when a current carrying loop is placed in a magnetic field which we shall discuss in Chapter 8.

Four Fundamental Forces of Nature

The potential energy forms are related to the four fundamental forces of nature from which all other forces are derived. In Appendix D a description of the four fundamental forces is provided.

1.6 ENERGY IN ELECTRIC FIELDS

An *electric field* surrounds a charge (created by the charge) and its strength falls off as the square of the distance ($1/d^2$) from that charge. The field vector (lengths represent field strength) at representative locations in a two-dimensional slice of space containing the charge are shown in the diagram below and, are drawn *from* (location P1) for positive charges and *to* (location P2, to avoid confusion showing the vector through the negative charge) for negative charges.

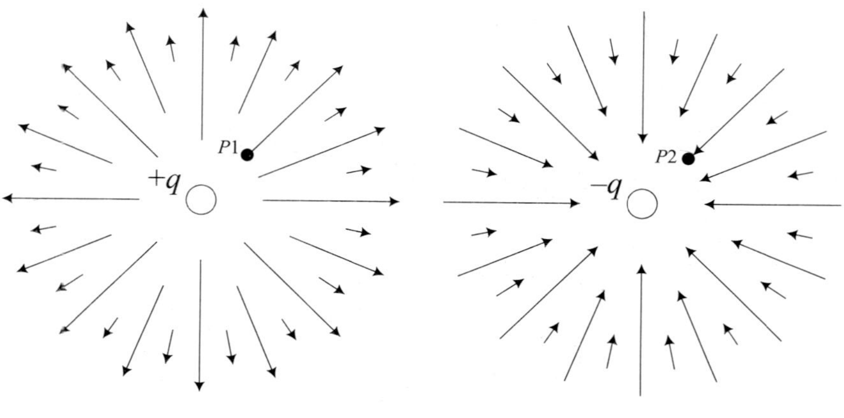

A correct visualization of the electric field surrounding a positive point charge is shown in the figure below that takes into account the fact that the electric field exists in *all* directions surrounding the charge and fills the entire universe. Visit www.falstad.com for a 3-dimensional view of the field.

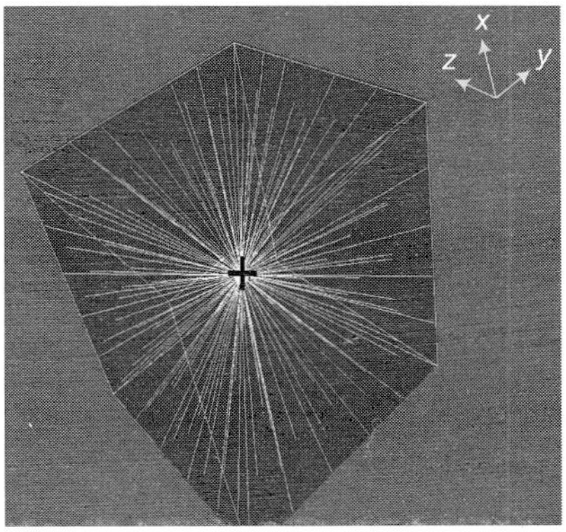

From Einstein's famous $E = \text{mc}^2$ equation, we can show that the energy contained in one gram of matter is 90×10^{12} joules ! This energy is equivalent to that produced by a 100 watt bulb operating for 30,000 years ! But we do not use tiny grains of matter at home and convert these to obtain such huge amounts of energy for use in our daily lives !

The picture associating energy with charges is also complex, and is the subject of ongoing study by physicists. Every proton and electron built by nature has an electric field, and the field surrounds the charge. The electric field, whether powerful or weak due to a charge that is large or small, exists over the entire universe. The field energy of an electron, computed by integrating its' field squared (E^2) over infinite space, is infinite due to the $1/r^2$ form of the electric field[3]. Imagine drawing several thin wires from a block of metal in all directions, making them thinner and thinner as you move away. The electric field around a point charge (such as an electron or a proton) is like the wires pulled from the block; but is special in that its spread is uniform (alteration of space) and not discrete. The rest energy (see Appendix B "Energy stored in a two electron configuration") of an electron is 0.511×10^6 eV, not infinity, and, relating this to the integral of E^2 is fraught with difficulties. In our study of circuits, the amount of energy of an individual point charge, be it an electron or a proton, is of little consequence as the following will suggest. Note that for individual particles we associate *energy*; not potential energy. The electric *potential energy* of a system containing *only one* point charge is "zero", as no energy is required to move the point charge from infinity to its location. But, we are rarely, (I don't feel like "superman", though I have zillions of these in my body !) if ever, required to deal with a single particle's huge energy and have to bring our focus to moving around prefabricated charges; not one, but several (electrons, say) and compute the *potential energy of their interaction*, which will be in more manageable amounts, of such *systems* of charges, obtained by summation using potentials or integration using the electric field (squared)! When we study electromagnetic enegy in Chapter 4, we will see that the manageable amounts of energy required in most circuit operations are transported by the electric and magnetic fields.

The electric potential energy of an electron and proton (or nucleus) system is due to the proximity of the electron to the proton (or nucleus) and the attractive force on the electron by the proton (or nucleus) as shown in the figure below.

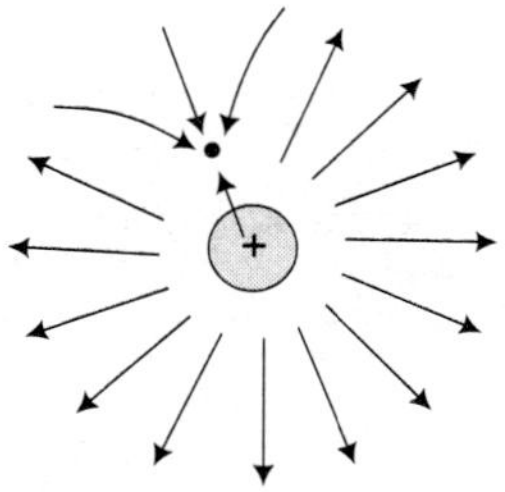

Take away the proton or nucleus and electric potential energy loses meaning. It is the electric field (negligibly small or large) of the two charges interacting that connected the two charges and potential energy is stored in the system of the two charges separated a distance.

Note: Physicists are prone to say the *potential energy of an electron* ………..when what they *really* means is the potential energy of the *electron and nucleus (or proton) system of charges*. See Appendix B "Energy stored in a two electron configuration" when two electrons interact at a distance of 10^{-10} metres.

We restrict our discussion of the electric field sourced by charges to its strength, direction and effect on other point charges at distances a little away from the point charges, which is sufficient for the study of most circuits.

The reason for this will become clear in Section 1.13 when we discuss the structure of atoms, Section 1.18 that describes how to visualize current and in Section 2.26 where neutrality in conductors carrying currents is dealt with.

The electric field that surrounds a point charge *does not* exert a force on that same charge.

In a system of charges that gained or lost potential energy, the charges do not change color, volume or mass !

The potential energy of the system is due to the relative positions the charges are made to occupy, stretching or compressing the invisible springs between individual charges.

When we bring an uncharged conductor near an isolated positively charged sphere then, charges will be induced on the surface of the conductor and, there will be a net attraction between the sphere and the conductor as seen in the figure below.

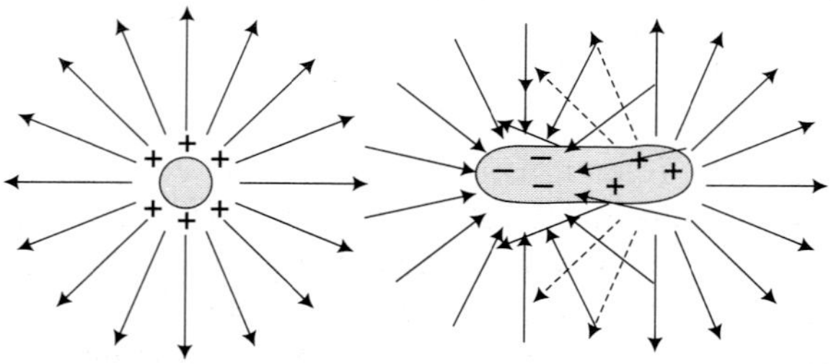

Positively charged sphere Conductor

When the conductor was uncharged and far away from the positively charged sphere, there was no electric field perceptible surrounding the conductor and we could say that it was *neutral*.

When the positively charged sphere was brought near however, the charges of the conductor separate, and one end is progressively positively charged while the other end is progressively negatively charged. In the figure above, the electric field lines are drawn crossing each other (which are never to be shown this way) merely to illustrate the intermediate stage of development of the resultant field around the system, the final stage of which is shown in the figure below.

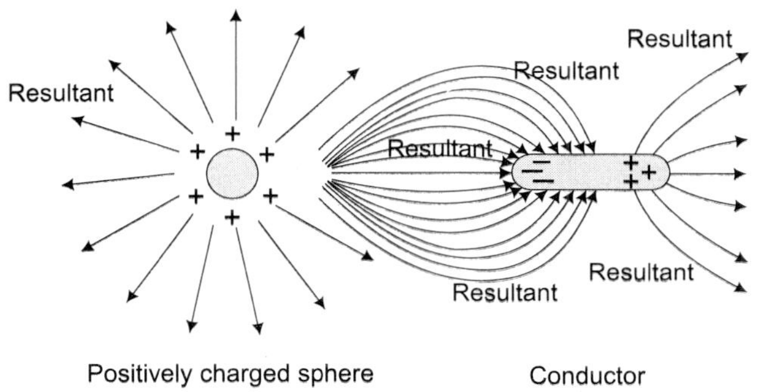

A resultant electric field that is non-zero appears surrounding the conductor.

The system did some work to cause the *separation* of charges when the conductor was brought near. The energy of the system has *gone down* in comparison to the energy in the field *before* the conductor was brought near the positively charged sphere.

Note very carefully that the electric field surrounding the system is due to the electric fields of individual charges and is a resultant computed from these fields.

The fields of individual charges *ever exist*. They existed since the time the charges existed and will ever remain with the charges.

What is computed (or measured) and shown of the system is a *resultant* of the fields due to all charges.

1.7 POTENTIAL ENERGY AND KINETIC ENERGY OF CHARGES

The relationship between work, kinetic energy, and potential energy for a system comprising charges is : Work done by an external force $W = \Delta KE + \Delta PE$ where KE = Kinetic energy and PE = Potential energy.[26]

Example

Two positively charged balls are tied together by a string. One ball has a mass of 30 g and a charge of 1 μC; the other has a mass of 40 g and a charge of 2 μC. The distance between them is 5 cm. The ball with the smaller charge has a mass of 30 g; the other ball has a mass of 40 g. Initially they are at rest, but when the string is cut they move apart due to mutual repulsion.

When they are a long way away from each other, how fast are they going?

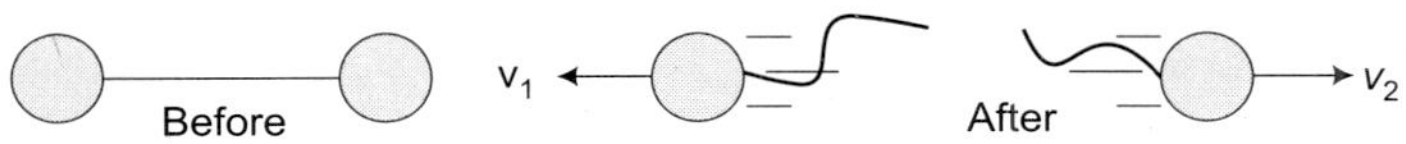

Let's start by looking at energy. No external forces act on this system of two charges, so the energy must be conserved. To start with, all the energy is potential energy; this will be converted into kinetic energy when the string is cut.

Energy at the start: $KE = 0$ and $PE = k\,q\,Q/\,r = (9 \times 10^9)\,(1 \times 10^{-6})\,(2 \times 10^{-6})\,/\,0.05 = 0.3596\ J$

where $$k = \frac{1}{4\pi\varepsilon_0} = \left(9 \times 10^9\ \frac{Nm^2}{C^2}\right).$$

When the *balls are very far apart*, the r in the equation for potential energy will be large, making the *potential energy negligibly small.*

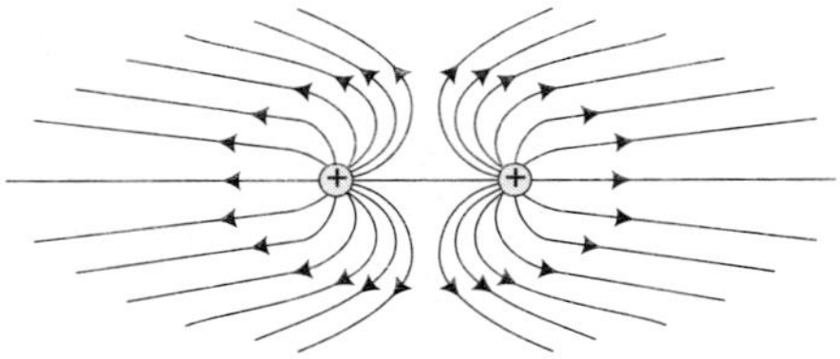

Before the string is cut

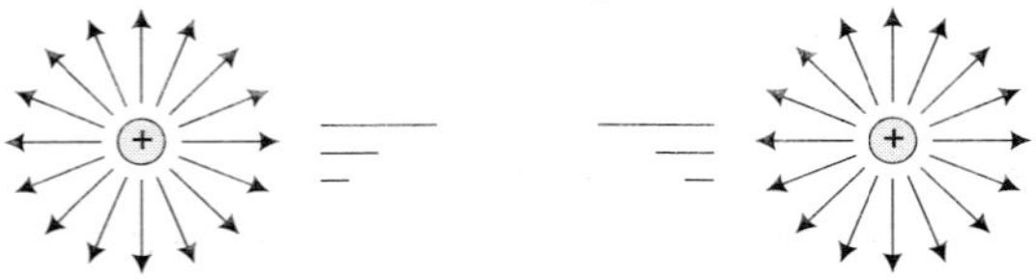

The charged balls are far apart

Energy at the end : $KE = ½\,(m_1 v_1^2) + ½\,(m_2 v_2^2)$ and $PE = 0$.

Energy is conserved, so the kinetic energy at the end is equal to the potential energy at the start:

$$½\,(m_1 v_1^2) + ½\,(m_2 v_2^2) = 0.3596\ J.$$

The masses are known, but the two velocities are not. To solve for the velocities, we need another relationship between them.

Because no external forces act on the system, momentum will also be conserved.

Before the string is cut, the momentum is zero, so the momentum has to be zero all the way along.

The momentum of one ball must be equal and opposite to the momentum of the other, so,

$$m_1 v_1 = m_2 v_2 \text{ and therefore,}$$
$$v_1 = (m_2/\,m_1)\,v_2 = 4/3\ v_2$$

Plugging this into the energy equation gives:

$$m_1(4/3\ v_2)^2 + (m_2\ v_2)^2 = 2 \times 0.3596$$

This gives $v_2^2 = 0.7192/[1.778\ m_1 + m_2] = 7.7057$ where $m_1 = 30$ g and $m_2 = 40$ g.

So, $v_2 = 2.78$ m/s and $v_1 = 3.70$ m/s. The speeds are tiny compared with the speed of light, about 2.99×10^8 metres per second. We shall see in Chapter 4 that the energy transferred *in electric circuits* does not lie in the kinetic energy of the tiny almost weightless moving electrons though numbering in the millions.

One joule in everday life is approximately the energy required to lift an apple one metre straight up.

In the next section, we shall see how a net electric field is produced between the terminals of a battery.

1.8 ACTION OF A BATTERY AND PRODUCTION OF A NET ELECTRIC FIELD

A battery is used as a device that separates charges by a chemical reaction as shown in Fig. 1.1

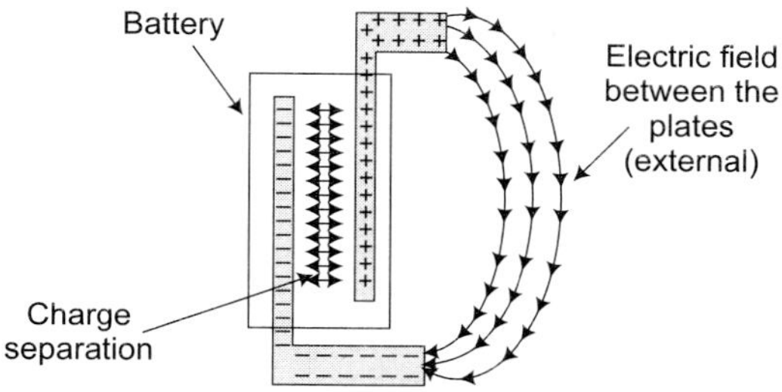

Fig. 1.1 A charged battery.

Charges are *separated* by chemical reactions. The exact process is quite complex and need not concern us. Positive ions form at the positive plate and, negative ions form at the negative plate.

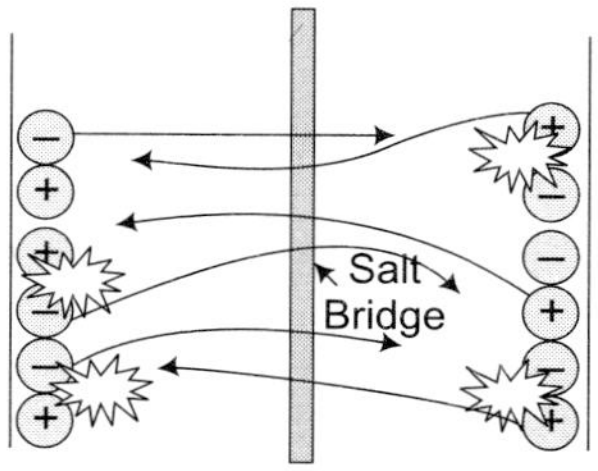

This separation is performed by a chemical process, resulting in a state of equilibrium with a certain number of electrons on the surface of the negative terminal and the same number of electrons missing on the surface of the other positive terminal.

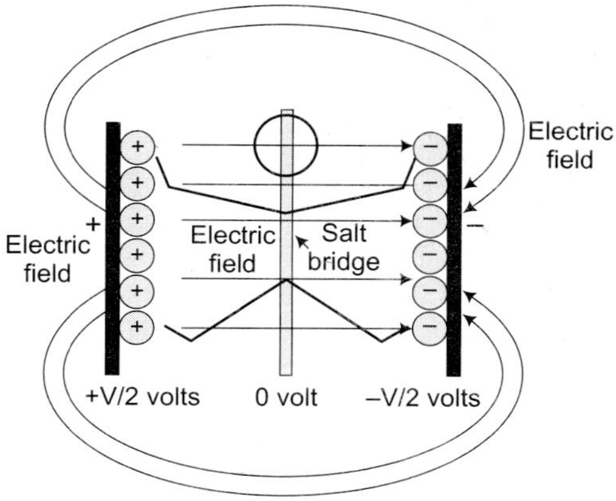

A dynamic equilibrium is reached when the attracting coulomb forces between these separated charges are equal to the separating chemical forces. There exists an electric field outside as well as inside the battery between the separated charges on the plates.

Potential energy is stored in the electric field of the *configuration* of separated charges!

Conservation of Charge[2]

The net charge of an object is the sum of all the positive charges minus the sum of all the negative charges. Usually, matter has the same number of protons as electrons, so its net charge is zero. *The net charge of a closed system does not change.* In an extremely wide variety of experiments, no one has ever observed electric charge to be created or destroyed. These observations are summarized by the principle called *conservation of charge*: the net charge of the universe or of any closed system, does not change. By a closed system we mean any region through whose boundaries charged particles neither enter nor leave. For example, the charges in a battery and on its terminals are conserved and constitute a closed system. For a V-volt battery not having any *external* excess charge, i.e charge not local to the closed system, at the physical centre in the space between its terminals, there will be an equipotential line of zero volt, with the negative terminal at –V/2 volts and the positive terminal at +V/2 volts produced entirely by internal charge separation. There are high-energy particle reactions that create or destroy charged particles, but always in a way that the *net* charge does not change.

Important: Whenever it is said that a material is electrically neutral, it is merely a statement of account of an equal number of positive and negative charges and that the conservation of charge is preserved in a closed system and, *not* a statement about how these are *distributed* in the system; as dipoles or as electron hole pairs in semiconductors *and* whether there is a resultant electric field around them, unless this is explicitly stated.

1.9 SURFACE CHARGES AND FIELDS IN AND AROUND CIRCUITS

Indicated below are typical questions students raise in the study of electric circuits and circuit behaviour reported by Hermann Haertel, a scientist at the Institute for Science Education, University of Kiel who studied how students attempt to understand the principles and processes of electrical circuits. Later sections develop answers to these questions. (Reproduced with permission from Hermann Haertel.)

Student Questions

The analysis of circuits in terms of surface charge provides answers to some profound questions students sometimes raise, *questions which may have no satisfactory answer within the context of traditional analysis in terms solely of potential.* Haertel reports students asking, *"What is the difference in the conditions at the two ends of a resistor in which there is a current?"*

It is quite unsatisfying to say merely that the potential is different.

Consider a battery connected to a bulb with a pair of wires as shown in the figure below from "Unified treatment of electrostatics and circuits" by Chabay and Sherwood[1].

Is electric current due to charges in and around the battery?

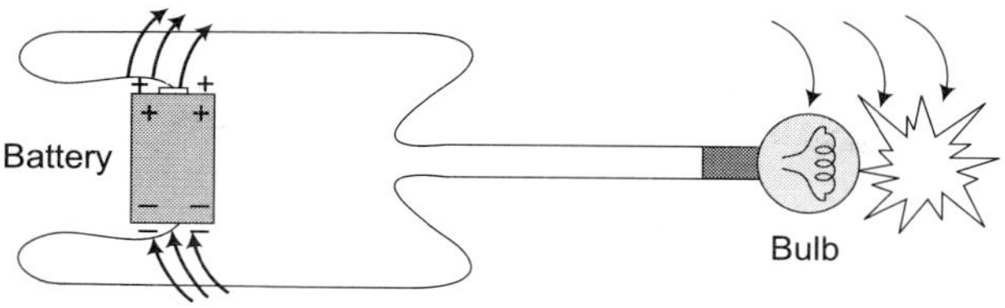

If we bring the bulb closer to the battery, does the bulb glow brighter?

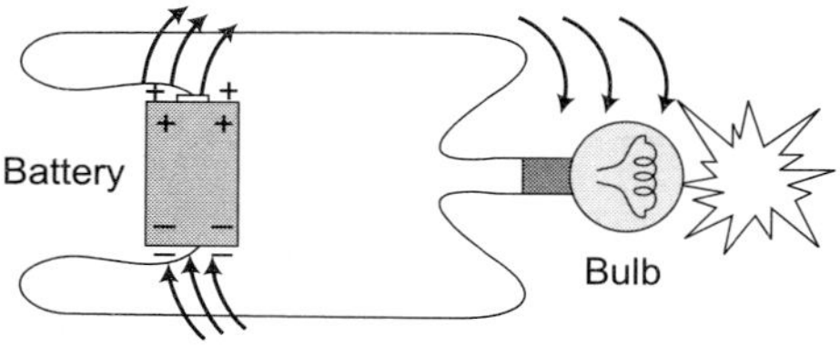

The bulb glows with the *same* brightness even when brought *closer* to the battery.

Coulomb's law predicts that the electric field is inversely proportional to the square of the distance from the source charge. If the bulb is brought closer to the battery, then there should be a corresponding increase in the electric field due to the charges on the battery that would exert a force on the conduction band electrons in the wires and in the filament. But, we observe that there is no increase in the current in the wires and in the brightness of the bulb.

Now rotate the bulb by 90 degrees.

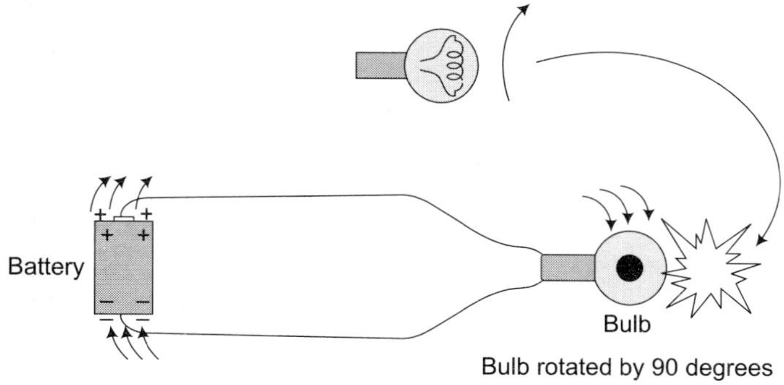

Bulb rotated by 90 degrees

Since the electric field is now perpendicular to the filament, the bulb *should* stop glowing.

The bulb does *not* stop glowing.

Therefore, the electric field from the charges on the battery has *no significant effect* on the bulb filament.

Evidently, charges in and on the battery *cannot* be the only source of the electric field inside the bulb filament.

It is therefore reasonable to conclude that some *other charges* in addition to those on the battery are responsible for producing an electric field that is *pushing* the electrons,

Or,

is our conclusion wrong and the flowing electrons are simply pushing each other through the wire?

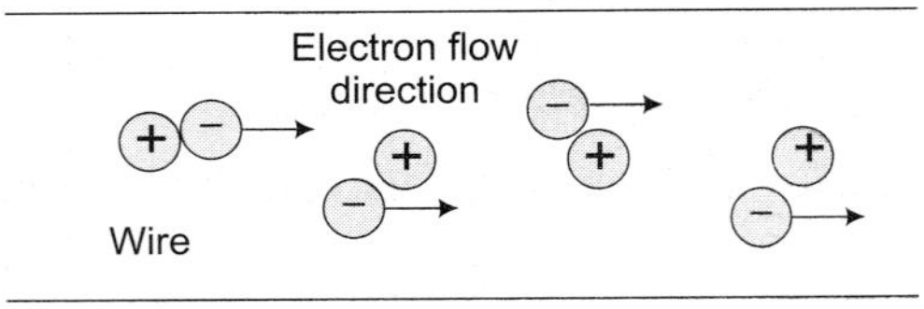

The flowing electrons inside a wire *cannot push each other* (like water molecules that are neutral inside a pipe) through the wire because, on an average the *repulsion* by any electron is *cancelled* by the *attraction* of a nearby positive atomic core.

One may think that the flowing electrons push each other through the wire as in the figure above, like water molecules pushing neighbouring water molecules through a hose.

However, this cannot be true because averaged over a few atomic diameters, the interior of the metal is everywhere neutral, and, on average, the repulsion between flowing electrons is cancelled by the attraction to positive atomic cores.

Some other charges in addition to those on the battery are responsible for producing an electric field that *pushes* the electrons.

This is one of the reasons why an analogy between electric current and the flow of water can be misleading.

It is easy to understand the constant current in terms of potential in a circuit comprising a battery, wires and a resistor. The potential drop across the resistor is always nearly equal to the emf of the battery, and the electric field inside the resistor is simply the gradient of the constant potential drop. Is this satisfactory?

We know that charges are the only source of electric field (in the absence of varying magnetic fields). Then, where are the charges that are responsible for the field inside the resistor?

Hermann Haertel published a paper "Constant Electric Current and the Distribution of Surface Charges1"[6] in which he detailed the process of conduction and showed how "surface charges" are responsible for the phenomenon of conduction.

We now introduce surface charges and the nature of the electric field due to surface charges, to gain an intuitive feel for its presence in wires in circuits and in the next section go into some detail about the configuration of the surface charges that produce it.

Surface Charges

(Reproduced with permission from Hermann Haertel)

According to Haertel, surface charges are present, whenever a wire carries an electric current and, are necessary for a causal explanation of how an electric current *actually* works.

Consider the case of a battery connected by wires to a resistor as shown in Fig. 1.2.

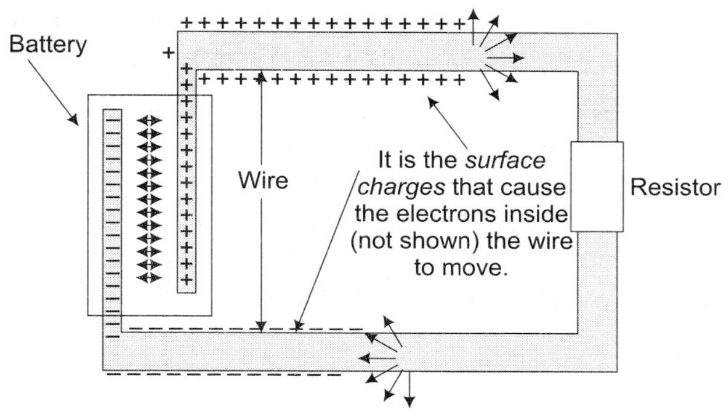

Fig. 1.2 Development of surface charges—intermediate stage.

When we close a circuit we are applying a potential difference across the circuit and the charge imbalance that exists between the two terminals of the voltage source causes a field to appear across the terminals of the circuit.

To deliver energy to the circuit surface charges appear and, the electric field due to these charges apply forces that cause the movable charges in the wires to begin a slow flowing motion: $F = qE$, where F is the force in Newtons, q is the charge in coulombs, and E the electric field in Newtons per coulomb. The electric field is also exactly equal to volts/metre. See the end of Appendix A for a derivation of the equivalence of the two units of the electric field.

The surface charges develop on the wire connecting the battery with the resistor. Shown in Fig. 1.2 is the intermediate development stage.

The way in which these charges are distributed is shown in Fig. 1.3.

In Fig. 1.3. we see the final stage of development of surface charges. The surface charges develop along the wire and positive surface charges build up at one end of the resistor while negative surface charges develop at the other end. The surface charge density gradient produces a powerful resultant electric field inside the resistor that pushes the conduction electrons to establish a current in the circuit. The surface charges around a circular cross-section wire is in the form of a ring and in Section 1.10 below we shall see how it produces an axially directed electric field inside the wire.

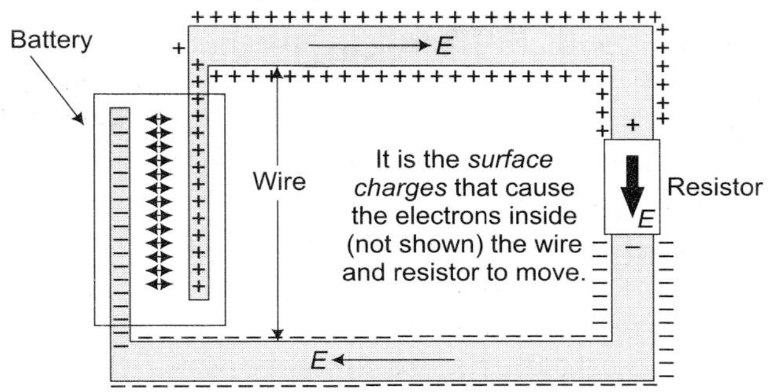

Fig. 1.3 Development of surface charges–final stage.

Another student question reported by Haertel is, *"How does the current know how to split when there are parallel branches?"* This question will be answered in Chapter 2.

Are electric fields really present in and around circuits?

The presence of electric fields in and around circuits can be demonstrated by circuits made out of conductive ink and placed on glass plates, photographs of which are shown in Fig. 1.4. The electric field in the space around the circuit due to surface charges is demonstrated with the aid of grass seeds strewn upon the glass plate. The grass seeds orient themselves in the neighbourhood of charges because they act like electric dipoles (analog to magnetic compass needles).

These demonstrations were made by Professor Oleg Jefimenko, West Virginia University, USA and published in a paper titled "Demonstration of the electric fields of current-carrying conductors", American Journal of Physics, Vol. 30, No. 1, 1962.

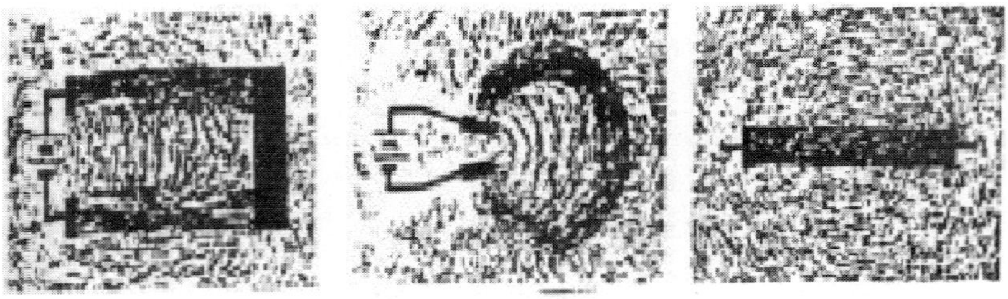

Fig. 1.4 Electric field lines around conductors.

At the end of Section 1.19, we will answer the question as to how the bulb connected using long wires to a battery does not increase in brightness when brought close to the battery or if the bulb were twisted to make the electric field from the charges on the battery perpendicular to the filament.

1.10 ELECTRIC FIELD DUE TO A RING OF CHARGE

It is instructive to get a feel for the electric field, its production and its strength due to a uniform charge distribution around a ring. Such rings of charges on surfaces of circular cross-section wires used in circuits, produce an electric field *inside* the wires that establish the currents.

In the real world of electricity and electronics, point charges are only electrons (that can never be stationary) and protons. But note how the study of electrostatics using stationary point charges, dipoles and other charge distributions enables us to understand circuit processes and mechanisms that depend on these electrons (though moving that form clouds of charge), protons and *vibrating* lattice ions. We give below the derivation of the electric field due to a ring of charge that causes conduction in round uniform wires and resistors.

To derive the expression for the electric field we make use of an important property of the electric field which is the *principle of superposition*. Superposition as a vector sum holds true for forces in general, so it also holds true for fields, because field is a force per unit charge. For a definition and description of this important principle see Appendix A "Superposition of electric fields"

The electric field strength E at a point is defined as $E = \dfrac{F}{q_{\text{test}}}$, where F is the force acting on the test charge q_{test} at the point where the field E is being measured. The electric field for points *along the axis* of a ring of charge is of interest in conduction processes and can be calculated using the principle of superposition, as shown in the figure below.

We consider a differential element dl located at the top of the ring. It contains an element of charge given by $dq = q.dl/2\pi r$ where $2\pi r$ is the circumference of the ring and q is the total charge on the ring. The element sets up a differential electric field dE at point P[13].

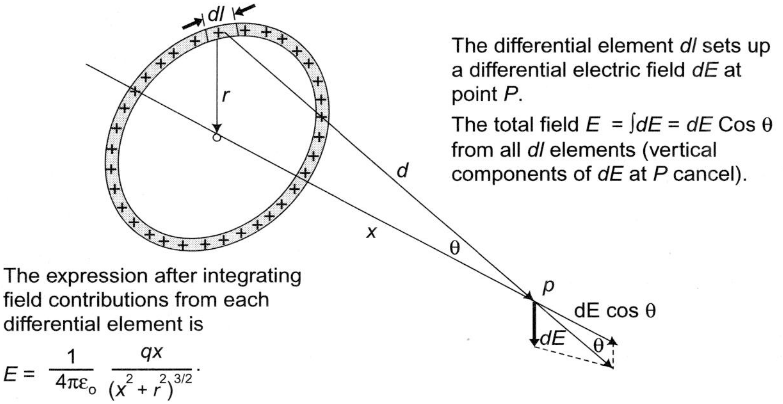

The resultant field E at P is found by integrating the effects of all the elements that make up the ring. From symmetry this resultant field must lie along the ring axis. The component established perpendicular to the axis is cancelled out by an equal but opposite component established by the charge element on the opposite side of the ring.

Thus, only the component of dE parallel to this axis contributes to the final result

$$\frac{1}{4\pi\varepsilon_0}\,\frac{qx}{(x^2+r^2)^{3/2}}.$$

See Appendix A for details of computing the field due to a ring of charge.

Nature of a Field

The nature of a "field" can be understood by analogy with a gravitational field that acts on all objects.

A field like a **gravitational field** is a ***special*** kind of field.

It applies a force ***continuously*** or rather relentlessly on you whether sitting, standing walking or sleeping for 24 hours and for years and on all objects all the time.

It does not stop ever applying a force. ***It never gets tired.*** That is the reason for the ***acceleration*** a field produces.

On the other hand, take a simple ***application of a force*** to an object ***once,* just once,** like ***hitting an object at rest once.*** The object will accelerate but **will settle down** to move with uniform velocity and will ***not accelerate any more***.

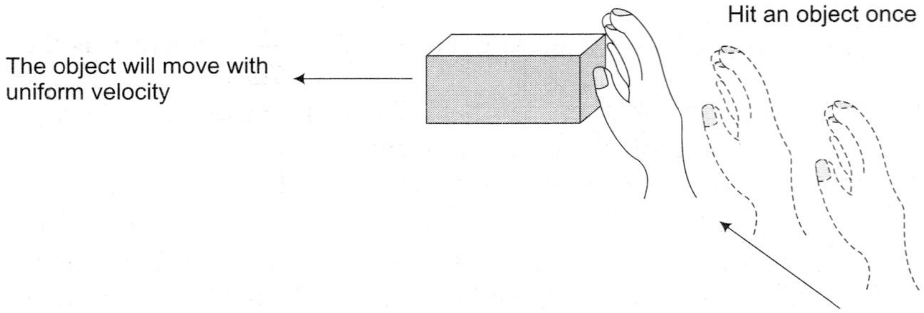

A gravitational *field* applies a force continuously on all objects and therefore, it accelerates the objects.

The gravitational force acting on the ball released from the top is continuous. It acts on it when held in the hand before release and during every instant of its movement downward after release and also after it hits the ground. Therefore, the ball accelerates, i.e. its velocity increases continuously. It does not remain uniform as it does when an object is hit just once.

An ***electric field*** like a gravitational field is another ***special*** kind of field.

It applies a force ***continuously*** or rather relentlessly on electric charges in its sphere of influence.

That is the reason for the ***acceleration*** it produces of *electric charges* within its sphere of influence.

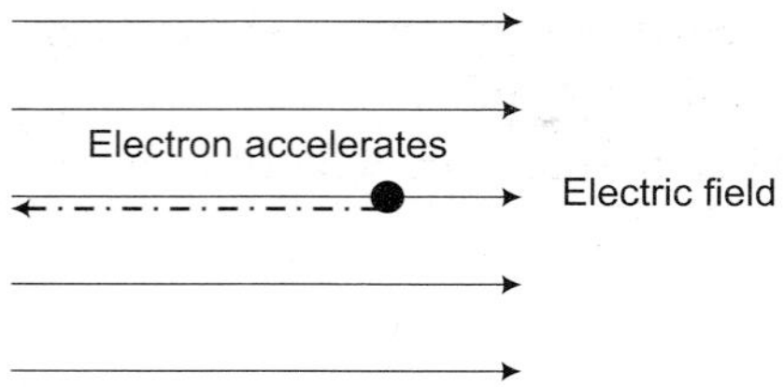

The electric field does not stop ever applying a force….***it never gets tired unless if it were reduced to an insignificant force due to its source charge unifying with an equal and opposite point charge.***

Separation of Charges

When unlike charges are separated (as in the formation of dipoles), *three* things happen:

1. A net non-zero *electric field* is produced in the space surrounding the charges.
2. A potential difference or *voltage* is created.
3. *Energy* (that caused separation) is stored in the electric field.

Every proton and electron built by nature has an electric field, and the field surrounds the charge. When the *unlike* charges were near each other before separation, the electric field in the space surrounding them is largely, yet not completely, "cancelled", and, after separation, it is as if the field was "uncancelled" and that produces a *significant* net non-zero field.

If pairs of equal and unlike charges were so distributed in space that around the charges there was a very *weak* resultant electric field (we say *resultant* because the fields due to the positive and the negative charges exists everywhere in space—at each and every point in space), we could treat the system of charges as being *neutral,* as we would a stable atom.

We can draw thick electric field lines originating on the positive charges and terminating on the negative charges if the charges are near each other. Then, we can draw less thicker and then thinner *appropriately* oriented field lines representing appropriate magnitudes and field direction to depict less strong and weak resultant electric fields at points a little away and at distant points respectively from the charges. Remember, an electric field due to a charge exists at all points in space, and resultant fields are produced at points where there are fields due to other charges.

Depicting an electric field using field lines can be deceptive because physicists and engineers use these to depict the most *dominant* field strengths. The key to understanding several electrostatic and circuit processes lies in the ability to visualize the *weaker* fields at points as either due to being very far away from source charges or, as a resultant of powerful (sometimes) yet nearly oppositely directed fields which *really should* be depicted by appropriately sized and directed field vectors, but is not done by the engineers and physicists for the sake of clarity.

Note: In this book, field flux lines (continuous) are used to portray the electric field in an expanse of space and will be used to convey a general feel for the field therein. For more accurate representations, field vectors whose length or thickness represents strength are used.

If now the unlike charges were to be separated further, a *significant* resultant field will begin to appear *around* the separated system of charges (where it was weak when the source charges were nearer each other before separation) and seemingly grow and spread in the space surrounding the charges. The energy that was given to cause separation will be stored in the electric field and the potential difference will increase.

To view the electric field that develops as a dipole is formed by separating charges, please visit the Masachusetts Institute of Technology webpage http://ocw.mit.edu/ans7870/18/18.013a/textbook/HTML/chapter28/section04.html and in the first applet, set *all* charges *except* $q1$ and $q2$ equal to zero and click "plot charges and field lines". Then, drag the charges $q1$ and $q2$ apart and observe the development of the resultant electric field around the dipole. If you drag them close to each other, watch the resultant electric field around the dipole collapsing to zero. See the "Links_to_animations_articles" pdf in the accompanying CD for another interactive webpage that shows the dipole and its electric field in 3D.

We shall see in subsequent sections that with real electrons and protons of atoms, it is not possible to make the resultant electric field around them zero.

1.11 DEVELOPMENT OF SURFACE CHARGE

If a current of 1 amp is flowing through a wire, there are about 10^{19} electrons passing through each cross-section within one second.

Because of the enormous strength of the Coulomb interaction and the very high mobility of electrons in metals, it takes only a few electrons at the surface of the wire to produce an electric field that can push 10^{19} electrons around in a circuit and to overcome the resistance of a metallic wire.

In this topic we shall study qualitatively various configurations and distribution of excess (local) surface charges of wires and the resulting electric field inside the wires recollecting from Section 1.10 how a ring of charge produces an axially directed electric field.

Let us first assume that the surface charges have a constant distribution[6] around a wire as shown in Fig. 1.5.

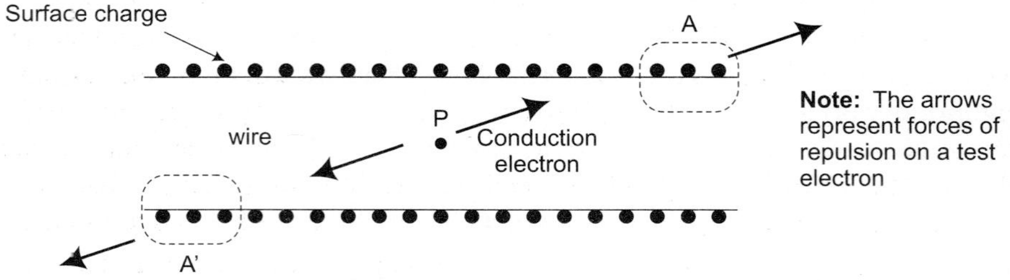

Fig. 1.5 Influence of a constant excess surface charge (for point P on the centre axis of the wire).

Along the wire, for every small surface area A containing a certain number of electrons, there exists another area A' at the opposite side with the same number of electrons which will balance the interaction with a test conduction band electron at point P. Therefore, no resultant force is produced on the conduction electron located at the point P and a constant excess (local) surface charge cannot produce a current in the wire.

A ring of surface charges has the strongest interaction with the free electrons within the same cross-section.

But because of the circular symmetry of the wire, however, the interaction of the surface charges cancel each other, and there is no net force on electrons situated at the same cross-section, as shown in Fig. 1.6.

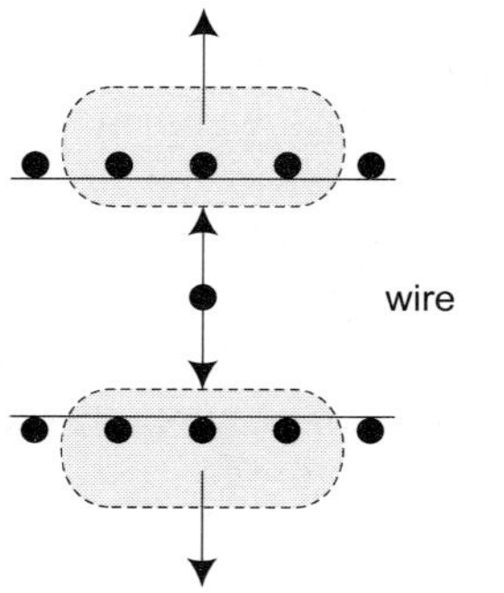

Fig. 1.6 Influence of opposite surface charges

For reasons of symmetry, it can be concluded that the interaction of equally distributed electrons on the surface of the round wire with centrally located electrons inside the wire cancels to zero. Therefore, no conduction mechanism is possible for centrally located electrons in the wire with such a constant distribution of surface charge.

We next consider the effects of a constant excess (local) surface charge distribution on electrons that are located a little away from the centre of the wire as shown in Fig. 1.7. The forces are resolved into their parallel and perpendicular components.

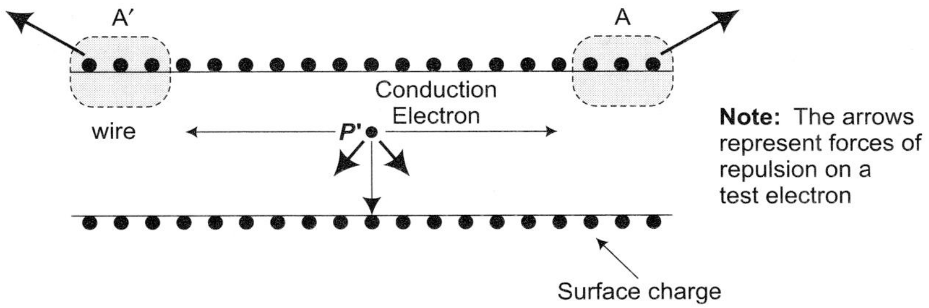

Fig. 1.7 Parallel and perpendicular (component) forces that act due to constant excess surface charge (for points P′ off the axis).

For electrons located off the centre axis, P′, the symmetry argument predicts for forces parallel to the axis of the wire, that the resultant force will be zero.

However, the component forces perpendicular to the axis will exert a net force on the conduction electrons at P′ and will initially cause the electrons to move to the surface, the bottom surface in Fig. 1.7.

This will cause a redistribution of the excess (local) electrons on the bottom surface of the wire as shown in Fig. 1.8 and, the forces due to these excess electrons on the bottom surface will cause the resultant force on the electrons at P′ to become zero. Again, no conduction mechanism is possible even for electrons located away from the central axis of the wire for a constant excess surface charge distribution.

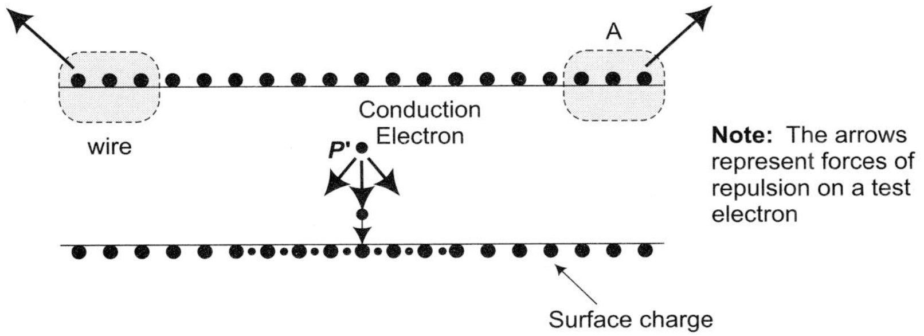

Fig. 1.8 Perpendicular (component) forces that act due to constant surface charge (for points P′ off the axis).

We have seen that *no* "net" electric field in any one direction can be produced *if* the distribution of surface charge is *spread uniformly* over the surface of conductors.

There has to be some kind of distribution that *can* produce a *non-zero net* electric field that will *force* the conduction band electrons in conductors into *motion*.

1.12 MECHANISM OF CONDUCTION

We now consider the case if there is a non-uniform distribution of excess (local) surface charges *due to the applied voltage* in a circuit that could produce a resultant electric field inside the wires.

In short, to break this equilibrium between the forces generated by the equally distributed surface charges as was seen in the Figs. 1.5-1.8 and to exert a driving force on the electrons within the wire, it is necessary to "change" this symmetric distribution of surface charges.[6]

On the surfaces of the wires, there have to be more excess (local) surface charges on one side than on the other side, so that the two interactions of opposite areas A and A′ do not balance.

DEVELOPMENT OF SURFACE CHARGE AND THE REAL MECHANISM OF CONDUCTION

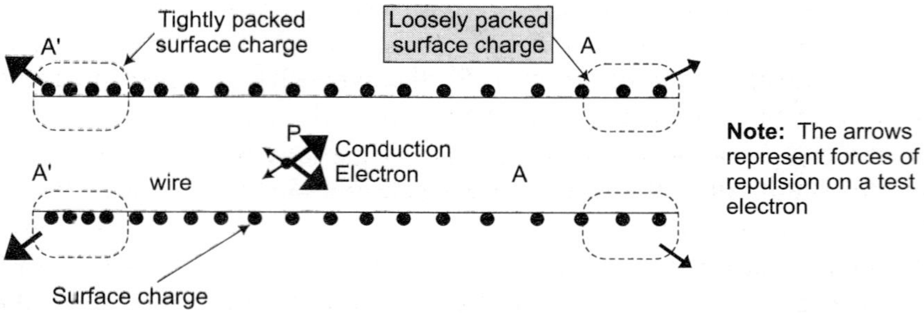

Fig. 1.9 Gradient in excess surface charge distribution.

The *net electric field* is due not just to nearby surface charges but also to ***all*** the other charges in the system, including surface charge on *distant parts of the circuit and charges on the battery.*

As shown in Fig. 1.9, there is a "gradient" of excess surface charge distribution that produces a resultant electric field on the electron located at point P and also on electrons located at other points. Notice that there is a strong force exerted by the charges in the area A′ while there is a weak force exerted by the charges located in the area A.

The two forces acting on the electrons at P cause a net force to act on the electrons and this causes the electrons to move under its influence.

This is the real mechanism of conduction.

One could imagine positive surface charges in the area A and negative charges in area A′ in Fig. 1.9 and such a surface charge distribution with charges of opposite signs can also produce the electric fields inside wires.

In Fig. 1.10 is shown a full view of the rings of the same sign distributed with a gradient on surfaces of wires, the electric fields that they produce and the resulting electric field E_{net} that enables conduction.

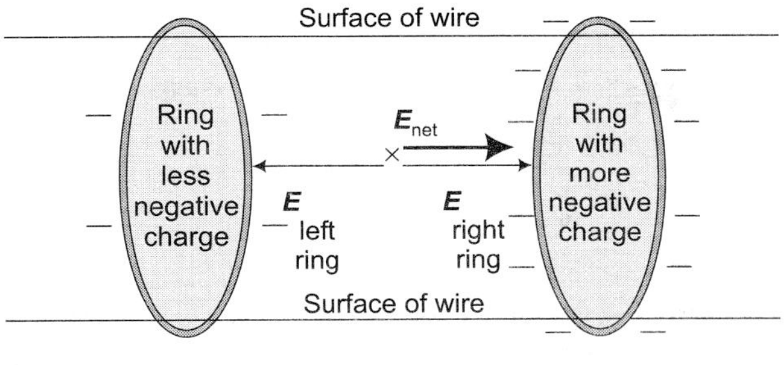

Fig. 1.10 At a location midway between two surface charge rings with different amounts of charge, the rings contribute a *non-zero* net electric field. Electric field inside the wire is associated with a *gradient of surface charge density.*

The setting up of the axial field due to the surface charges of a ring (at points along the axis, as was described in Section 1.10) is a very complex phenomenon since it involves the electric fields due to millions and millions of electrons and lattice ions. Adding to the complexity to graph the setup of the axial field, is the retardation effect because of the finite (speed of light) travel speed of changes in the electric field. These phenomena cannot be observed by the human eye and some scientists have made computations for simple circuits that are very complex and an overview of the results of these computations is given in Section 2.6.

It is once again brought to the attention of the reader that surface charge rings with varying density (*gradient*) would contribute to the *net* electric field (Fig. 1.10) inside the wire.

Rings of equal charge density (and the same sign) contribute zero electric field at a location midway between the two rings, whereas rings of unequal charge density (or different sign) contribute a non-zero field at that location (Fig. 1.9).

In one particularly simple geometry, that of a long straight coaxial cable driven by a battery at one end and short-circuited at the other end, the uniform electric field inside the centre conductor is actually produced by a constant gradient of surface charge density.

In other circuit geometries, the distribution of surface charge may be arbitrarily complicated even though the pattern of electric field they produce is simple.

The analysis of surface charge geometries and their resultant electric fields inside wires of varying sizes and arranged in different layouts was studied by a scientist A. Walz.

According to studies in experiments conducted by him, in some simple cases the dominant contribution to the electric field inside a wire may be made primarily by nearby surface charges, because the $1/d^2$ character of the Coulomb interaction may make the contributions of distant charges quite small. To the extent that we can neglect the effects of distant surface charges, we can make qualitative analyses in terms of local causes and local effects.

In Fig. 1.11 is sketched the results of an analysis by Walz of the relative importance of nearby and far away charges when there is a constant gradient of surface charge along a straight wire. One finds that the peak contribution dE to the field comes from surface charges at a distance of $1.4r$, from which Walz concluded that E is due to mainly nearby charges.

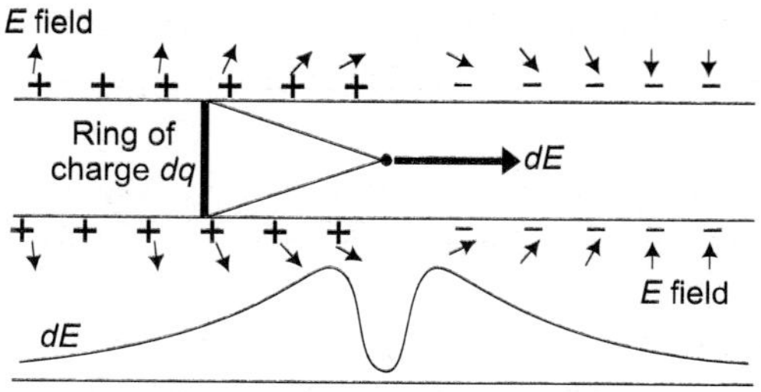

Fig. 1.11 According to Walz, most of the contributions dE to the field E come from nearby rings of surface charge (constant gradient).

However, with a constant gradient of surface-charge density (dq proportional to xdx) a distant ring contributes an amount dE proportional to $(xdx)/x^2 = dx/x$, so the contribution of distant charges along an infinite straight wire diverges logarithmically.

In a real circuit geometry this logarithmic divergence will be cut off due to bends in the wire. While Walz's analysis suggests that nearby charges make a major contribution to the electric field in simple geometries, it is easy to find examples where this approximation breaks down.

For example, if a wire crosses itself or there are multiple loops as in a coil, longitudinally remote but geometrically nearby sections make large contributions to the field.

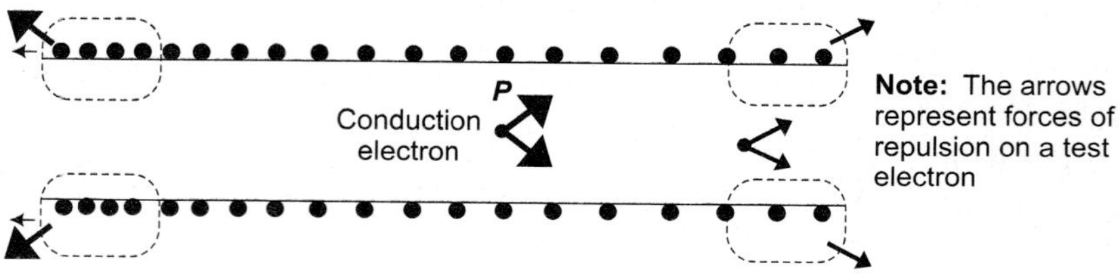

Fig. 1.12 Influence of surface charge at different distances.

Going back to the situation (see Fig. 1.12) where we have a surface charge gradient, when moving away from this cross-section the components in the direction of the wire do not cancel any more; and the resulting forces increase. They reach a maximum at a distance two to five times larger than the diameter of the wire. At larger distances, the interaction vanishes rapidly because the distance increases.

It is, therefore, the influence of a particular surface charge at distances comparable to the diameter of the wire for a long straight wire that exerts maximum force on the free electrons.

For a long and thin wire, it can be shown by calculation that a uniform decrease of surface charges in the direction of the axis (a surface charge with a constant gradient in the X-direction) results in a uniform and *constant force for all the millions and millions of electrons in the direction parallel to the axis of the wire, everywhere inside the wire.*

The electric field due to the surface charges *fills the wire* along its entire length.

It is interesting to see how the excess (local) surface charges arrange themselves when there is a bend in the wire.

If the wire is curved or if other wires with surface charges are closer together, the "gradient" is no longer constant and the circular symmetry of the charge distribution around the wire may be broken.

A few excess charges will bunch close together on the outer side of the bend and cause to exert a force on the conduction electrons so as to turn them and make them follow the bend. This non-linear distribution of surface charges is shown in the figure below.

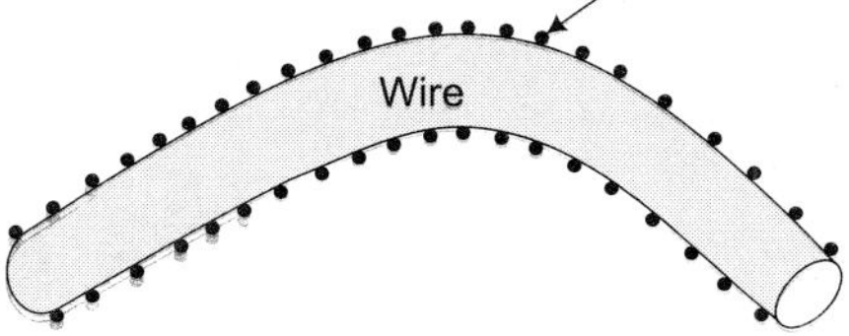

Non-linear distribution of surface charges in a wire with a bend.

1.13 STRUCTURE OF ELECTRICAL CONDUCTORS

In this section, we attempt to visualize the interior of a conductor in order to gain more insight into the process of conduction.

We first begin with a simple description of an atom, "neutral" atom to be precise. For an interactive applet of how an atom can be made, visit http://www.colorado.edu/physics/2000/waves_particles/wavpart2.html.

Most of us were probably introduced to atoms and that these are neutral because they consist of an equal number of positive (protons) and negative (electrons) particles. But that is not the end of the description. In addition, the following are inherent in neutral atoms: (see Section 5.23 also).

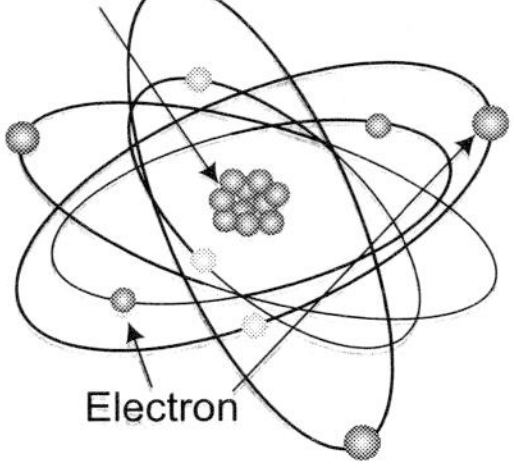

A neutral atom

(i) If the potential energy is set to zero at infinity, the usual convention, then bound electron states have negative potential energy which means some amount of energy has to be supplied to the *single* isolated atom to shift out the (outermost) electrons (say, by application of an electric field) from their orbital space and produce a dipole. Also, refer Section 9.7 on potential energy reference convention.

(ii) A nucleus consisting of neutrons and protons with electrons occupying orbital spaces very precisely; the fields due to the protons and electrons always exist, penetrating everything, but they cancel each other so well that there is hardly a net electric field around the atom!

Visualize *thick and hard* springs bonding the inner electrons and protons together in a single isolated atom, making it difficult to stretch it. See Appendix B for a comparison of gravitational and electric forces, notes on *stable atom* and "is an atom a dipole?". Some of the external forces that can lift or shift an outer electron (visualize *thin* springs holding it with the nucleus) to higher energy levels are heat, light and voltage.

If we visualize as we had done in Section 1.4, that the unlike charges, positive nucleus and electron of an atom, are connected by *a* spring then, the analogy with stretching a spring to store potential energy by stretching the bond is quite appropriate !

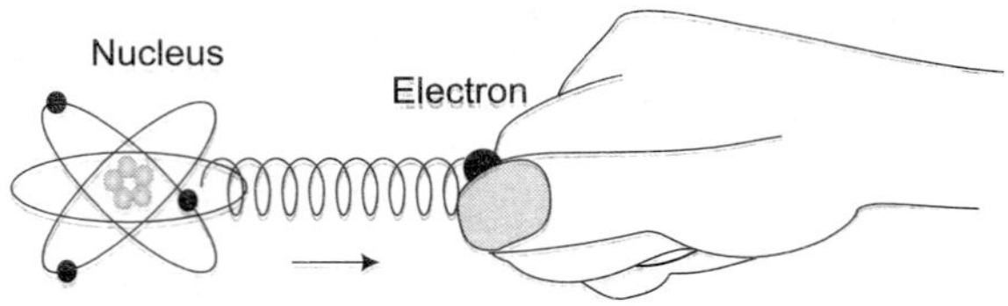

There is one difference though. In the real world of charges comprising electrons and protons and neutrons, the shifted electrons will continue to occupy slightly different orbital spaces when pulled away by supplying energy to them either in the form of heat, light or voltage and when they are released (by removing the heat, light or voltage) they will return to the stable orbital spaces. We shall see in later sections that it is very difficult to stretch the bonds of inner electrons.

See the end of Appendix A "Energy stored in stretched springs and unlike charges that are separated" for similarities in computation of energy stored in springs that are stretched and in bonds of attraction between electrons and positive charges that are stretched.

See Appendix B "Thermionic emission" for a description of how electrons can be extracted by emission from the surfaces of metals.

Most good conductors of electricity are metallic and commonly used ones in electrical and electronic circuits are copper, aluminium, gold, silver and silver-cadmium. Among the good non-metallic conductors are carbon, graphite and ionic solutions and several more.

In electrical conductors (metallic), comprising *millions* of very closely spaced atoms, visualize that the springs holding the *outermost* (valence orbit) electrons and the protons of individual atoms in the conductor have become ***thin and soft*** (the electron energy level occupancies expand towards *less* negative) due to the strong electric interactions (repulsion between electrons) with adjacent atoms, making them easier to shift away from the nucleus. See figure below and also "Metallic bonding and Metallic structures" in Appendix B.

In electricity and electronics, all that matters is the outermost orbit called the *valence* orbit. The electron occupying this orbit controls the electrical properties of the atom.

The outermost valence electrons can easily gain kinetic energy even at room temperature or by applied electric fields of moderate magnitudes such as found in electronic and electric circuits to become free and enable conduction which is why we call the valence electron a *free electron*; the inner electrons are *not affected easily* by moderate electric fields.

We will call the assembly of the neutrons and the protons *and the inner* orbit electrons "the positive atomic core" or "the lattice ions".

Consider the atoms of copper, a metallic conductor. Each atom of copper consists of 29 protons, 29 electrons and 35 neutrons as shown on the left in the figure below.

The lattice ion is the assembly of 29 protons, 28 (inner) electrons and 35 neutrons. The sum of the charges on the protons and electrons of the lattice ion is +29 (protons) –28 (electrons) = +1 and therefore, the lattice ion has a net charge of +1 as shown on the right in the figure below.

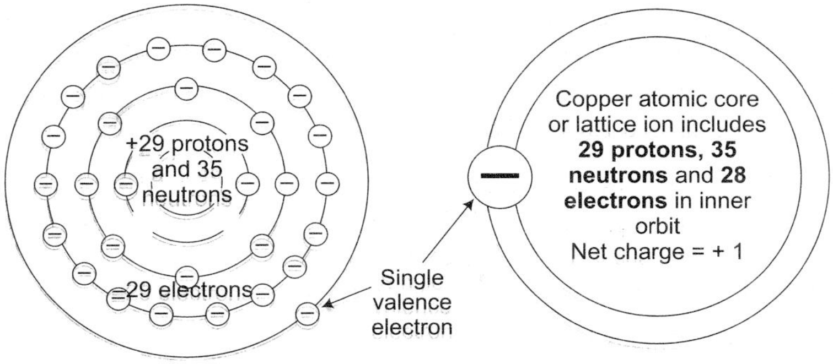

Copper atom

The outermost electron, the valence electron has a charge of –1 as seen on the right in the figure above. If we take the lattice ion and the valence electron together, the net charge is "zero".

The *outermost* (valence electrons) one (in copper) or two or three electrons (in certain other metals) that are quite loosely bound to the nucleus, enabling them to easily gain energy and go into larger orbits in which they become free electrons, roam freely around the lattice ions and form the "electron sea" as seen in the figure below.

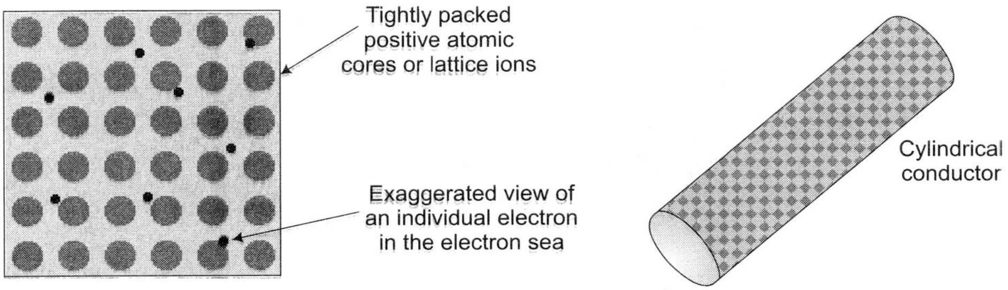

A view of the lattice in a metal and a cylindrical conductor with positive atomic cores and the electron sea.

A simple and useful visualization of a conductor is provided below:

Visualize a long rectangular box or cylinder

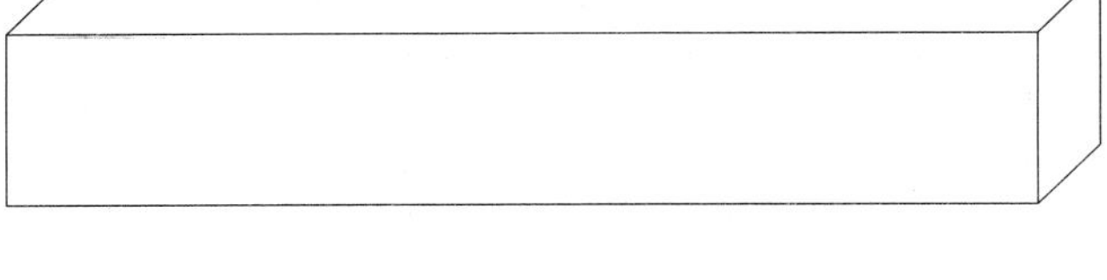

Visualize lattice (window grille) rectangular pieces

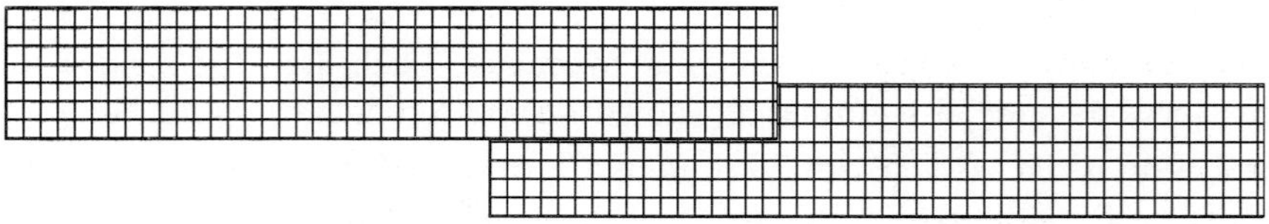

Visualize fixing these lattice strips in the box arranging them in layers....

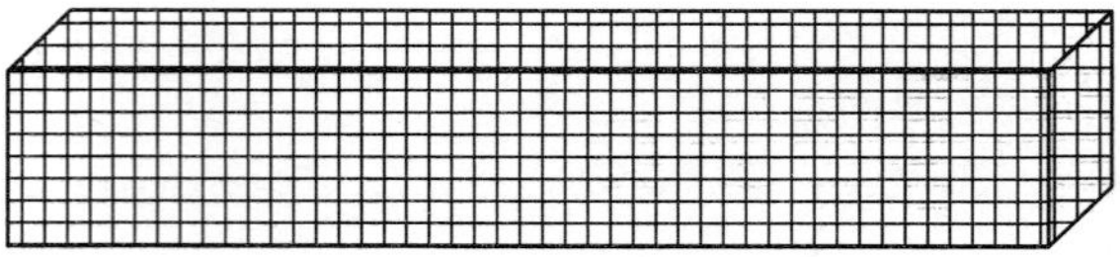

...adding several more until the box is tightly packed. What we are visualizing is a conductor with the positive atomic cores (lattice ions) located at the intersection points of the lattice.

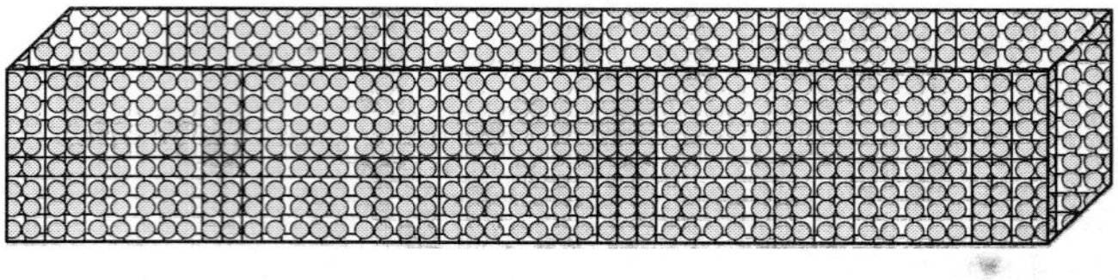

....these are composed of neutrons and protons and include the inner electrons.

Now pour some sand in and that is the conductor composed of the lattice ions and the sand is the *electron sea of the outermost electrons!*

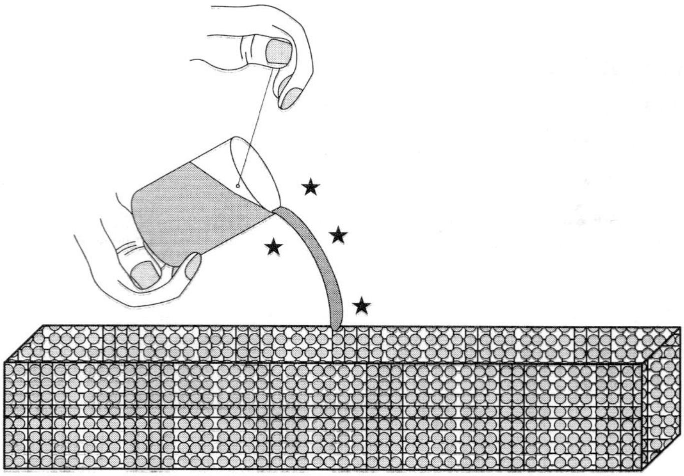

How is a metal "neutral" despite it being composed of ions (positive atomic cores or lattice ions) and electrons, each ion and electron possessing a huge amount of energy and whose individual potential close to it (volts with respect to a point at infinity) will range in the millions of volts?

When a metal is formed by packing atoms tightly together, then, at distances close to a lattice ion, the potential (volts with respect to a point at infinity) varies rapidly with distance and is extremely large (several kilovolts) *very close* to the lattice ion.

There is a very powerful electric field between the nucleus and the electrons orbiting it that may be indicated in figures of atoms by drawing closely spaced thick field lines originating from the nucleus and terminating almost entirely on the inner electrons.

However, due to the lattice ion's electron cloud and the electron clouds of nearby atoms, the net field at distances just outside the atom is very tiny and the voltage *rapidly diminishes* in the region where the free outermost electrons find themselves and this region is an equipotential region of near zero volt and the metal is therefore, *"neutral"*.

See http:// faculty.wwu.edu/vawter/PhysicsNet/Topics/ElectricPotential/PointCharges.html to view the high potentials very near point charges. Scroll to the bottom of the webpage and click "Show Topics Menu Frames" to view more on Electric Potential.

The regions near the lattice ions of rapidly varying potential represent a very *small* portion of the total volume of the metal, because the electron cloud occupies the *more significant portion* of the volume and, we can assume that the field distribution within the metal is equipotential and the free electrons are subject to no forces whatsoever.

While the mass of the proton is 2000 times the mass of an electron, the diameter of the proton is only 1/100000 as big as the diameter of the electron cloud, which is itself only about 2×10^{-10} metres ! An atom consists of a large volume of empty space ! The force of attraction is so huge between the *inner* electrons and the nucleus, that even metallic tools *deform* when attempting to separate these.

See Appendix B on metallic structures and how these are held together by the valence or delocalized electrons.

Between atoms in most metals, however, the forces are more reasonable permitting metallic ingots and sheets to be sheared and bent by cutting tools and presses.

The Electric Field Inside a Metal

(Chabay & Sherwood, 1995, used by Permission)[2]

When a metal or a conductor is subjected to an electric field then the electron sea shifts under the influence of this field as shown in the Fig. 1.13 and surface charges appear.

Electric fields can be present ***inside the matter*** as well as in empty space.

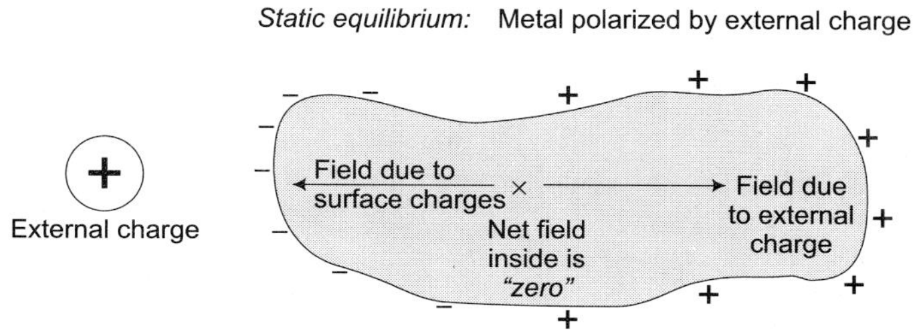

Fig. 1.13 Polarization of a metal.

An external charge polarizes an isolated block of metal completely. The free electron sea shifts until enough surface charge builds up to cancel the influence inside the metal of the external charge.

If this cancellation weren't perfect, there would be a force on free electrons inside the metal which would shift the electron sea until cancellation would be achieved.

Notice that the electric field of the external charge penetrated through the metal block. This is a property of the electric field.

In terms of electric field, we see that at any location inside the metal (marked **x** in the diagram), the electric field of the external charge is cancelled by the electric field produced by the surface charges.

A condition of *static equilibrium* then exists and, the net electric field is zero inside a metal as a result of the field due to external charges and the field due to the surface charges.

It may be noted that there was a movement of the free electrons under the influence of the external electric field prior to the condition of static equilibrium.

1.14 ELECTRIC FIELDS IN MATTER

Polarization means the displacement of positive and negative electric charge to the opposite ends of a nuclear, atomic, molecular, or chemical system, especially by subjection to an electric field or in simple terms, "separation of charges". Atoms and molecules have some inherent polarization. The polarization principle in action in conducting metals is reiterated in the next section. Here, we will study how metals and insulators behave when they are subjected to an external electric field. We will also see the effect of electric fields on atoms.

This is important from the viewpoint of putting together the processes involved in conduction, charging and discharging of capacitors and also in the co-occurrence of non-coulomb electric fields with time-varying magnetic fields (Faraday's law).

In a solid metal the atomic cores are packed close together, so that their electron clouds *nearly* touch their neighbours (Chabay & Sherwood, 1995, used by permission).[2] In such metals that are conductors the conduction electrons can *move freely* throughout the material.

The *sea of free electrons* is nearly *uniform* throughout the solid. When a positive charge is brought near the conductor, its electrons sea *shifts* under the influence of the electric field of the external positive charge as shown in Fig. 1.14.

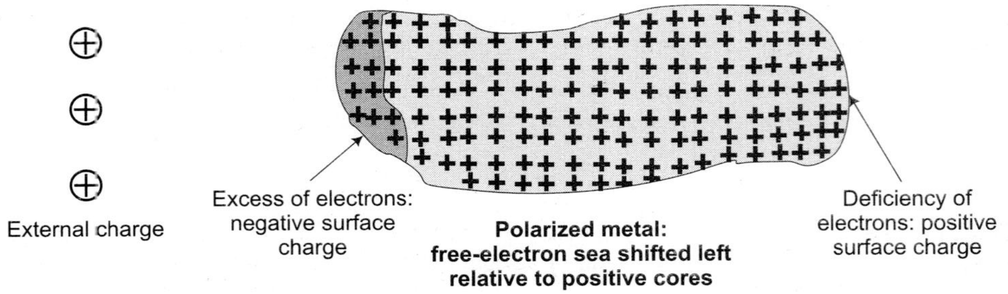

Fig. 1.14 Shifting of the electron sea in a metal under the influence of an electric field.

The shifting stops when the electric field due to the (separated) shifted sea and the exposed positive cores cancels the field of the external charge. In the *interior of the metal*, there is essentially *no change and no excess (local) charge*–we still have a uniform sea of electrons filling the space around the positive atomic cores.

At the *surface there is some excess charge*, so we can represent a polarized metal as having *thin layers of charge on its surfaces* but being unpolarized in the interior, unlike an *insulator that is polarized throughout its interior*.

It is important to note that in static equilibrium the achievable shift in the electron sea is extremely small, typically a tiny fraction of one atomic diameter!

We are unable to apply electric forces so strong as to shift the electron sea very much, because even a tiny shift leads to the appearance of enough surface charge (and accompanying electric fields) to cancel out the forces we're applying.

Polarization of Atoms under the influence of an Electric Field

(Chabay & Sherwood, 1995, Used By Permission[2])

In an atom, the electron cloud is *not rigidly connected* to the nucleus. The electron cloud and the nucleus can move relative to each other. If an external charge is nearby, it can *exert forces on the electron cloud* and on the nucleus which will cause them to *shift position*....creating ....a *dipole*. This is shown in the Figs. 1.15 and 1.16.

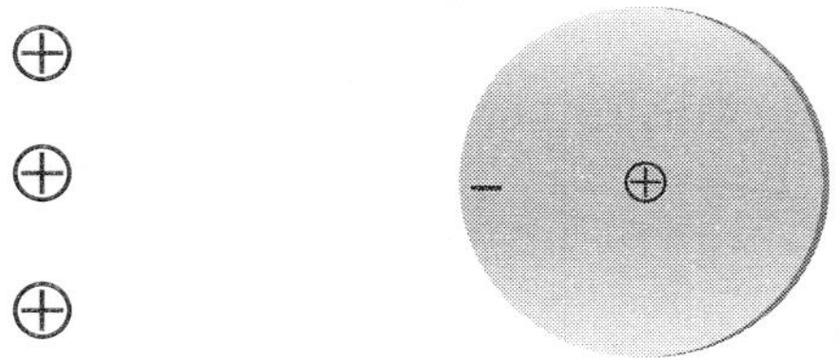

Fig. 1.15 An external charge near a neutral atom.

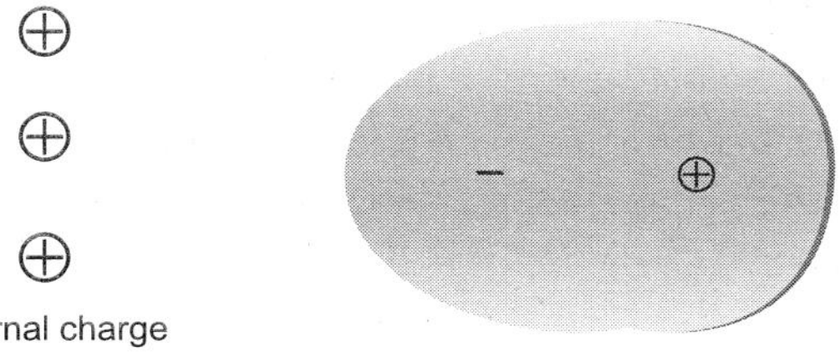

Fig. 1.16 Creating an atomic dipole.

The outer regions of the cloud are affected the most by the external charge because, in the outer regions the electron is attracted less to the nucleus and can be influenced more by the external charge.

Note: Certain molecules occur in nature as dipoles with inherent polarization and are called "permanent" dipoles, for example, water molecules and HCl molecules.

We have seen that an *individual* atom or molecule can be *polarized* by an external charge, but this polarization only produces an *induced dipole*.

When the electron cloud of an atom that was neutral, that is, an *insignificant* electric field surrounds it, is shifted away by an external charge, then, a *significant* resultant electric field will be produced around it due to the separation of the positive atomic core and the outermost electron. A computation of the electric field of induced dipoles is given in Appendix A.

In Fig. 1.17 is shown a piece of dielectric or insulator which, has no conduction band electrons, subjected to the electric field of an external charge.

In insulators, charged particles are ***not free*** to move through the material. All of the electrons are firmly bound to the atoms or molecules making up the material. Commonly used insulators in electrical and electronic circuits are vinyl (polyvinyl chloride or PVC) and rubberised polymers coverings on wires and paper, mica, polycarbonate, polystyrene and Teflon as dielectrics in capacitors.

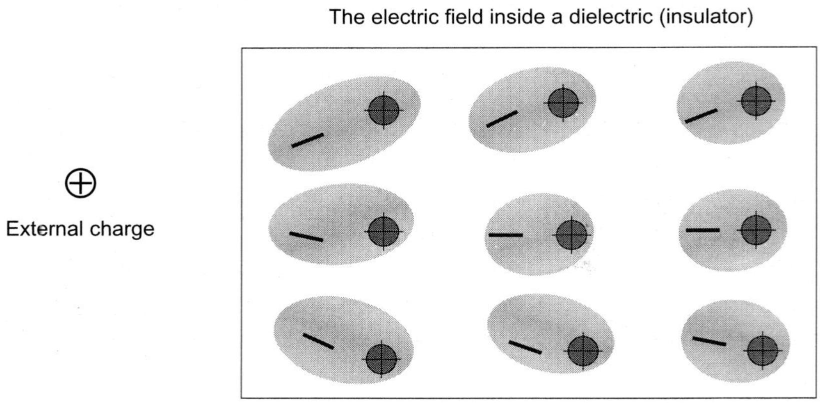

Fig. 1.17 Polarization of atoms in a dielectric.

The *electrons in the atoms or molecules of an insulator shift position slightly*, but remain bound to the molecule; they are *not ripped out* of the molecule. And, of course, protons cannot be ripped out of the nucleus of an atom by an external charge; it would take a *huge force to overcome* the strong force holding the nucleus together.

We see that induced dipoles are a result of separation of unlike charges by external "inducing" charges and these produce electric fields that can exert forces on the inducing charges and the inducing charges also exert forces on the dipole charges.

We can thus summarize the effects of fields due to external charges on conductors and insulators:

1. For metals (conductors):

   One property of metals (and of the $1/d^2$ property of the electric interaction) is that any *excess charge* on a piece of metal is always *found on an outer or inner surface.*

2. For insulators (dielectrics):

   This is different from an *insulator*, where *excess charges can be located in the interior*, or *bound to a particular spot on a surface without spreading out along that surface.*

The effects of electric fields on conductors and insulators are shown in Fig. 1.18.

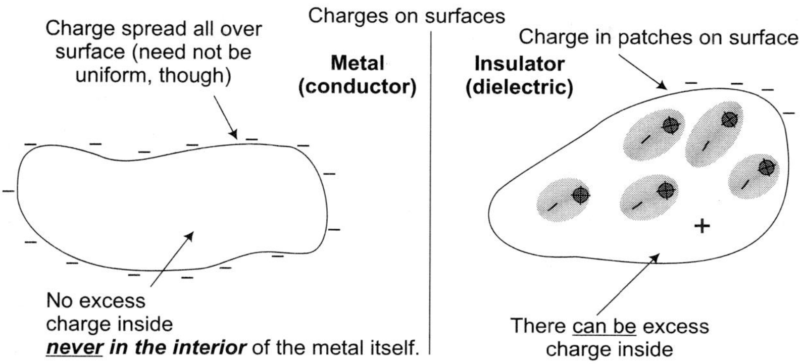

Fig. 1.18 Charges on the surface of conductors and insulators.

Polarization of Ionic Solutions (Chabay & Sherwood, 1995, used by permission)[2]

Ionic solutions, such as a solution of *sodium chloride (table salt)* in water, are conductors. In a solution of sodium chloride, the mobile charged particles are Na^+ ions, Cl^- ions, and the few H^+ and OH^- ions that are present in the solution.

When such a solution is subject to an external electric field, polarization into ions and their subsequent migration to the edges of the container results, as shown in the Fig. 1.19.

Bring a charged object near a beaker of sodium chloride solution (salt water) as shown in Fig. 1.19a.

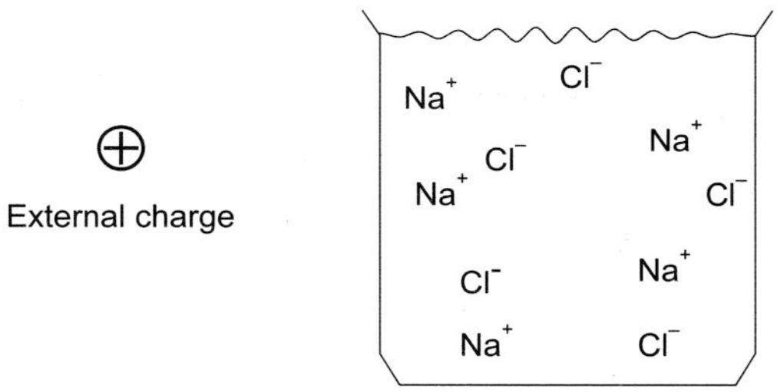

Fig 1.19a External charge brought near an ionic solution.

Polarization of the ions occurs and the *chloride ions* move close to the side of the beaker that is near the positive external charge and the *sodium ions* move towards the opposite side as shown in Fig. 1.19b.

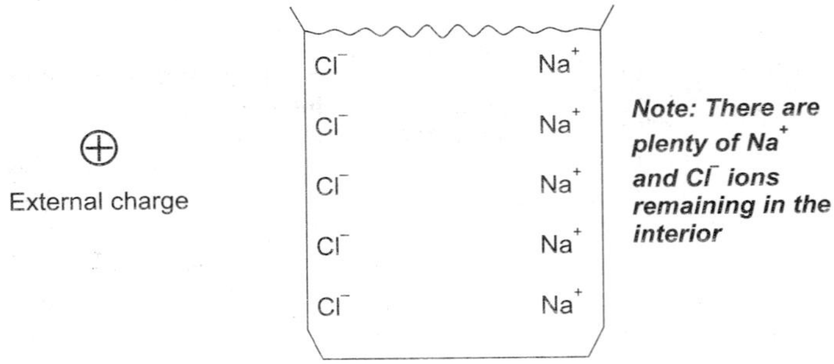

Fig 1.19b Polarization of ionic solutions.

The Electric Field of a Dipole

The figure below shows the electric field of a dipole. Picture courtesy: www.falstad.com reproduced with permission.

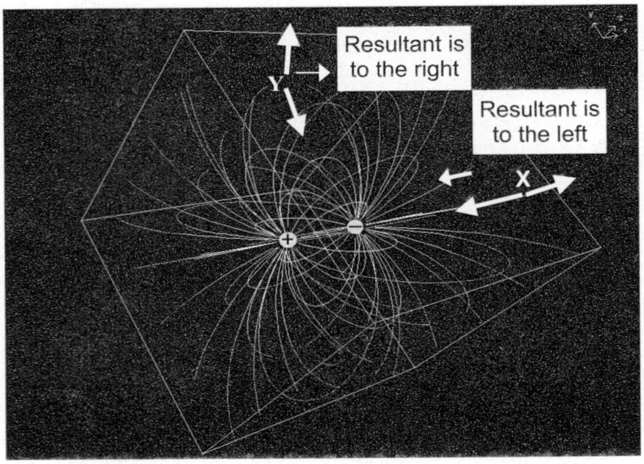

At y: the vertical components cancel and the resultant *E* field is to the right

At x: the resultant *E* field is to the left

The electric field at points **far away** in space where distance x (or y) >> **s** (where *s* is the separation between the individual charges of a dipole) is important in the study of circuits.

$$\text{Electric field at point } x = \left[\frac{1}{4\pi\varepsilon_o}\right]^2 \frac{2qs}{x^3} \quad \text{and, at point y} = \left[\frac{1}{4\pi\varepsilon_o}\right]^2 \frac{2qs}{y^3}$$

The electric field of an isolated point charge drops away at the rate of $1/d^2$ with distance. Note how on the other hand, the electric field of a dipole drops away at the rate of $1/d^3$... that is much *steeper*. See Appendix A for details of the computation of the field of a *permanent* dipole. Millions of point charges whose electric fields drop away at the rate of $1/d^2$, on the surfaces of conductors, add their tiny electric fields to produce large fields that cause sizable currents in circuits, just as millions of tiny droplets of water form the oceans!

Interactions of Electric Charges

A simple experiment that explains the interactions of charges can be performed using scotch tape (cellotape). Use a strip of tape about 20 cm or 8 inches long (about as long as the width of an A4 size paper) (Chabay & Sherwood, 1995, used by permission).[2]

Fold over one end of the strip to make a non-sticky handle. See figure below.

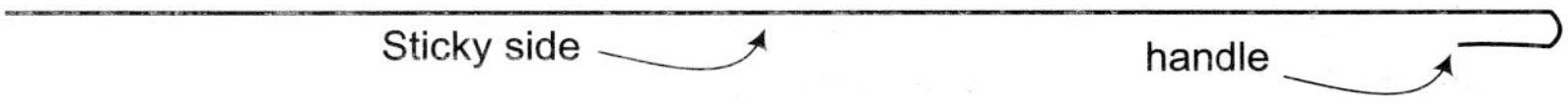

Now, stick a strip of tape with a handle down onto a smooth flat surface such as a desk.

Smooth this tape down with your thumb or fingertips. We will call this the base tape. Stick another tape with a handle down *on top of* the base tape. See the left figure below.

Smooth the upper tape down well with your thumb or fingertips. With a quick motion, pull the upper tape up and off the base tape, leaving the base tape stuck to the desk. See the second figure above. Mark the tape you pulled off "U-tape".

Your Hand and the U Tape

Hang the U tape vertically from the edge of the desk, and bring your hand near the hanging tape.

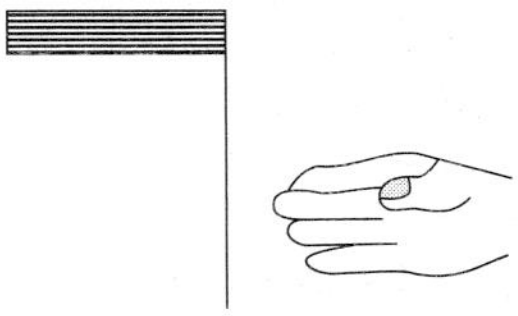

If the tape is in good condition and the room is not too humid, you should find that there is an attraction between the hanging strip and your hand when you get close to either side of the tape.

If you don't see an attraction, remake the tape. You can reuse the same piece of tape by again sticking it down on the base tape and then, running your finger along the base tape that is on the desk, sticking the U-tape down on the base tape and, smoothing the tape down well.

Now, jerk the U-tape very quickly off the base tape.

The tape should attract to your hand when hung vertically from the desk.

Why does the U-tape get attracted to your hand?

Your own body consists mainly of salt water, including the blood and the insides of cells. One effect of polarization of the salt water inside your finger is that there will be a shift of Na^+ and Cl^- ions in the blood and tissues. This shift may be a larger effect than the molecular polarization.

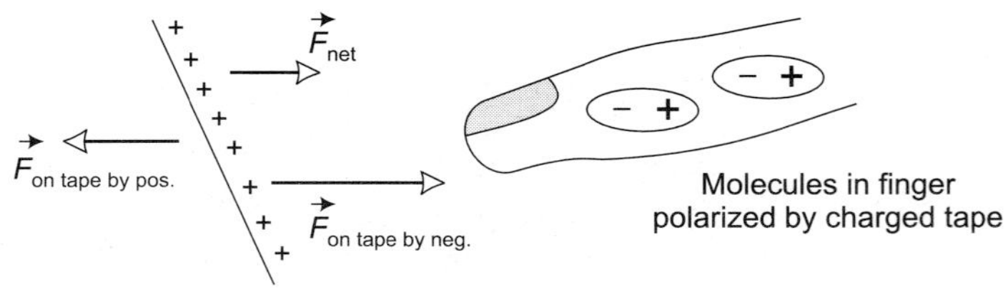

The tape holds a positive charge as seen in the figure and due to these charges, the tape polarizes molecules in the finger. Negative ends of molecules end up closer to the tape. The tape is attracted to the negative ends, and repelled less strongly by the positive ends because they are farther away, and force is proportional to $1/d^2$. The net force on the tape by the finger is the superposition of these forces, and is towards the finger, resulting in the attraction of the tape towards the finger.

Unless the external charge is just a few atomic diameters away, the shift in the electron cloud is typically so tiny that it can't be seen in a picture. But, if all the molecules in your hand are polarized even very slightly by a charged tape, the tiny interactions with the huge number of molecules can add up to an observable effect.

The tape is shown to hold an excess of positive charge but that could have been negative either.

The polarity of the charge would depend on the type of tape you have used. See Section 1.25 for an explanation of how the tape is charged.

Electric and Gravitational Fields Penetrate Intervening Matter

Observe that when bringing your hand towards either side of the charged hanging tape, the tape is attracted to your hand, despite the charges being on just one side of the tape. Intervening matter does not "screen" or "shield" the electric field, just as your desk does not "screen" or "shield" your book from the gravitational field of the Earth! Else, you would find the book floating in space! It is evident that the gravitational field penetrates intervening matter.

The Electric Field Inside Metallic Wires

During the *polarization* process that we saw earlier in a metal, leading to static equilibrium, charges are *moving* and the electric field is *not zero.*

Non-equilibrium: In an electric circuit powered by a battery, the free electron sea in the body of the wires and resistances is *continually shifted* through the wires by a non-zero applied *electric field inside the wires which applies a force on the electrons*.

It is to be noted that this is not a state of static equilibrium.

This is also known as a ***steady-state***.

1.15 THE NATURE OF SURFACE CHARGES IN CIRCUITS

We shall now see how when we apply a voltage to a simple circuit consisting of a battery and wires, an electric field is established by the development of surface charges due to the applied voltage. Refer to Fig. 1.20 that shows a simple circuit consisting of a battery and wires.

For the purpose of the analysis we shall ignore the fact that if we connect a circuit in this fashion the battery is effectively short-circuited and that the wire will heat up.

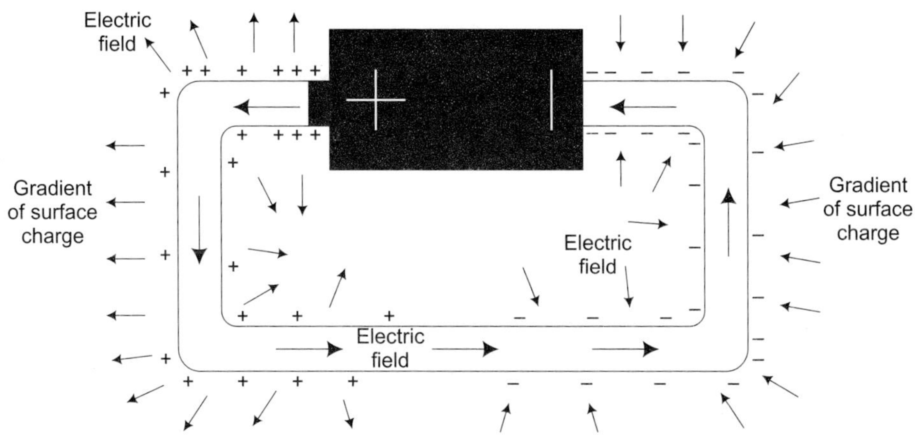

Fig. 1.20 Surface charges in a circuit.

There must be *positive and negative charges* on the surface of the wires *near* the positive (+) and negative (-) plates of the battery, and in the steady state the surface *charge must vary smoothly* from positive to negative around the circuit, possibly with extra charge on the bends in the circuit, to produce non-zero resultant electric fields inside the wire and in the space surrounding the wire and the circuit (not forgetting that an electric field is an invisible *alteration of* space). If the wire is uniform, the field inside will be uniform too. The field is *not* uniform in the space surrounding the circuit, as there are no mobile carriers there that can produce a uniform field.

Electrons in the free-electron sea inside the wires will move due to the fields from the battery. Imagine that you could take a "snapshot" of the surface charge after the electrons inside the wire have moved just a little bit (that is, after a very short time). Electrons moving toward a positive region of the surface will make the surface charge in that region less positive (negative point charges), while electrons moving away from a negative region of the surface will make the surface charge in that region less negative. A + sign on the surface does not represent some exotic positive charge. Rather, a + sign indicates a region where there is a deficiency of electrons, leaving partially exposed the metal's positive atomic cores, or positive point charges.

If the ends of a wire are connected to a battery, an electric field E will be set up at every point *within* the wire. If the potential difference maintained by the battery is 10 volts and, if the wire (assumed uniform) is 5 metres long, the strength of this field at *every point* will be 2 volts/metre. This field will act on the electrons and will give them a resultant motion in the direction of $-E$. In Section 1.23, we will verify the field strength value by applying the law of conservation of energy.

Some electrons will *leave the negative terminal because of Coulomb repulsion and enter the wire*, other electrons will *leave the wire and enter the positive terminal*, charging the wire negative near the negative terminal and positive near the positive terminal.[3]

Thus, a *gradient* of charge density (in the shape of *rings* for circular cross-section wires) will form along the wire, and the field inside the wires from these charges, as well as those on the terminals, will cause a current of density J, given by,

$$J = \sigma E$$

where, J is in amperes per square metre or, current 'I' moving past a unit cross-sectional area in a conductor whose total area is 'A' ($J = I/A$) of conductivity σ mhos per metre and E is the applied field due to the battery. The current I is shown in Fig. 1.21.

We can state a rule thus: When a dc circuit is closed, positive charges will enter the wire near the positive terminal of the source and negative charges will enter the wire near the negative terminal.

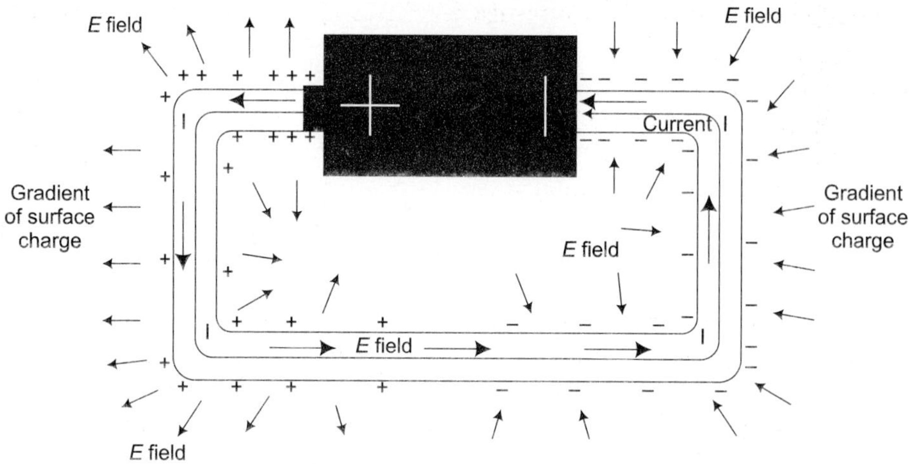

Fig. 1.21 Uniform electric field inside the wire and current I.

"Steady state" means that some charges *are moving*, but their velocities at any location aren't changing, and there *isn't any change in the deposits* of excess (local) charge anywhere. The circuit achieves and maintains this condition by a continuous feedback mechanism between the electric field and the surface charge and the moving conduction band electrons in the wire.

Feedback Leads to Current Conservation

(Chabay & Sherwood, 1995, used by permission)[2]

Consider a straight section of circular cross-section wire that has *more* incoming electron current than outgoing electron current (shown by the unequal white block arrows at the entrance and exit of the wire in Fig. 1.22) *leading to* a build-up of excess (local) *negative* charge on the surface of this section of wire. This could happen if there was an increase in the applied voltage.

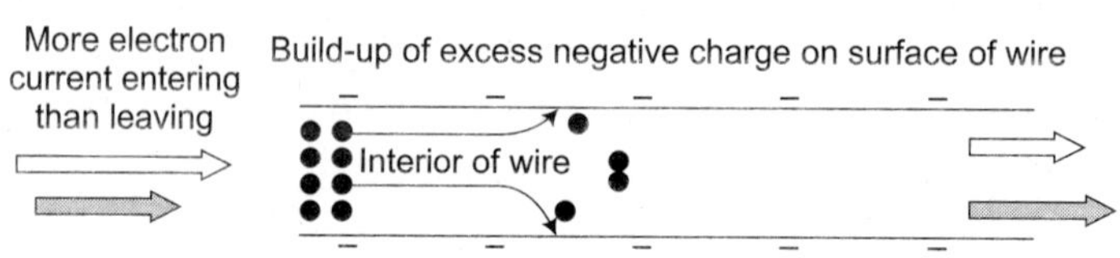

Fig. 1.22 Surface charge and feedback to maintain steady state.

The surface charge *build-up* (in the shape of rings) will tend to *equalize* the incoming and outgoing electron *currents*.

The outer surface of this section of wire becomes more negative, and this *negative surface charge tends to impede* the incoming electrons and tends to speed up the outgoing electrons, thus making the incoming and outgoing electron currents more nearly equal to each other.

This *feedback process* will continue until the two currents are *exactly equal* to each other. At that point there will be no further change in surface charge. This is shown by the shaded block arrows.

Now, consider that there is *more* outgoing electron current than incoming electron current say, due to a decrease in the applied voltage (shown by the unequal white block arrows at the entrance and exit of the wire in Fig. 1.23) *leading to* a build-up of excess (local) *positive* charge on the surface of this section of wire.

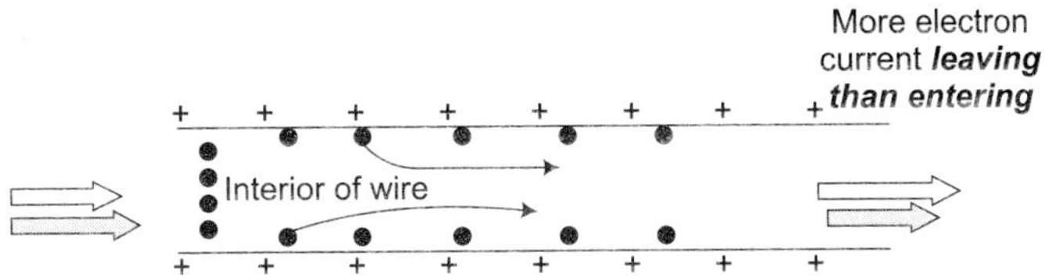

Fig. 1.23 Surface charge and feedback to maintain steady state.

The resulting *surface charge build-up* (in the shape of rings) will tend to *equalize* the incoming and outgoing electron *current* by speeding up the entry and slowing down the exit of electrons. Eventually, the entry and exit electron currents are equal as shown by the shaded block arrows.

So, *whichever way there is an imbalance* between incoming and outgoing currents, the two currents will quickly become equal to each other due to the *automatic feedback mechanism* of changes in the surface charge.

In their paper "A unified treatment of electrostatics and circuits", Chabay and Sherwood indicate that for a straight coaxial cable 30 cm long whose central conductor has a diameter of 1 millimeter, driven by a 3-volt battery at one end and shorted at the other end, the maximum linear surface charge density on the central conductor is *only a few million* electrons per centimeter.

Note: A 1 centimeter long 1 millimeter diameter copper wire contains 6.6×10^{14} *million* free electrons and therefore, the surface charges are indeed very few in number compared to the free electrons in the wire.

Connecting a Circuit: The Initial Transient

When you connect a circuit by making the final connection, *feedback* forces a *rapid rearrangement* of surface charges leading to the steady-state.

This *period* of adjustment before establishing the steady-state is called the "*initial transient*".

Surface Charge and *Feedback* to Maintain Steady-State

The wires in a circuit with steady currents are electrically neutral (neutrality will be explained in Section 1.18) in their interiors, but have *small amounts* of charge on their surfaces.

These surface charges are critical to the behaviour we expect from circuits, for they make the *internal electric field* parallel to the wires and ensure that the *same* amount of current flows in different parts of the circuit. *If*, the electric field and surface charges do *not* create the conditions for steady-state currents, rapid changes occur in the surface charge distribution until these conditions *are* satisfied.

There is feedback because the electric fields *change* the surface charges and the surface charges *produce* the electric field.

This *feedback* between charges and the electric field is not instantaneous, because changes in the field propagate at the speed of light *c*.

What happens if you bend a straight wire while the current is running?

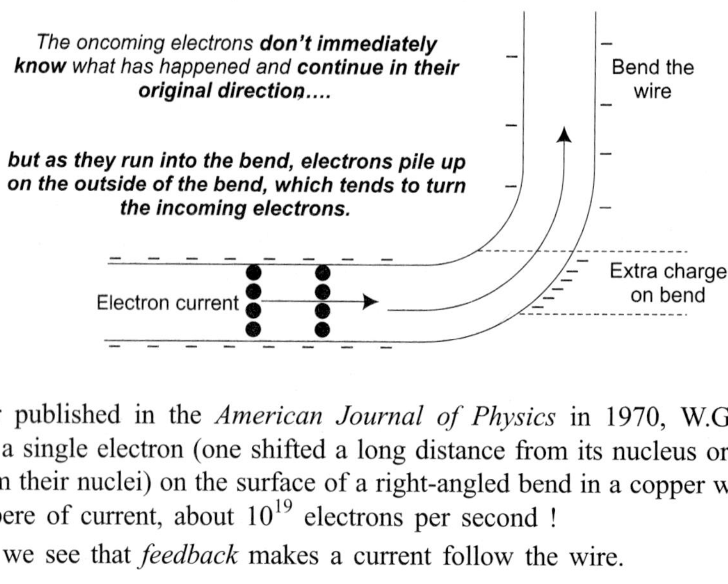

In a paper published in the *American Journal of Physics* in 1970, W.G.V. Rosser has calculated that a single electron (one shifted a long distance from its nucleus or several shifted marginally from their nuclei) on the surface of a right-angled bend in a copper wire is sufficient to turn an ampere of current, about 10^{19} electrons per second !

Therefore, we see that *feedback* makes a current follow the wire.

1.16 THE SPEED OF SURFACE CHARGE BUILD UP

The build up of charges on the surfaces of conductors in circuits occurs very rapidly of the order of 2 to 5 times the *light crossing time* of the circuit.

The *light crossing time* of a circuit is the time to cross a physical circuit at the speed of light. In typical electronic circuits the light crossing time will be of the order of a few ***nanoseconds***.

The time required to *move* the electrons *towards or away from the surface is completely negligible*, because the electrons *only* have to *move* an *infinitesimal distance* in order to establish a significant amount of surface charge! Surface charges are point charges that set up to produce a resultant field.

In the simple circuit of Fig 1.24, the *electron-sea is continuously shifting* under the influence of a battery.

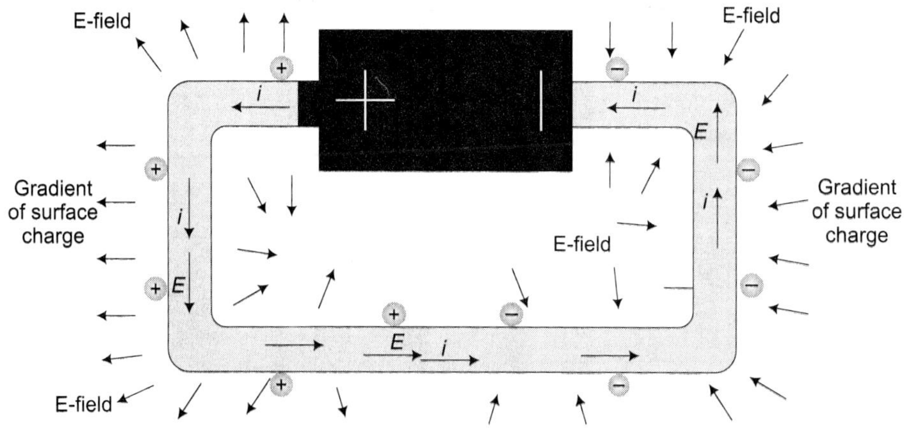

Fig. 1.24 Speed of surface charge "development".

There is *no excess (local) charge anywhere inside the wire* – to every free electron there corresponds a singly charged positive atomic core – an atom minus one electron which has been released to roam freely in the free electron-sea. The conservation of charge principle is upheld when we consider that the battery and wires constitute a larger closed system; never mind that the charges were separating or depleting (inside the battery) and moving as inside the wires.

Averaged over a few atomic diameters, the *interior of the metal is neutral*, and the *repulsion between the free electrons* is *balanced* out on the average by the *attractions of the positive cores*; at the surface, the elcctrons *only* have to *move* an *infinitesimal distance* in order to establish a significant amount of surface charge rings because, in reality, surface charges are point charges, that create an electric field in the interior that will extend around the wire.

In Fig. 1.24 surface point charges of both signs (see Section 1.15) are depicted. In a few nanoseconds, the *rearrangement* of the *surface charges* will extend *all the way* around the circuit, through slight expansion or contraction of the mobile-electron sea (compare with the molecular stretch associated with polarizing an insulator, which is less than the radius of a single proton!). When the circuit is completed with the battery, excess negative charge expands the mobile-electron sea so that some of the sea peeks out at the surface, giving a negative surface charge. Excess positive charge contracts the electron sea, allowing positive cores to peek out at the surface.

Summarizing the Effects of Feedback

Feedback in *static electricity situations* leads to static equilibrium ($E_{NET} = 0$ inside a metal).

Feedback *in a circuit* leads to surface charges and steady-state current ($E_{NET} \neq 0$ inside a metal).

Never forget that a **NET** field implies the superposition of all the fields that are alive and existing even as you are computing or visualizing the NET field and that is the reason for the resultant field to be called a **NET** field. See Appendix A, "Superposition of electric fields".

We had stated that the fields due to charges always exist, then how can we speak about its travel at the speed of light ? Assume that you have been observing the electric field due to a stationary charge 30 feet from you. If the charge moved, it would take a finite time (30 nanoseconds, because light travels 1 foot per nanosecond) before the new magnitude of electric field due to the new *position* of the charge could be observed simultaneously because, both light and the *change* in the electric field travel at the speed of light. It would be wrong to say that the electric field of a charge travels at the speed of light. What the observation proves is that, it is the *change* in the electric field due to movement of charges that travels at the speed of light. According to J.D. Jackson[10], surface charges perform the following major roles in circuits:

1. ***maintain the potential*** around the circuit
2. ***provide the electric field*** in the space around the circuit
3. assure the ***confined*** flow of current.

Keeping in view, the conduction mechanism in circuits and the processes that enabled the maintenance of static equilibrium in conductors it should be possible to answer the question:

Does a voltage develop across an open switch?

We know V = IR therefore, if there is no I there cannot be a V.

Wrong!

The presence of surface charge should dispel that notion.

1.17 CONDUCTION – A MICROSCOPIC AND MACROSCOPIC VIEW

We have seen in Section 1.13 how in metals, the valence electrons are not attached to individual atoms but are free to move about within the lattice and are called conduction electrons.

In copper, there is one such electron per atom, the other 28 remaining bound to the copper nuclei to form lattice ions or atomic cores.

The figure below attempts to depict the random motion of the electron in the metal.[13]

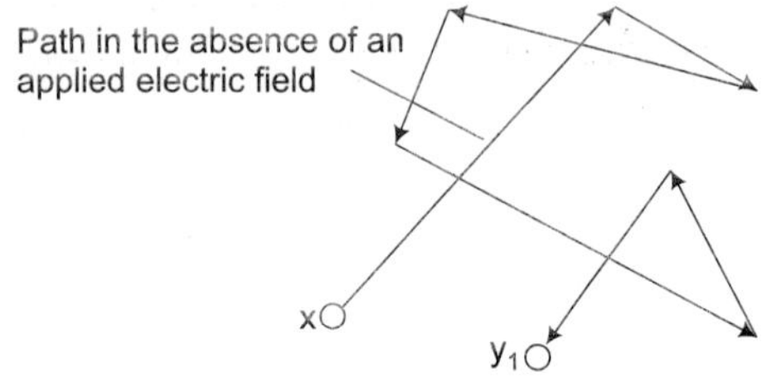

The solid lines depict a possible random path of an electron in the *absence* of an applied electric field, proceeding from x to y_1, making six collisions on the way, bouncing about in an atomic lattice at a speed of about 1/200 the speed of light. The electron collides constantly with the atomic cores (lattice ions), often suffering sudden changes in speed and direction.

In an isolated metallic conductor with no electric field applied, there is no net motion because at any cross-section and for a given interval of time for as many electrons that would have moved to the right, an equal number of electrons would have moved to the left and the current is construed to be zero. The dashed lines in the figure below show how this path is altered when an electric field is applied.

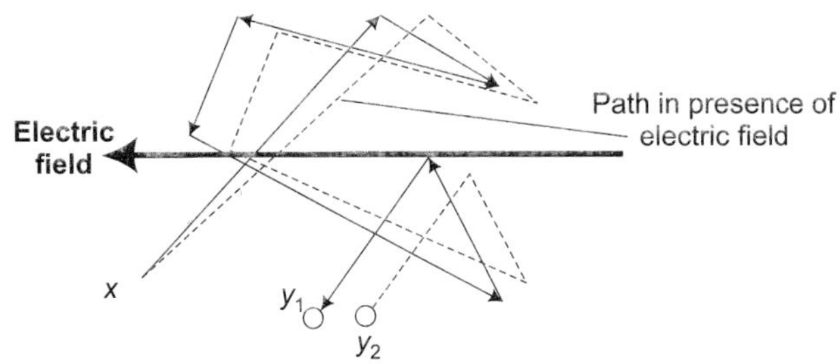

The electron drifts towards the right with a speed much less than a snail's pace. Note that the electron drifts steadily to the right, ending at y_2 rather than at y_1. When a field is applied to an electron in the metal, it will experience a force qE which will impart to it an acceleration. It soon collides with a lattice ion and loses its tendency to drift and will have a truly random direction after the collision. If the applied electric field is steady, the electron accelerates after the collision and again collides and moves to a point a little beyond the point y_2, and so on. A net motion to the right would have been established. In the next section, we shall see how the continuous collisions the electron makes with the lattice results in a slow drifting motion of the electrons. The number of collisions per second is a stupendous 10^{14} *per second*, and the mean free path between two collisions is about 10 diameters of a copper atom.

When the electric field is impressed on a conductor, however, the conduction band electrons are accelerated and their velocity (drift) increases in proportion to mobility. The number of collisions remains close to 10^{14} per second but the *intensity* of the collisions increase. At a macroscopic level we say that a current density is established.

The quantity J denotes current *density,* a number representing flow of charges per unit time in elemental volumes, and because the electric field penetrates through matter and due to surface charge effects, acts with the *same* strength on *all* the charges in the volume uniformly, it also refers to the number of charges that cross a *unit area* per second. If the current density is uniform and flows straight into a wire of cross-sectional area A then, the current $I = JA$.

In the figure below a pictorial view of the total current I and current density J is given. For ease in visualization, a rectangular cross sectional wire is assumed.

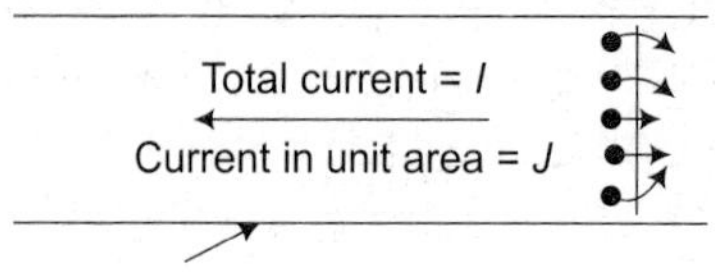

Rectangular cross-sectional wire

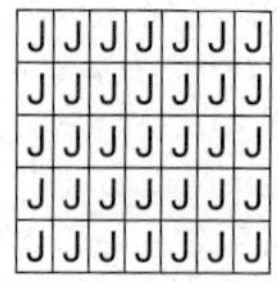

Total current I is sum of the current densities (J) in each elemental area

There are two views of conduction and its causes—the microscopic and the macroscopic.

Microscopic	**Macroscopic**
The drift velocity $v = \mu E$	The current density $J = \sigma E$
and, $I = qnAv = qnA\,\mu E$	and, $I = qnAv = \frac{1}{R}\lvert\Delta V\rvert$
where, v = drift velocity μ = mobility (cm^2/V-s) E is the applied electric field A = cross-sectional area n = number of free electrons per unit volume	where, σ = conductivity = $qn\mu$, q = magnitude of charge of an electron in coulombs.

In the *microscopic* view, we focus on individual electrons whose drift speed is caused by the electric field produced by external charges (*nearby surface charges, surface charge on distant parts of the circuit and charges on the battery).*

Typical values of mobility for electrons are: silicon 1350, germanium 3900 and copper 100.

The smaller the mobility, μ, the *tighter* the collisions and resulting movement of the electrons within the lattice (more lattice defects can produce this effect, see "Resistivity of metals" in Appendix B) and larger their mobility μ, the more loose or *freer* their movement, as is evident from the equations for conductivity $\sigma = nq\mu$ and the drift velocity $v_d = \mu E$.

In the *macroscopic* view, we measure conventional current in the every-day units of amperes, and this current is numerically proportional (with proportionality constant $1/R$) to the potential difference, which we measure in the everyday units of volts.

It is useful to gain a feel for the mechanism of current flow. Some scientists have captured the flow and these are given in the next section. These should be only treated as aids to visualization and though not accurate by all means because of the myriads of particles involved at the atomic level, they serve to capture the essence of the movement.

1.18 VISUALISING CURRENT AND NEUTRALITY IN CONDUCTORS

Consider a simple circuit comprising a battery, connecting wires and a resistor that is closed. After the initial transient, lasting a few picoseconds, all the electrons, the millions everywhere in the wires gain a net movement in one preferred direction, and we call this "current".

Since fast running water has kinetic energy which has to leave the system when the *waterflow* is switched off and the kinetic energy of *electrons*, however, is negligible compared to all other interaction processes, the *analogy* of *fast flowing* water does not hold for conduction processes.

A better and more *suitable analogy* is a closed water circuit under *high pressure but with a very low flowing rate.* A current can be switched off with almost no energy dissipation.

Electric current is actually a flowing motion of charged particles.

When charge flows in a battery and resistor circuit, the current is *sustained* by an "applied" e-field in the steady state like a steady and continuous application of force to maintain motion which otherwise will cause the current to stop because of the dissipation or energy exchange in the resistive portion.

Consider a section of metal wire through which the free-electron sea is *continuously shifting to the left*, with a "drift speed" v as shown in the figure below. Averaged over a few atomic diameters, the interior of the metal is *neutral.* A detailed discussion of neutrality inside current carrying conductors is dealt with in a subsequent paragraph in this section.

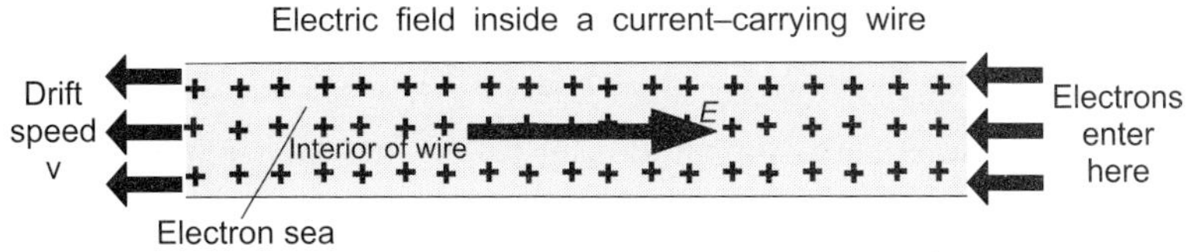

If there is an electric field inside the wire, *wouldn't an individual electron accelerate rather than drift at a constant speed*? After all, Newton's second law says that an electron should experience,

an acceleration, $$a = \frac{\sum F}{m} = \frac{-eE}{m}$$

A free electron in the metal *does accelerate* and gain energy, but then it *loses that energy* by *colliding with the lattice* of *atomic cores, which is vibrating because of its own thermal energy.* (Chabay & Sherwood, 1995, used by permission).[2] In general, the cooler the metal, the less lattice vibration there is and the fewer collisions between electrons and the lattice. After collision, an electron *again* gets *accelerated*, and *again collides*. This process is what makes the metal filament in a light bulb or the nichrome wire get hot. The figure below shows a graph of this start-stop sequence.

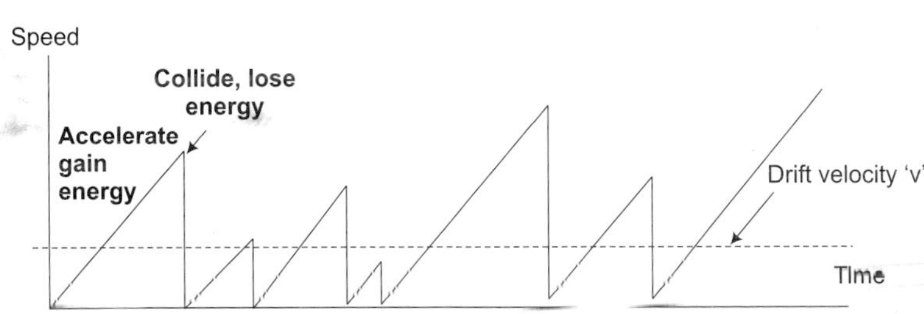

Electric current is a very *slow* flow (drift) of charges produced by the start-stop motion of millions and millions of electrons already existing in the wires, as shown in the Fig 1.25.

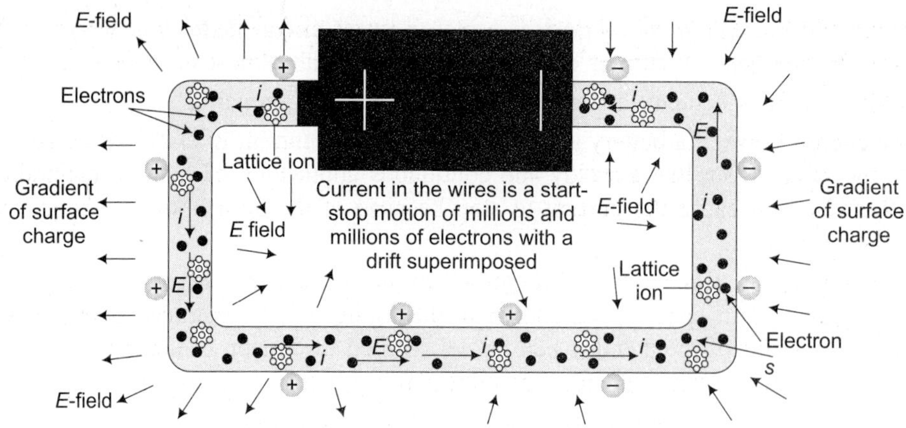

Fig. 1.25 Electric current in a simple circuit.

The average speed in this start-stop motion is called the *"drift" speed v*, and we say that the electrons *"drift"* through the metal. Suppose you hook a battery to a light bulb. Electric current appears in the connecting wires. Now break the circuit. The millions of flowing electrons everywhere in the wires, the bulb filament and through screw terminals, solder joints, crocodile clips and plug and socket, do not zoom back into the battery; instead they halt where they are and, begin executing the random movement with zero drift in the absence of the electric field as described in the previous section.

If the applied voltage is increased, the start-stop motion of the millions and millions of electrons becomes more vigorous but remains *tight* within the confines of the lattice making intense collisions, and there is a consequent increase in the drift velocity and the current accordingly increases. For a constant applied voltage (therefore, constant field) and in the steady-state, the drift velocity is constant.

There are several analogies described in textbooks; for example of the motion of objects dropped in a tank of liquid and that attain a constant "terminal velocity" *when* the gravitational force is balanced out by the viscous forces.[13] The viscous forces of the liquid cancel out the acceleration produced by gravity leaving the object to move with a steady velocity, the "terminal velocity". Note that the viscous forces will be there as long as the object moves, and will disappear if the object is held steady. Therefore, the object moves with a *terminal velocity*.

Analogously, in a conductor subjected to a steady applied voltage (field) the electron collisions with the heavy lattice ions, "cancels" out in a "discretized" manner in a manner of speaking, their acceleration produced by the steady applied field, resulting in a steady velocity, the "drift velocity". There is one difference between the force and acceleration component cancellation of the object in the tank and that of the electrons in the conductor; the viscous forces on the object are produced on its *surface* and yet cancels out the acceleration of the object as a whole, while the acceleration of the millions of electrons is impeded by their collisions with the millions of lattice ions everywhere *in* the conductor, as would ball bearings (electrons) moving down a ramp studded with a forest of nails (lattice).

Such movements are described by equations where velocity and not acceleration (!) is proportional to applied force. However, I strongly recommend visualizing the actual movement of the electrons as described above and this should be sufficient and accurate enough for the visualization of the process of conduction.

If you *take away* the lattice ions and leave a vacuum and apply the field, the electrons will accelerate continuously and their motion is termed "ballistic". This is what happens to *liberated* electrons in vacuum tubes and oscilloscopes.

If the voltage applied to the conductor is changed, it is the *change* in the electric field that travels at the speed of light and *not the electron*. The *conduction band* electrons (in conductors, which are giant crystals made from atoms) accelerate to very large velocities that are yet small compared to the speed of light, and gain a huge amount of kinetic energy. But they soon decelerate due to the collisions. If they did not gain such large kinetic energies due to their high velocities, their collisions would be analogous to a tiny marble (electron) colliding against a cricket ball (lattice ion) when little energy can be transferred to the cricket ball (lattice ion). At higher velocities, with their large kinetic energies however, they transfer *most* of the energy they gained *between* collisions to the lattice ions and heat them up, like shooting swarms of high velocity marbles at cricket balls to make them move.

Since a coulomb is approximately equal to $6.24150948 \times 10^{18}$ elementary charges, one ampere is approximately equivalent to $6.24150948 \times 10^{18}$ (6.2 million-million-million) elementary charges, such as electrons, moving past a boundary in one second.

Note that in the description of the ampere it is the number of charges that move *past a boundary* in one second that matters and there is *no requirement* for the charges to travel all the way around the circuit to qualify for labelling as a current.

If the electrons in a conventional circuit have a start-stop motion moving a few micrometers, there will be a net transport of electrons past one cross-section (boundary), and there are some other electrons (not the same) that will be crossing a boundary at another cross section (boundary) a few micrometres away and different electrons crossing at any boundary, all around the circuit, at the same instant, with the same number flow rate.

That is sufficient to be registered, as a current....by an ammeter.

Figure 1.26 gives a more detailed view of the electron movement in a small section of a conductor. This view was made based on the above descriptions of the start-stop motions and does not claim to be accurate.

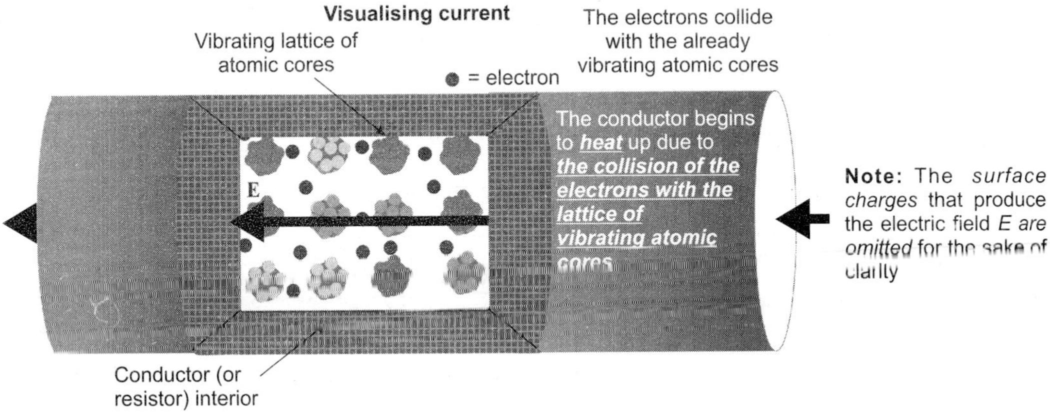

Fig. 1.26 Action in one small slice (a few atomic diametres wide) of a conductor impressed by an electric field E in a circuit.

With no electric field applied, the electrons gain energy from their collisions with the rocking lattice ions. The lattice ions, even in the absence of a current, vibrate on their own accord due to temperature. To gain an intuitive understanding of temperature, please read "What is temperature" in Appendix B. To view the rocking motion of the lattice ions in a metal, visit http://www.chem.iastate.edu/group/Greenbowe/sections/projectfolder/flashfiles/thermochem/eqilibrium-v1.html and raise its temperature to increase the rocking intensity.

It is instructive to calculate *how long it takes an electron to get from a car battery to the starter motor after the ignition switch is turned.* Assume that the current flowing is 115 A, and that the electrons travel through copper wire with cross-sectional area 31.2 mm^2 and length 85.5 cm.

What is the *current density* in the wire?

$$3.69 \times 10^6 \text{ A/m}^2$$

The number density of the conduction electrons in copper is 8.49×10^{28} /m^3.

Given this number density and the current density, what is the *drift speed* of the electrons?

0.00027 metres/second or 0.97 *m/hour* = nearly *forty thousand times slower* than the speed of the fastest runner in the Olympic games.

Is it anywhere near the speed of light? c = 1080 × 1000000 km/hour.

How long does it take for an electron starting at the battery to reach the starter motor? A length of 85.5 cm or nearly a metre.

52.5 minutes (or nearly an hour!)

In that time, the slowest snail has been recorded to move a distance of 12 meters!

Comparing this drift speed (0.00097 km/hr) with the instantaneous speed of electrons in copper at room temperature which is about 5760×10^3 km/hr (only 0.005 times the speed of light), we see that the drift speed is indeed very very small.

We see that the electron moves only a small distance inside a conductor in which there is a current. Therefore, current in the wires is a start-stop motion of millions and millions of conduction band electrons everywhere within with a drift superimposed (see Fig. 1.25). This raises a question as to how energy is transported in circuits and its speed. This question will be answered in subsequent chapters.

Neutrality in Current Carrying Conductors

Conductors carrying currents *remain neutral* in the interior.

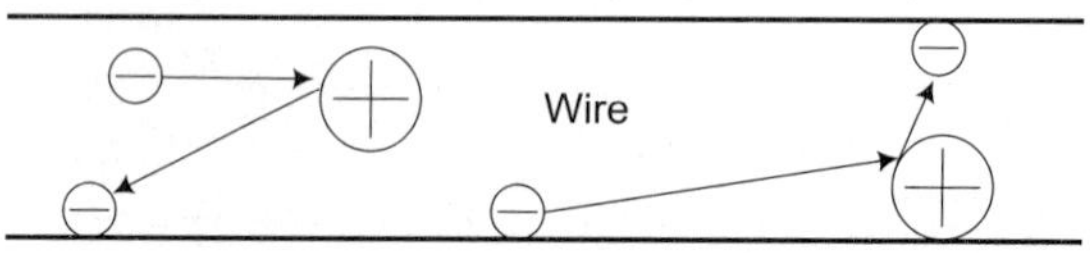

Fig. 1.27 Conduction mechanism in metal.

The mobility of electrons in metals like copper is many orders of magnitude larger than in a resistor. Within the copper wire, the electrons experience a very small force; and it is therefore possible to postulate this concept of a mechanism of conduction where the single electrons are accelerated under the influence of the interaction with the surface charges and collide with some atoms within the lattice as shown in Fig. 1.27. The surface charges are not shown in the figure.

As we stated for conductors, the number of collisions per second is 10^{14} per second, and the mean free path between two collisions is about 10 diameter of a copper atom. (Note: the world's fastest drummer has been recorded to make *only* 20 strokes per second and the fastest rotating microfabricated gas turbine reached *only* 33,333 revolutions per second). I urge the reader to pause for a moment and visualize the hundred million million collisions per second each electron makes with the lattice ions!

The electrons can, therefore, be regarded as independent particles which are accelerated for a short moment and then stopped by collision. With this approximation, the interaction *between* the free electrons is neglected. In equilibrium, when the distribution of the surface charges is stable and the electric current is constant, this approach is correct because the interaction between the electrons and the enforced increase of the velocity of the electrons is so very small.

Therefore, the density of the electrons is practically *not changed*, and the demand for **neutrality** is always fulfilled in the interior, despite the presence of the applied electric field. Whenever it is said that a material is *electrically neutral*, it is merely a statement of account of an equal number of positive and negative charges of a closed system. If we treat the interior of the wire (*excluding* the surface) as a closed system, the neutrality condition is fulfilled *and also* the atomic stretch during the current is very small (no significant dipoles are created) and the interior may be considered to be "truly" neutral.

1.19 MECHANISM OF CONDUCTION IN RESISTORS

What is a ***resistor*** ? What is ***resistance***?

We understand *"resistance"* in several ways. Here are a few samples.

Electrical resistance is a measure of the degree to which *a body opposes the passage of an electric current.* The SI unit of electrical resistance is the ohm. Its reciprocal quantity is electrical conductance measured in siemens.

Here is another: The *restriction to the flow of electrons.*

And another: The *difficulty* in moving electrical current *through a conductor to which voltage is applied;* expressed in ohms.

Resistor: This is a *circuit element* which *dissipates energy* in the form of heat when current flows through it. Example: an electric duct heater, in which ribbons or coils of *metal having a fairly low conductivity (high resistivity)* are heated to a high temperature by passing current through them.

An appropriate definition of resistance: It is the *property of a conductor that determines the current produced by a given difference of potential.* Ohm is the unit of resistance.

The definition enables us not to forget that a resistor is made from conductive material.

The following description of the function of a resistor was adapted from the article "Constant Electric Current and the Distribution of Surface Charges" by Hermann Haertel and reproduced with permission.

The function of a resistor is, in general, to hinder the flow of charge carriers and to *exchange energy*. There are different ways to accomplish this function. One possibility is to reduce the diameter of the conducting wire without changing the material.

In most practical cases, a resistor is introduced by choosing a different material. Such a material possesses a higher density of obstacles (lattice imperfections such as rows of missing atoms and displaced atoms) to electron flow, and/or it offers a smaller concentration of free electrons. Both effects normally appear together within manufactured resistors and can cause a dramatic change in the mobility for the electrons. The resistivity of a metal can be increased by adding small amounts of impurities or, by straining it severely by drawing it through a die to increase the number of lattice imperfections.

The resistivity of metals is due to lattice imperfections and the vibrations of the lattice ions at elevated temperatures. For a description of the phenomenon of resistivity and its variation with temperature, please see the Appendix B "Resistivity of metals".

If a uniform conductor and a uniform resistor are soldered together, there exist two thin areas or volumes within which the change of conductivity occurs. These areas are represented in the picture (Fig 1.28) on an exaggerated scale.

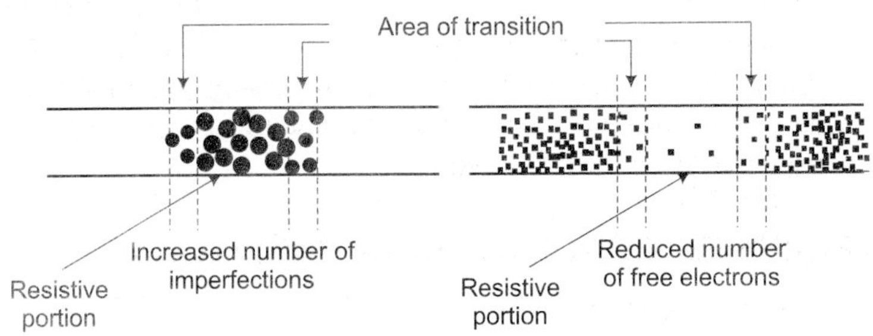

Fig. 1.28 Different mechanisms for electrical resistance.

In the left picture (increased number of imperfections), the filling represents the density of obstacles (lattice imperfections) for the moving electrons in the resistive portion. In the right picture, the reduced number of free electrons is represented in the resistive portion.

In both cases, the function will be the same. Within the area of transition from high to low values for the mobility, one side there will be a slightly higher concentration of electrons than neutral and a slightly lower one on the other side. These extra charges, which are spread out within two thin volumes, will have two effects: on one hand, they oppose the flow in conductors and on the other, they increase the force on the free electrons within the resistor.

In dynamic equilibrium the same number of electrons will flow through any cross-section in any given time. Due to the extra charges and the different gradient of the surface charge distribution, the driving force within the conductor and the resistor are, however, quite different.

According to Haertel, to treat the effect resulting from these transition areas is important for two reasons. First, the principle that nature changes gradually and continuously in space and time is basic for qualitative reasoning. Therefore, the dramatic change in mobility should not be treated as a step function with no further consequences.

Second, this area of transition plays an important role when studying the conduction mechanism in semiconductors such as diodes and transistors.

A last possibility to resist an electron flow is through magnetic interaction from the outside. In this case, it can be assumed that an additional force (or induced emf that will be dealt in Chapter 5) from outside is holding back parts of the electron distribution within the circuit.

This leads to a redistribution of the surface charges within the other parts just as if an ordinary resistor would have been introduced.

The effect of resistance in respect to electron flow can easily be explained through geometric arrangement. The surface charges A, B, C and D (see Fig. 1.29) which always exist in the presence of a changed cross-section, will oppose the effect of the original charges on the battery terminals and will thus, reduce the intensity of the current flow in the conductor.

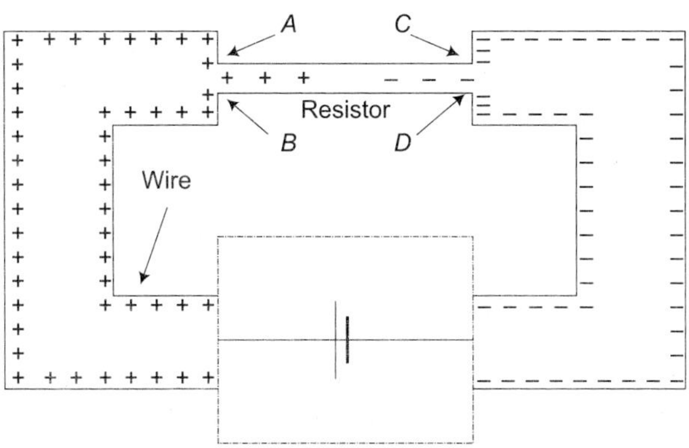

Fig. 1.29 Resistance through reduction of cross-section.

The gradient of the surface charge distribution (point charges set up by the shifting electron sea as described in Sections 1.15 and 1.16) is larger along the thin wire than along the thick ones. Therefore, a stronger force due to a stronger field is exerted on the free electrons within the resistor leading to a higher drifting speed. The longer the thin wire, the smaller the gradient and resulting axial electric field will be inside the wire, and this will lead to a smaller current.

When equilibrium is reached, the same number of electrons will pass through any cross-section of the system in any given time. The drifting speed, however, will be different at different cross-sections.

Resistance Combines Conductivity and Geometry

Instead of saying, "the cross-sectional area goes up, the length decreases, and the mobility increases" when comparing two different circuits, it is useful to have a *single quantity* to *describe a resistor* that *encapsulates both the properties* of the material namely, *conductivity (mobility, free-electron density)* and the *geometrical properties (cross-sectional area, length)* of the particular object.

Such a quantity is called "resistance". And is given by

$$R = \rho \frac{l}{A}$$

where, ρ is the resistivity of a conductor expressed in ohm-meter; l = the length of the conductor; and A = area of cross-section.

ρ is itself given by: $\frac{E}{J}$ where E = the electric field in the conductor; and J = the current density.

Now consider a circuit consisting of a voltage source, connecting wires and a resistor as shown in Fig. 1.30.

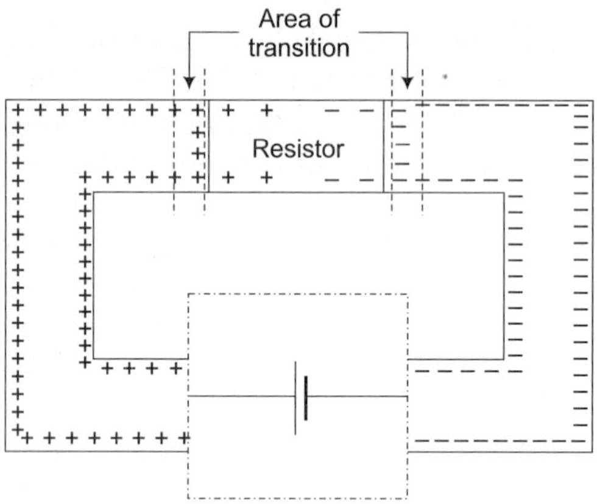

Fig. 1.30 Surface charge distribution (circuit with one resistor)

For any of the different resistors either with more obstacles (more imperfections in the lattice) or fewer movable charges (fewer number of free electrons), a similar surface charge (point charges set up by the shifting electron sea as described in Sections 1.15 and 1.16) distribution will occur. The figure indicates the principal idea, which is the establishment of a *steeper* gradient of surface charge across the resistor and a more powerful electric field inside it than inside the wires; it does not represent reality in all respects.

The scale of the wire and resistance in the Fig. 1.30 was chosen to optimize visibility. Moreover, the gradient in the surface charge distribution on the copper wires is not shown because it is too small compared with the gradient along the resistor. To view a live demonstration of surface charge phenomenon and its mechanical effects in circuits, please visit http://www.matterandinteractions.org/Content/Materials/materials.html.

Recollect from Section 1.9 where we sought an answer to the question as to how the electric bulb does not increase in brightness when brought closer to the battery or, when the bulb is twisted so that the filament is perpendicular to the electric field due to charges on the battery. The answer is the production of a uniform electric field in the filament due to the development of surface charges in combination with the charges on the battery and at the interface of the wires connecting to the bulb. Continuous feedback between the electric field and surface charges ensures that the brightness of the bulb does not vary as it is moved around or twisted.

1.20 A MECHANICAL BATTERY THAT SIMULATES THE ACTION OF A CHEMICAL BATTERY

We now describe both qualitatively and quantitatively, the action of sources of emfs in simple circuits, the conduction mechanism and the current determining parameters.

The source of emf, a battery, is depicted using a mechanical battery that consists of a conveyor belt driven by a motor which conveys electrons from one plate of the battery to the other.

This analogy fairly accurately captures the mechanism of a real battery in which the process of separation of charges is accomplished by chemical reactions. The chemical reaction considerations will assume importance in a real battery when it discharges to levels that will affect its emf. The description of the mechanical battery is based on the concept used by Chabay and Sherwood in their book "Electric and Magnetic Interactions", 1995, John Wiley, USA; used by permission.

In the mechanical battery shown in Fig. 1.31, the conveyor belt applies a *non-coulomb force* $\mathbf{F}_{NC}$ to extract an electron from the positive plate of a large parallel-plate capacitor and pushes it onto the negative plate.

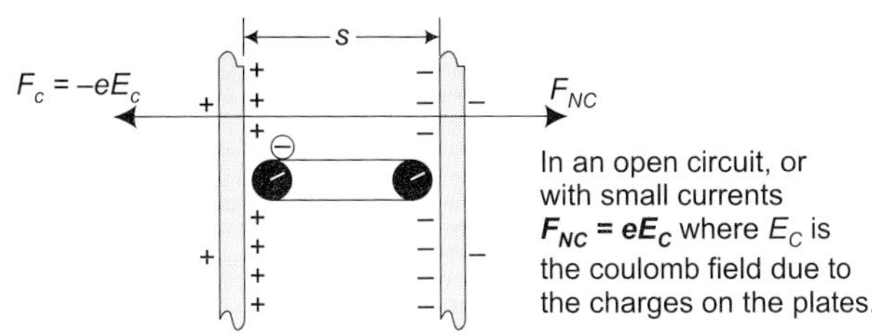

Fig. 1.31 An isolated mechanical battery.

Charge builds up on the plates of the battery until the *coulomb force* eE_c is equal to the *non-Coulomb force* F_{NC}. The separation s is small compared to the size of the plates.

Note on Non-Coulomb Force

We call the force of separation by the conveyor belt a non-coulomb force because, it is due to the *mechanical force* exerted by the belt on the electron as it drags it away from the positive plate.

This is in contrast to the electric force due to electric fields set up by point charges and which is called a coulomb force.

Energy Conservation and Electron-current Conservation

In a *circuit* containing a mechanical battery with the plates separated by a distance 's' and a series of wires:

Energy conservation leads to an equation relating the amount of work done on an electron by the mechanical battery to the energy dissipated in the wires, where E_i (newtons/coulomb) is the electric field in a wire of length L_i.

$F_{NC}s = \Sigma eE_iL_i$ and e = magnitude of the charge of an electron in coulombs.

Current conservation Because charge is conserved, in the *steady state* (drift speeds and excess charges not changing with time), the amount of current entering any portion of a circuit must equal the amount of current leaving that portion of the circuit.

If there were any *imbalance*, charge would build up inside this portion of the circuit, and steady state would not be achieved.

From the electron current conservation equation:

$$i = nAv = nA\mu E$$

where, n = number of free electrons per unit volume, A = cross-sectional area, v = drift velocity, and, μ = mobility in $cm^2/V\text{-}s$.

A battery puts ***energy*** into a circuit.

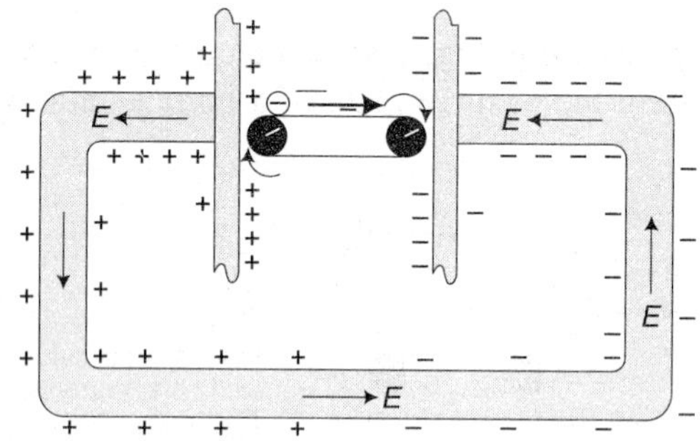

A battery puts energy by separating unlike charges and depositing them on the plates and then holding them apart. If charges are continuously being depleted on the end plates (terminals) after the circuit is closed and a current is established, the battery works to replenish the charges on its plates continuously, thereby maintaining the voltage across its terminals.

In the mechanical battery, the motor ***puts*** energy into the circuit.

Within the circuit, the moving electrons collide with the atomic cores and heat the metal. The warm metal ***heats*** the surroundings, and in the ***steady-state*** the energy input of the battery must be equal to the heat transfer output to the surroundings.

The amount of energy the mechanical battery ***must expend*** to move one electron from the positive plate to the negative plate is:

$$F_{NC}s$$

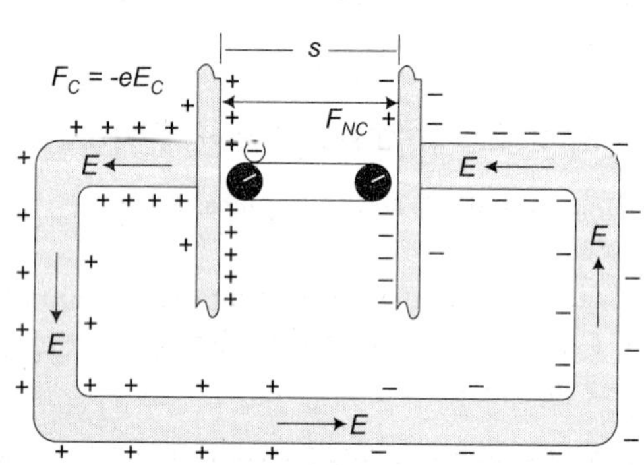

Fig. 1.32 A simple circuit showing energy and current conservation.

This electron is *pushed* through the wire (shown uniform in Fig. 1.32) by the electric force eE inside the wire. The work done on the electron in *forcing it through* the entire length of the wire L is eEL. This work goes into heating the wire.

Due to energy conservation, $F_{NC}\ s = eEL$ *in the steady state.*

The above formula is *unaffected* if the wire is *bent* into some complicated shape because, no matter how many twists and bends there are, the *surface charge* will *arrange itself to make the electric field be uniform* in magnitude, and make the electric field follow the direction of the wire.

Hence, the work done on the electron by the surface charges is *eEL*, *independent of* the shape of the wire.

Joules and Electron Volts

Physicists and engineers prefer to use the electron volts abbrev. eV when dealing with electron energies (1 eV = 1.6×10^{-19} joules) because an electron's potential energy is so tiny that it makes little sense to express this routinely as so *many million million millionths* of a joule! In the operation of electronic circuits, electric lamps and motors, millions and millions and millions of electrons in tiny energetic amounts (expressed in electron volts), harmoniously convert energy that is measured in joules! Therefore, physicists and engineers switch between electron volts and joules in the descriptions of microscopic and macroscopic aspects of circuit processes. More details on the mechanism of energy tranfer in circuits are provided in Chapter 4.

1.21 DISTINGUISHING INTERNAL ENERGY MECHANISMS AND EXTERNAL ENERGY MECHANISMS

Note that in considering the energy balance, we don't include a term $-F_C s$ corresponding to the coulomb force F_C acting "on" the electrons that are being moved from the "+" ve plate to the "–"plate.

That force, and the associated work or energy, are *internal* to the circuit (but numerically equal to the work done on the electron by the motor $F_{NC}s$, if there is no internal resistance of the battery).

Similarly, the work done on the drifting electron by the electric field (due to the charges located on the ends of the battery and on the surfaces of the wires) is *internal* to the circuit but numerically equal to the heat transfer to the surroundings.

The only *external* energy *input* to the circuit is the work done by the motor, and the only *external* energy *output* from the circuit is the heat transfer to the surroundings.

The "emf" of a Battery

What is the significance of the term $F_{NC}s/e$ (e = electron charge in coulombs) in the energy conservation equation? Chabay & Sherwood, 1995, used by permission.

Since $F_{NC}s$ is the amount of work done by the motor in moving one electron through a mechanical battery, $F_{NC}s$ is the *work done per unit charge* (joules per coulomb or volts).

This quantity is called the *"emf"*— the *energy input* to the circuit – or the circuit's *"starter"*.

In a chemical battery, the "emf" is a measure of the chemical energy per unit charge expended by the battery in moving charge through the battery and is *nearly constant*, independent of how much current is running through the battery (if the current is not too large) and approximately independent of how long the battery has been used.

It is the approximate constancy of the "emf" that makes this a useful concept, and which makes it useful to anlayze circuits in terms of "emf" and the potential differences in the circuit elements connected to the battery.

1.22 DIFFERENCE BETWEEN EMF AND POTENTIAL DIFFERENCE

It is important to keep in mind that although the units of "emf" are volts, the "emf" is not a potential difference.

"Potential difference" is a path integral of the electric field made by real charges.

The *"emf"* is the non-coulomb work per unit charge and might in principle be gravitational or nuclear in nature.

The *"emf"* of a flashlight battery is 1.5 volts, and the associated *charge separation is due to chemical reactions* in which electrons are reactants or products.

A Simple Circuit and Calculation of the Field Inside the Conductor and Resistor

In Fig. 1.33 is shown a battery connected to a thin wire by using 2 thick wires of equal length.

We have to calculate the electric fields E_1 and E_2 inside the thick and the thin wires respectively.

The joule value $F_{NC}\,s$ given for the battery is for a typical 1.5 volt battery and F_C represents the coulomb force acting "on" the electrons that are being moved from the "+" ve plate to the "–"plate.

Also given for the circuit are n = number of free electrons per unit volume of the wire, A_1 and A_2 are the cross-sectional areas of the thick and thin wires respectively, and, μ = mobility in $cm^2/V\text{-}s$.

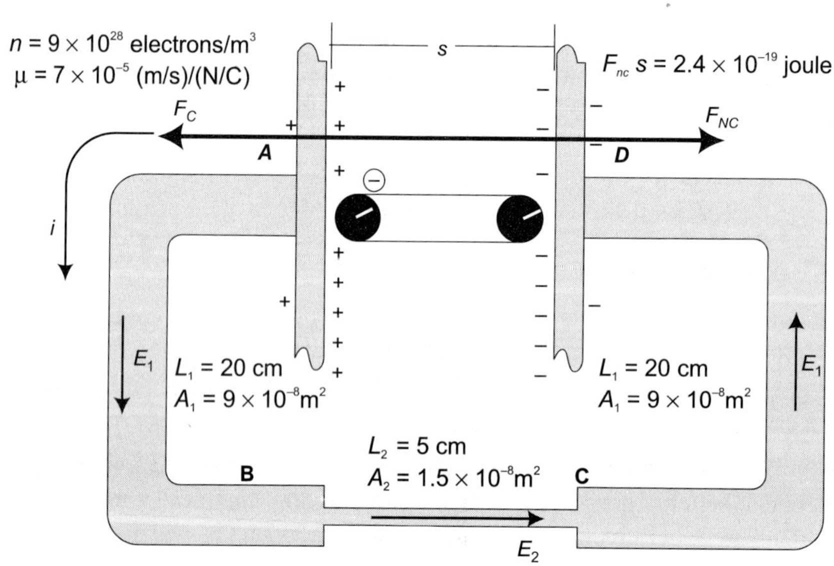

Fig. 1.33 A simple circuit consisting of a mechanical battery and a thin wire resistor.

Calculation to obtain the electric field inside the thick (conductor) and thin wire (resistor)
From the energy conservation equation, we know

$$F_{NC}s = \Sigma eE_iL_i \quad ...(1)$$

$$\therefore \quad F_{NC}s = e(2E_1L_1 + E_2L_2) \quad ...(2)$$

From the electron current conservation equation

$$i = nA_1\mu E_1 = nA_2\mu E_2 \quad ...(3)$$

From (3): $$\frac{E_1}{E_2} = \frac{A_2}{A_1} = \frac{1.5\times10^{-8}}{9\times10^{-8}} = \frac{1}{6} \quad ...(4)$$

Or, $$E_2 = 6 \times E_1$$

From (2): $$2.4 \times 10^{-19} = 1.6 \times 10^{-19} (0.4 \times E_1 + 0.05 \times E_2)$$

Substituting $E_2 = 6 \times E_1$ from (4):

$$1.5 = 0.4E_1 + 0.05 \times 6 \times E_1 = 0.7E_1$$

Therefore, $$E_1 = 2.1 \text{ V/m and } E_2 = 12.6 \text{ V/m}$$

Verification that the result conforms to Ohm's law, i.e., voltage drop across the thick and thin wires = 1.5 V.

The potential difference between points B and C (across the resistor) = $E_2 \times L_2$ = 12.6 × .05 = 0.63 V and, the drop across the wires = E_1 x 2 x L_1 = 2.1 x 2 x 0.2 = 0.84 V

Since, 0.84 V + 0.63 V = **1.47 V** = **(approx. 1.5 V)** ...(5)

the relation (5) establishes the conformance of the circuit to Ohm's law.

Note that the electric field in the thin conductor (resistor) is *6 times stronger* than that of the field in the thick conductor.

It is this powerful field in the resistive portion that guarantees a constant current through the entire circuit.

Calculations Based on Resistance and Voltage

Let us check our results by calculating the resistance of the wires given that the battery whose energy is 2.4×10^{-19} joule corresponds to a 1.5 volt battery (1.5 V = 2.4×10^{-19} joules/1.6×10^{-19} coulombs).

Conductivity of the wires $\sigma = qn\mu = 1.6 \times 10^{-19} \times 9 \times 10^{28} \times 7 \times 10^{-5}$ = 100.8×10^4 mho per metre.

Resistivity ρ = 1/ Conductivity = $1/\sigma$

Therefore, the resistivity of the material is 1/ 100.8×10^4 = $9.9 \times 10^{-7} = 990 \times 10^{-9}$ = (approx) $1000 \times 10^{-6} \times 100 \times 10^{-3} = 100 \times 10^{-6}$ = **100 microohm-cm.**

Consider the wire of length L = 20 cm and area of cross-section A = 9×10^{-8} m^2.

Resistance of the wire = resistivity × L/A = $990 \times 10^{-9} \times 20 \times 10^{-2}/9 \times 10^{-8}$ = $19800 \times 10^{-11}/9 \times 10^{-8}$ = 2200×10^{-3} = **2.2 ohms.**

Consider the wire of length L = 5 cm and area of cross-section A = 1.5×10^{-8} m^2

Resistance of the wire = resistivity × L/A = $990 \times 10^{-9} \times 5 \times 10^{-2}/1.5 \times 10^{-8}$ = $4950 \times 10^{-11}/1.5 \times 10^{-8}$ = 3300×10^{-3} = **3.3 ohms.**

The total current will be 1.5 V/ resistance of the wires.

Therefore, the current = 1.5 V/ (2 × 2.2 + 3.3) ohms = 0.19 Amps.

The voltage in the thick sections is 0.19 A × 2 × 2.2 = 0.85 V.

The voltage in the thin section is 0.19 A × 3.3 = 0.627 V.

The sum of the two voltage drops is = 0.85 V + 0.627 V = 1.477 V = **approx 1.5 V** ...(6)

This is the same result that we obtained with calculations (see relation (5)) based on electric fields in the wires.

1.23 FIELD STRENGTH INSIDE THE WIRES AND THE CONSERVATION OF CURRENT

It is useful to keep in mind that the electric field strengths inside wires in most applications where currents do not exceed a few amperes will be in the range 0.1 V/m to 20 V/m.

If we plot a graph as in Fig. 1.34, of the electric field and drift velocity inside the wires and the resistor, we will observe a large drift velocity of the electrons inside the resistor portion in comparison with the low resistance wire portions.

This is rightly so because if it were not, then, excess (local) charges (neglecting surface charges) will build up inside the circuit in violation of the charge neutrality condition that should always be maintained.

Because charge is conserved, in the steady state (drift speeds and excess charges not changing with time) the amount of current entering any portion of a circuit must *equal* the amount of current leaving that portion of the circuit and is simply called the *conservation of current*.

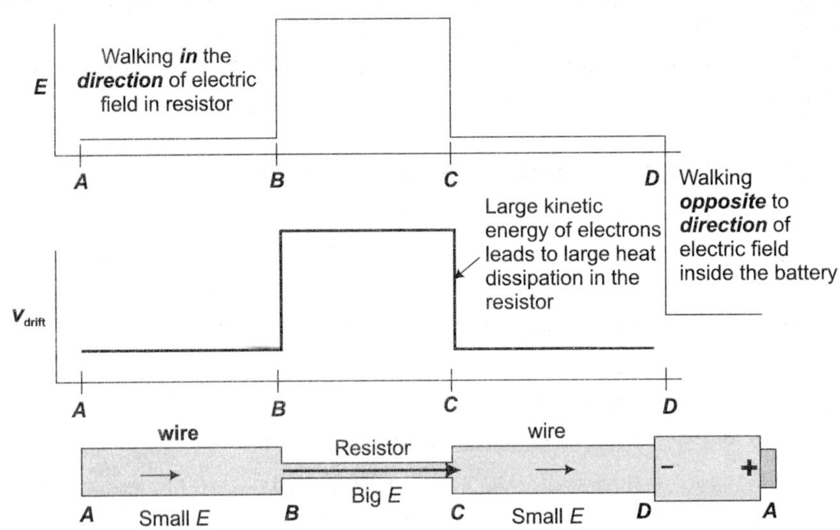

Fig. 1.34 Graph of electric field and drift velocity inside the wires and the resistor.

It is once again emphasized that there is an ever vigilant feedback mechanism between the excess surface charges and the electric field that ensures a constant and steady current leading to steady-state conditions, after Chabay & Sherwood, 1995.

The millions of conduction band electrons in the bulk of the wires are extremely busy: even as they are accelerated by the applied electric field due to the voltage source, only to collide with the lattice ions, they move only infinitesimal distances and this maintains charge *neutrality* in the bulk of the wire.

In response to the applied field that seems to have disturbed their otherwise pre-programmed movement, they make adjustments to the *surface* charges to preserve voltage conditions on the surface and within the wire and, more importantly, ensure that the conservation of current law is not violated at all times.

Notice the large kinetic energy imparted to the electrons by the electric field in the resistor leading to large heat dissipation in it.

How is heat produced by an electron collision with a lattice ion? Please see Appendix B for a description of the mechanism of heat production in conductors and resistors.

The conduction band electrons are busy in the resistor as well. Here, they continuously make adjustments to the surface charges to ensure a uniform voltage drop along its length (provided, of course, the resistance is uniform along its length) (see Fig. 1.35 and the potential graph between points B and C), maintaining charge neutrality in elemental volumes and that the required field strength is available that will ensure that the conservation of current law is not violated at all times. Neutrality is maintained *because* the interaction between the electrons and the enforced increase of the velocity of the electrons due to the applied field is so very *small.*

In the bulk of the resistor (conductor) not counting the surface charges, the conduction band electrons **do not** create *significantly strong* **dipoles** (charge centre separation between free electrons and lattice ions) that could produce *opposing* internal electric fields of significance. But the discrete nature of the electrons (charge carriers) and their own random (because of the movement of millions of atoms) thermal agitations, should be borne in mind to understand the origins of certain types of internal sources of extremely *tiny* emfs[21] (dipole effects) "noise" (in fact, using sophisticated instrumentation, very tiny "noise" emfs can be observed even across *open-ended* resistors), that superimpose on the voltages of interest across conductors.

A plot of the potential inside the wires and the resistor is shown in the Fig. 1.35 (Chabay & Sherwood, 1995, used by permission)[2] and indicates a large potential drop across the resistor.

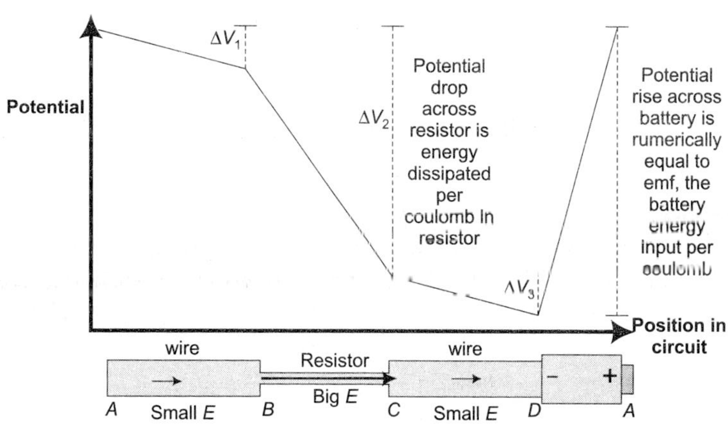

Fig. 1.35 Graph of potential inside the wires and the resistor.

The potential drop is caused by the steep surface charge gradient over the resistor caused by a buildup of positive charges at its positive end and negative charges at its negative end.

How much current if we double the length of the wire (after Chabay & Sherwood, 1995).

If we double the length of the wire:

The *amount of charge* on the *endplate* of the battery *stays the same.*

The *rate of variation of the surface charge* on the wire *decreases;* the progression from "+" to "–" surface charge *spreads out over twice the length.*

Since, $eEL = F_{NC}s,$

Therefore, $E = F_{NC}s/eL$

So if we *double the length L*, we'll get *half as big an electric field E* (since the other quantities *do not change*).

$v = \mu E$ and Electron current is nAv.

With half the field we get half the drift speed and *half the electron current.*

How much current if we double the cross-sectional area of the wire?

If we double the cross-sectional area of the wire:

The *amount of charge* on the *endplate* of the battery *stays the same.*

The *electric field* inside the wire *hardly changes* at all.

The *drift speed* of individual electrons is *unchanged.*

The free *electron density is unchanged* because the wire is made of the same material

$v = \mu E$ and electron current is nAv.

If n and v are the same, only A matters. Therefore, doubling A doubles the current.

With twice the cross-sectional area, we get *double the electron current.*

In Section 1.15, we stated that the electric field inside a wire 5 metres long and of uniform cross-section connected to a 10 volt battery is 2 V/metre.

Using the energy conservation equation:

$F_{NC}s = \Sigma eE\ L$ with, $F_{NC}s = 16$ joules corresponding to a 10 volt battery, $L = 5$ metres, we can verify that the electric field is indeed 2 volts/metre. Any other value for the electric field will violate the law of energy conservation.

1.24 PROCESS OF ENERGY SUPPLY TO MAINTAIN THE CURRENT

As the electrons in the wires and the resistor are being *drawn* from the negative plate towards the positive plate there is a *depletion* of charges on the endplates of the battery as electrons from the wire near the positive terminal neutralize the positive charges (deficiency of electrons) on the positive terminal by uniting with them …in reality…..the electrons settle into convenient orbital spaces surrounding the nuclei, producing neutral atoms….(see Section 1.13 on how atoms are neutral) and we call the process neutralizing or uniting with positive charges.

This in turn *slackens* the strength of the electric field due to the endplate charges *and* the surface charges in the entire circuit momentarily and, the charges on the endplates and the surface charges are immediately replenished (by the process of charge separation) by the motor in the mechanical battery or the chemical action in a commercial battery.

The slackened field is strengthened by the replenished charges and the process is a *seemingly continuous though intermittent* one occurring at nearly the speed of light...depletion of charges on the plates of the battery and their replenishment by internal charge separation....the kinetic energy of the electrons is lost in the resistive portion of the circuit continuously as they collide and pick themselves up under the influence of the electric field exerting a relentless pull on them.

1.25 WHY DOES A CHARGED PLASTIC PEN OR COMB ATTRACT TINY BITS OF PAPER? AND WHY DOES IT LOSE ITS CHARGE QUICKLY?

(Chabay & Sherwood, 1995, used by permission)[2]

The comb is made up of large organic molecules and when rubbed in your hair, the organic molecules of the surface strands of your hair break and electrons are transferred from them to the plastic comb. Large organic molecules in the plastic and/or your hair break at their weakest bond in such a way that negative ions (negatively charged fragments) are deposited on the comb and/or positive ions (positively charged fragments) are deposited on your hair. The plastic comb becomes negatively charged.

Molecular breakage or electron transfer provides a rough explanation as to why the plastic comb and the U-tape when pulled off the base tape (see Section 1.14) got charged, but such details as to why the comb rather than your hair becomes negative are the subject of research by physicists and chemists. Part of the complexity of these phenomena is due to the fact that they are surface phenomena and the special nature of intermolecular interactions at the surface of a solid are generally less well understood than those in the interior. The U-tape used in the experiment in Section 1.14 could have charged positive or negative and this would have depended on the type of the tape. When the charged comb is held near tiny bits of paper, induced dipoles (or separated charges formed by the shifting of the electron cloud of the outermost four carbon atoms by a tiny amount, less than 10^{-15} metres) are formed in the paper by their interaction with the electric field of the charged comb. And, since there are millions of atoms, the net force of attraction between the near side positive nuclei and the negatively charged comb is sufficient to lift the paper against the force of gravity. See Appendix A for a calculation of the amount of electron cloud shift of the carbon atoms in the bits of paper attracted by the comb.

Different atoms of different elements will have different strengths of interactions with the applied field. When a small electric field E is applied to an atom or molecule that is not a permanent dipole like a water molecule, it is found experimentally that for the charge separation s produced, $p = qs$, which we define as the dipole moment, is proportional to the applied field E. It is an excellent approximation to say that the dipole moment $p = qs = \alpha E$, where α is a constant that expresses the "polarizability" of the particular kind of atom or molecule. See the Appendix A for details of the calculation of the electric force of attraction by the pen to lift the bit of paper.

Water molecules are permanently polarized. The water molecule (H_2O) is not symmetrical but has both hydrogen atoms off to one side of the oxygen atoms. Slight shifts of the electron cloud to one side of the molecule leaves that side a bit negative as compared to the other side. So the water molecule is a permanent dipole. When water molecules in the air (when the atmosphere is humid) strike a surface they sometimes become attached to the surface, probably because the charged ends bind to the surface. A film of water builds on all surfaces. Pure water is a poor conductor but does contain small amounts of mobile H^+ and OH^- ions.

More importantly, the water dissolves surface contaminants such as salt, and the impure water provides an effective path for charges to spread onto neighbouring objects. After a while a charged surface loses its original charge, which is why the charged pen loses its charge (by spreading over large surface areas), quickly. When grounding is discussed at the end of Chapter 2, this aspect will become clearer. So, experiments with charged objects work better when the humidity is low.

I conclude the chapter, with a note on the visualization of atoms in conductors and surface charges that enable conduction.

I find it useful to visualize the nucleus and electrons (cloud) as if being held together by *tiny rubber bands*. Visualize these as having different stiffness for different atoms. The stiffer they are, the tinier the stretch when influenced by external electric fields; the smaller their stiffness as in conductors, the rubber bands can be stretched more and can eventually be broken and result in producing isolated charges!

In this way, it is not too difficult to visualize the extremely thin rubber bands (representing the metallic bonds) between the outermost electrons and the nuclei stretching and snapping to produce the electron sea. At the surface, the sea thins (creating positivity) or condenses (creating negativity) as the circuit warrants!

In Appendix C, a qualitative description of the effects of an electric field on insulators, their surface voltages and field strengths and descriptions of the internal mechanism that enables these to prevent electric shocks from live conducting parts is given.

Questions

1. Explain what is meant by a "closed system" and how does the conservation of charge principle apply to such a system?
2. What is an electric field? Describe its nature and its penetration through matter and action on charges. What is the resulting velocity when an object is hit once? Why does an electric field produce an acceleration of charges in its influence? Does an electric field penetrate all matter?
3. Describe how there must be an electric field inside the wires of a simple dc circuit that lights a bulb and how the charges on the battery alone cannot be responsible for lighting it because the brightness of the bulb does not change (which should, if we apply Coulomb's law) when brought closer or taken farther away from the battery.
4. What is polarization? Discuss the effect of an external electric field on atoms and molecules. Describe what happens when the external electric field is removed.
5. What does a battery do? Describe its operation and also explain the conservation of charge principle in the universe.
6. What are permanent dipoles? Describe with the help of examples.
7. Why are the energies of electrons (at different levels) in atoms considered negative? Are these energies potential energies? Explain. Why do the electrons not fall into the nucleus of stable atoms? Explain.

8. What are surface charges and how do they enable the process of conduction? Is electric current a result of pushing of charges as water molecules do when flowing in a pipe? Describe how electric currents are different.
9. Derive an expression for the electric field at a distance x from a ring having a charge q that is uniformly distributed around it.
10. Describe qualitatively what happens when unlike charges are separated to form dipoles.
11. Describe with the help of sketches the surface charge distribution on a round wire that would enable the conduction process. Will surface charge distributions be like rings for conductors of other geometries and when they are crossed.
12. Describe metallic structures and relate to the type of bonding that holds the metal together.
13. Is an atom a dipole? Describe how a stable atom is treated as being neutral and is there zero electric field around it?
14. Describe the atomic structure of metallic substances and how a piece of metal is neutral for practical purposes though it contains thousands of millions of electrons and protons whose potentials range in the thousands of kilovolts
15. How does the magnitude of the electric force binding the hydrogen atom together compare to the gravitational force exerted by the entire earth on the hydrogen atom? Would you conclude that electric forces are indeed much more powerful than gravitational forces?
16. How much force does an ordinary chisel apply when hit by a hammer and explain whether it can prise electrons from a metal.
17. Draw the electric field of a dipole lying centred on the origin along the x-axis consisting of unlike charges $+q$ and $-q$ separated by a distance 's'. Derive an expression for the field at points far away in space along the x-axis and along the y-axis.
18. Describe the structure of metals with reference to metallic bonding and how does the structure differ from other structures formed due to ionic and covalent bonding.
19. With the help of neat sketches describe what happens when metals (conductors), insulators (dielectrics) and ionic solutions are subject to an external electric field. Do insulators have a voltage? Explain with the help of neat sketches.
20. When a metal is subject to an external electric field what is the amount of shift of the electron cloud that produces the field necessary to cancel the influence of the external electric field?
21. Do electrons that form currents move all the way around a closed circuit? Is an electric field necessary to establish a current? Describe qualitatively currents in circuit wires and the relation between drift velocity and mobility.
22. A wire is connected to the two terminals of a battery. What is the rule for the very initial charge movement whenever an electric circuit is made? Describe how feedback enables the current to remain constant in a circuit in which the incoming current were to exceed the outgoing current and when the outgoing current were to exceed the incoming current.

23. What are resistors? Describe qualitatively their lattice structures and defects in them mentioning those that affect resistivity and the effect of temperature on resistivity and currents in them. Is an electric field necessary to establish a current in them?
24. How do conductors and resistors remain *neutral* in the interior in circuits (not the surface charge) despite a current and the electric field that motivates it? How is this so? Describe the processes that keep it so.
25. What is meant by "initial transient" and "steady-state". Describe the two terms with respect to a simple series circuit consisting of a battery, wires and a resistor and describe the conservation of current principle. What is the speed of surface charge build up in a typical circuit?
26. What are induced dipoles? A dipole is induced of an atom centred at the origin as a result of polarization by a charge $+Q$ located along the x-axis at a distance d which is much greater than 's' the distance of separation of the dipole. Derive an expression for the force F on the point charge $+Q$.
27. With a mechanical battery made using a large capacitor with plate separation 's' fitted with a motor and conveyor belt and given the non-coulomb force exerted by the motor to be F_{NC}, derive an expression that balances the energy input and output when the terminals of the battery are connected using a wire of length L.
28. What order of velocity does an electron reach in the conduction process? Is it close to the velocity of light? or is it non-relativistic ?How is it instrumental in the dissipation of heat in a conductor?
29. With a mechanical battery made using a large capacitor with plate separation 's' fitted with a motor and conveyor belt and given the non-coulomb force exerted by the motor to be F_{NC}, derive an expression that balances the energy input and output when the terminals of the battery are connected using a wire of length L. What will the current be in the wire if its length is doubled? Give reasons.
30. With a mechanical battery made using a large capacitor with plate separation 's' fitted with a motor and conveyor belt and given the non-coulomb force exerted by the motor to be F_{NC}, derive an expression that balances the energy input and output when the terminals of the battery are connected using a wire of length L. What will the current be in the wire if its cross-sectional area is doubled? Give reasons.
31. Explain how a charged plastic pen is able to lift tiny bits of uncharged paper from a height of 1 cm and the amount of shift of the electron cloud necessary to achieve this.
32. Explain how heat is produced inside resistors when an electric field from a voltage source is applied to it and a current is established.

2

Capacitors in Circuits

How Current Splits in Parallel Branches; Introduction to Magnetism and its Interaction

Chapter 2 commences with a description of the role of surface charges on the discharging and charging of capacitors. This is then followed by a description of the behaviour of RC (resistor-capacitor) circuits to pulse inputs and the phenomenon of how the voltage across a capacitor cannot change instantaneously.

Next, the application of a sinusoidal input to an RC circuit is presented with input and output waveforms, and, the fringe field that plays a role in modulating the current is also presented. The way the current leads the voltage in RC circuits is presented next. In this section, the difference in the qualitative nature of resistance and reactive capacitance is presented. I felt that the time was appropriate to introduce in this section of Chapter 2, the division of currents in resistive parallel circuits.

The chapter progresses with a description of the concept of voltage and how it can be visualized and an answer to "where is the voltage" is provided. It is then followed by calculations of the capacitance of a single conductor such as the earth.

This is followed by the all important concept of potential and, in particular, the process that is involved in charging up a single wire connected with one terminal of a battery. In the following sections, the concepts of neutrality and equipotential regions in conductors is explained. In the concluding sections, a description of how capacitors share charges and the processes involved before steady state or equilibrium is reached. Finally, the concept of "grounding" of circuits is described.

2.1 DISCHARGE OF A CAPACITOR

Why does a capacitor discharge?

Consider that a capacitor has been charged by connecting it to a voltage source like a battery and this would have caused electrons from one plate to be displaced to the other plate leaving one plate positive with a deficiency of electrons and the other plate negative with an excess (local) of electrons. Now, the capacitor is disconnected from the source and a wire is used to short-circuit the two plates of the capacitor to discharge it. This will cause the electrons from the negative plate to neutralize the positive charge on the positive plate.

Why should a capacitor discharge? Imagine the capacitor with all its charges is kept ***inside a bag and tied*** at two open ends (open-circuit) as shown in Fig. 2.1.

From: "Electric and Magnetic Interactions" by Chabay and Sherwood.

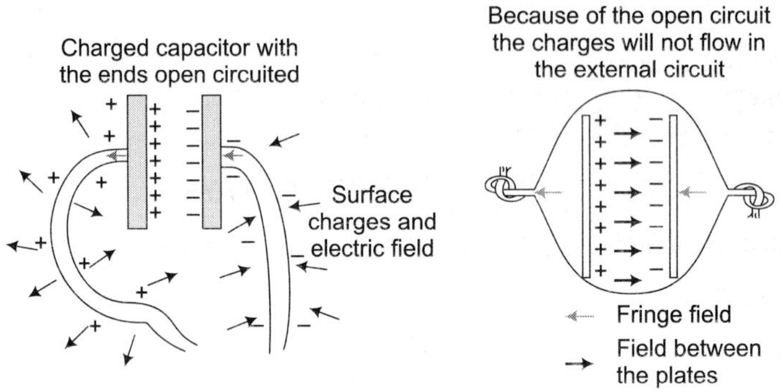

Fig. 2.1 Discharge of a capacitor.

The positive charges on one plate are attracted strongly to the electrons on the other plate and the electric field is strong and uniform between the plates between the opposite charges on the plates. The opposite charges on either plate seem to hold each other tightly. The field is not so strong to cause the electrons to cross the gap and reunite with the positive charges. There is a fringe field at both ends ready to burst the bag, but the open circuit does not allow charge flow in the external circuit. The electric field of surface charges that move onto the wires from both the plates (We will learn more about this phenomenon in Section 2.21 "How is there a voltage or potential for a single wire connected to one terminal of a battery") and the fringe field will, during the initial transient when the wires are attached, combine to make the resultant electric field in the wires, *zero*. The fringe field is *different* from the field *between the capacitor plates.*

2.2 THE FRINGE FIELD AND ITS ORIGINS

There is nothing mysterious about the origin of the fringe field.

The fringe field (see Fig. 2.2) exists slightly *external* (marked 'x') to the plates and is a result of the uneven effect or *contributions of the two powerful fields set up by the opposite charges* on the capacitor plates that are separated by a ***"tiny"*** distance '*s*'.

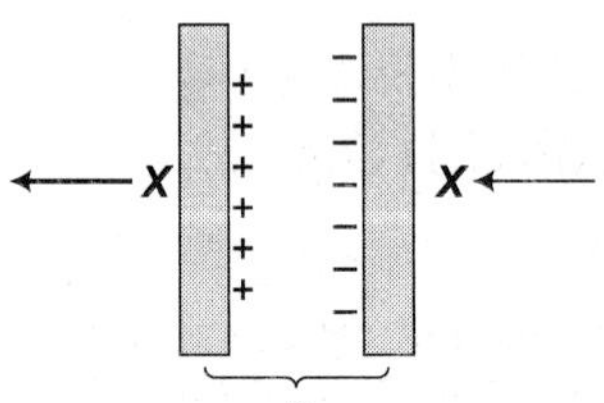

Fig. 2.2 Fringe field and its origins.

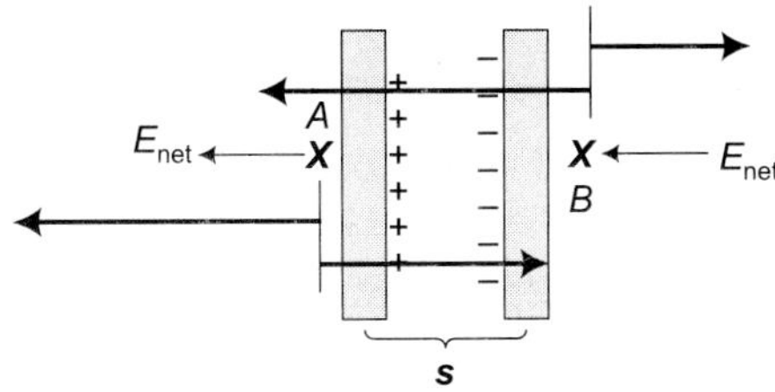

At point A, the field due to the positive charges is *slightly more powerful* (indicated by the longer length of the arrow) than that due to the negative charges because the point A is exactly length s away from the negative plate. The *net* field is the algebraic sum of the two *nearly yet not equal* powerful fields and is *very small.* The reader should note many Physics textbooks indicate that the net (fringe) field *outside* the capacitor is zero but this is not valid. For disk type parallel plate capacitors with separation s, radius R, area of plates A and carrying a charge Q, it is given by:

$$\frac{Q}{2A\varepsilon_o}\frac{s}{R}$$

See Appendix A for details of calculating the field between two charged plates and the fringe field. Also see http://caccphysics.cacc.cc.al.us/phy-webpages/phy214/214workbook/4-Wkb3ch3c&schap%27s3&4.htm.

A similar argument holds for the point B across and away from the negative plate.

It takes a little getting used to …to *estimate* the e-field at points in space near an assembly of charges; *the Principle of superposition* built-in by nature enables the estimation to be performed easily.

1. Remove all source charges except *one* and estimate the field due to it at the target point.
2. Next, remove all source charges except the *next one* and estimate the field due to it at the target point and so on.
3. Vectorially add the contributions and you have the field due to the complete assembly of charges at the target point.

Notice that in the process we learn to *see* the e-field in space as being due to several contributory source charge fields…maybe some of these are on the other side of plates…or other objects…and so on…but the field *penetrates through matter...so no bother !*....remove one set from the source charge assembly…….and you have the contributory effect of the *e-fields* of the rest……remaining.

In short, we learn to *see and feel* how distant source charges influence the strength of fields at points in space, with or without objects in between the target and the source charges.

2.3 VISUALIZING AN ELECTRIC FIELD AT POINTS IN SPACE

One of my favourite ways of *visualizing* electric fields in space is to imagine a *force or wave* (as in a sea)…*filling the space*....around and near the target point of interest….with a sound …. a gentle…zzzzzz or mmmmmm….

....exerting a *relentless force....never tiring unless destroyed*....on some charges at the points of interest in space.....keeping a mental image of the *source charges*.....

I then imagine the *magnitude and assign a direction* to this force or wave that suits the situation, that is for the individual parts of the assembly of source charges.

I imagine the magnitudes in relative terms to the situation.

I do find it difficult to stick in *absolute magnitudes*, relative magnitudes work fine for me.

2.4 THE PROCESSES WHEN A CHARGED CAPACITOR DISCHARGES

When the charged capacitor is short-circuited using a wire, the effect is to open the tied ends of the bag and the charges are released.

When the circuit is closed, the effect is as if the closed bag is opened and the charges are released.

Closing the circuit between the two plates....

...has the effect of opening the tied ends and the charges are released.

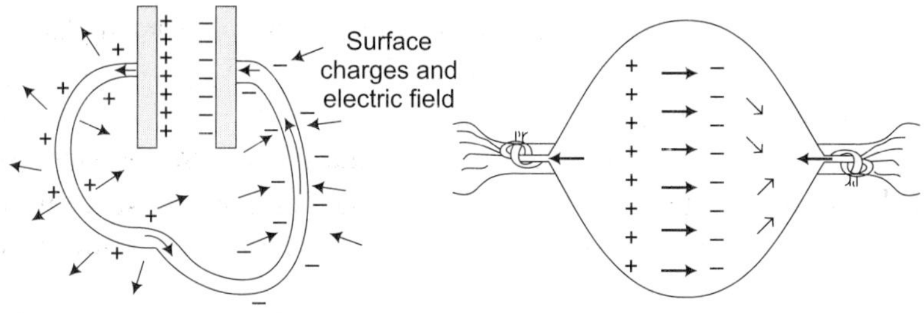

If there were *no fringe field* there would not result a movement of charge. Nature takes care of everything!

→ Flow of electrons

→ Field between the plates

The **fringe field** causes the ***initial crucial*** release of charges for further flow of the charges....and development of surface charges that will enable the charges to flow around the wire and neutralize each other. After all the charges on the plates are neutralized, the circuit attains a state of static equilibrium, a natural relaxed state of equilibrium.

2.5 DISCHARGE OF A CAPACITOR USING A WIRE WITH A BEND

Supposing a charged capacitor is short-circuited with the intention of discharging it using a wire with a bend as shown in Fig. 2.3 after Chabay & Sherwood, 1995.

On closure of the circuit, charges, according to rule (see Section 1.15), move out from the positive plate on the wire and from the negative plate on the wire. The electric field due to these charges and the charges on the plates of the capacitor fills the space surrounding the capacitor and because it penetrates through matter it sets up the field as shown in the Fig. 2.3.

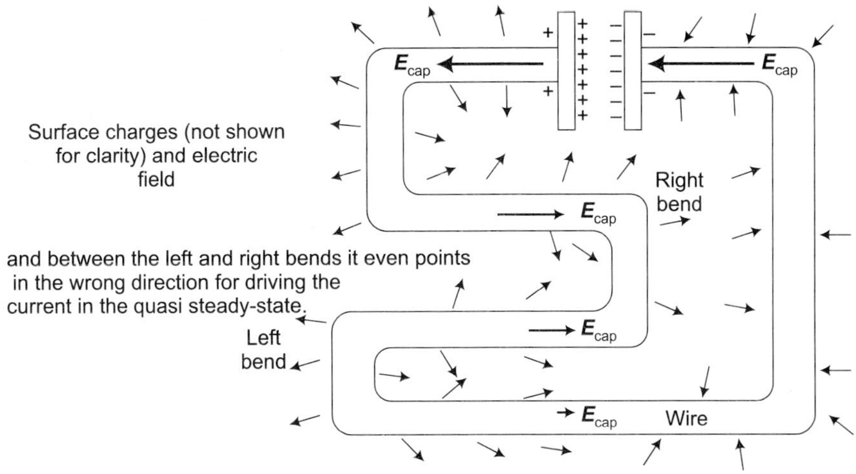

Fig. 2.3 Discharging a capacitor using a wire with a bend.

The capacitor looks approximately like a dipole, so E_{cap} falls off *rapidly* ($\sim 1/r^3$) with distance from the capacitor, as may be seen by the length of the E_{cap} field arrows near and away from the capacitor. The *electric field* E_{cap} due to the charges on the capacitor is *not uniform* in magnitude and *between the left and right bends it even points in the wrong direction* for driving the current in the quasi steady-state, that is, having some resemblance to a steady-state. There is no discharging current yet.

But now something remarkable happens.

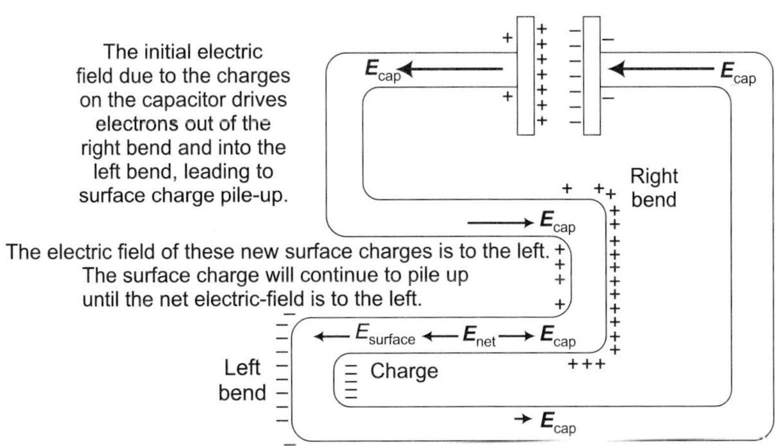

Fig. 2.4 Electric field being reoriented in the bend.

It is useful to review the nature of the electric field as described in Sections 1.4 and 1.11. The electric field is an invisible alteration of space that exerts a force continuously on charged particles (like electrons) causing them to accelerate. When an electric field is applied to wires, it produces a drifting motion superimposed on the rapid random movement of the electrons because of their collisions with the lattice ions or atomic cores, and this drift results in a current.

After a very short time the non-steady-state configuration of electric field causes surface charges to appear on the wire. What contribution to the net electric field do these charges make? In the branch between bends, where the electric field initially points the "wrong" way (to the right), the new surface charges make an electric field in the "correct" direction (to the left).

If there is insufficient surface charge on the bends to make a net electric field to the left, of the approximate magnitude, pile-up of surface charge will continue until the net electric field does point to the left, with the appropriate quasi steady-state magnitude.

Ultimately, this elegant negative feedback mechanism will result in the production of an arbitrarily complex distribution of surface charge all over the wire such that the electric field due to all the charges (surface charge plus the charge on the capacitor plates) is uniform throughout the wire as seen in Fig. 2.5 and has the magnitude predicted by conservation of energy.

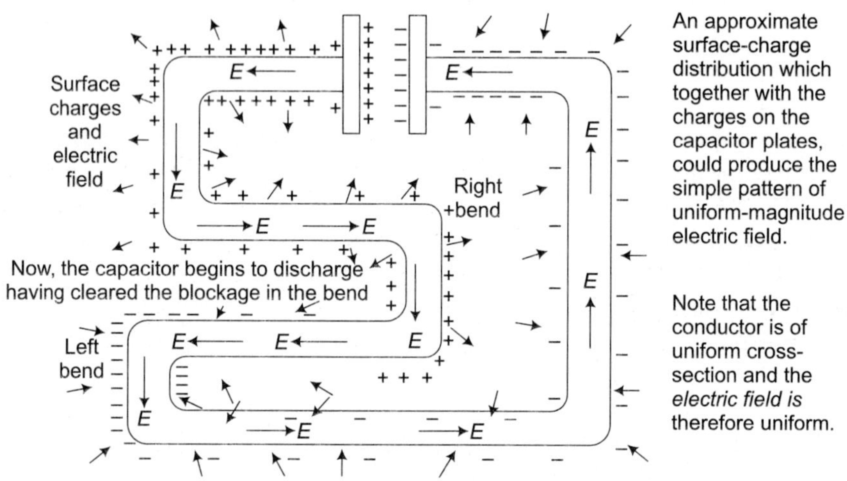

Fig. 2.5 Electric field in the quasi-steady state.

Note that a capacitor cannot ***sustain*** a flow of charge like a battery because there is no mechanism to supply energy to the circuit once the capacitor plates lose their excess charge to the opposite plate.

The key element that enabled the process of discharging was the *fringe field* of the capacitor outside the capacitor plates.

The fringe field of the capacitor (plus the electric field of the charges on the surfaces of the wires) *drives* current (the millions of conduction band electrons) in such a way as to *reduce* the charge on the capacitor plates (and to reduce the charges on the surfaces of the wires). As a result of these reduced charges, the electric field inside the wires is reduced, and so the current is reduced.

The process continues with *ever-decreasing current* until all the charges have been neutralized.

The system reaches "static equilibrium".

As a matter of fact, it should be noted that after the capacitor is *fully discharged no surface charges* remain either on the capacitor or on the wires.

In the next section we shall see the results of a quantitative study of how during the initial transient phase of the discharge of a capacitor the fields evolve and establish the quasi steady-state condition.

2.6 THE EVOLUTION OF THE FIELDS AND SURFACE CHARGES OF AN RC CIRCUIT

The evolution of the fields and surface charges of an RC circuit were simulated and studied by Norris W. Preyer, Department of Physics and Astronomy, College of Charleston, Charleston, South Carolina, USA and the results[9] are shown in Figs. 2.6 and 2.7 (reproduced with permission from Norris W. Preyer).

The circuit consists of a charged capacitor connected by a single fat wire of high resistance with charges of equal and opposite sign placed on the plates at $t = 0$. The diagrams in Fig. 2.6(a) are at 0 and 15 picoseconds (ps) after the system is allowed to evolve in time. **Note:** 1 picosecond (ps) = 10^{-12} seconds.

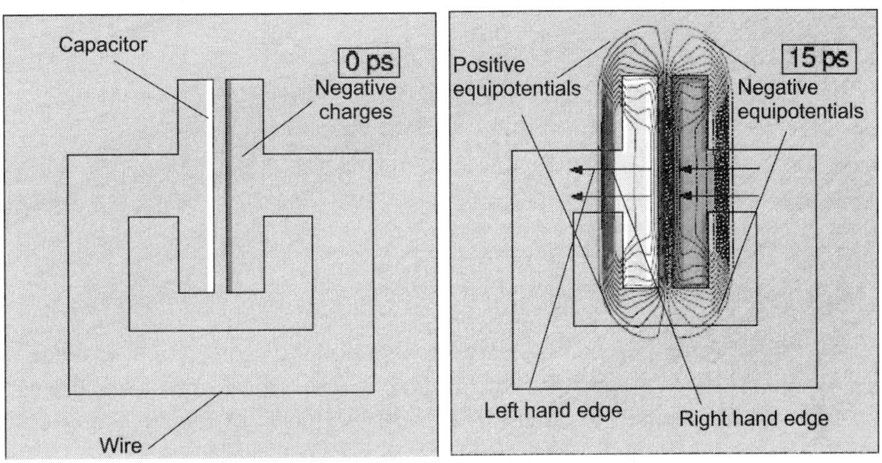

Fig. 2.6(a) Discharge of a capacitor: very intial transient.

The diagrams are a cross-section through the midplane of the circuit and the white shade on the left plate ($t = 0$) indicates the initial positive charge and the dark grey shade on the right plate indicates initial negative charges.

The various grey shades represent the excess charge density, ranging between +/–5000 e(charge of an electron)/mm^3. The charge density on the capacitor plates exceeds this range, which was chosen to illustrate prominently the surface charges on the wires. The arrows plot the square root of the electric field. The large field vectors between the plates have been omitted.

Also shown are the equipotentials from –5 V to +5 V in steps of 0.4 V; positive equipotentials using a solid line and negative equipotentials using a dashed line. As one moves *across* these lines along a conductor or resistor, the potential changes and, this change is what is commonly called "voltage drop".

Note the tight bundle of equipotentials to the left of the positive plate when the elapsed time is 15 ps in the Fig. 2.6(a). The left-hand edge is caused by the new positive charges, while the right-hand edge is caused by the new negative charges (the two together give a much smaller field than a single plate). The width of this region of very strong electric fields is just about the width of the gap between the capacitor plates.

In Fig. 2.6(b), the situation shown is at 30 ps. The steep potential gradient (closely spaced equipotential lines) and large electric field (shown by the long field vector lines) at the left-hand wire upper are entirely due to the charges on the left-hand positive plate and is the chief cause of rapid removal of charges from the plate.

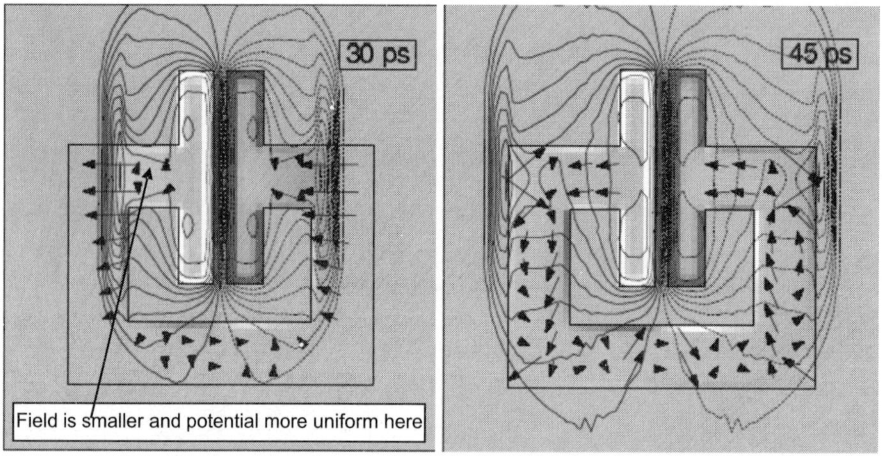

Fig. 2.6(b) Discharge of a capacitor: very intial transient.

Notice how at 30 ps the potential is more uniform and the electric field is much smaller to the right of this point due to contributions from both plates. This is the classical weaker fringe field due to charges from both the plates. The pattern extends all around the circuit.

The electric field's vectors are drawn 5 times larger in the last three diagrams (at $t = 45$, 80 and 160 ps) to display small features more clearly. Notice the surface charge density gradient from the various grey to light shades on the edges of the wires. Each cubic centimetre of the connecting wires (assumed made of copper) contains 8.4830×10^{22} free electrons and the charge density at the surface of the plates and on the connecting wires does not exceed 5×10^{6} charges per cubic centimetre indicating that only a very tiny fraction of the total number of free electron point charges $\frac{1}{6 \times 10^{17}}$ $(= 5 \times 10^{6}/8.4830 \times 10^{22})$ move to the surface of the wires and the capacitor plates. In other words, a small number of surface point charges are sufficient to produce a significant potential (a potential variation from – 5V to + 5V across the plates of the capacitor = voltage of 10 volts) difference. There are plenty of free electrons still in the conducting wires and plates. The fraction of surface charge to free charge in the wires will be tiny even if the wires were assumed to be thinner.

At 45 ps (see right diagram in Fig. 2.6(b)), the electric field is seen to reach the first turn in the wires from both the plates and, upto this time there was no significant surface charge to modify the large electric field of the closest plate of charge. It is after this time that we see surface charges building up on the wires as shown in Fig. 2.7, at $t = 80$ ps.

Note on Equipotential Lines, Surfaces and Regions

Equipotential lines are fictitious lines that are drawn to provide a visual representation of the behaviour of the potential at points around electric circuit diagrams and around distributions of charges. An equipotential line usually carries a number that represents the potential (with respect to a point at infinity) on *all* points on the line. Unlike electric field lines which represent a physical quantity namely, the electric field (an alteration of space) that causes a force to act on charged particles, an equipotential line is not a physical entity like a force or a vibration. An equipotential line is simply a graphical representation of the locus of points where there is an electric field and at which points, the *total work done*, (by computing the integral of the dot product $E.dl$ ($\int E.dl$) along infinitesimal distances on the chosen path[13]) to move a unit positive charge from a point at infinity where the potential is assumed to be zero, is equal. The *equipotential* lines are always perpendicular to the electric field lines. More descriptions will be provided in Section 4.5 "Power flow when a capacitor discharges into a resistor" on equipotential lines.

Their usefulness in the study of circuits lies in their meaningful visual representation of potentials at points that lie on *different* equipotential lines and highlight the amount of voltage drop (potential difference) between these points. In Section 2.17 "what is voltage" we will learn that voltage is a way of using numbers to describe an electric field.

Equipotentials can occur on surfaces and in regions. It must be appreciated that in a real conductor or wire (in 3-dimensional space) the equipotentials can be on surfaces (for example, of isolated charged conductors) and in regions (for example, inside a conductor in the interatomic spaces). We will learn more about this when we describe the average field in matter in Section. 2.26.

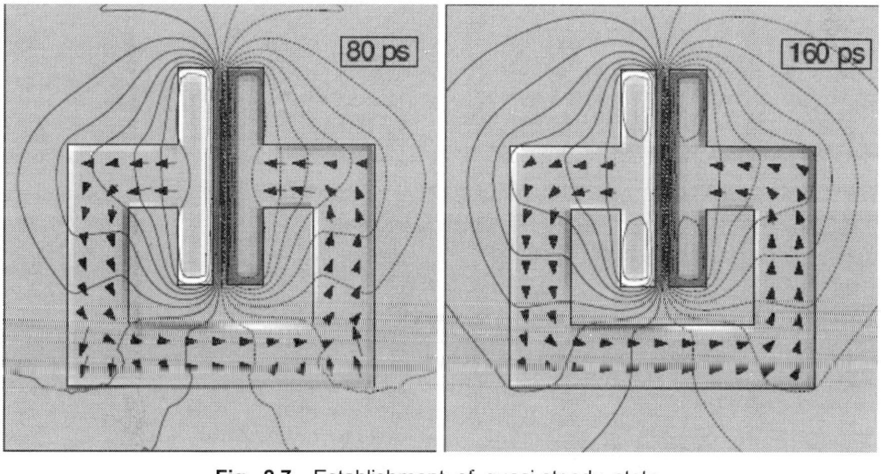

Fig. 2.7 Establishment of quasi-steady state.

It takes a light-crossing time (the time taken for the electric field to cross the length of the circuit), before the electric field vectors begin to point correctly to reach quasi steady-state.

Note that after 160 ps, the electric field vectors are being tipped towards vertical from being horizontal at the corners of the wires. This is due to the feedback mechanism that begins to take effect when the electric field and surface charge development becomes coordinated. As the charge develops, the field advances. This in turn causes further charge development and the charge is advanced further. There is a further final rearrangement of surface charges and the resulting electric field becomes uniform along the wire (see Fig. 2.7, 160 ps) assumed to be of uniform cross-section.

After this transient period of 160 ps, all the conduction band electrons motivated by the field set-up in the wires, move and begin the discharge cycle. The charges would have been moving during the transient also, but ever so slightly.

While the charges and fields do form a feedback loop beginning around 160 ps, initially there is no feedback. At $t = 0$, charges were placed on the capacitor plates and, the changing electric field from the sudden appearance of the charges on the capacitor plates, races away at the speed of light and a bit after the $t = 45$ ps has moved off the plotted area, never to return (there's no rebound). As this advancing electric field reaches new pieces of neutral conductor, the charges respond by separating. This change in the charge distribution creates a new propagating electric field which, in time, modifies all the other charges slightly.

The only thing that has happened at 45 ps is that, the first electric fields have finished sweeping across the circuit and separating charges. From here on what we see is feedback; the newly separated charges are not an equilibrium configuration and the charges move (and change the fields) until a quasi steady-state equilibrium is reached by 160 ps = roughly 2 or 3 light-crossing times.

For the above computations, six 400 MHz Pentium II computers were run in parallel and each integration step took about 5 minutes.

2.7 CHARGING A CAPACITOR

We now study the development of surface charge and the mechanism of feedback to maintain steady-state when charging a capacitor (after Chabay & Sherwood, 1995).

What happens when an uncharged capacitor is charged?

Shown in Fig. 2.8 is a circuit used to charge a capacitor comprising a battery, connecting wires and a bulb.

During the first few nanoseconds (Chabay & Sherwood, 1995, used by permission)[2], while the surface charges were arranging themselves, it is as though the *capacitor weren't there* – as though there were a continuous wire with no break in it.

There is *no electric field* E_{cap} *and fringe field* because there is *no* initial charge on the capacitor.

On closure of the circuit, the *net field* is due just to battery and surface charges.

Note: The voltage across the capacitor is zero at this instant of turn on and therefore the *voltage of plate B is equal to the voltage of the battery positive terminal.*

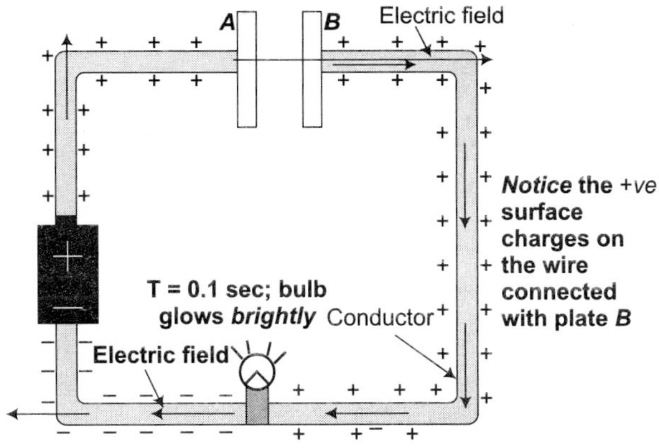

Fig. 2.8 Charging a capacitor.

On closure of the circuit, the *net field E* is due just to battery and surface charges. The field sourced from charges on plate *A* which is in contact with the battery positive terminal, *penetrates* the gap between the plates and then *penetrates* the plate *B*.

Surface charges appear on the plate *B* and further, appear on the wire to the bulb.

The capacitor is not "charged" at this instant, yet.

Then, the free-electron sea *shifts* in plate *B* due to polarization and negative charges *begin* to appear on the inner side of plate *B* as more and more electrons travel along the wire connected to the bulb.......and charges will run in the wires, governed by the relation $J = \sigma E_{net}$.

The voltage *across* the capacitor is zero at this instant of turn on and therefore the voltage of plate *B is equal* to the battery voltage.

During the *first few nanoseconds* while the surface charges are arranging themselves, the *capacitor* makes *no* contribution to the electric field in the wires because there are no charges on its plates yet. It is as though the capacitor weren't there – as though there were *a continuous wire with no break in it.*

The situation is shown in Fig. 2.9a where the plates of the capacitor are shown distorted to afford a clear view. The positive charges appear on plate *B* initially and these gradually give way to the negative charges as polarization progresses

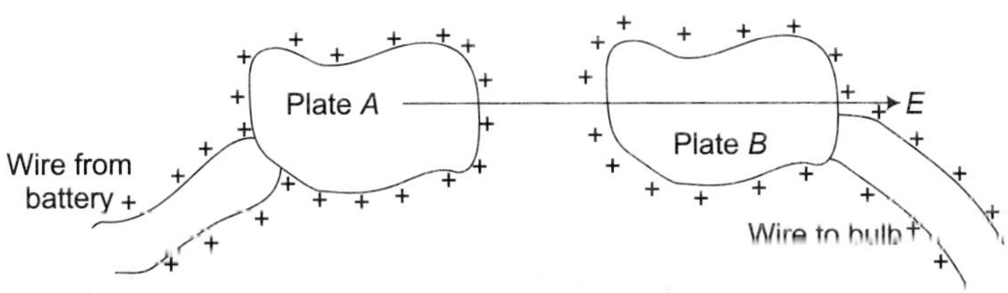

Fig. 2.9(a) The RC charging circuit – Initial transient.

Soon, the polarization of plate *B* and the connecting wire begins and negative charges appear on the inner surface of the plate and the resultant field is smaller as shown in Fig. 2.9b. The field vectors between the plates of the capacitor are not shown for reasons of clarity.

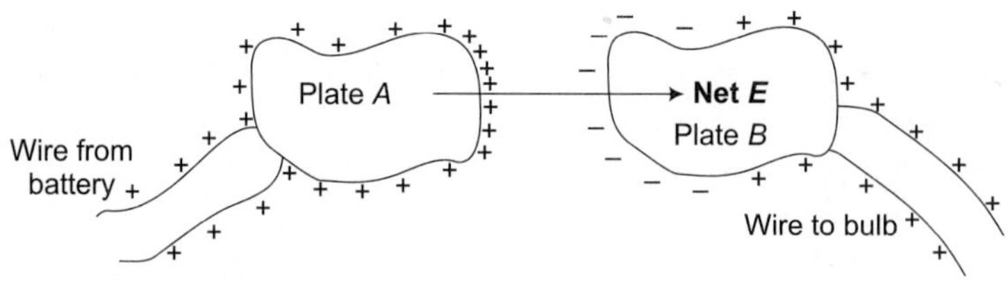

Fig. 2.9(b) The RC charging circuit – polarization of plate.

If, the initial electric field in the wires is due entirely to the surface charge on the wires and on the battery plates, with *no contribution* from the capacitor, then the initial *current* should be the *same* as it is in a circuit with a bulb and batteries but no capacitor. So, the current is maximum (as shown by the large block arrow in the wires in Fig. 2.10) at turn ON and the bulb will be bright initially.

That is why at turn ON of the circuit, the electron current is a maximum that causes *polarization* of the plates *A* and *B*.

Note that polarization occurs at *both* plates. And, the electrons *displaced* from plate *A* cause a *corresponding number* to make their presence on plate *B*; the electrons appearing on plate *B* move from the wires and are not the ones that left plate *A*.

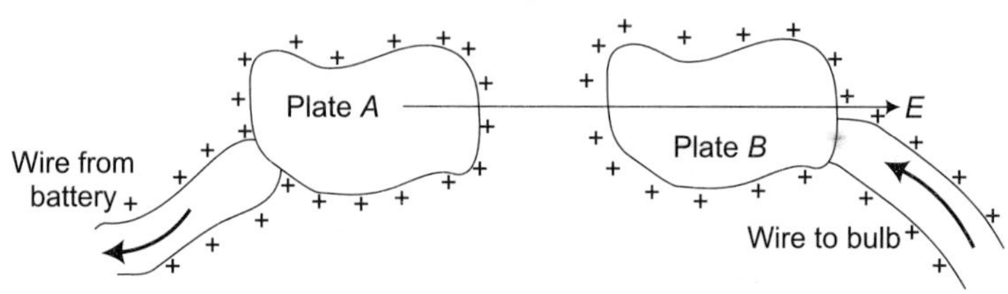

Fig. 2.10 The RC charging circuit – initial transient and maximum current at turn on.

After the initial transient is over, the charging of the capacitor progresses and now that there is some charge on the capacitor, the *net* field is *smaller* due to the fringe field of the capacitor; and this will result in a smaller current. See Fig. 2.11.

On the wire connecting the lamp and the capacitor note the gradual reduction in positive charges.

The net field which is a result of the vectorial addition of the applied field and the growing fringe field gradually begins to reduce in magnitude.

The initial voltage recorded at plate *B* gradually begins to reduce in magnitude as the capacitor charges and there is a reduction of the net field everywhere in the wires.

Visualizing the movement of charges (electrons) during charging, in the wires, inside the material of capacitor plates and, on the surfaces of capacitor plates.

One must be cautious when visualizing the movement of electrons in the wires and on the surfaces of the capacitor plates. As was seen in Chapter 1, current is not a zippy movement of electrons from one terminal of the battery to the surface of the capacitor plate. It is a *start-stop movement of millions of electrons that already exist* everywhere inside the wires and in the material of the capacitor plates, *with a drift superimposed.*

At every instant during charging, it is only the electrons inside the *last layers* of the material of the plate B that shift and rest on its surface. A corresponding number of electrons that leaves from the *last layer* of the material on the surface of plate A, similarly *shift and move into the wire*. Do remember that the conductors (wires, capacitor plates) with current in them are everywhere "neutral" inside (of course, leaving the surface charges that also determine the voltage) which means that dipoles are not formed within them. See Section 1.18 "Neutrality in current carrying conductors".

As the last layer of electrons shift, the resultant electric field everywhere in space due to the charge on the surface of the plates, reaches out and around and *slaps* onto the opposite charges on the other plate; the lines of the electric field can be drawn originating on the positive charges of plate *A* (visualize the positive charges as lattice ions *shining* through the shifted and pulled cloud of orbiting electrons) and terminating on the electrons resting on the surfaces of plate *B*.

Do not visualize the electrons as having left the plate *A* and as having moved all the way around to plate *B*. It is only the last layer of electrons that have moved. After all, there are millions available everywhere in the circuit, inside the wires, inside the battery terminals, and inside the material of the capacitor plates and all are alike!

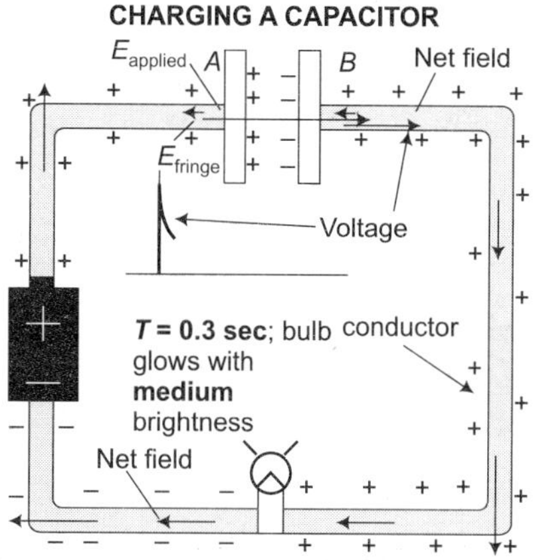

Fig. 2.11 The RC charging circuit – after 0.3 second.

As more and more charges accumulate on the capacitor, *net field* is still *smaller* due to growing fringe field of capacitor; therefore, smaller current results.

Visualizing the circuit process: One problem most students experience in visualizing circuit processes is the deeply ingrained idea that current flow is sequential. The reader may have by now observed with circuits that one has to think about what happens to *every* element in the circuit all at the *same* time; it normally causes problems if one focuses in on one part and then another. The presence of surface charges should make the visualization comparatively easier.

The applied field (due to the battery) *feels* an opposing force due to the fringe field. The applied field does not reduce in magnitude because the battery holds it as is its duty, but the net field reduces in magnitude because of the opposition.

The Action of a Resistor and a Capacitor in Electric Circuits

A resistor *physically impedes* the flow of charge carriers in a circuit. It never opposes the applied voltage (field). That job of opposition to applied voltages (fields) is for capacitors and inductors (which we shall study in Chapter 5).

A resistor *absorbs* all the voltage applied to it. It has a stiffness to charge movement that is uniform throughout its length if it is geometrically uniform throughout its length and the entire applied voltage *drops* along its length overcoming this stiffness equally within all its elemental segments. It does not form dipoles inside it....not counting the surface point charges....but those belong to the applied field. A resistor will oppose an externally applied field only if it were to be *isolated* as we saw in Section 1.14 (See Figs. 1.13 and 1.14). In a circuit, the applied field *stokes* a movement of charges *continuously* inside the resistor, which we call *current*. In the relation $J = \sigma E_{net}$, a resistor has *no* influence on E_{net}.

A capacitor does not have the kind of stiffness that a resistor has. A capacitor opposes the applied voltage (field) directly by its fringe field which is due to the charges that accumulate and form dipoles on its plates. In the relation $J = \sigma E_{net}$, a capacitor *affects* E_{net}.

Going back to the RC charging circuit, there is a further *reduction* of positive surface charges between the lamp and the capacitor.

The bulb glows with less than medium brightness due to a reduction in the current as a result of the overall reduction in the net field.

This situation is shown in Fig. 2.12.

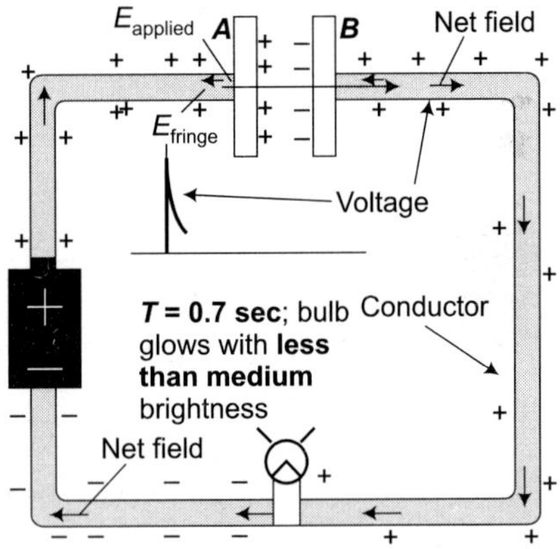

Fig. 2.12 The RC charging circuit – after 0.7 second.

It is instructive to *see and feel* the electric fields.....and to not see them as mere vectors...as described in Section 2.3 on the fringe field and its origins.

The battery source field *penetrates* the space between the plates ......*assisted* by surface charges on the wires.......then, surface charges develop on the *plates*....these charges also (never mind if these were induced) ....are surrounded by their own electric field.........all the fields have their own source ...and strengths....their contributions *add up vectorially* at target points.....in the wires for example........ knock off one source charge and the field at the target point is the one due to the ***other*** source charges.

As charge accumulation continues on the capacitor, *net field* keeps reducing and the current becomes smaller and smaller and the voltage of plate B is moving closer to the battery negative voltage.

Note there is a still further *reduction* of positive surface charges between the lamp and the capacitor.

The situation 1 second after closing the circuit is shown in Fig. 2.13.

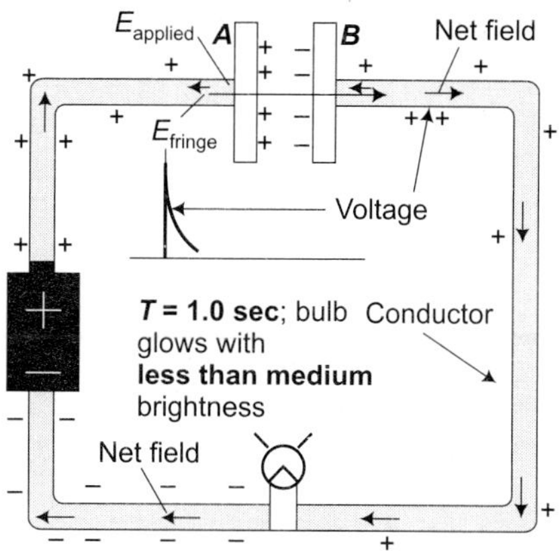

Fig. 2.13 The RC charging circuit – after 1 second.

Finally there is so much charge on the capacitor plates that the fringe field "cancels" the fields due to the battery and surface charges.

The situation after 50 seconds is depicted in Fig. 2.14. Note the appearance of a *dash of negative* surface charges between the lamp and the capacitor.

The voltage of plate B is nearly 0 volts and the voltage across the capacitor is nearly equal to the battery voltage.

The bulb *stops* glowing.

To "charge up" a capacitor, electrons are removed from the positive plate and "carried" to the negative plate. (**Note:** In a real charging circuit, it is only the last layer of electrons that is moved). In doing so, the agent charging the capacitor (the battery) has to *fight* against the electric field, which is pulling them back towards the positive plate and pushing them away from the negative plate.

For the calculation of the amount of energy stored when a capacitor is charged, see Appendix A "Energy stored in capacitors".

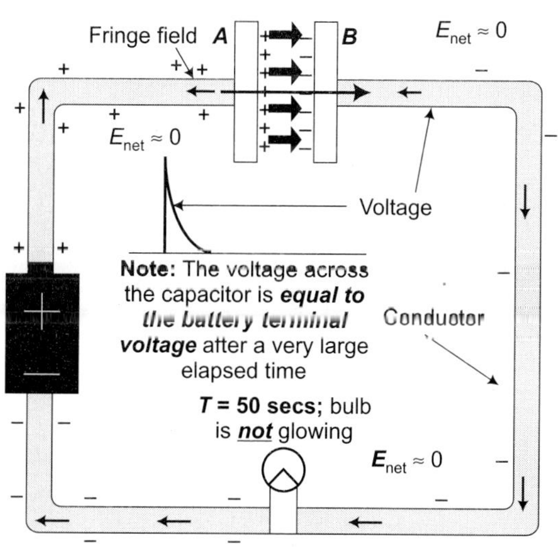

Fig. 2.14 The RC charging circuit – after 50 seconds.

As time further elapses the capacitor is fully charged and there is no net field and no net current.

The potential of the plate A will be the same as that of the battery positive terminal and the potential of plate B will be the same as that of the battery negative terminal. If they were not, static equilibrium will not have been attained and a current will result until static equilibrium is achieved.

Notice the very powerful field shown by thick field vectors (see Fig. 2.14) that has come into existence between the plates of the capacitor. This is due to charges of opposite sign tightly packed on the inner surface of each capacitor plate. One of the specialities of this field is that it is *uniform between the plates* and does not reduce with distance moving from one plate to the other in the gap.

Expressions for the field are given in Appendix A and I urge the reader to derive the expressions for the field of two uniformly charged disks beginning with the electric field of a dipole.

The final situation is depicted in Fig. 2.15.

However, let us not forget that the fields from the battery and the surface charges are still *real and existing* and so are the fringe fields of the capacitor that are *alive and existing as well.*

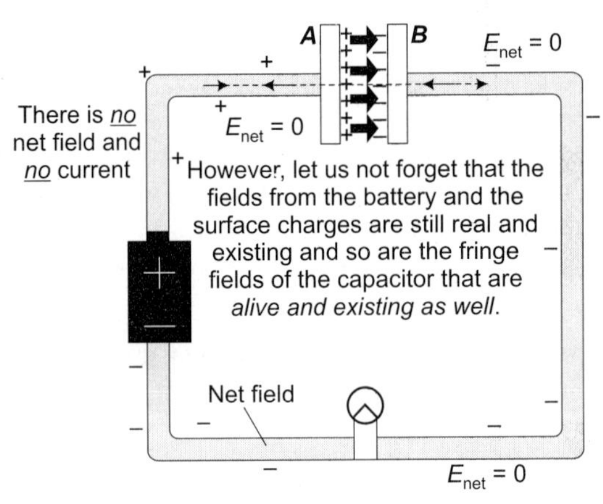

Fig. 2.15 The RC charging circuit – finally after a very long time

The fields, the applied and the fringe are in a tense standoff in the wires as shown in the figure below.

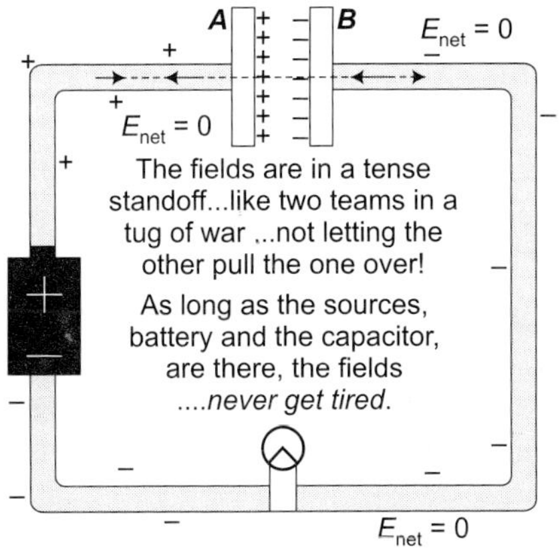

The fields are in a tense standoff

Note that the charges on the plate *B* were *sourced* from the plate *A* of the capacitor, therefore, the charging current in the circuit is termed *displacement current.*

As a matter of fact, after a capacitor has been *charged, some* surface charges exist on the wires connecting the capacitor and these are responsible for *holding* off the capacitor from discharging.

Here is One Way to Visualize the Charging of a Capacitor

Whenever we "charge" a capacitor, the *path for the current* is *through* the capacitor and back out again. Of course, between the plates no charges ever cross over; but an ammeter connected to either plate of the capacitor will indicate the *same* current.

Why? How?

A certain amount of charge has been forced to flow momentarily during charging, and a *rising* potential difference has appeared. *But the* current is directed *through* (only apparently, not really) the capacitor and the *incoming* electrons on one plate force *other electrons to leave* at the same time.

Every bit of charge that is injected into one terminal must be *forced* out of other terminal at the same time. From: http://amasci.com/emotor/cap1.html.

The amount of charge *inside* the capacitor *never changes.*

The net charge on each plate is *cancelled* by the opposite charge on the other plate. When charging a capacitor, a *momentary current* causes the voltage to rise. Volts times the electron-flow equals *energy-flow* (V × I − P). Therefore, during a momentary current through a capacitor, there is a joules-per-second *transfer of energy from the power supply to the capacitor.*

During the *charging* process, *energy* is placed in the capacitor.

The energy is *transferred from the battery to the capacitor* at the moments when the charge, driven by the applied field, overcomes the fringe field and accumulates on the plates of the capacitor.

Capacitors store ***energy***, **not** charge. When we "charge" a capacitor, we *give it a charge of energy!*

Charging a capacitor means *injecting* electrical *energy into* the device.

Notice how when there is enough charge on the capacitor plates to produce a large enough fringe field to counteract the field made by all the other charges, the current stops completely.

The initial establishment of surface charges takes a few nanoseconds, while it takes several seconds to reach static equilibrium.

During the *several seconds while the bulb is glowing*, the *current* is not constant, it *changes so slowly* that it makes sense to describe the situation as a **quasi steady-state** in which *quasi* means "having some resemblance usually by possession of certain attributes".

During every tiny interval of time when, the voltage across the capacitor (and, the current in the circuit) is changing, the surface charges rearrange themselves to maintain appropriate electric field conditions around the circuit and to uphold the conservation of current law, in a sort of steady-state during each of those tiny intervals, or quasi steady-state.

Now for some Mathematics

If the instantaneous current through a capacitor is i, then the change in voltage across the capacitor in time t_l is $\frac{1}{C}\int_0^{t_1} i.dt$. Practically, zero resistance is never achieved, therefore, the current is always of finite magnitude, then the above integral approaches zero as $t_l \to 0$.

Hence, it follows that the voltage across a capacitor cannot change instantaneously, *provided that the current remains finite;* charges cannot move with the speed of light (because no object or particle is known to move faster than the speed of light) to charge up the capacitor to the applied voltage, which we reiterate is always so in practical circuits because zero resistance is never achieved in electrical and electronic circuits.

The above reasoning is in line with what we observed in the action of charging a capacitor and the descriptions given with reference to Figs. 2.9 and Fig. 2.10. The output voltage and the current is given by the expressions $Ve^{\frac{-t}{RC}}$ and $\frac{V}{R}e^{\frac{-t}{RC}}$ respectively, where V is the battery voltage and R is the resistance of the bulb. These may be easily derived from $V - v_c = RC.dv_c/dt$ written around the loop beginning with $i = C.dv_c/dt$. Refer to "Electrical Technology" by Edward Hughes for this derivation.

The waveform of the output voltage is shown in the left figure below.

Voltage across the resistor

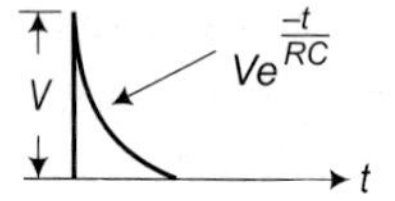

Voltage across the capacitor

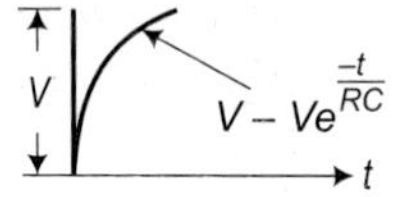

As charges accumulate on the plates of the capacitor, the voltage across it grows. And this is indeed so. Its voltage is plotted in the right figure above and is given by

Voltage across capacitor v_c = Battery voltage – the output voltage across $R = V - Ve^{\frac{-t}{RC}}$

The output voltage decays exponentially with respect to time and note the dependence of the rate of decay on the product *RC*. The greater the product, slower is the decay and, the smaller the product, faster is the decay.

Why does the voltage decay have to be "exponential"?

***'e'* is the base amount of growth (or decay) shared by all continually growing (or decaying) processes just as Pi or 'π' is the ratio between circumference and diameter shared by all circles.** The constant '*e*' allows you to extend a simple growth rate such as the end of year interest added to a savings bank account and produce a compound, continuous growth at much faster rates down to fractions of a second. '*e'* shows up whenever systems grow continuously and is used in the field of many natural growing processes: population, radioactive decay and so on. Even jagged systems that don't grow smoothly can be *approximated* by *e*.

Pi or 'π' is a fundamental ratio inherent in *all* circles and therefore impacts any calculation of circumference, area, volume, and surface area for circles, spheres, cylinders, and so on. Pi shows that all circles belong to one family, not to mention the trigonometric functions derived from circles (sin, cos, tan). Similarly, '*e*' shows that all growth and decay belong to *one family,* that of '*e*'.

Just like every number can be considered a *scaled* version of 1 (the base unit), every circle can be considered a *scaled* version of the unit circle (radius 1), and every rate of growth (or decay) can be considered a *scaled* version of '*e'* (the *unit* rate of growth or decay).

To gain an intuitive understanding of the constant '*e'*, and why it has the value '2.718', please visit http://betterexplained.com/articles/an-intuitive-guide-to-exponential-functions-e/.

2.8 APPLYING A VOLTAGE PULSE TO A CAPACITOR–RESISTOR CIRCUIT

Assume we apply a voltage pulse to a capacitor –resistor circuit. This is shown in Fig. 2.16.

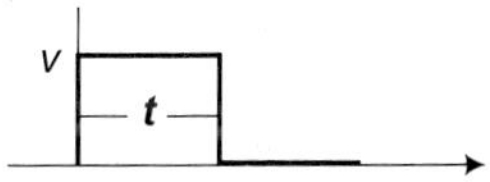

Applying such an input is like connecting a battery of voltage '*v*' for a time '*t*' and then "*shorting*' the capacitor resistor series combination subsequently...while disconnecting the battery from the circuit.

This is shown in Fig. 2.16.

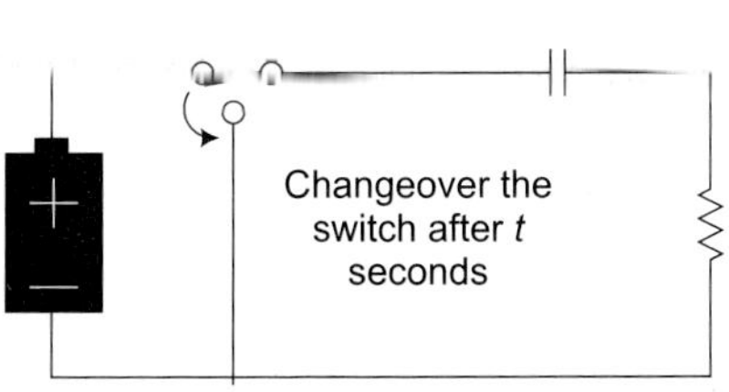

Fig. 2.16 Applying a voltage pulse to an RC combination.

If the product of the capacitance and the resistance is ***large... (RC >> t),*** that is, either one of the two is ***large* or** maybe even ***both are large,*** the capacitor will take a ***long time*** to *charge up*; because the ***resistor restricts*** the charging current, if it were to be large, else, if the capacitance were to be large ....the capacitor will *thirst* for a *huge* amount of charge.....Q being = CV.

In either case, the output voltage will initially *jump to* '*v*' as described earlier and then will **reduce *slowly*** towards the negative voltage of the battery. This is shown in Fig. 2.17.

Due to the slow passage of current and resulting *slow* charging of the capacitor, the *strength* of the fringe field *grows slowly*.

The output voltage *lingers* and reduces slowly from its initial jump to the input voltage '*v*'. The electric fields inside the wire are shown only near the capacitor for clarity though the fields exist everywhere in the connecting wires.

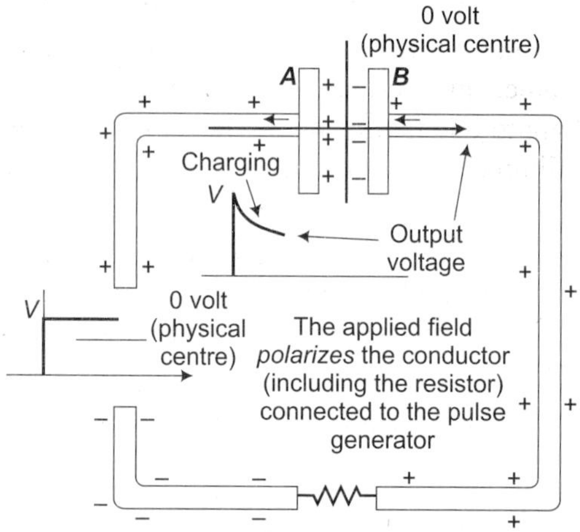

Fig. 2.17 Applying a pulse input to a CR combination.

The electrons in the wires connecting plate B slowly *neutralize* the positive surface charges and make the wire less positive—polarization progresses as seen in Fig. 2.17.

Notice, how the applied field is *driving* charge (sourced from opposite plates) inside the wires onto the plates negative on B and positive on A despite the growing fringe field.

At the physical centre of the battery, in the space between the terminals, the potential is zero with reference to a point at infinity as we shall see in Section 2.21 "Potential of a point charge relative to infinity". Also see Section 1.8 "Conservation of charge" and the example of battery terminal voltages.

At the physical centre of the capacitor, in the space between the plates of the capacitor, the potential is zero with reference to a point at infinity. These are shown in Fig. 2.17 above.

These potentials are a natural consequence of separating unlike charges, a speciality in electricity.

All of a sudden, the *applied voltage is made zero...from 'v'.*

It is like shorting out the CR series combination and, causing the capacitor to *discharge* through the resistor.

The situation is shown in Fig. 2.18.

Note: The voltage of plate B is still short of becoming *equal* to the pulse generator's negative terminal voltage.

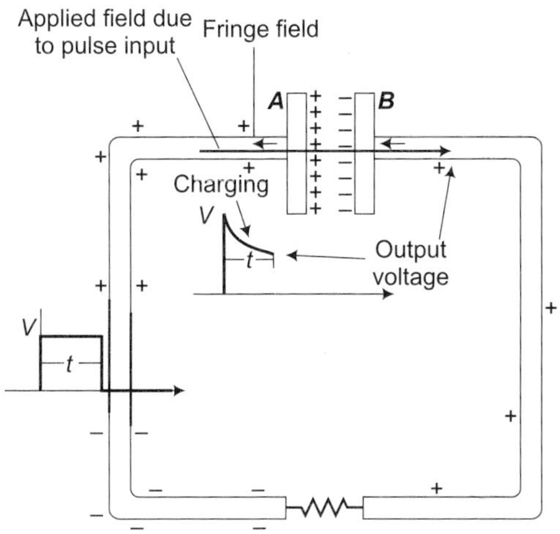

Fig. 2.18 Pulse input – input voltage is made zero.

How does the fringe field *oppose* the electric field due to the applied input voltage?

After all, we had stated that the fringe field is the vectorial sum of slightly unequal though large capacitor plate fields?

As a consequence, the fringe field cannot be very large but only a *very tiny fraction* of the individual fields on the plate.

How strong is the fringe field?

Let us put in a few numbers to check what the *real value* of the strength of the fringe field is.

In Chapter 1, we had seen that the applied field to a resistive circuit for a typical 1.5 volt battery is 2.1 volts/metre in the wires…and for a 10 volt supply it will be, say 21 volts/metre.

Now, consider that a 10 Vrms, 1 kHz wave is applied to a capacitor of value 0.01 microfarad through a 15 kiloohm resistor.

The voltage will be 7 volts *across* the capacitor, and since the capacitor is 0.01 microfarad, then, assuming a typical air insulated disk plate construction…with area $A = 110\ \text{cm}^2$, the airgap is 1 mm ("*s*") and radius is about 6 cm.

The resulting field strength is 7 V/1 mm = $\mathbf{7 \times 10^3}$ **volts/metre** *between* the plates or 7000 volts/metre *between* the plates…and the fringe field at a point a little distance away from one plate ….outside the capacitor is; (See Appendix A for the expression for the fringe field)

The fringe field is …(s/2R) times the "field inside" – 1 mm/2 × 60 mm or 0.008 × 7 × 10^3 volts/metre = **56 V/metre.**

We see that the fringe field is *comparable* with the field in the wires which was 21 V/metre.

And it is *in the wires* that the fringe field is *called on duty to oppose the applied field.*

We have indicated only sample values for illustrative purposes.

How strong is the fringe field?

Recap our illustration of the fringe field in Section 2.2...noting the relative strengths of the fields...

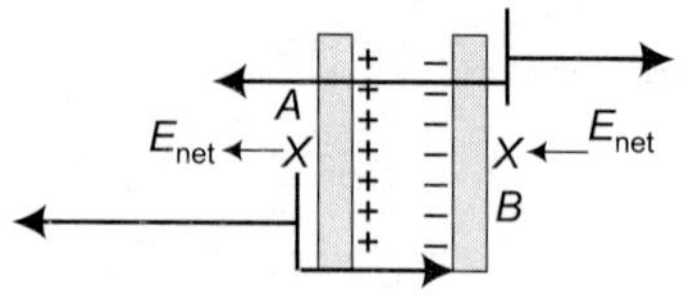

How can this be so? When the applied input field is so *"weak"*?

Applied field = 21 volts/metre

Between the plates = 7000 volts/metre

Fringe field = 56 volts/metre

Are we to believe this?

Yes. We must be careful not to forget that the charges on the plates of a capacitor are extremely tightly packed together....therefore, the **"*strong* 7000 V/m"** field strength ! Giving us our tiny fringe field to *oppose* the small input field.

The voltage *across the capacitor* in our example maybe a *mere 7 volts* but the field is very powerful.

So, when you visualize the charging of a capacitor...

..do not forget

.......to visualize the small applied field...with a tiny sound of zzzz... or mmmm ..... pushing some charge on the plate...source charges for the capacitor field...

...the comparable small fringe field outside the plate with a tiny sound of zzz... or mmmm.... which is due.....

...to the powerful fields between the charged capacitor plates with a loud sound of zzzz... or mmmmm...

...and the charging process with the applied field...

....pushing charges against the fringe field

Charging and Discharging a Capacitor–Applying a Pulse Input-contd.

Refer to Fig. 2.18. When the pulse input goes to zero volt, the effect is to *remove the applied field*....of applied voltage '*v*'... by which it was *holding up* the voltage of plate *B* and its connecting wire.

What *remains* is the fringe field and the polarization of the plate *stops* and the incoming electrons are *reversed* in direction. This situation is depicted in Fig. 2.19.

The fringe field takes charge of the tug of war !...and electrons from plate *B* make the plate *B* voltage *negative*... down instantaneously by the amount of voltage '*v*' by which they were held up by the now "removed" applied field.

In Section 2.26 "Applying a pulse to an RC circuit:- Another close look", this aspect will be dealt with in some more detail. Also see the topic "Superposition of Potential" which is a consequence of the "Superposition of electric field" principle in Section 3.1.

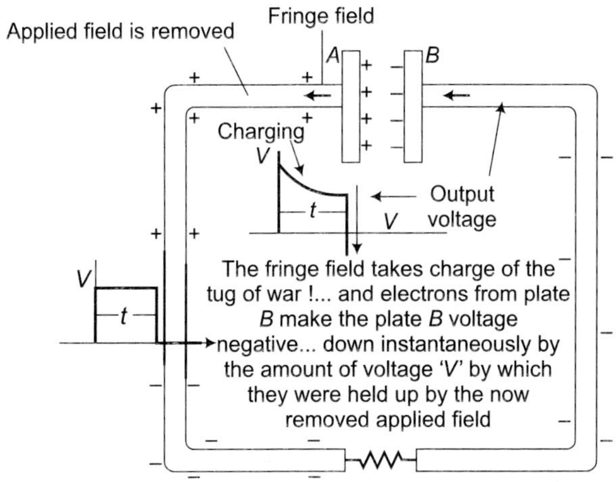

Fig. 2.19 Pulse input – input voltage is made zero.

Gradually, more and more *excess* (local) electrons from plate *B* move towards plate *A* as the capacitor begins to *discharge*.

This is depicted in Fig. 2.20.

The voltage of plate *B* of the capacitor *rises* towards "zero" from its negative value and that of plate A *falls* towards zero.

The surface charges simply *disappear* and so do the charges on the plates of the capacitor and finally, the fringe fields disappear.

Energy is simply radiated away in the resistor, not to return. A resistor does not *store* energy. It takes it away from a circuit.

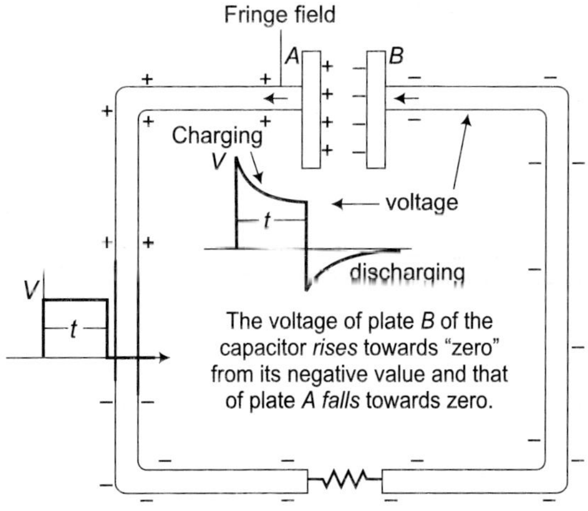

Fig. 2.20 Pulse input – the capacitor discharges.

If the product of the capacitance and the resistance is *small and RC* << *t,* that is, either one of the two is *small* or maybe even *both are small*: the capacitor will take a *very short time* to *charge up* because, the *resistor allows a large* charging current....if it were to be small else, if the capacitance were to be small it develops a voltage with small amounts of charge.

In either case, the output voltage will *quickly* reduce from the value '*v*' to which it had jumped initially and move towards zero *much before t* when the input voltage goes to zero.

When the input voltage is reduced to "zero", the output voltage will jump to "–*v*".

The capacitor will then take a ***very short time*** to discharge because, the ***resistor allows a large*** discharging current.

In either case, the output voltage will *quickly* rise from –*v* to "zero" as the capacitor is *fully discharged.*

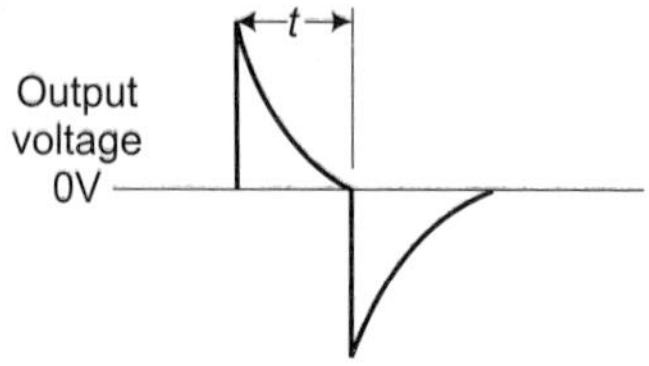

Summarizing, for a pulse input of width '*t*' to a C-R circuit, if the *product* of the capacitance and the resistance......

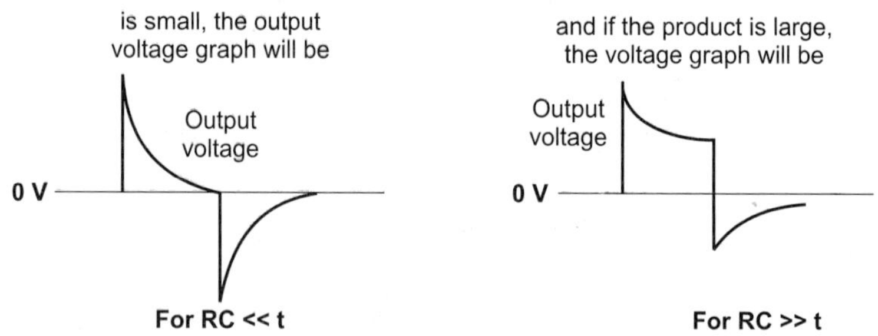

Voltage Changes Across Capacitors

Provided that the current remains finite, *the voltage across a capacitor cannot change instantaneously*. What this means is, as seen in Fig. 2.20, at $t = 0$, when the input voltage changes discontinuously or instantaneously, the capacitor does not charge instantaneously and instead seemingly "passes" the voltage to appear across the resistor. In reality, the electric field *penetrates* the gap between its plates. The capacitor begins to charge only *after* this instant.

The same comments are applicable at the trailing edge of the pulse input, when the output voltage becomes negative and *then* the capacitor begins to discharge.

It should be noted that a sinusoidal voltage does not change instantaneously and has instantaneous-like properties only at very high frequencies.

Important: Energy considerations when charging a capacitor using a battery (application of a step voltage).

When a capacitor of capacitance C is charged using a battery of voltage V and a switch, the battery expends an amount of energy QV while the energy stored in the capacitor (after charging) is $QV/2$ where Q represents the amount of charge transferred when the capacitor voltage is V. Note that with a potential difference V, a capacitor C will store charge Q, but the energy stored is *not* $Q \times V$. Why not? Because, as the capacitor charges, both Q *and* V increase, so we have not moved *all* the charge with a potential difference of V across the capacitor (Fighting against the growing fringe field to transfer energy from the battery), which the battery did. See http://230nsc1.phy-astr.gsu.edu/hbase/electric/capeng.html.

Where is the missing energy? $(QV)/2$ from the battery?

Directly connecting the battery to the capacitor (application of a step voltage) results in an instantaneous energy *input* of QV to the capacitor, which is capable of *receiving only an amount* $(QV)/2$. The circuit shakes away excess energy by: heating up the parasitic resistance such as the capacitor equivalent series resistance or wiring resistance; arcing across switch contacts; or, by electromagnetic radiation (in digital circuits this disturbs operation of nearby circuits). This method of charging is analogous to standing on a table and lifting a weight from the floor by jerking a rope tied to the weight with the intention of placing it on the table. The jerk causes the weight to accelerate, and make it land on the table with a 'thud', dissipating "excess" kinetic energy in the form of frictional heat. If however, the weight were to be pulled with a slow and steady speed, it can be placed gently on the table surface without loss of excessive energy ! Likewise, if the capacitor is charged using a *ramp* voltage with a uniform rate of rise, energy losses could be kept to a minimum. See also Section 2.23 "The process of capacitors sharing charges".

2.9 APPLICATION OF A SINUSOIDAL VOLTAGE TO A CAPACITOR *C*

A sinewave depicts movements in the form of waves. It has a datum, rapid, slowing and steady growth in one direction for a quarter cycle with reference to a datum (a reference) and then, slow and faster decay in one direction for the next quarter cycle and, all these elements again in the opposite direction (reversal) of the forward movement for the next half cycle. It has peaks and valleys. In essence, the sinewave is a perfect embodiment of oscillatory movements like springs and quantities, like a voltage. It retains its waveshape when added to another sinewave of the same frequency and arbitrary phase and is the only periodic waveform that has this property. When a sinusoidal input voltage $v(t)$ is applied to a capacitor, then, during the positive half-cycle, notice that in the *absence of a resistance* the capacitor *instantaneously* attains the voltage of the input.

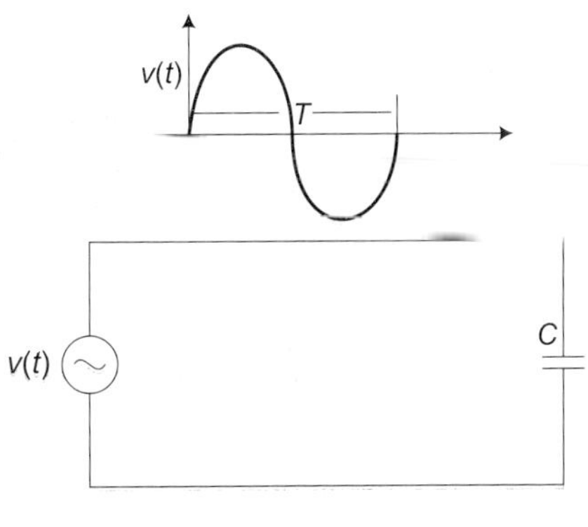

In short, the voltage *across the capacitor* changes by *charging* and then *discharging* as a half-sinusoid *following* the input voltage *at every instant*.

A sinusoidal voltage generator operates by separating charges for the positive half-cycle in one direction initially rapidly and slowing down and momentarily halting when its voltage is at the positive peak and then, gradually reuniting the separated charges to make the separation zero; when all charges re–unite, the generator voltage is zero.

The negative half-cycle now commences when, the generator operates by separating charges in the *opposite* direction initially rapidly and slowing down and momentarily halting when its voltage is at the negative peak and then, gradually reuniting the charges to make their separation zero; when all charges reunite, the generator voltage is zero. It then begins the next positive half-cycle followed by the negative half-cycle and the process repeats.

If connected to a resistor or a capacitor then, charges (surface) move onto the wires and their number and orientation changes at *every* instant of time. Remembering that surface charges are point charges (produced by a shift in the sea of electrons), it may be noted that the surface charge changes that occur at nearly the speed of light, produce the varying electric field inside the wires. The voltage at the source and the load terminals if plotted, produces a sinusoidal waveform.

The situation when the positive half of the sinusoidal voltage is applied is shown in Fig. 2.21.

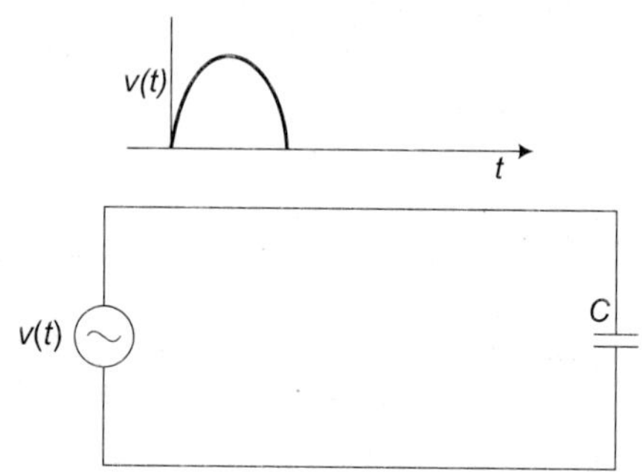

Fig. 2.21 Positive half sinusoid applied to a capacitor.

Notice the subtlety in this, the capacitor can charge instantaneously following the input because *there is no (or rather near zero) resistance* and develop the *fringe* field almost instantaneously to closely *match* the e-field of the applied input voltage...of course due to surface charges. See Fig. 2.22.

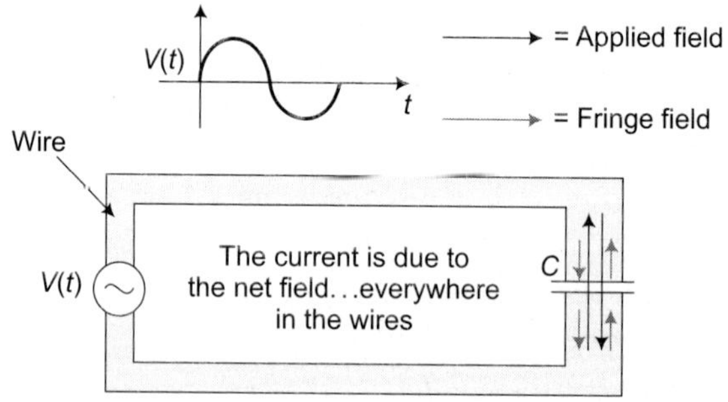

Fig. 2.22 Capacitor charges instantly following the input voltage.

The input voltage rises...the e-field *slams* into the capacitor plate...and through it.

There is *no* fringe field... and the resistance being near "*zero*"...the current zooms instantly... governed by the relation $J = \sigma E$, and, *positive charge* is *sprayed* on to the capacitor plate...(shown is one of the pairs of the capacitor plates)......the current is large...!

The situation is depicted in Fig. 2.23.

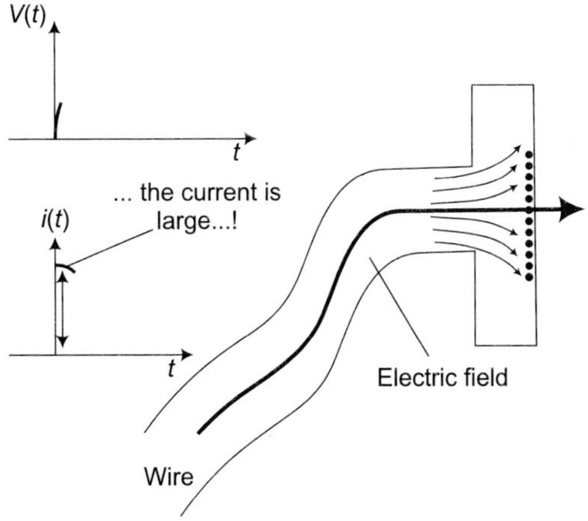

Fig. 2.23 Charge movement when a sinusoid is applied to a capacitor.

From the movement of charge depicted in Fig. 2.23 it is easy to understand the e-field *due* to the applied input. It is instructive to see how the *fringe field* develops which is described next.

In Fig. 2.24 the action at both the capacitor plates is depicted.

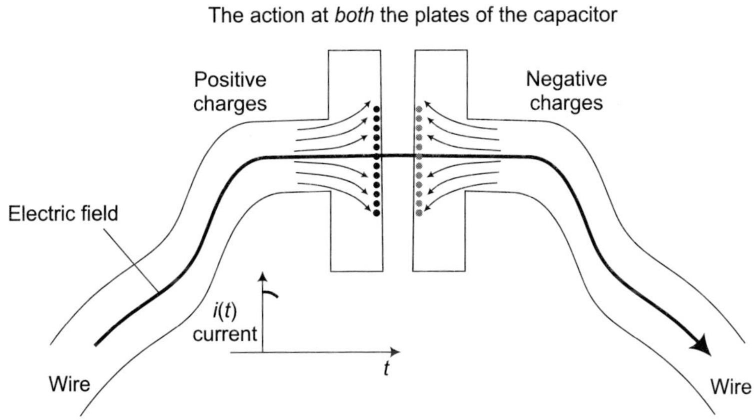

Fig. 2.24 Charge "spray" on both the plates of a capacitor.

The electric field, remember, always "fills" the wire and should not be visualised as being present in a narrow central portion of the wire.

At the initial instant, positive charge is sprayed on to the left capacitor plate and even as the charges are being sprayed on the capacitor plate, a fringe field develops and this is in opposition to the applied electric field due to the sinusoidal signal generator.

This is shown in Fig. 2.25.

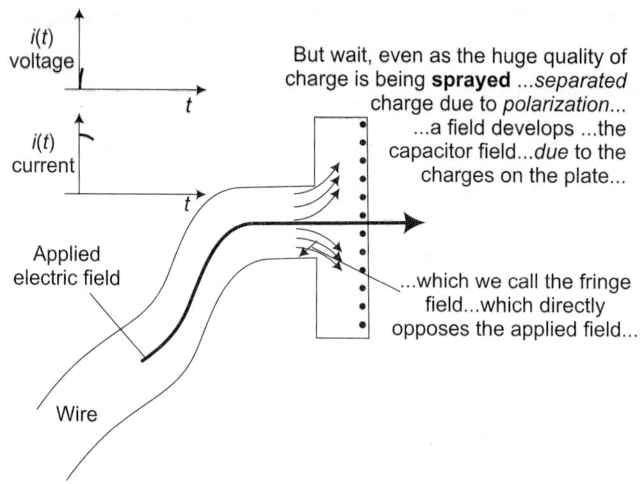

Fig. 2.25 Fringe field development.

The opposition force of the fringe field to the force of the applied field can be likened to the opposing force felt when compressing a ballpoint pen spring.

As time advances, the input voltage increases but, with a smaller rate of growth. Accordingly, the electric field (applied) also increases but with a smaller rate of growth. The force on the electrons in the wires increases but with a smaller rate of growth.

The charge *spray* and therefore, the **current** is reduced in magnitude due to the slow growth rate of the motivating applied field of the input voltage giving more time for charges to accumulate on the capacitor plates and allowing a stronger growing fringe field to *oppose* the applied field.

Never forget that the current in the wires is a start-stop motion of millions and millions of electrons everywhere within with a drift superimposed.

Therefore, as the applied voltage increases but with a *smaller* growth rate, the current ***reduces…to zero***.

This is shown in Fig. 2.26.

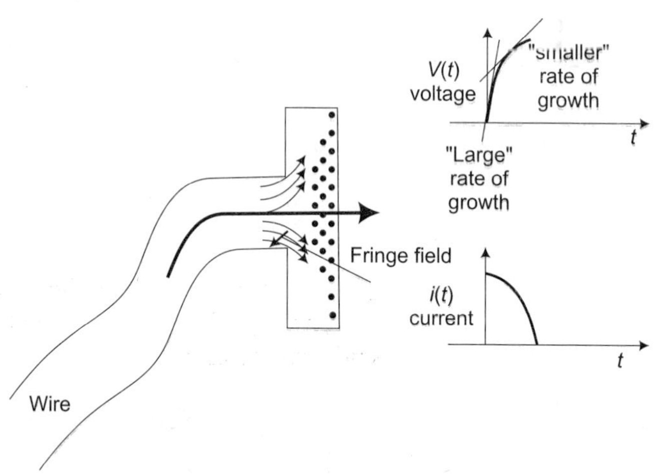

Fig. 2.26 Reduction in current due to the fringe field.

In Fig. 2.27, is shown how the fringe field *brakes* the oncoming charges during the positive half of the sinusoid.

Note the production of current the charge "spray" or current can be huge—yet cannot be infinite....

...because, the fringe field the spray (on both the plates) develops, is in direct opposition to the applied field, matching the applied field at every instant.

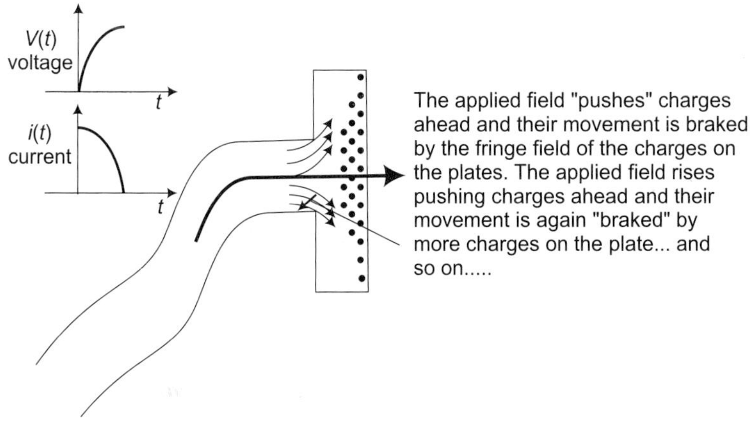

Fig. 2.27 Braking action of the fringe field.

The actions are described considering positive charge movement for simplicity and clarity in understanding the circuit processes. This does not in any way detract from the accuracy of description of the process or its mechanism, being considered opposite to the movement of electrons.

As time advances further and the input voltage *reduces* as a sinusoid the applied voltage causes the applied field to *reduce* in magnitude and the positive charges on the capacitor plate are *dragged* away from the plate and move into the wire under the influence of the *dominant* fringe field and the capacitor discharges.

This will cause the current direction to *reverse*....and its magnitude *grows*, in reverse.

This is shown in Fig. 2.28.

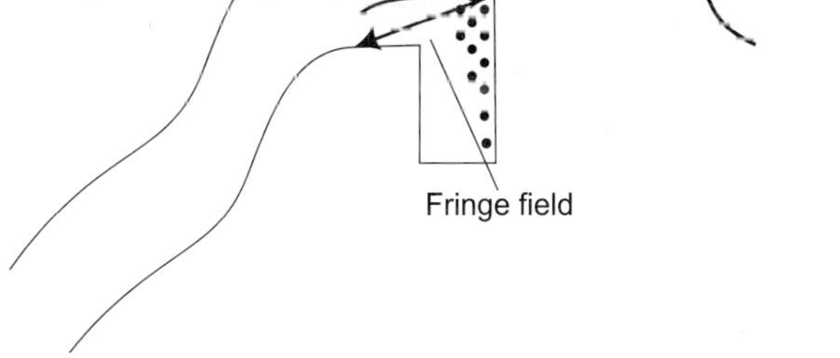

Fig. 2.28 Current reversal led by the fringe field after the input voltage has peaked.

The current reaches a *maximum* in reverse when the applied voltage is "zero". The capacitor is now fully discharged and its voltage is zero, as well.

When the applied voltage reverses or becomes *negative*, the applied field changes direction and this causes more positive charges to be *dragged* away from the capacitor plate.

The capacitor plate now becomes negative since positive charges are being dragged away into the wire after the capacitor voltage was zero.

A fringe field develops this time in the opposite direction to that when the charge on the plate was positive. It again ***opposes*** the applied field reducing the magnitude of the current.

This is shown in Fig. 2.29.

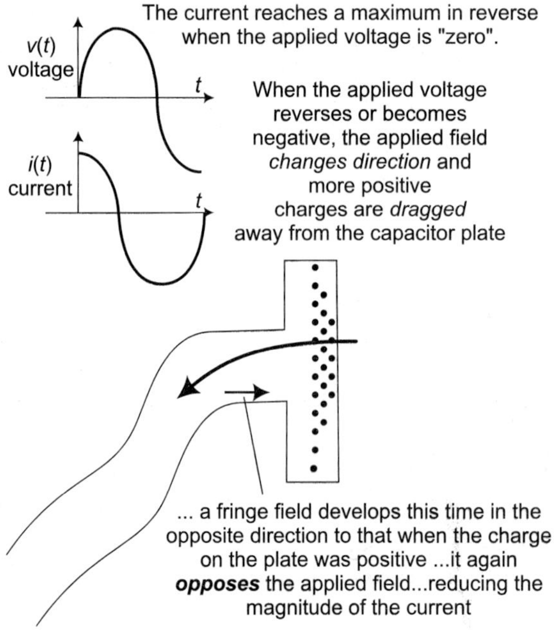

Fig. 2.29 Current reduction after reversal.

The reader is advised not to forget that the processes (the applied and fringe field development) and effects (the deposit and withdrawal of charges on the plate) were described in relation to one plate, but identical though reversed processes occur and inverse effects produced simultaneously at the other plate as well (see Fig 2.24). This is how the charges on one plate are sourced by the other plate.

Notice that the current is a *maximum* and *reducing* even as the voltage is increasing (or decreasing through "zero"). This is due to the *slow* growth rate of the applied field as time progresses allowing time for the fringe field to develop and *oppose* the applied field more.

The effect is that the current is large when the input voltage is small and rising and the current is zero when the input voltage peaks.

We say that the current *leads* the voltage by 90 degrees for a sinusoidal input.

This is shown in Fig. 2.30.

The current leads the voltage by 90 degrees for a sinusoidal input starting with a bang and ending in a whimper. Remember, for a capacitor

$$i_c = \frac{C\,dv_c}{dt}$$

Notice that the current is a *maximum* and *reducing* even as the voltage is increasing (or decreasing through "zero"). The is due to the *slow growth rate* of the applied field as time progresses allowing time for the fringe field to develop and *oppose* the applied field more.

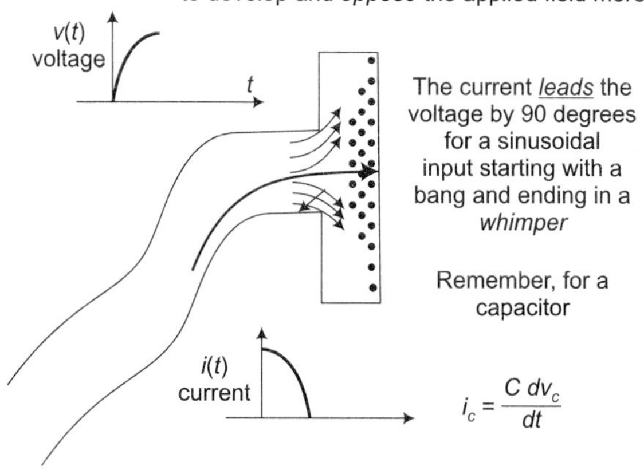

Fig. 2.30 Current "leads" the voltage for a capacitor.

That the current formula for a capacitor is correct can be observed when a ramp voltage is applied to a capacitor.

The *rate* of growth of the motivating applied field remains *constant* (unlike in a sinusoid where it changes continuously) and the *net* field which is the difference between the applied and the fringe field is constant and the charge spray is *continuous*.

The applied field always makes the motivating leap ahead of the fringe field.

The current is therefore *constant* and is given by $J = \sigma E_{NET}$ as shown in Fig. 2.31. Evidence that the current formula for the capacitor is correct can be obtained by observing what happens if the applied voltage were to be a "ramp" voltage. The rate of growth of the motivating applied field remaining constant (unlike in a sinusoid where it changes continuously) the net field is constant and the charge spray is continuous and successive batches of charge spray are of equal magnitude The current therefore, given by $J = \sigma E_{NET}$ remains constant and does not reduce.

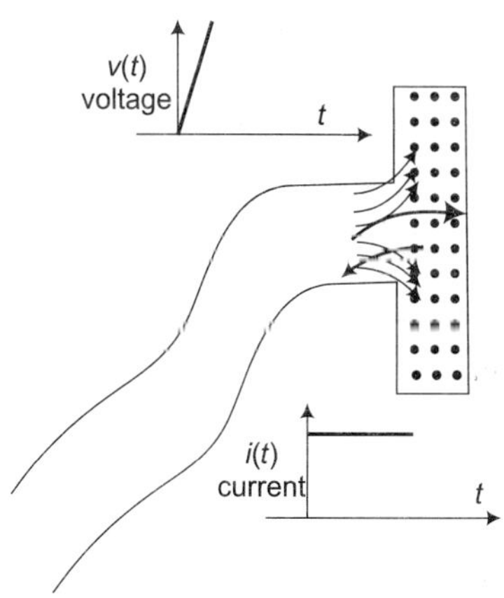

Fig. 2.31 Current remains constant for a ramp input voltage.

Visualizing the Charging of The Capacitor

One very powerful visualization of "charging" a capacitor—
one of my favourites is this.

For a "steady" current in a resistor, one has to apply a
steady voltage... DC

On the other hand... apply a "steady voltage" to a capacitor and
the current rises sharply. Charges "accumulate" instantly on
the plate "*opposing*" more charges from accumulating due to
the e-field these set up and the current is "zero" almost immediately.

"Increase" the voltage "*steadily*" (*and "continue"
to increase*) to the capacitor and the applied field is able to
"overcome" the "resisitive" effect of the fringe field due to the
accumulated charges, resulting in a "net field" whose
value is just "steady enough" so that the capacitor current
(charge spray) is steady or ***constant current.***

That is the reason the ***current*** in the capacitor is constant or
steady when a applied sinusoidal voltage "goes through
zero..." where it resembles a ramp.

"Reduce" the rate of growth of the applied field...

and the "*net field*" (i.e applied minus the fringe field)
"rate of growth" is *reduced*...

…never forgetting that the accumulating charges introduce
a field that is ever in "*opposition...*"

...the current is "reduced..." in magnitude

...make the applied voltage "*steady*" or constant and quickly
the accumulated charges develop a fringe field that directly, exactly
and continuously "opposes" the applied field... the net field is "zero"….

...and the current is "*reduced to zero*"

Why does the current remain steady and maximum when the applied sinusoidal voltage crosses zero when swinging from a positive to a negative magnitude or the reverse?

The rate of change of voltage is quite similar to a *ramp* voltage at this instant of time and applying the reasoning for determining the current in a capacitor subjected to a ramp voltage described above (See Fig. 2.31), the current is seen to remain steady and at a maximum at this instant.

This is depicted in Fig. 2.32.

This is quite unlike the magnitude of current if this voltage were to be applied to a resistor, where there is no fringe field to oppose due to accumulated charges and the current will be zero when the voltage is zero.

VISUALIZING THE CHARGING OF THE CAPACITOR... ONE MORE AID

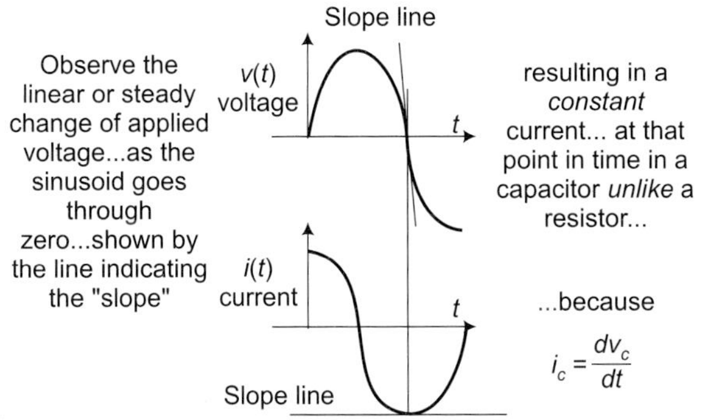

Fig. 2.32 Current is constant at the zero crossing instant of applied voltage.

Note that it is at the *zero crossing point* in a sinusoid that the rate of change of the input voltage is a *maximum*.. $C\,dv_c/dt = C \cos \omega t = C \cos 0° = C$ and so the current is also a maximum and naturally also *steady*.

2.10 CURRENT IN A CAPACITOR CIRCUIT – SINUSOIDAL INPUT

By a little stretch of imagination, it is possible to see that the *current* in the circuit is a sinusoid and *leads* the sinusoidal voltage *across* the capacitor ....which is also the applied voltage .....by 90 degrees as shown in Fig. 2.33.

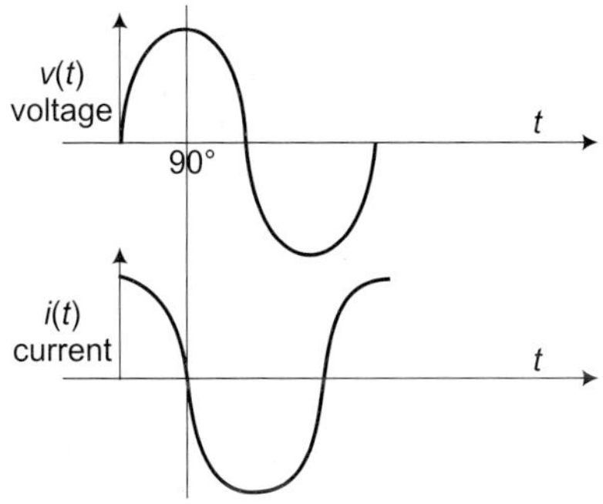

Fig. 2.33 Current leads the voltage for a sinusoidal input.

The existence of a *sinusoidal* current resulting from the application of a sinusoidal voltage to a capacitor is a characteristic of fringe field development due to charges on the plates of the capacitor.

The relation between the "RMS" (square root of the mean of the squares) current and voltage of a capacitor is given by:

$$\frac{V}{I} = \frac{1}{2\pi fC} = \text{capacitive reactance.}$$

where f = frequency of the sinusoidal voltage and C is the capacitance of the capacitor.

The capacitive reactance is expressed in ohms and as we saw in Section 1.2 "current", is useful to determine steady-state current values to *sinusoidal* voltage inputs. It is to be noted that, though the value of the reactance is expressed in ohms, it was not obtained by computing a resistance, as of resistors.

The current is due to the resultant electric field E_{NET} (resultant of the applied field and an opposing electric field, the fringe field), obtained by applying the relation $J = \sigma E_{NET}$. The implication of the capacitive reactance relation will become clear when we study in Section 2.11 the circuit processes when the frequency of the applied voltage to a capacitor is varied.

The full current and voltage waveform of a capacitor subjected to a sinusoidal voltage is shown in Fig. 2.34.

What is meant by steady-state in a capacitive circuit subject to a sinusoidal voltage input?

Imagine that you and your friend are playing a game of "swing". When your friend sits on a stationary swing and you begin pushing it, it will take a few hard pushes initially to overcome inertia when the swing moves with a small displacement. You then synchronize your pushes by progressively moving slightly away from the stationary position of the swing, initially, pushing gently, and then pushing harder as you move away from the central stationary position of the swing.

The point of pushing the swing is usually at the top of the swing cycle at one end. It takes a while of pushing before you are able to establish a rhythmic swing.

The period before the rhythmic swing is established is the "transient" and the rhythmic swing that is established after the few transient cycles elapse is the "steady-state".

This is analogous to establishing the steady-state in a capacitive circuit subject to a sinusoidal input. The capacitor being initially uncharged, will cause the current during the transient period to assume a value that will be quite different from that at the same voltage angle after steady-state is established. It will take a few cycles before steady-state is established and, the current will then begin to follow a pattern that is rhythmic. There are some differences with the swing analogy though. The person pushing the swing can adjust the "push" rhythm initially so that the swings increase in amplitude gradually, while a sinusoidal applied input voltage is normally of fixed amplitude and there is no adjustment possible once the circuit is switched on.

The descriptions of circuit behaviour that we consider in these and subsequent sections assumes that *the steady-state has* been established.

These remarks on transient and steady-state are applicable to inductive circuits that will be described in Section 5.17 "Current in an inductive circuit-Sinusoidal input".

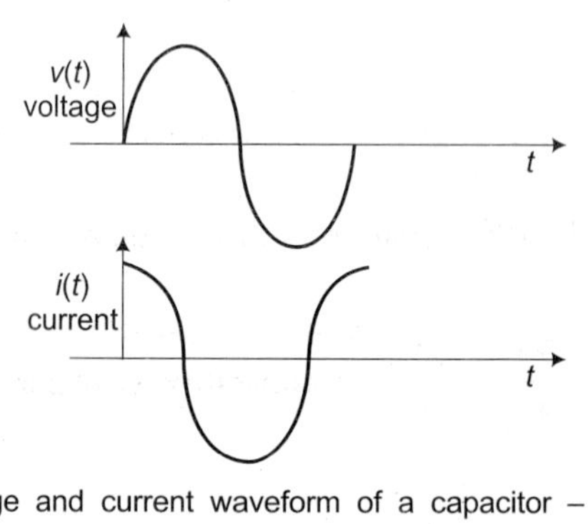

Fig. 2.34 Voltage and current waveform of a capacitor – sinusoidal input.

I urge the reader to visualize all the circuit processes that lead to sinusoidal current when a sinusoidal voltage is applied to a capacitor....the *changing* rates of change of the applied voltage ....the surface charge set up changing at every instant....the applied field changing in the wires ....and the current at each and every instant in time...the charges packing on the plates forming a *dipole*....the fringe field development....dismantling of the charges on the plates as the applied field reduces ...dominance of the fringe field when the applied field reduces....and so on. One critical aspect of the visualization, the one that many among us miss...when we plot the current at the zero-crossing point is its *reversal*....the charges that were moving rapidly..in one direction slow down..as if braking....then, come to a standstill...when the current is zero and then reverse..to move in the opposite direction as if at the end of a pendulum's swing.

A resistor obeys Ohm's law: $V = IR$ and for a sinusoidal applied voltage, the current is always in phase with the applied voltage. The interior of a resistor carrying a current is always neutral despite the presence of the applied field and the surface charges; no dipoles are formed in the interior, therefore there is no opposing field and the current simply follows the voltage in phase. See Section 1.23.

A capacitor is different. At low frequencies the capacitor "fills up", but at high frequencies it begins discharging before "filling up" as we shall see in Section 2.11.

The current *leads* the voltage; current flows *before* the charges/potential difference on the capacitor appear.

If the capacitance of the capacitor C is made large, then the fringe field does not build as fast as it would have if C were to be smaller. With a large C, the charge sprays on the plates do not result in developing a large voltage in a given interval of time as evident from the capacitor voltage-charge relation $Q = CV$. The fringe field is smaller and the net field consequently is greater. Therefore, at a fixed frequency, the current increases ($J = \sigma\ E_{NET}$ in the wires) as the size of the capacitor is increased and this is encapsulated in the capacitive reactance expression, $\frac{1}{2\pi fC}$.

2.11 VARYING THE FREQUENCY OF INPUT SINUSOIDAL VOLTAGE TO A CAPACITOR

If we increase the *frequency* of the **input voltage,** the *rate* of change of the input voltage and the applied field is *greater than* the rates obtained with applied voltages at lower *frequencies.* At high frequencies, energy from the source is transferred at quicker rates to the circuit as if being hurled into it like a javelin by a javelin thrower! As the frequency increases, the applied electric field is near and at the maximum and minimum for smaller lengths of time, as may be seen by the steep ascent and descent near the wave crest and trough of the high frequency sinusoidal input voltage in Fig. 2.35.

At high frequencies charges are moved in larger quantities within small intervals of time in the wires to the plates of the capacitor but do not give enough time for the plates to accumulate charges in large numbers (as would have if the frequency was smaller) due to the rapid forward and reverse movements of charges. There is therefore, a smaller fringe field built that could oppose the applied voltage (field). E_{net} therefore, is greater at higher frequencies in the wires.

The result: The current is *greater than* that at lower frequencies of input voltage as seen in Fig. 2.35.

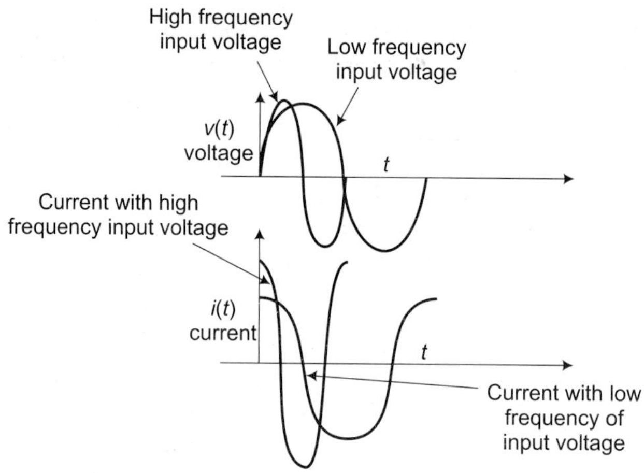

Fig. 2.35 Current when the sinusoidal input voltage frequency is increased.

$J = \sigma E_{NET}$ *and* $\dfrac{V}{I} = \dfrac{1}{2\pi f C}$, **which one among the two relations is satisfied in a capacitor circuit?**

It may appear that there are two relations (see the previous section) that the current seems to be required to satisfy: $J = \sigma E_{NET}$ *and* $\dfrac{V}{I} = \dfrac{1}{2\pi f C}$. This is always a source of confusion in most students. The answer to this is: *both* relations are satisfied. Between two circuits whose resistances are insignificantly small (one cannot realize "zero resistance" in any practical circuit) yet different, the E_{NET} change in the wires of the two circuits is so *tiny* that the capacitive reactance relation will appear not to have been violated, and becomes an *ideal approximation* (Refer "Ideal Approximation" in "Electronic Principles" by Paul Malvino[20]) while noting that E_{NET} regulates the total charge transfer and withdrawal from the capacitor plates, and is not easily measured nor computed in a practical circuit. If the resistance were to assume larger values then, the capacitive reactance relation does not apply to the circuit as a whole and the circuit becomes a capacitor-resistor circuit which is described in Section 2.12.

In the solutions to problems of circuits subject to sinusoidal voltages in the steady state, Euler's identity $e^{j\theta} = \cos(\theta) + j\sin(\theta)$ which is the constant 'e' with an imaginary exponent, is frequently used. I urge the reader to visit https://ccrma.stanford.edu/~jos/mdft/Proof_Euler_s_Identity.html for a description of how imaginary exponents came about and for a proof of this important identity.

2.12 APPLICATION OF A SINUSOIDAL VOLTAGE TO A CAPACITOR AND RESISTOR COMBINATION

When a capacitor is connected in series with a resistor R, and a sinusoidal voltage is applied to the combination, then the current is *diminished* in magnitude in comparison to the current that would result if the resistor were not there. This is to be expected, since a part of the input voltage will be used to push charges through the resistor and is shown in Fig. 2.36.

When a capacitor is connected in series with a resistor *R*, then the current is diminished in magnitude and the applied voltage *v*(*t*) is split between the capacitor and the resistor in a slightly more complex way.

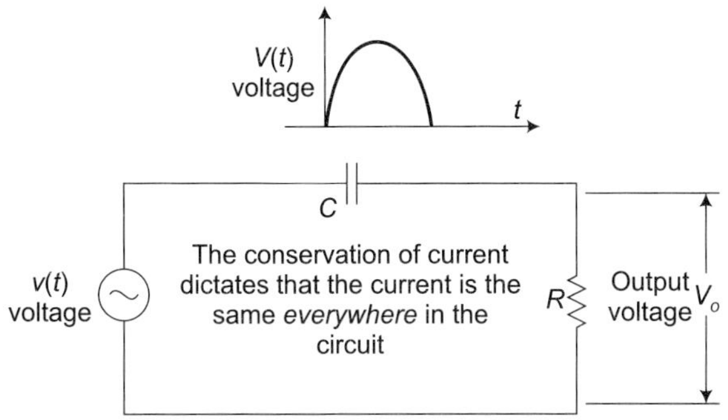

Fig. 2.36 Current when a sinusoidal voltage is applied to a capacitor resistor combination.

Surface charges appear and *bunch up* at the junction of the capacitor and the resistor.

The ***initial*** current is not as large as that in the capacitor circuit. It is restricted by the resistor in series.

All the processes that lead to the current *leading* the voltage *across* the capacitor ….mind …..despite there being a *resistor in series*….by 90 degrees….hold.

But, due to the resistor, the voltage *across* the capacitor is not the same as the input voltage….it is smaller….overall, smaller in magnitude.

This is shown in Fig. 2.37.

The *initial* current is not as large as that in the capacitor circuit. It is *restricted* by the resistor in series.

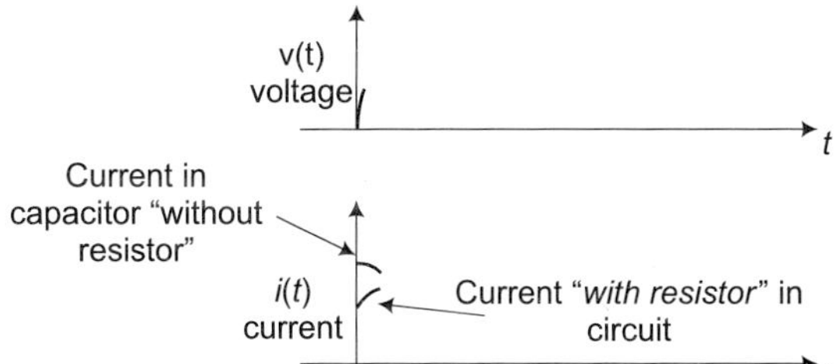

Fig. 2.37 Comparison of currents in the capacitor circuit (Fig. 2.22) and in a capacitor-resistor combination circuit subjected to a sinusoidal voltage.

It is to be noted here that we are considering the situation when steady state is reached and in its previous cycle the voltage across the capacitor was negative….and the input voltage is now rising from zero. See Fig. 2.41.

The ***initial*** current grows with the applied input voltage ...and the *capacitor voltage* rises in accordance… in the *positive* direction.

The current levels off and remains constant as the voltage across the capacitor reaches zero and is just becoming positive. The instant this occurs is before the applied voltage *peaks*.

Remember, the current is a constant when the *rate* of change of the voltage is constant across a capacitor.

The situation is shown in Fig. 2.38.

At this "*instant*", the rate of growth of the input voltage and the rate of growth of the capacitor voltage result in a "*constant*" *net* electric field in the wires that maintains the current constant before the input voltage has reached its peak.

Also note that at *this instant*, the voltage across the resistor which has the same phase as the current is also "constant though for a very brief period..." (see Fig. 2.41). Why does the current peak and remain "*constant*" briefly "*before*" the applied voltage has "peaked"? While this happens in the capacitor circuit (see Fig. 2.32) when the applied voltage was crossing zero?

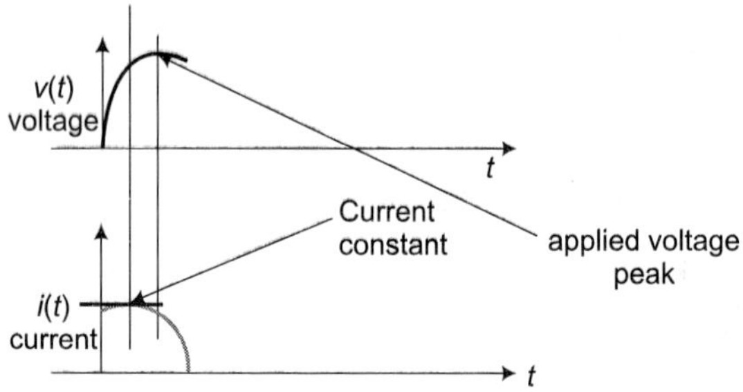

Fig. 2.38 Constant current in a capacitor-resistor combination circuit subjected to a sinusoidal voltage.

When there is a resistor, the capacitor has a smaller voltage across it because the current is smaller, fewer charges are transferred in a given time interval and so, the rate of change of its voltage is also smaller, (draw a couple of sine waves of the same frequency one with a "large" amplitude and the other with a *smaller* amplitude both starting at $t = 0$ and compare the slopes at the same instant of time, any time).

You will note that the capacitor voltage's *zero crossing* rate of change will *exactly match* the rate of change of the input voltage at an instant of time *later* than when the input crosses zero…(slide around the smaller amplitude waveform on the time axis to convince yourself of this point). Closer observation reveals that the rate of change of the input voltage and the rate of change of the voltage across the capacitor is the *same* at *this instant* of time. See sketch below and Fig. 2.41.

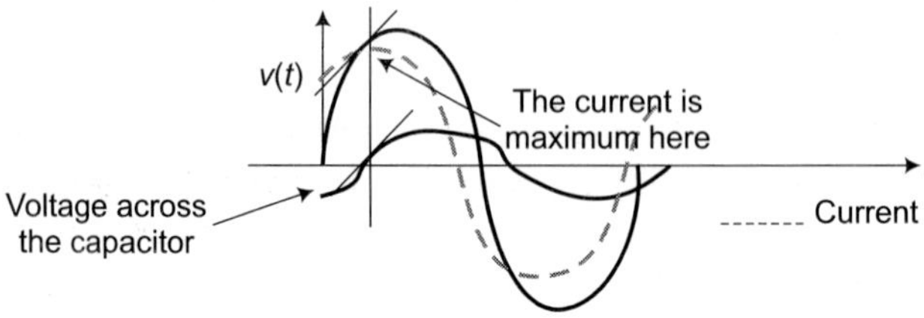

In other words, the rate of growth of the applied input field and the rate of growth of the fringe field (at different magnitudes) is the "*same*" at this instant of time, as we shall soon see, resulting in a *constant* net field, therefore, the current remains "*steady*" since $J = \sigma E_{NET}$ = a constant.

In the capacitor circuit, this instant occurs when the input voltage and so the capacitor voltage (being the same as the input voltage) is going through "zero"...and their rates of change match at every instant.

As the input voltage reduces in magnitude, the current crosses "zero" and reverses under the influence of the *dominant* fringe field. At this time, energy will be returned to the source. (The transport of encrgy and the processes involved are described in Chapter 4.)

The fringe field has assumed control at some point "*ahead*" of 180 degrees when the input voltage is zero. This is shown in Fig. 2.39.

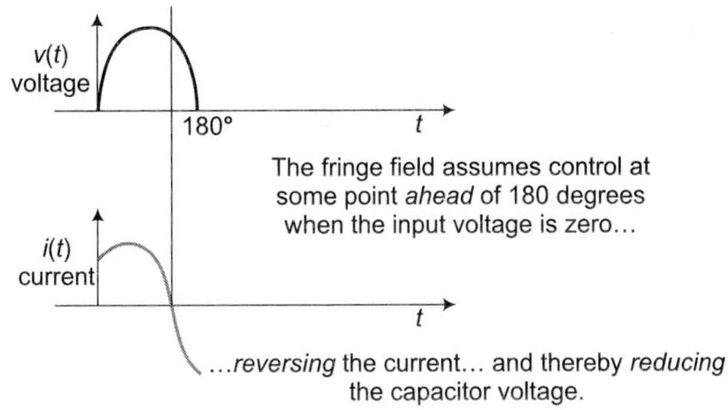

Fig. 2.39 The current in the capacitor resistor combination circuit subjected to a sinusoidal voltage.

The current in a circuit comprising a capacitor with a resistor is smaller and *leads* the input voltage by an angle "*less than*" 90 degrees.

This is shown in Fig. 2.40.

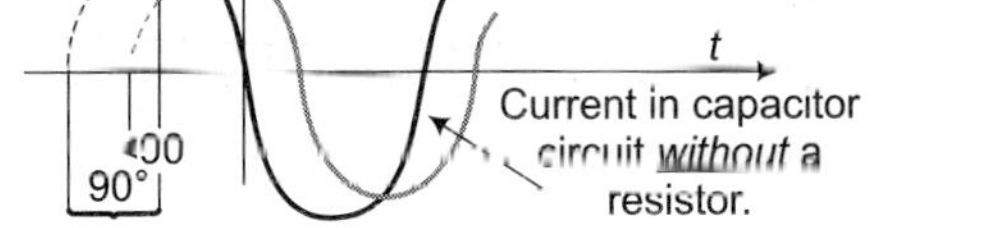

Fig. 2.40 Comparison of the current in the capacitor resistor combination circuit with that in the capacitor circuit (Fig. 2.22) subjected to a sinusoidal voltage.

The output voltage V_o is measured across the resistor.

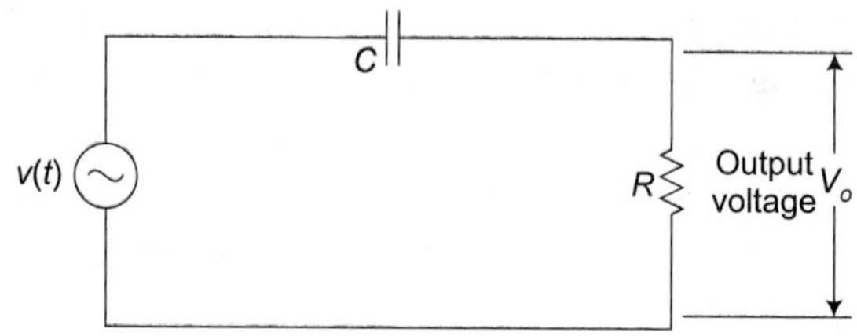

The output voltage across the resistor *leads* the input voltage by the same amount as the current because they are in phase for a resistor.

The current which has to be the same everywhere in the circuit leads the voltage *across* the capacitor by 90 degrees—a rule that is *never violated.*

Note that the *slope* of the capacitor voltage change *at the zero crossing point* is the *same* as the slope of the input voltage change before the input peaks, where the current is constant.

Observe that the output voltage is *smaller* in magnitude than the input voltage, as it should be.

The waveforms of input and output voltages and the voltage across the capacitor are shown in Fig. 2.41.

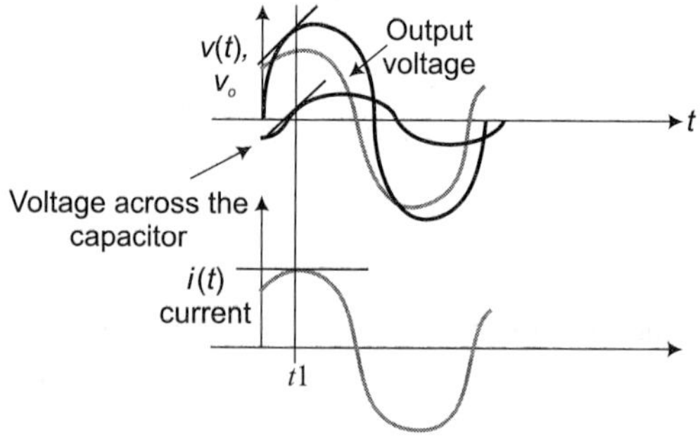

Fig. 2.41 Voltage waveform of input, output and the current in a CR combination circuit subjected to a sinusoidal voltage.

Note equal slopes of input voltage and voltage across the capacitor at time *t*1 when the current is "steady" at its peak.

2.13 APPLICATION OF SINUSOIDAL VOLTAGES AT VARIOUS FREQUENCIES TO A C-R COMBINATION

What happens when the frequency of the input sinusoidal voltage is increased?

If we *increase the frequency*, the fringe field becomes weaker. Due to the rapid charge forward and reverse movements caused by the high frequency positive and negative cycles of the applied voltage, the charges do not get to separate and stick to the plates of the capacitor—these being buffeted in both directions, and so the fringe field is weaker. The current therefore increases and as the frequency is further increased is limited entirely merely by the resistor...and the input voltage drops almost entirely across the resistor.

This resistor limited current is not enough to *dump charge fast enough at such high frequencies and of sufficient quantity* to produce any significant opposing fringe field.

Therefore, for a given RC combination the output voltage picked across the resistor is able to reproduce the input signal with *less attenuation.*

We say that the capacitor *bypasses* the high frequencies .....in reality, the electric field of the input voltage passes "through" the capacitor with almost no opposition.

As the frequency of the applied voltage (e-field) is increased, the applied voltage (e-field) is not opposed as much by fringe fields that are weaker (thereby enabling the applied e-field an easy penetration through the capacitor) and, the RC circuit *impedance* (see Section 1.2 "Current") will be *small.*

The output voltage is almost *in phase*...with the input voltage.

The waveforms for the circuit are shown in Fig. 2.42. Also depicted are the voltage and current waveforms for input voltages of a lower frequency.

CR COMBINATION - SINUSOIDAL INPUT - HIGH FREQUENCY

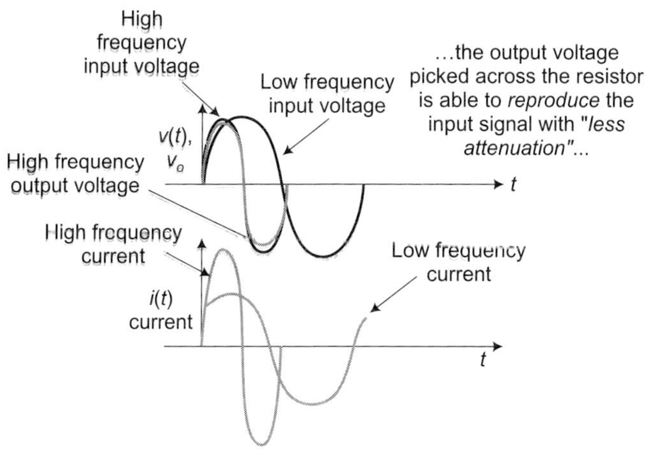

Fig. 2.42 Voltage waveform of input, output and the current in a CR combination circuit subjected to a high frequency sinusoidal sinusoidal voltage.

The Capacitor and Resistance Combination – More on Fringe Field

Do not forget that the fringe field is *due* to the charges on the capacitor plates and the charges *get there* by the current.

The current is limited by the resistor. The oncoming charges cannot move at the speed of light and are *opposed* by the fields of charges that had earlier *accumulated* on the plates.

And that is the reason why at high frequencies the fringe field is unable to build fast enough to *oppose* the applied field.

Therefore, we say the capacitor acts as a "short circuit" at high frequencies.

What happens when the frequency of the input sinusoidal voltage is decreased?

If we *decrease the frequency* of the input voltage there is sufficient time for the fringe field to develop because charges have *ample time* to accumulate on the plates in relation to the voltage change and the current is *small*. See Fig. 2.43.

The output voltage is unable to follow the input signal being increasingly blocked by the capacitor's fringe field and therefore, the *attenuation* is also high.

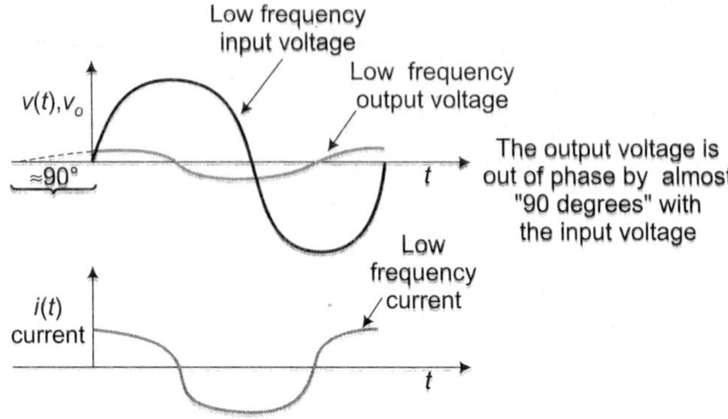

Fig. 2.43 Voltage waveform of input, output and the current in a CR combination circuit subjected to a low frequency sinusoidal voltage.

The waveforms of output voltage and current, when a low frequency sinusoidal voltage is applied, are shown in Fig. 2.43.

At low frequencies, the circuit has sufficient time to *push* the charges on the capacitor plates, and the input voltage changes *slow* enough to allow the charges to accumulate on the plates. The fringe field develops and is able to oppose the applied voltage on time.

The potential difference across the capacitor is frequency dependent and current flows *before* the charges/potential difference on the capacitor appear; therefore, the current *leads* the applied voltage by a phase angle greater than 0° but less than 90°.

So, we say, the capacitor acts as an *open circuit to low frequency input voltages*.

As the frequency of the applied voltage (e-field) is reduced, the applied voltage (e-field) is opposed by fringe fields that are *stronger* (thereby making it difficult for the e-field to penetrate through the capacitor) and, the RC circuit *impedance* (see Section 1.2 "Current") will be *high*.

The Capacitor and Resistance Combination – As a High-Pass Filter

We have seen from the sinusoidal inputs to the CR combination that...

the circuit "passes" high frequency sinusoidal signals easily with "little attenuation" though with "almost zero phase shift..."

...on the other hand, the circuit is "*unable*" to pass low frequency sinusoidal signals and the phase shift is "almost 90 degrees leading" between the output and input voltage.

This leads us to make an important observation regarding C-R circuits: A capacitor-resistor circuit is a high pass *filter* of input signal voltages. The capacitor effectively blocks low frequency and dc signals and transmits the high frequency content of signals that are a combination of several frequencies, to the resistor.

If one is interested in the low frequency content of the signal, this can be picked across the capacitor.

The Capacitor and Resistance Combination – As A Low-Pass Filter

Referring to Fig. 2.41, we observe that the voltage across the capacitor has a very small rate of rise and fall and if the capacitance is made large enough then the voltage across the capacitor will indeed vary very slowly and may be used as a filter to suppress "ripples" in rectified AC voltages to produce DC voltages for power supplies in electronic circuits.

One such filter circuit using a capacitor is shown below. Note that the output voltage is picked across the capacitor.

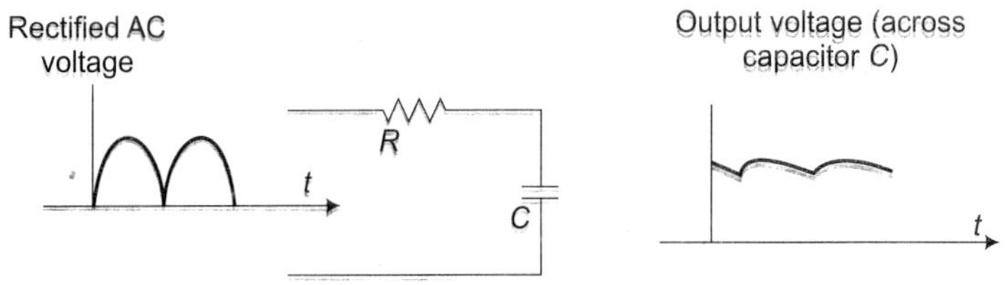

There are some interesting observations we can make to understand and get a feel for the mechanism of capacitive circuits and capacitive resistive circuit operation.

2.14 FOR CHARGE TO FLOW IN A CIRCUIT – THERE HAS TO BE A CLOSED PATH – IS THIS NECESSARY?

From: http://amasci.com/emotor/cap1.html

No!

There must be a closed path **ONLY** if the charges must ***flow continuously*** in the circuit forever (DC).

But if the charges flow briefly, or if they flow back and forth (AC), then an ***open path*** is "acceptable".

For example, with ***a single metal rod 1 metre in length***, it is possible to create a standing wave of ***charges oscillating lengthwise in the rod*** with frequency 150 MHz. Inside the rod, ***the charges move back and forth*** (this resembles the motions of the compressible air inside the pipe of a church organ).

The charges within the rod are acting like a *"compressible"* fluid.

At lower frequencies, the electric current can still be large if the ends of an open-circuit are joined to the plates of a capacitor.

On the metal surfaces **of the plates** of a capacitor, the ***quantities of charge carriers behave as a "compressible fluid",*** while the **charges within a wire** behave as an ***"incompressible fluid".***

Note on incompressible and compressible fluids.

An incompressible fluid flows so that the net flux of fluid into or out of a given volume within the fluid is zero.

If the fluid is compressible, however, and its density $\rho(r)$ varies with position because of pressure or temperature variations, the net outward flux of mass from some small element is determined by divergence(ρv), and this must be related to the rate at which the density of the fluid within is changing.

Suppose we ***force*** (use a Non-coulomb force!) some charge ***into one plate***, but we ***do not connect the other plate to anything.***

In this case, the **lines** of electric force extend outwards from both plates to the earth. Important question: what is the capacitance of the "capacitor" formed by one plate and the distant earth? Very small.

Very small, possibly millions or even billions of times smaller than the number of microfarads existing between the two plates. Therefore, if we connect ***both capacitor terminals to a circuit*** and force charges "through" this capacitor, the ***voltage*** across the plates ***will rise slowly***, but if we instead try to force the same current ***into only one plate*** (and leave the second plate disconnected), then the ***voltage*** between the capacitor and earth will ***rise millions or billions of times faster*** for the **same current**.

In electronics we *usually ignore the "one-plate"* phenomenon entirely, and imagine that capacitors are a ***strange kind of two-terminal conductor.***

2.15 WHAT ARE SURFACE CHARGES REALLY?

Surface charges are point charges produced by the shifting of free electrons (see Section 1.15 and Section 1.16) as if compacted or dispersed and arranged in the form of rings in long wires of circular cross-section, seen to guide the motion of charges and create the appropriate electric potential and Poynting vectors for the circuit, and hence are an important factor in the study of circuit theory.

Charges on charged capacitors are *excess* charges local to that capacitor plate and that are sourced from the other capacitor plate (the two plates form a closed system, refer section 1.8 "Conservation of charge").

Surface charges perform three roles in circuits

They appear upon closure of electric circuits

1. To *maintain the potential* around the circuit
2. To *provide the electric field* in the space around the circuit
3. To assure the *confined* flow of current

In Fig. 2.44 is shown the surface charges that are present in a simple series circuit.

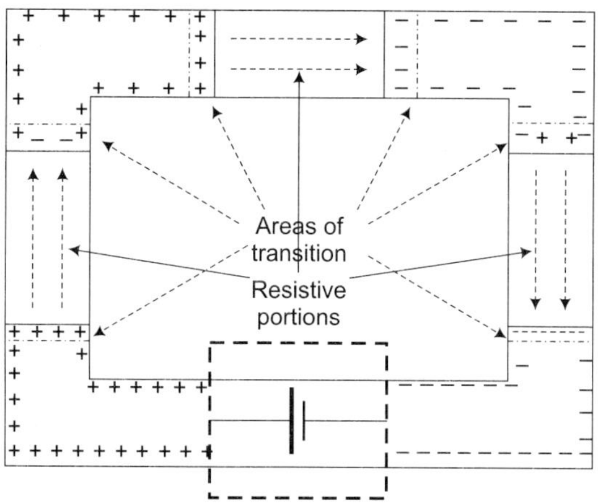

Fig. 2.44 Surface charge distribution in a series circuit.

The figure is not to scale and the gradient of surface charge density along the wires is not shown, because it is quite small.

The presence of the interface charges where the resistive portions join the wire provides a steep surface charge gradient and a resulting powerful field in the resistive portions.

In Fig. 2.45 is shown the interface charges that are present at the junction of two materials that carries a current. (From: Massachusetts Institute of Technology lecture notes)

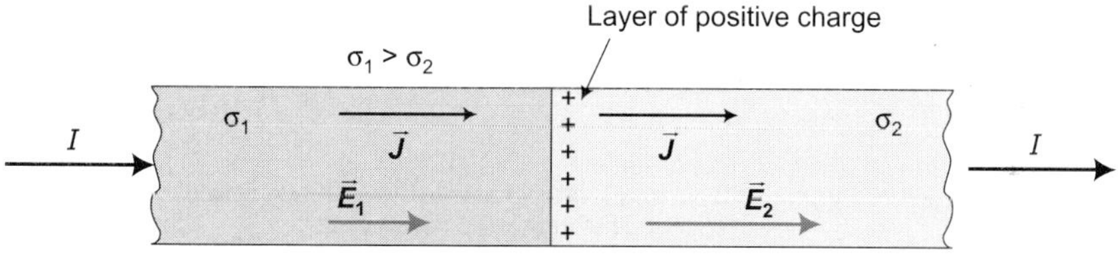

Fig. 2.45 Interface charges at the junction of two materials.

The above figure shows another example of the charge at the ***junction*** of two materials joined to form a series resistance and through which a current is established by a source (not shown).

It is to be noted from Fig. 2.45 that in a steady-state of current flow, the normal component of current density J must be the same on both sides of the junction. Since the current is the same because the area of cross-section is the same, *the layer of charge develops a field* that is "different" in the two resistor materials of different conductivities.

Note: $J = nqv = nq\mu E = \sigma E$, where n = number of free electrons per unit volume, v = drift velocity, μ = the mobility for the material, and σ = conductivity of the material and, $\sigma = nq\mu$.

In Fig. 2.46 is shown the surface charge distribution in a parallel circuit comprising two resistances in parallel.

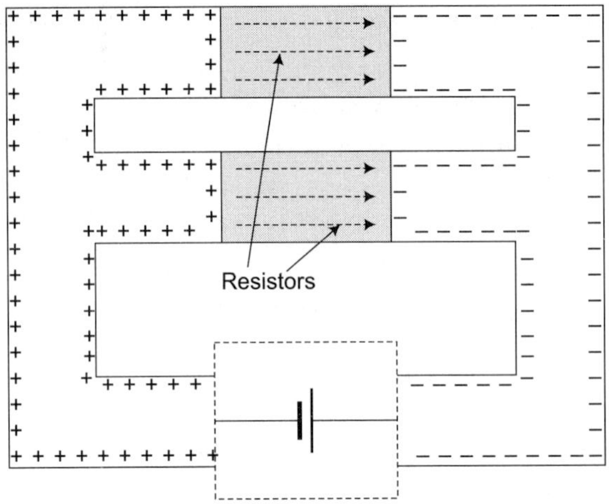

Fig. 2.46 Surface charge distribution in a parallel circuit.

We will now study the mechanism by which current is appropriately steered down a parallel path that is introduced in a simple series circuit as an answer to the question: "How does the current know how to split when there are parallel branches?"

2.16 CURRENT SPLITTING IN PARALLEL BRANCHES

(Chabay & Sherwood, 1995, used by permission)[2]

When a dead end branch is added to a series circuit as shown in Fig. 2.47 then, initially electrons *flow into both branches*, but the surface of the dead-end branch becomes charged so negatively, that no more electrons can enter.

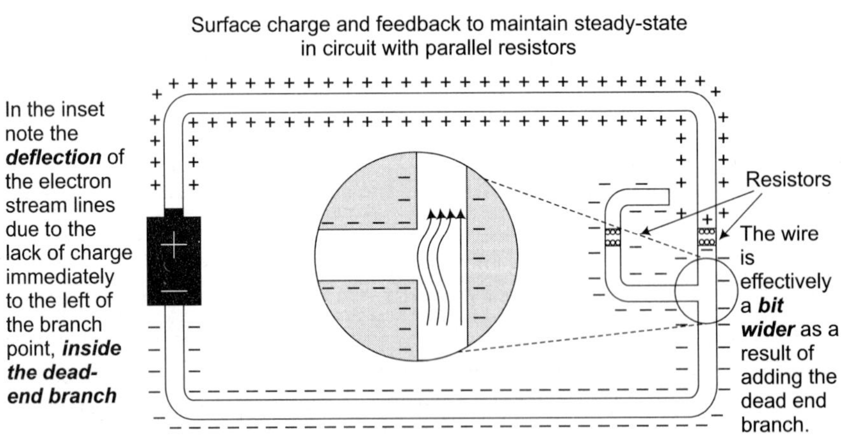

Fig. 2.47 Parallel circuit—dead-end branch.

Also the wire is *effectively* a bit wider as a result of the dead-end branch where conduction electrons make an attempt to enter the dead end branch but are deflected away by the highly negatively charged branch.

We now ***complete*** the parallel connection, which leads to a rearrangement of surface and interface charges.

The new arrangement is shown in Fig. 2.48.

During the initial transient phase, current may run equally down two parallel branches, but different resistance in the two branches leads to different surface-charge build-up along the two branches, which in the steady state will ***steer*** appropriate amounts of current down each branch.

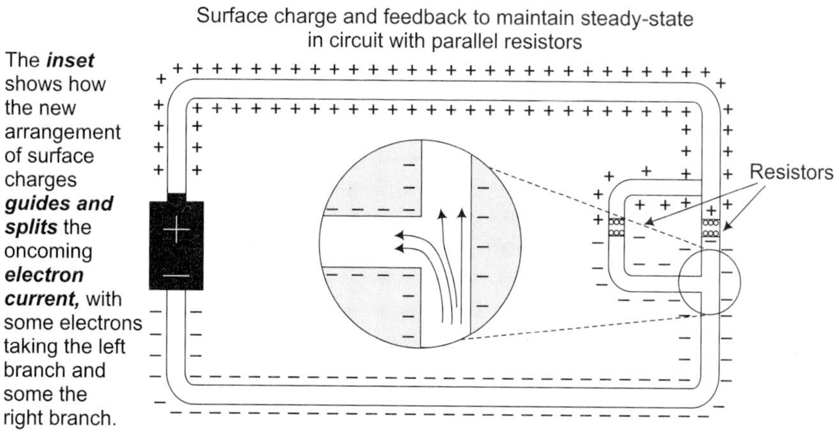

Fig. 2.48 Parallel circuit-closed.

We will recapitulate why the current is maximum at the instant of closure of the switch of a capacitive resistor circuit. This is being done with a view to get a feel for the circuit processes with the electric fields included in the mechanism.

We saw how when we closed the switch in a circuit with a parallel *dead-end branch*, the circuit initially was not "aware" of the presence of the dead-end and caused the charges to flow into the dead-end which then accumulated there and *prevented further movement* of electrons in the dead-end branch.

There must be *positive and negative charge* on the surface of the wires near the "+" and "–" plates of the battery, and in the steady state the surface charge must *vary smoothly* from positive to negative around the circuit, possibly with extra charge on the bends in the circuit.

When we close the switch in a charging circuit of an uncharged capacitor, similarly, the surface charges, developed by the e-field which is produced by the charges from the battery, initially appear around the circuit. The electric field produced by the surface charges on the wires, is ***unopposed*** by any fringe field on the uncharged capacitor and the charges also cause an electric field to appear in the space surrounding the circuit, in, ***"through"*** the capacitor, and around the wires in the circuit as if ***"unaware"*** of the presence of the capacitor.

Remember: *An electric field penetrates through matter.*

In the circuit used to charge a capacitor from a battery with a bulb, at the instant of turn on, surface charges appear causing an electric field that penetrates through the plates of the capacitor and appears in the connecting wires.

The field also fills the space surrounding the capacitor and the wires. The situation is shown in Fig. 2.49.

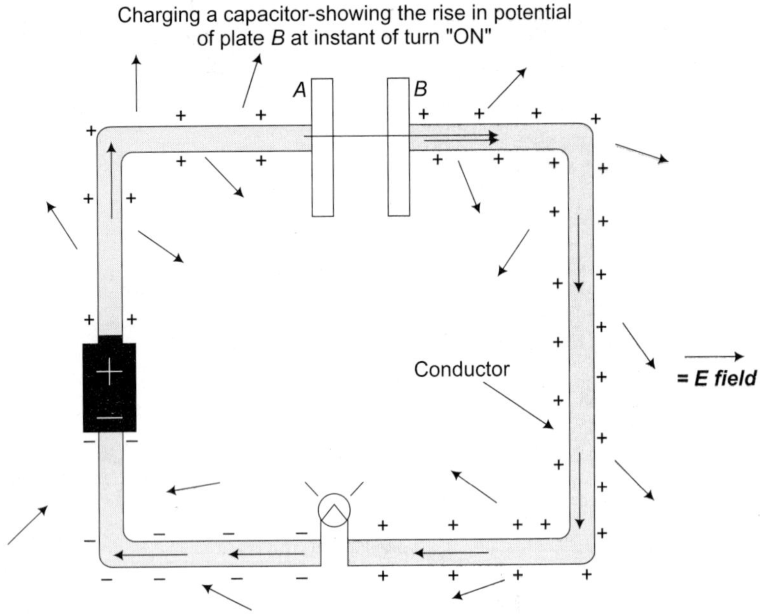

Fig. 2.49 The voltage across a capacitor cannot change instantaneously.

Positive surface charges will appear on plate *B* and cause the plate *B* of the capacitor to attain a potential difference (p.d.) = the voltage of the battery at the instant of turn on.

The current is limited solely by the resistance of the bulb and is therefore a maximum at that instant.

As the *fringe field* develops and grows in magnitude on the capacitor plates, the capacitor charges and stores energy. The fringe field *opposes* further flow of charges and the current gradually decreases in magnitude.

2.17 WHAT IS VOLTAGE?

We now turn our attention to answering qualitatively the question: "What is voltage?"

This is an important concept that needs to be addressed because voltage is intimately linked to electric fields. The transport of energy in circuits is through the electric and magnetic fields and therefore a qualitative understanding of these is very important.

Voltage is a way of using numbers to *describe an electric* field. Electric field is the force that acts on a charge placed in a field.

Electric fields or "E-fields" are measured in volts over a distance; volts per centimetre for example.

"Voltage" and "e-fields" are two ways to describe the *same basic concept.*

Consider the potential energy of objects in a gravitational field.

A stone on the ground has a lower potential energy (PE) in comparison to the PE it possesses when raised some distance above the ground.

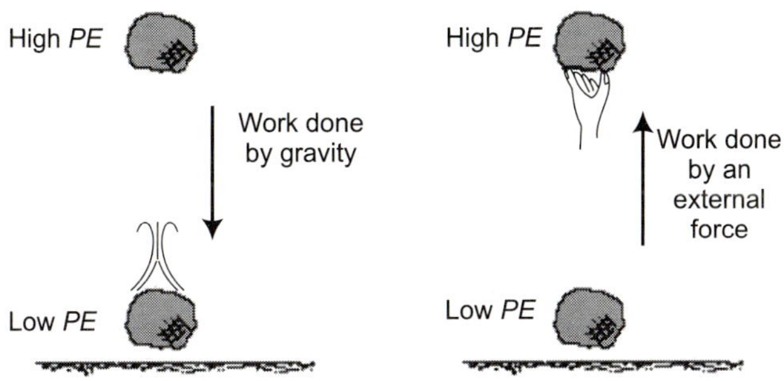

It has gained this energy by virtue of the work done by an external agency in raising it. Note that no matter what path is taken to move the stone from the ground to its final position, potential energy gained (or lost) will remain the same. Also note that if the stone is returned to its original position after raising it gently so that there is no conversion to kinetic energy, the potential energy gained is zero. We say that the field is *conservative* because a round trip path integral of work done is zero. In quite analogous fashion, it can be shown that the electric field of a point charge is conservative. A positive point charge has a field pattern as shown in the Fig. 2.50.

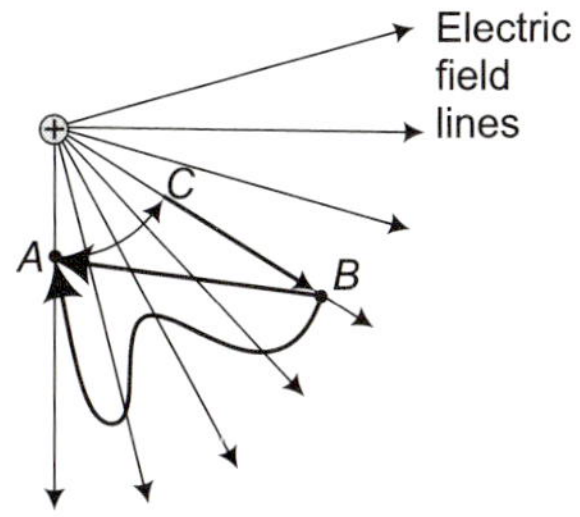

Fig. 2.50 The concept of electric potential being independent of path.

The potential difference between points **A** and **C** is given by ΔPE/charge where ΔPE is the change in potential energy when moving the charge from location **A** to location **C**. The field is said to be "conservative" because, the work done to move a test charge from **A** to **C** to **B** and back to **A** is "zero".

In other words, the ***round-trip path integral*** of the electric field is "zero" ***no matter which path is taken.*** Intuitively, this can be understood by considering that the potential energy gained when moving a test positive charge *against* the field is lost when moving *with* the field measured in small incremental movements with direction taken into account. The analogy with the stone moved against the gravitational field and gently lowered (so as not to convert into kinetic energy) is fairly obvious.

Note: Care should be exercised to ensure that the force applied to move the test charge towards the point charge is ***equal to qE** and **not greater*** because if it is, then, the charge will *accelerate* and the work done to push will be ***wasted as kinetic energy.***

Hence, the potential difference between two locations near a point charge is ***independent of the path.***

We say that a property of the electric potential due to a point charge is *path independence* and the electric field due to point charges is *conservative.*

In Section 2.19, we shall see that voltages are associated with electric fields due to several charges that are separated. This is due to the production of resultant electric fields in the space surrounding the separated charges as was described in Section 1.10 "Separation of charges".

2.18 WHAT IS ELECTRICAL ENERGY?

Electrical *kinetic energy* (KE) appears whenever electrons ***flow*** through the lattice of atomic cores in conductors. We call this *"electric current"* and, it causes *magnetism.*

Electrical potential energy appears when positive charges are ***yanked*** away to a distance from their corresponding negative charges (separation of charges). We call this "net electrostatic charge" and, it causes *voltage.*

Electrical kinetic energy is associated with current, and *electrical potential energy is associated with voltage.*

This will be dealt with in some more detail in Chapter 4.

2.19 CONSERVATIVE AND NON-CONSERVATIVE ELECTRIC FIELDS

Visualize an electric circuit as an area in space in which the battery provides the + and the – between which electric field vectors are spread out around the space and the *round trip integral* around this of the electric potential is *"zero"*. We will shortly explain why.

Before we do that, let us invert the argument and study whether it is possible to produce an electric field pattern that does *not* have the divergent spread pattern associated with point charges.

In the light of path independence of electric potential, "*Is it possible* to set up a *configuration of charged objects in such a way that the electric field along a circular path* looks like this?"

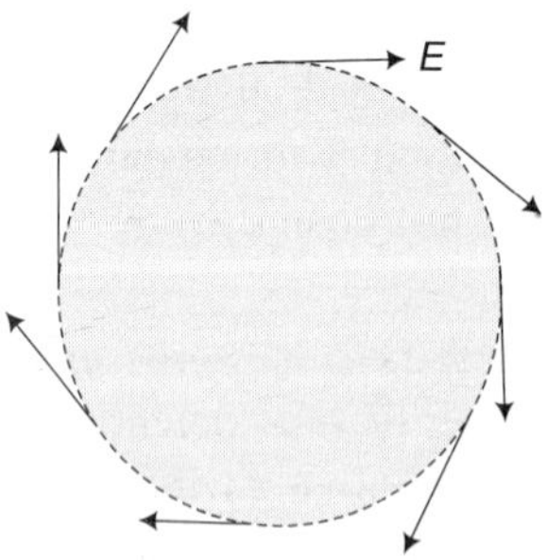

Calculating the round-trip potential difference from one point back to the same point travelling all the way around the circle:
All along the path (of length $2\pi r$):
E is in the direction of the path, so
$E_{parallel} = +E$ and,

$$\Delta V\ round\text{-}trip = -\int_{initial}^{final} E.dl = -E.dl = -E(2\pi r) \neq 0$$

We state one important attribute of charged objects: For any collection of point charges, $\Delta V_{\text{round trip}}$ of a round trip ***must be zero.***

So, this curly field cannot be produced with charged objects.

That is the reason we say, that the *curl* (wherein the electric field vectors will close on themselves) of any static charge distribution (including surface charges in circuits) as seen in Figs. 1.24, 2.5 and 2.50 is always "zero". The electric field vectors *diverge* but do not close on themselves. See Section 5.11 "Conservative fields" also.

Allowable Patterns of Electric Fields Using Charged Objects

It ***might seem*** that an electric circuit has a curly pattern of electric field as if the field vectors close on themselves. But examine the diagram below ***carefully***.

Here you will see that ***inside the battery the electric field is in the opposite direction to the electric field in the wire*** **and the round trip potential-difference is therefore "zero".**

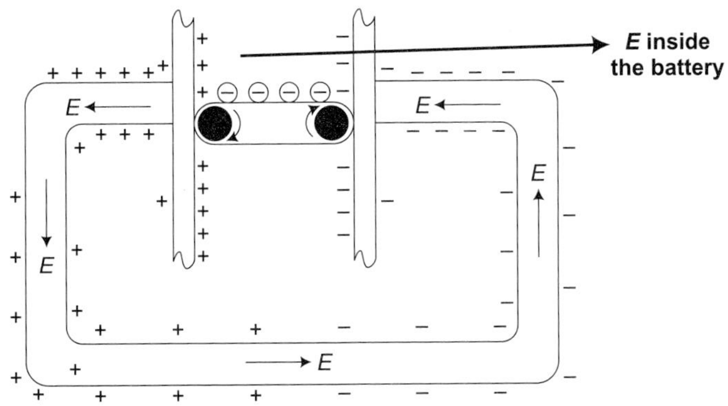

This is also indicated in the voltage "drop" waveform plotted in the circuit shown below.

The electric field in the battery is labelled E_{NC} to indicate a non-coulomb electric field that is produced as a result of charge separation owing to a chemical reaction.

Round-trip path integral in circuits

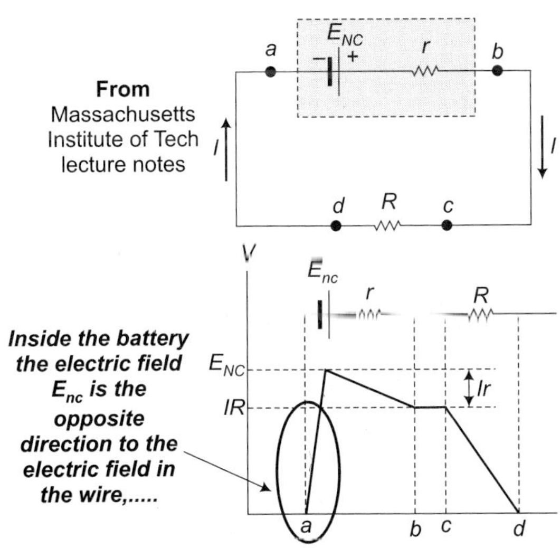

$$\Delta V = -\int \vec{E} \cdot \vec{ds} = 0$$

Closed path

The following description of the non-conservative electric field in a circuit containing a battery is from a paper titled "Charge distribution on DC circuits and Kirchhoff's laws" by W R Moreau, Dept. of Physics, University of Canterbury, New Zealand in *European Journal of Physics*, Volume 10, 1989.

The electric field in the vicinity of a conducting DC circuit can be expressed as:

$$E = E_c + E_{nc}$$

where, E_c is a conservative field with zero curl and E_{nc} is a non-conservative field with zero divergence (and non-zero curl).

Mathematicians call E_c irrotational and E_{nc} solenoidal, but we prefer to call them *conservative and non-conservative* to emphasize the fact *that only the conservative field is related to electric potential.*

Consider the circuit shown in Fig. 2.51.

With the switches open $E_c = E_{nc}$ and so the net field $E = 0$ and there is no current.

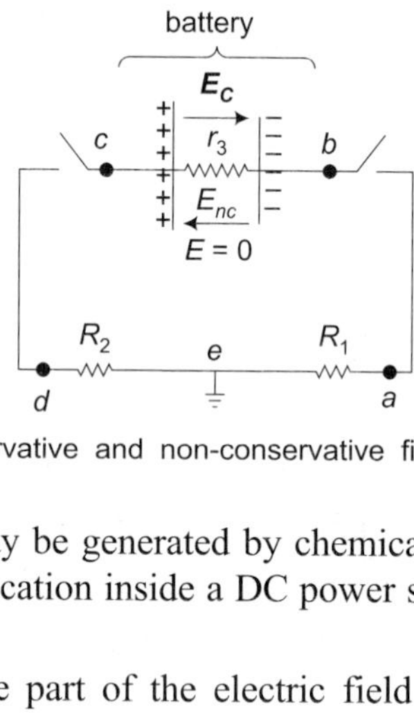

Fig. 2.51 Conservative and non-conservative fields (switch open).

The *non-conservative* part may be generated by chemical action inside a battery or a time-varying magnetic field with rectification inside a DC power supply and is usually non-zero only inside these devices.

We associate the conservative part of the electric field with a charge distribution on the circuit.

When the switches are closed, the charge density on the plates has *decreased* from the open-circuit condition, thus reducing E_c while E_{nc} remains the same.

In the battery the total electric field is now non-zero and directed from the negative plate to the positive plate and, net total electric field now *forces charge through the internal resistance of the battery* constituting a current.

This is shown in Fig. 2.52.

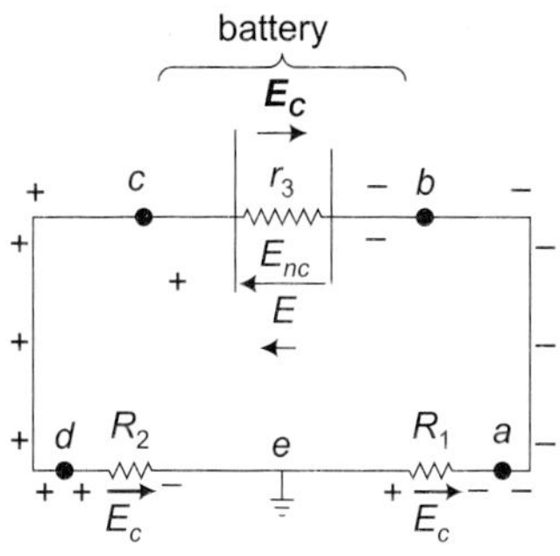

Fig. 2.52 Conservative and non-conservative fields (switch closed).

Effects inside a source of emf - battery

What is the "*direction*" of electron flow?

"*Inside*" a battery?

Source: http://ocw.mit.edu/high-school/physics/electric-circuits/steady-state-direct-current-circuits-batteries-resistors/8_022_fall_2004_lecture8.pdf and www.amasci.com

When a battery is assembled, charge *"pumps"* are located on the *surface* of the battery plates…one on each plate. These are called the "half-cell reaction sites".

Electrons and positive ions are provided by the *negative* plate reacting with the electrolyte. When a number of these are produced, the electric field they give rise to, increases in magnitude, large enough to "*block the migration"* of positive ions to the positive plate. An equilibrium is reached and separation stops when an electric field of sufficient strength is *established to oppose the reaction; an emf has developed between the plates.*

These cells are wired in series, so their voltages add together. The *surfaces* of the metal plates act as the true energy-producing "batteries". The *conductive liquid acts as a wire.* It connects these two "batteries" together.

When a circuit is completed with the battery, then the charges on the *surfaces* of its plates are *"depleted",* because these react at the positive plate in the presence of positive ions provided by the electrolyte and rejoin to become "neutral". The electric field between the plates *reduces* in magnitude and, as a result of this, the *chemical reaction restarts, with electrons* via the external circuit that *reappear* at the *negative* terminal.

Separation of charges *is resumed* and the depleted charges are *continuously replenished.*The direction of electron flow is *towards* the negative plate inside the battery while the direction of electron flow is *away* from the negative plate in the external circuit.

When the battery is delivering energy to the circuit, the separation of charges is a continuous process and the depleted charges are *continuously replenished, restoring the electric field and the "emf"*. All these effects take place at nearly the speed of light. All sources of emf behave in this fashion. We will be seeing another example when we discuss generators.

2.20 VISUALIZING VOLTAGE

How can we visualize voltage?

Considering the simple series circuit in Fig. 2.53(a) consisting of a battery, connecting wires and a resistor, how can there exist a *potential difference* between *A* and *B* and, practically *no potential difference* between *B* and *C*?

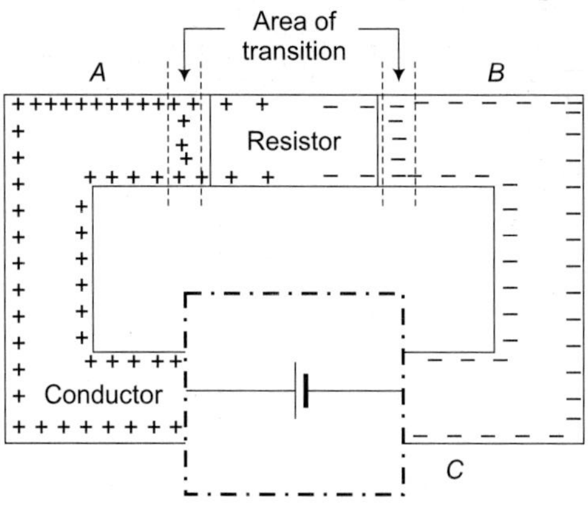

Fig. 2.53(a) Simple series circuit and the existence of a potential difference.

The three cross-sections at *A*, *B* and *C* seem to be identical on a microscopic scale. Also, the same amounts of electrons per second are passing through as follows from Kirchchoff's first law.

Most textbooks explain that there is a potential difference between the points A and B. But what corresponds to this difference in microscopic dimensions?

There is an electric field, alright. It is constant within the resistor. But what causes this field?

The problem is, of course, not one of the physical theory.

In the case of flow of fluid in pipes there is a pressure difference that can be directly measured at different points but inside a conductor or resistor nothing seems to change over its length. There is a condition of conservation of charge resulting in charge neutrality in every elemental volume within the conductor.

Voltage is intimately connected with electrical energy. So is magnetism. We can say that electrical energy is the fundamental object, while voltage and magnetism are the two faces it displays to the outside world.

Another analogy: in mechanical physics, both the kinetic energy (KE) and the potential energy (PE) are part of matter: relative motion of an object has kinetic energy, and stretched or compressed objects (for example, springs or rubber bands) have potential energy.

In a similar way, electrical kinetic energy appears whenever electrons flow through (really, around) the atomic cores. We call this "electric current," and it causes magnetism. Electrical potential energy appears when positive charges are yanked away to a distance from their corresponding negative charges. See Section 1.15 "The nature of surface charges in circuits". We call this "net electrostatic charge," and it causes voltage.

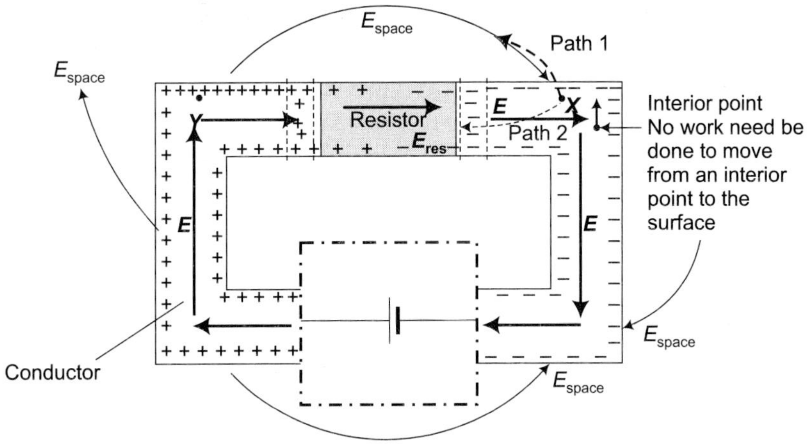

Fig. 2.53(b) When charges are present there must be e-fields and $V = -E \int dx$.

Therefore, the answer to our question "Where is the voltage" is:

The voltage or potential difference is *entirely due to the existence of surface charges*. Surface charges accompany any current flow within a conductor as in the Fig. 2.53(b). The work done to move a unit charge from one point *X* in the circuit to another point *Y*, whether via the interior of a field-filled connecting wire (path 2) or, via the field-filled space (path 1) will be the *same* and thus, the *potential difference* between these two points will also be the same whichever path one takes. See in Section 1.16, the three roles of surface charges. For voltage measurements in circuits, we *always* consider that the points for computing potential differences are located *outside* the surface charges with their small electric fields. See Appendix B and D on the force exerted by a voltmeter probe on a wire surface. If considering points lying *interior* (of wires and resistors), these are always *outside* atoms since dipoles do not exist there (neutrality conditions prevail despite the current and the driving field) and these points are equipotential with the nearest surface.

Therefore, for the path of a test unit charge there is no work done (and so, there is no voltage drop to consider) to move from an interior point to the nearest surface or to move from the nearest surface to the interior point.

The reason why most textbooks ignore this simple fact of the existence of surface charges could be that the surface charges are rather small and normally not detectable (the electric field in space due to them in simple low voltage circuits will not make the hair on your hand stand as a charged plastic comb can as described in Section 1.3!) unlike the fluids in pipes where most parameters associated with its flow such as pressure, velocity can be easily measured using suitable instruments and meters. In the case of fluids, the energy associated is carried along in the fluid. This is not the case with the flow of charge in a wire.

Electrical KE is associated with current, and electrical PE is associated with voltage. If electrical energy is the same as electromagnetism, then maybe we should be more sensible and name it "VoltageCurrent-ism." Is voltage a type of potential energy? Think of it like this. If you roll a big boulder to the top of a hill, you have stored some potential energy. But after the boulder has rolled back down, the hill is still there. The hill is like voltage: the height of the hill has "Gravitational Potential." But the hill is not made of Potential Energy, since we need both the hill and the boulder before we can create potential energy. The situation with voltage is similar. Before we can store any electrical potential energy, we need some charges, but we also need some voltage-field through which to push the charges.

A charged metal object is like a flat-topped mountain, where the altitude of the mountain top is like potential of the charged object, and the mountain slopes are like e-fields in the space around the object. On one side of the mountain, the slope might be very gentle. And, on some other side, steep. Yet no matter how gentle (or steep), we arrive at the same altitude when we walk to the top.

If a charge lands on the mountain top where the potential is positive, after a long and arduous climb along its slopes, it is well received and given a free pass to roam around the top without losing energy! The "equipotential" inside a conductor (of any shape) is like the mountain top.

The free electron sea shifting inside the conductor, in every elemental volume, keeps the entire conductor an equipotential "pushing" any excess charge all around to its surface that will maintain its voltage condition and darting around and stirring up the charges within to maintain equipotential conditions inside yet maintaining neutrality in the bulk of the conductor. Neutrality is maintained because the interaction between the electrons and the enforced increase of the velocity of the electrons is so very small.

This shows that, if a conductor has a potential which is not zero, there must always be an e-field surrounding that conductor and extending to infinity, even if that e-field is very weak. No matter how weak, it must still add up to the same potential at the object surface. And, since e-field flux lines must terminate upon charges, therefore the presence of the e-field requires that surface charges must be present on that conductor.

Recollect from Section 1.5 that the potential energy (stored energy) is proportional to the number of separated charges and the distance separating them. The same remark applies to voltage when we saw in Section 2.7 "Charging a capacitor" that the voltage across the capacitor grew in magnitude as the number of separated charges closely *packed* on its plates increased as time elapsed, the distance they were separated remaining fixed being the distance between its plates.

A complete circuit with an indication of surface charges could be represented as depicted for the circuit on the left of Fig. 2.54 by Hermann Häertel, in his paper "**Constant Electric Current and the distribution of Surface Charges**[1]**"**. (Reproduced with permission from Hermann Haertel).

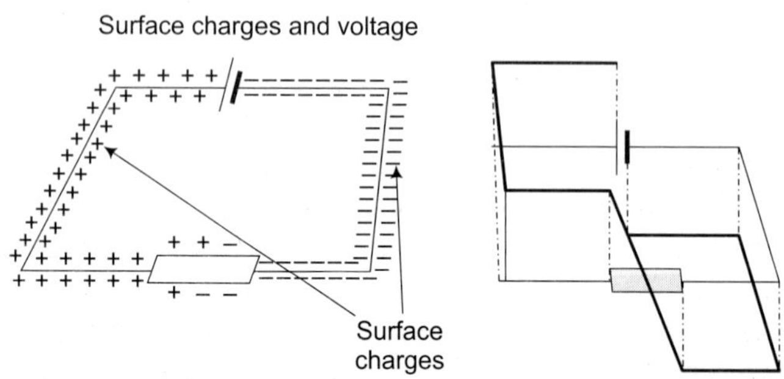

Fig. 2.54 Different possible respresentations of voltage.

A different representation could be chosen by representing different values of surface charges, or better: different values for potential differences by vertical lines of different length.

In Fig. 2.54 on the right, positive surface charges are indicated as lines in the upper direction and negative charges as lines downwards. This representation is more abstract but better to read, and the gradient of the surface charge distribution (potential difference) is immediately visible. This same information can be presented by choosing one side as zero (same height as the circuit) as shown in the next Fig. 2.55.

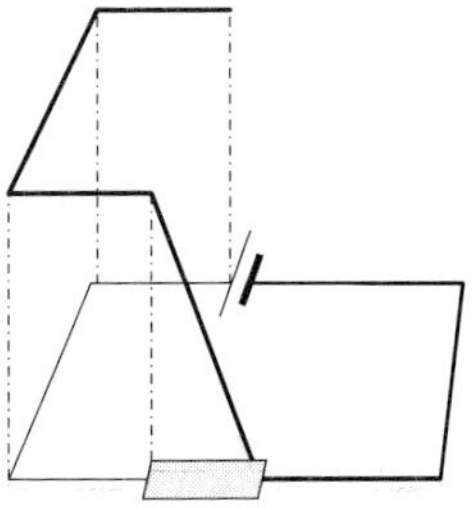

Fig. 2.55 Representation of voltage (change of reference point).

It is not the absolute value of the surface charges (potential) but ***only the difference*** in density between the two points (potential difference) that causes a net driving force on the mobile electrons within the conductor and the resistor.

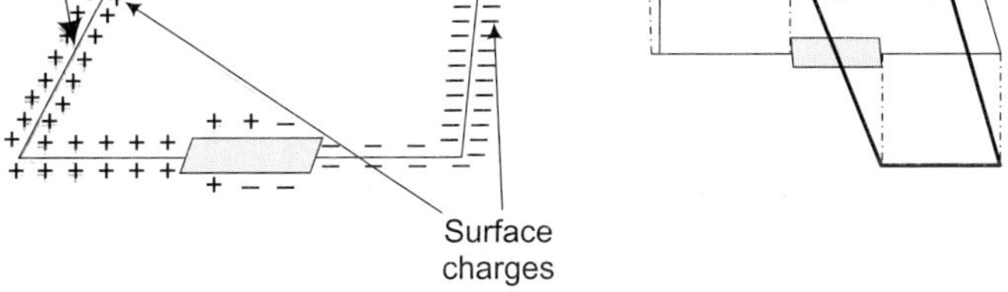

Fig. 2.56 Measurement of voltage.

When probes to measure voltage using a voltmeter are touched (see Fig. 2.56) at two different points in a circuit, then surface charges appear on the probe wires. The surface charges enable a tiny current in the wires and this current deflects the coil in the meter and the pointer attached to the coil swings to indicate the voltage. Note that a voltmeter probe merely comes *very close* (where the e-field is quite strong) to the surface charges when it touches a terminal and it *never* really touches neither the valence electron nor gets inside a weak atomic dipole. Refer Appendix B. In electric and electronic circuits the electric field drops away so steeply with distance from the wires into space that no voltage is recorded in voltmeters if its probes are held even micrometers away from the wires.

The experiment and the explanation above of how a meter measures voltage raises another interesting question, as discussed in the next section.

2.21 HOW IS THERE A *VOLTAGE* (OR POTENTIAL) FOR A *SINGLE* WIRE CONNECTED TO ONE TERMINAL OF A BATTERY?

Before we answer this question, it is necessary to understand how charges spread when placed on an insulated conductor and the concept of capacitance applicable to a single conductor. When an excess charge is placed at random on an insulated conductor (Fig. 2.57a), it will set up electric fields inside the conductor.[13] These fields act on the charge carriers of the conductor and cause them to move, that is, they set up internal currents. These currents redistribute the charge in such a way that the internal electric fields are automatically reduced in magnitude. Eventually the electric fields inside the conductor become zero everywhere, the currents automatically stop, and electrostatic conditions prevail as shown in Fig. 2.57b. The potential energy after the charges redistribute themselves will be *less* than that when the charges were placed; the difference will appear as heat in the sphere or radiated away.

Fig. 2.57(a) Excess charges placed on a conductor **(b)** the charges redistribute themselves.

Note the existence of electric fields and tiny currents inside the conductor for a short duration that caused the redistribution of the excess charges even though there was no voltage maintained as by a battery between any two points on the conductor. The fields of the excess charges exist, but their resultant or net inside the conductor is zero.

Next, we consider that there are two *uncharged* conducting spheres and these are separated by pulling one sphere far away to infinity. See Fig. 2.57c.

The potential of the conductive sphere at infinity may be assumed to be zero. That is a convention. Thus, we can introduce a notion of the concept of capacitance of a single conductor with reference to a spherical conductive shell at infinity.

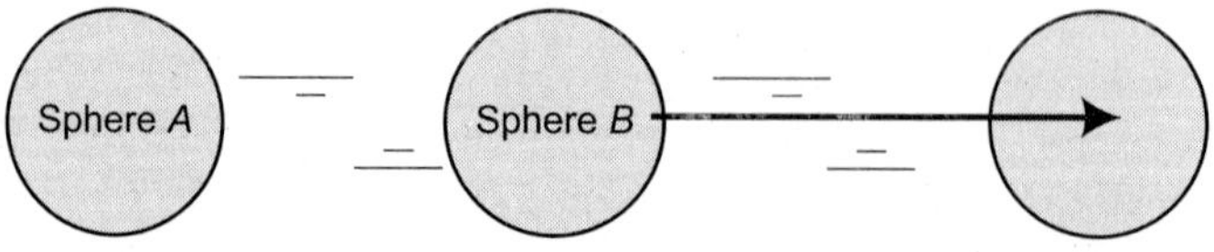

Fig. 2.57(c) Two conductive spheres near to each other are separated and one is taken away to infinity.

It is reasonable to assume that if some positive charge is placed on the conductor nearby then electric field flux lines will originate on the surface of the conductor and terminate on "some induced negative charge" on the conductive shell placed at infinity and that is how we would normally depict the electric field flux lines of isolated positive charged spheres: lines extending in all directions towards infinity. See Fig. 2.57d. It is as if the positive charge is in search of a companion negative charge wherever possible....infinity. We can thus consider and compute the capacitance of a single conductor with reference to a spherical conductive shell at infinity.

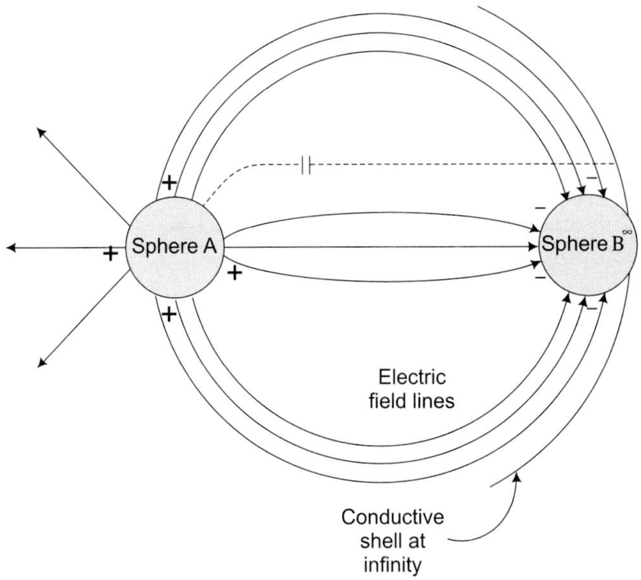

Fig. 2.57(d) Positive charges placed on the conductive sphere nearby will induce negative charges on the sphere at infinity.

All matter in the universe may be considered to be neutral. This is the natural and neutral world we are familiar with and used to in our daily life.

Whenever an electric charge is isolated, it is always in search of a companion opposite charge and an isolated charge searches for its companion opposite charge because its electric field extends in all directions in space. If it finds one, it terminates its electric field on it.

It should be noted that the field lines between the pair of opposite charges show the direction of the electric field strength that is a resultant computed from the fields of the opposite charges whose fields *always* exist in all directions in space. The "termination" of lines of a positive charge on a negative charge, is simply an "illusion" created by the *resultant field* of the two charges.

If the isolated charge were free to move, it will move towards that opposite charge and produce a neutral atom. It seems natural that this should happen, and all the observations in electrostatic experiments confirm it.

Therefore, it seems reasonable to define a capacitance for single conductors that will tell us what potential that conductor will (which it does, practically) attain if some charge were to be pumped on to it with reference to another conductor placed at infinity (which, of course, we cannot). This is the next topic of this section.

Potential of a Point Charge Relative to Infinity

It is often useful to *choose one reference position* (Chabay & Sherwood, 1995, used by permission),[2] and measure the potential at all other locations relative to that reference position.

When dealing with individual point charges, it turns out to be particularly convenient to *select infinity as the reference* position, infinitely far away from all charged particles. The rationale for this is that all points at infinity (that is, very far from all charges) have the same potential.

If we go from one distant point to another distant point along a distant path, along which the electric field is negligible (proportional to $1/d^2$ and because we are ***faaaa....r*** from the electric fields due to dipoles whether permanent or induced and multipoles of manmade capacitors, batteries, generators and all the not so perfectly neutral atoms of all matter with their lattice ions consisting of protons and inner electrons and outer electrons of, the earth, atmosphere, planets, stars, galaxies, asteroids, comets and so on), the potential difference between two different *distant* points is "zero". See Note in Fig. 2.58 and "Is an atom a dipole?" in Appendix A.

So all locations at infinity, in all directions, are at the same potential namely, *"zero"*, and is an ideal approximation ("Ideal Approximation" in "Electronic Principles" by Paul Malvino[20]). So is the case of the isolated point charge; it is just that, an ideally approximated *isolated* point charge.

The potential due to the point charge can be computed as shown in Fig. 2.58.

A test charge q_0 is moved by an external agent from point A to point B in the field set up by a point charge q.

$$V_B - V_A = -\int_A^B E.dl = \int_A^B Edr$$

The field E is given by

$$E = \frac{1}{4\pi\varepsilon_0}\frac{q}{r^2}$$

Combining the above two equations, the potential difference between A and B:

$$V_B - V_A = -\frac{q}{4\pi\varepsilon_0}\int_{r_A}^{r_B}\frac{dr}{r^2}$$

$$= \frac{-q}{4\pi\varepsilon_0}\left(\frac{1}{r_B} - \frac{1}{r_A}\right)$$

Choosing reference position A at *infinity* and choosing $V_A = 0$ ***at this position*** and dropping the subscript B leads to

$$V = \frac{1}{4\pi\varepsilon_0}\frac{q}{r}$$

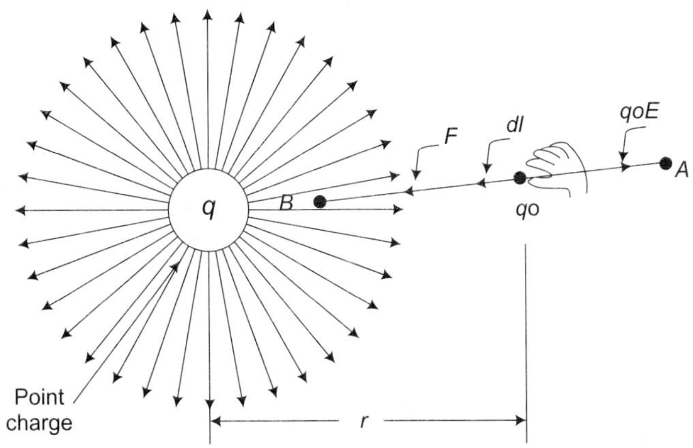

Note: The expression for $V_B - V_A$ also bears out the ideal approximation that the potential difference between two different distant points at infinity is zero because as both r_B and r_A tend towards infinity, we can replace these two variables by a single variable r and so $\left[\frac{1}{r} - \frac{1}{r}\right]$ becomes '0'.

Fig. 2.58 Potential due to a point charge.

From the above computation we see that the potential due to a point charge with reference to a point at infinity is given by

$$V = \frac{1}{4\pi\varepsilon_0}\frac{q}{r}$$

Using the above formula it is possible to calculate the capacitance of the earth. Refer to the figure below.

Assume that the earth is a conductive sphere of radius R (=6400 km) and that it carries a charge Q. Then its potential (at the surface or inside the sphere) is given by

$$V = \frac{1}{4\pi\varepsilon_0}\frac{Q}{r}$$

We assumed above, when computing the electric field due to a conductive sphere, that, it can be approximated as a point charge.

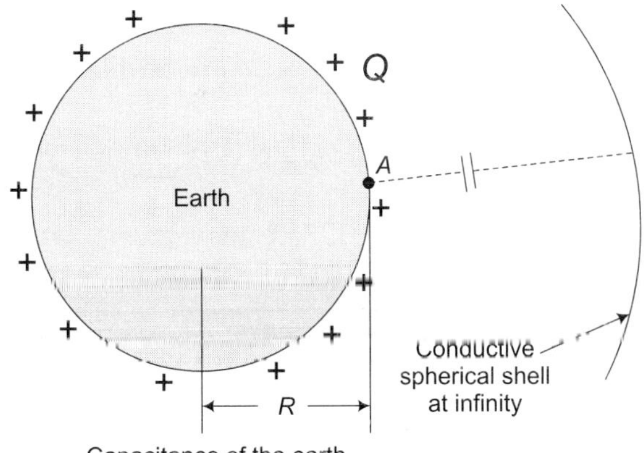

Capacitance of the earth

We now assume that a ***"second"*** conductive shell is located *at infinity and carries an equal and opposite charge* ***Q***, therefore the capacitance of the earth is given by rearranging the above equation to give $C = Q / V$.

$$C = \frac{Q}{V} = \frac{6.4 \times 10^{6}\,\text{metre}}{9 \times 10^{9}\,\text{nt} - \text{mt}^{2} / \text{coulomb}^{2}} = 711 \times 10^{-6}\ \text{F} = 711\ \mu\text{F}.$$

In electromagnetism and electronics, **capacitance** is the ability of a body to hold an electrical charge. There are therefore, types of capacitance other than the simple capacitor consisting of two conducting plates separated by a distance, as we saw in the above example.

Self-capacitance is the amount of electrical charge that must be added to an isolated conductor to raise its *electrical potential* by one volt; the reference point for this potential is a theoretical hollow conducting sphere, of infinite radius, centred on the conductor.

Any two adjacent conductors can be considered a capacitor, although the capacitance will be small unless the conductors are close together for long. This (unwanted) effect is termed *stray capacitance*. Stray capacitance can allow signals to leak between otherwise isolated circuits (an effect called crosstalk), and it can be a limiting factor for proper functioning of circuits at high frequency. Some authors also refer to the stray-capacitance to denote an equivalent capacitance; for example, the capacitance of an inductor between its ends, between turns and between the inductor and turn-to-core capacitance are all combined under certain specified conditions to form one expression that is taken as the stray-capacitance.

With small or no stray capacitances (as if the conductor is *very far* from other conductors or isolated in an ideal approximation), surface charges can accumulate rapidly at low to medium frequencies (due to the large times available for setting up the charges) enabling signals to be channeled strongly with little weakening of electric fields and consequently little loss of signal power to the loading end. With higher signal frequencies, in the presence of stray capacitances (whether small or significant) however, surface charges do not accumulate rapidly enough (due to the very small times available for setting up the charges) causing weak signal voltage and losses in the power transmitted due to losses in the stray capacitance charging processes.

We extend the concept of the capacitance of a single conductor to explain the potential of a single wire connected to a battery.

I owe gratitude to William Beaty of the University of Washington for providing the answer and which I have explained below with figures.

"Is there a voltage (or potential) for a single wire connected to one terminal of a battery?"

If the battery's net charge is zero, then for a 12 V battery, one terminal would be at a potential of +6V with respect to a conductor at infinity, while the other terminal would be at –6 V. *At the physical centre in the space between the two terminals of the battery, the voltage is zero.*

If we then connect a small wire to, say, the positive terminal, a few excess point charges will move onto the wire, and that wire will now be at the same potential as the battery terminal: +6 V. *The wire will be surrounded by an e-field,* and since the lines of electrostatic flux must always terminate upon charges, the wires *must possess an excess surface charge.*

Actually the wire upsets the picofarad values of each battery plate.

This is shown in Fig. 2.59.

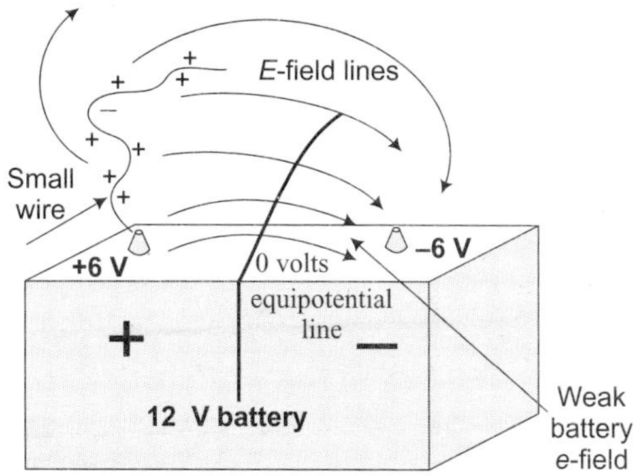

Fig. 2.59 Potential of a single wire connected to a battery.

Case 1: Treating the battery as an inert object with no net charge

We need to know if the net charge of the battery as a whole is zero. If we treat the battery as an inert conductive object hanging in space, it will have a capacitance of perhaps a few picofarads.

For such single conductors we assume that there exists a conductive shell surrounding it at an infinite distance as described in Section 2.23 with reference to Fig. 2.57d.

In other words, does the battery have an e-field which stretches to infinity? Or is the battery a neutral object, where the only e-field is the weak one produced by the potential across the battery terminals?

Actually, the small wire would upset the picofarad capacitance values of each battery plate, so the voltage at the wire might only be +5.99 V, while the other battery terminal would be –6.01 V.

Effect of Picofarad Capacitance on the Potential with Respect to a Conductor at Infinity

If we connect a very long wire to the battery positive terminal, the positive battery terminal must supply significant excess charge in order to electrify the wire (electrons move from the wire into the terminal). The battery then pumps charge through itself in order to maintain the 12 V difference. When things settle down, perhaps the wire is only charged to +1 V with respect to infinity, while the battery's negative terminal is now at –11 V.

This is shown in Fig. 2.60 with different values of the potential with respect to infinity: +1V, +5.99 V and 6 V for different picofarad values of the wire.

But the battery always maintains a 12 V difference across its plates by charge separation.

Potential difference between the plates is always 12 V.

$+1\text{ V} - (-11\text{ V}) = +6\text{ V} - (-6\text{ V}) = 12\text{ V} = +5.99 - (-6.01)$

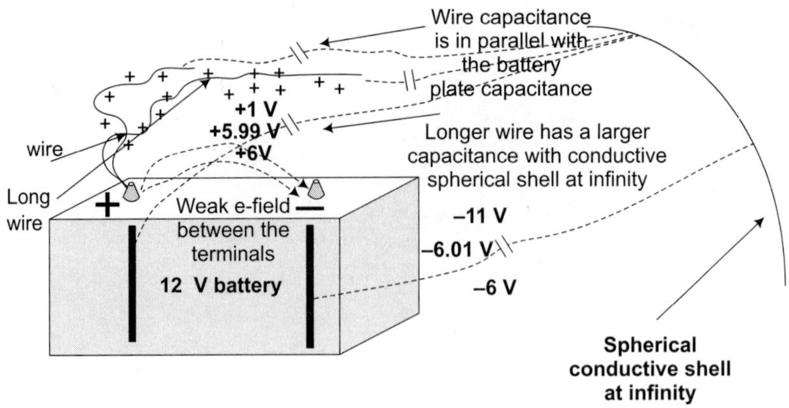

Fig. 2.60 Potential of a single wire connected to a battery.

Thus, the values would depend on capacitance measured with respect to the earth, so it would depend upon the size of the wire, the size of the battery, and upon the proximity to any other objects such as the earth's surface.

Case 2: Now supposing we connect a long wire to the battery with a "net charge"

We assume the net charge of the battery was not initially zero. Perhaps someone with rubber-sole shoes carried the battery while walking on a wool carpet or perhaps the battery was charged to –5000 V, and would deliver a tiny spark if touched. See Fig. 2.61. (In that case one terminal would still be 12 V different from the other terminal, perhaps -5000 V on the positive terminal and –5012 V on the negative).

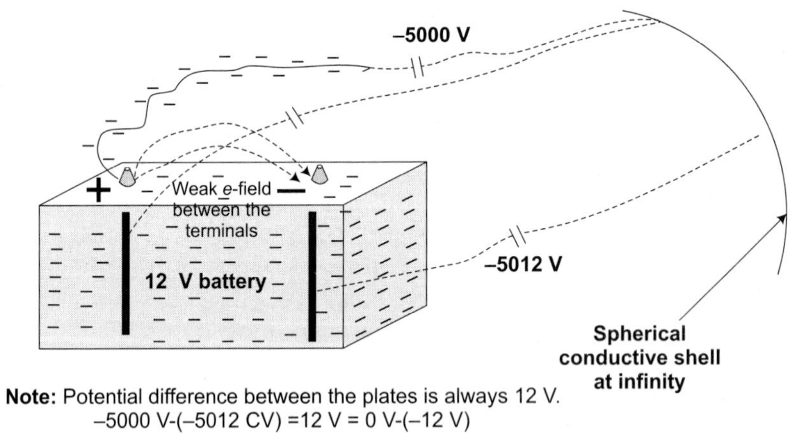

Fig. 2.61 Potential of a single wire connected to a battery having a net charge.

Or, perhaps the battery had a much smaller net charge. If we wish, we can deposit just enough charge onto both the terminals, onto the battery as a whole, so that one terminal will have a potential of zero with respect to infinity (or with respect to the earth) while the other terminal is 12 V different as shown in Fig. 2.62.

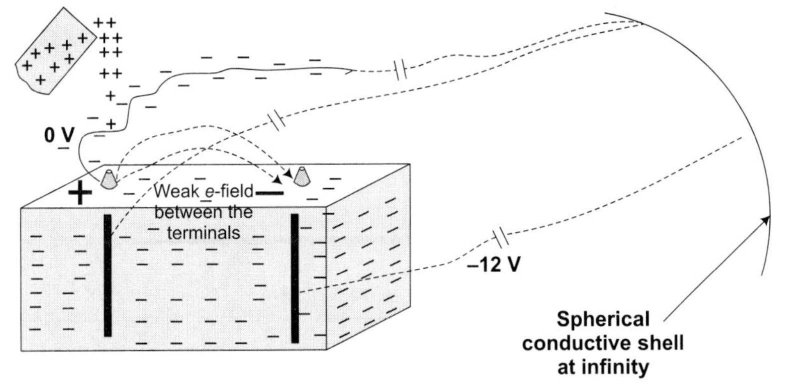

Note: Potential difference between the plates is always 12 V.
0 V-(–12v) =12 V

Fig. 2.62 Cancelling the excess charge of the battery.

By depositing a net charge upon the battery as a whole, we can choose to place one battery terminal at 0V with respect to infinity. If this terminal happens to be connected to a long wire, then there will be no e-field surrounding most of that wire, and most of the wire will have zero surface charge excess.

Thus, we see that the *voltage* of a single wire connected to a positive (say) terminal of the battery will be that of the terminal, and this is due to excess charges on the wire surface that the battery will *pump (in reality, electrons enter the terminal from the wire to the positive terminal and neutralize positive charges there, exposing positive lattice ions in the wire and causing the battery to pump up the depleted charges on its plates)*, in order to maintain its difference of potential with respect to the other (negative) terminal.

The *potential* of the wire will depend on the *state* of the battery potential (taken as a whole with respect to a point at infinity).

However, the difference of potential between the terminals and between the connected wire and the other terminal will never be different from the emf of the battery which in the example above we have assumed to be 12 volts.

The potential of the wire itself (and therefore of the terminal to which it is connected) depends on its capacitance and the capacitance of the terminal and the amount of excess charge that is already present.

What are Currents and What is Voltage?

It is wrong to associate currents with voltage.

Currents don't have Voltage

Voltage is *not a characteristic* of electric current. It's a common mistake to believe that a current "has a voltage" (and this mistake is probably associated with the 'current electricity' misconception, where people believe that 'current' is a kind of substance that flows).

Voltage and current are *two independent things*. It is easy to *create a current that lacks a voltage*: just short out an electromagnet coil.

It is also easy to create a *voltage without a current*: flashlight batteries maintain their voltage even when they are sitting on the shelf in the store.

Water analogy: Think of water pressure without a flow. That's like voltage alone. Now think of water that's coasting along; a water flow without a pressure. That's *like electric current alone.* Refer:http://amasci.com/emotor/cap1.html.

2.22 POTENTIAL ENERGY OF ELECTRIC FIELDS

How is the potential energy of an electric field gained and how is it lost? Of course, in systems of charges.

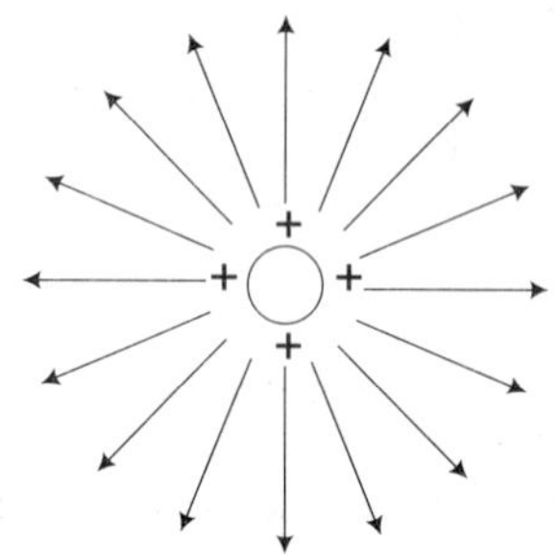

An electric field surrounds an isolated positively charged sphere

An *uncharged* conductor is brought near the positively charged sphere. Charges will be *induced* on the surface of the conductor and there will be a net attraction between the sphere and the conductor. See Fig. 2.63. The system did some work to cause the *separation* of charges when the conductor was brought near. The energy of the system has gone down in comparison to the energy in the field "before" the conductor was brought near the positively charged sphere. Look at it this way: If you were to stand very far away the net charge *will appear* like a *smaller* point charge.

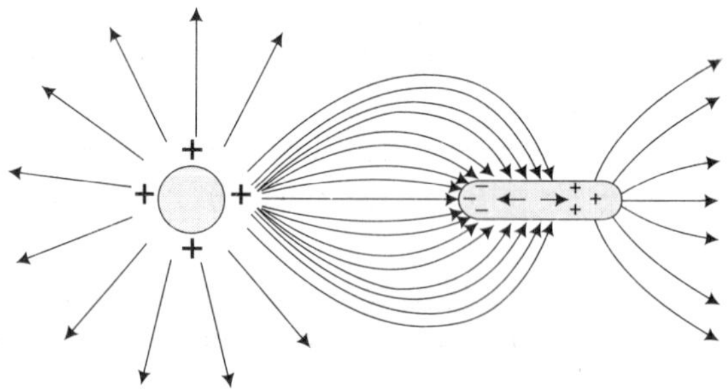

Fig. 2.63 Potential energy of electric fields.

There is a *slight loss in energy* in the conductor *during* the charge separation.

This is *carried away* in the form of electromagnetic (EM) radiations, heat and other EM waves.

Of course, the *bulk* of work done by the positively charged sphere is *stored* as electric potential energy in the *separated charges* of the conductor.

A graph of the potential of the system is plotted in Fig. 2.64a. The potential of the system *before* the conductor is brought near is shown with a dashed line. The altered potential *after* the conductor is brought near is shown with a full line. Far away from the system, the *electric field* strength will become smaller due to the polarization of the conductor. Between the conductors the electric field has *increased* due to the presence of the image charges and so the potential *decreases more rapidly* as compared to the isolated sphere.

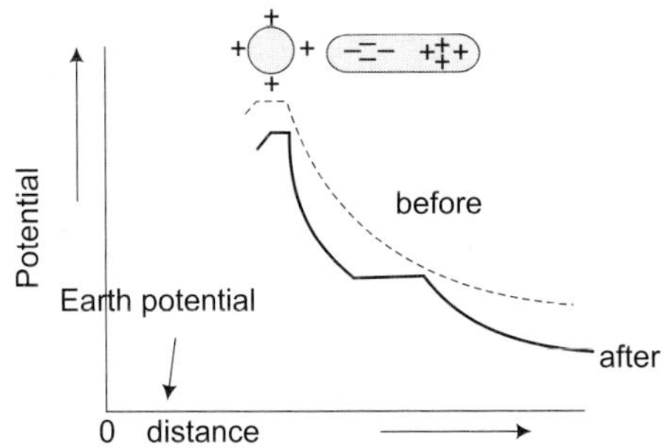

Fig. 2.64a Graph of the potential of the system.

If we move the conductor nearer the charged sphere, the potential graph will shift lower and lower and the potential near the left end of the conductor becomes sharply negative as seen in Fig. 2.64b.

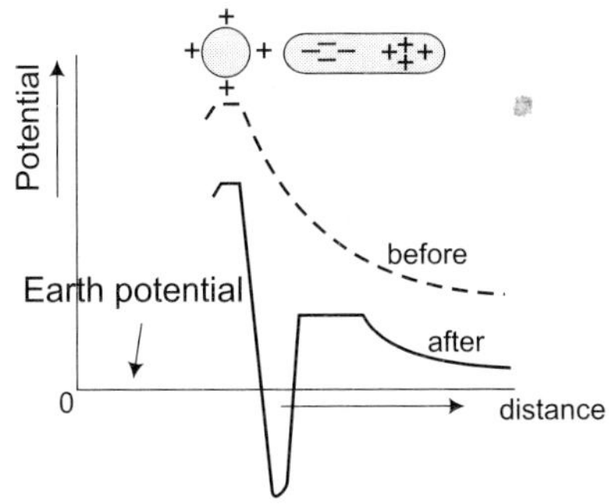

Fig. 2.64b Graph of the potential of the system.

The figure below shows the electric field flux lines of the system of Fig. 2.64 comprising a positively charged sphere and the uncharged conductor brought near it. A charged metal object is like a flat-topped mountain, where the altitude of the mountain top is like the potential of the charged object, and the mountain slopes are like e-fields in the space around the object. On one side of the mountain the slope might be very gentle as the fields we see surrounding the positively charged sphere to the left, top, below and, to the right of the conductor.

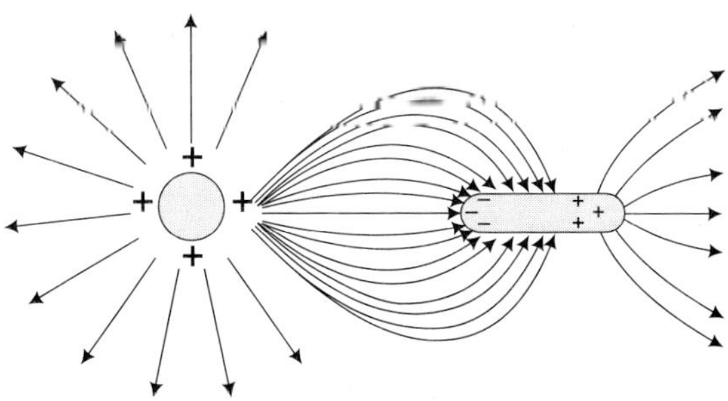

On some other side, the slope might be very steep, as seen *between* the charged sphere and the conductor.

Yet no matter how gentle (or steep), we arrive at the same altitude when we walk to the top as seen in the flat portions of the potential graphs of Figs. 2.64 for the sphere and the conductor.

Many students try to develop analogues in other areas such as mechanics between electrical forces and gravitational forces. But there are differences that complicate this. Coulomb's law equation for electrical force bears a strong resemblance to Newton's equation for universal gravitation.

$$F_{elect} = \frac{k * Q_1 * Q_2}{d^2} \quad \text{and,} \quad F_{grav} = \frac{G * m_1 * m_2}{d^2}$$

$$(k = 9.0 \times 10^9 \text{ N-m}^2/\text{C}^2) \quad \text{and} \quad (G = 6.67 \times 10^{-11} \text{ N-m}^2/\text{kg}^2)$$

The two equations have a very similar form. Both equations show an inverse square relationship between force and separation distance. And both the equations show that the force is proportional to the product of the quantity that causes the force – charge, in the case of electrical force and, mass, in the case of gravitational force. Yet, there are some striking differences between these two forces.

First, a comparison of the proportionality constants — k versus G — reveals that the Coulomb's law constant (k) is significantly greater than Newton's universal gravitation constant (G). Consequently a unit of charge will attract a unit of charge ***with significantly more force than a unit of mass will attract a unit of mass.***

Second, gravitational forces are only ***attractive***; electrical forces can be ***either attractive or repulsive.***

2.23 THE PROCESS OF CAPACITORS SHARING CHARGES

In this section, we study the processes using the concepts that we have learnt to understand how capacitors connected in parallel, share charges.

When capacitors connected in parallel share charges, the actual processes involved *before* steady-state occurs are obscured and, it is only when the *energy balance equations* are written do these *become more apparent.*

A capacitor C_1 is charged by a battery and, the battery is disconnected and the capacitor then connected to another *uncharged* capacitor C_2. (Refer to "Physics" by Resnick and Halliday).

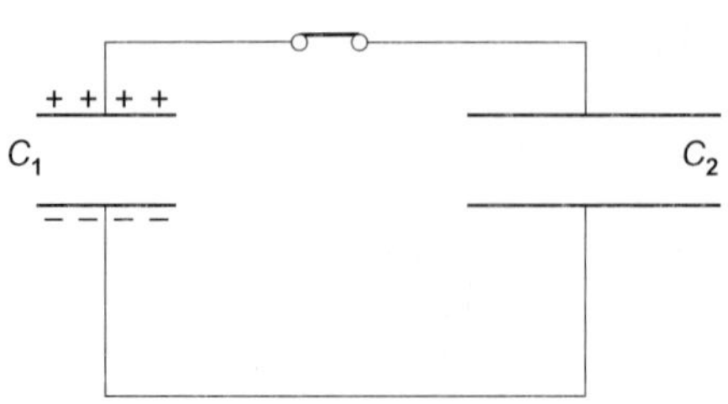

The charges *migrate* from capacitor C_1 to the plates of capacitor C_2.

How long should they migrate?

They will migrate until the *potential difference* between the plates of the respective capacitor are the *same* when further movement stops, that is *the rule.*

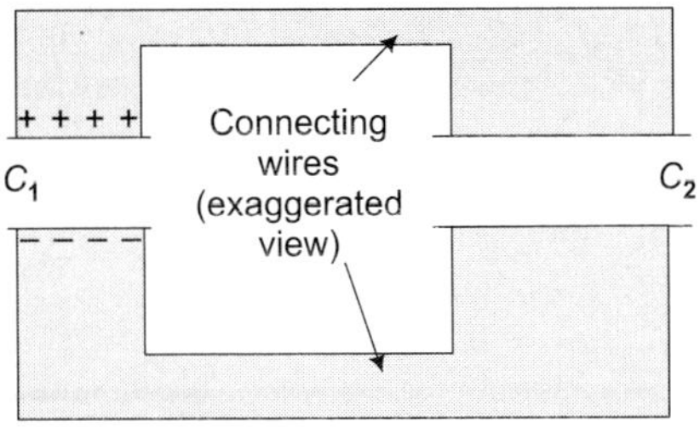

Consider the following example:

A capacitor C_1 is charged to a potential difference V_0. The charging battery is then removed and the capacitor is connected to an uncharged capacitor C_2.

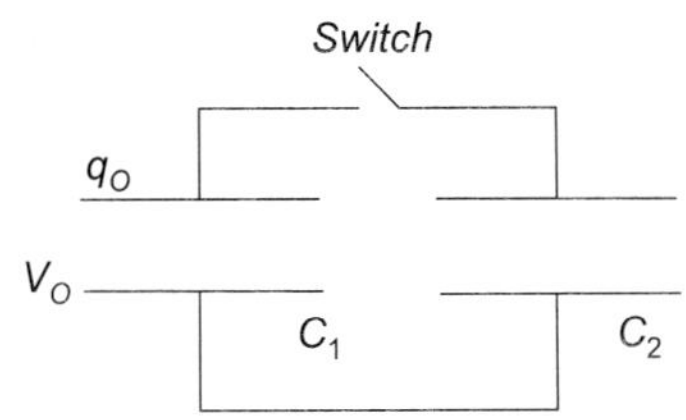

What is the potential difference V across the combination?

What is the stored energy before and after the capacitor C_1 is connected to capacitor C_2?

The original amount of charge on capacitor C_1 is q_0. After the two capacitors are connected the charge q_0 is shared by the two capacitors. Thus,

$$q_o = q_1 + q_2$$

Applying the relation $q = CV$ to each of these terms yields

$$C_1V_0 = C_1V + C_2V$$

or,

$$V = V_o \frac{C_1}{C_1 + C_2}$$

The initial stored energy is $U_o = \frac{1}{2}C_1V_o^2$ (see Appendix A "Energy stored in capacitors").

The final stored energy is

$$U = \frac{1}{2}C_1V^2 + \frac{1}{2}C_2V^2 = \frac{1}{2}(C_1 + C_2)\left(\frac{V_o C_1}{C_1 + C_2}\right)^2$$

$$= \left(\frac{C_1}{C_1 + C_2}\right)U_o$$

Final stored energy (in the two capacitors) = initial stored energy in C_1 multiplied by $\left(\frac{C_1}{C_1 + C_2}\right)$.

Thus, *U* is less than *U*$_o$!...

Final stored energy in both the capacitors is ***less than*** initial stored energy of capacitor C_1. This means that there was a loss of energy when the potentials of both capacitors became equal.

What is the form of this lost energy and where is this energy?

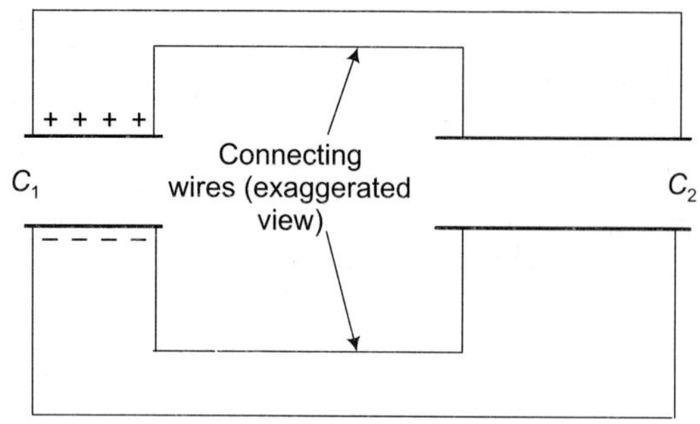

Intuitively, many of us then try to figure out what will happen if the resistance of the wire were to be *made zero.*

Assuming you can achieve zero resistance, you will find however that there is **still an impedance** due to the self inductance of the wire leads.

Even if this self inductance is infinitesimally small it will result in the capacitor actually "overcharging" to a voltage greater than that of the battery. Think of the inductance as an *"effective inertia"* in the flow of the current.

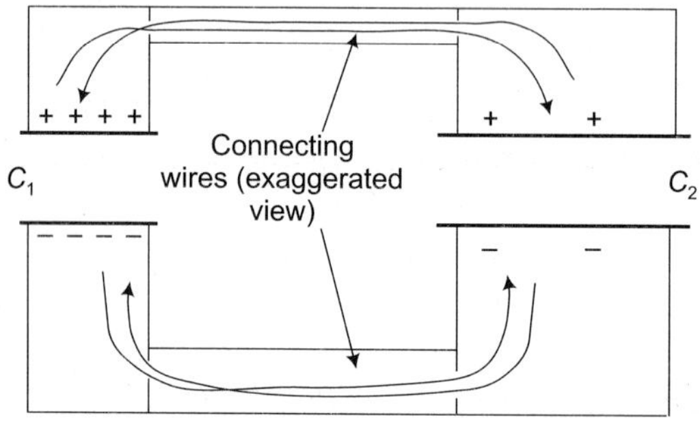

The circuit will then oscillate at a frequency √{C/L}between the initial zero charge on the capacitor and a maximum greater than the voltage of the battery.

If the circuit is not shielded the difference in energy will ***radiate away*** as electromagnetic (em) waves and the voltage of the capacitor will be a decaying oscillation centred around the battery voltage.

Eventually it will settle down to the static case and you'll find that the *missing energy is exactly the energy of the radiated waves.*

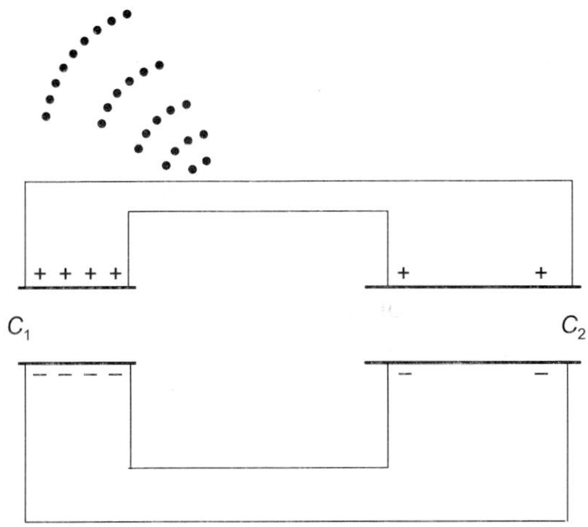

The amount of *missing* energy dissipated is *independent* of the impedance value and so is the same in the limit of zero impedance. The only variable is the *complex phase* of the impedance (how much is real = resistive and how much is imaginary = inductive)....and this tells you how the energy is dissipated, either as heat or as E-M radiation.

If the resistance is large, the ***missing*** energy is ***dissipated*** as heat in the wires!

The *em-waves/photons* of the "*missing energy*" result from *accelerating* charges. This is a fundamental property of a charged particle; it radiates energy when accelerated. The collisions essentially result in acceleration of the charges, and some of the energy is *lost directly* as photons or electromagnetic wave-packets.

Some of it results in energy being directed to the lattice. However, since the *lattice* points are also charges, they too undergo *acceleration*. There may also be collective mode of *charges sloshing* back and forth in the solid (also called *plasmons*), or lattice vibrations (called *phonons*). But the essential picture remains the same... charges are jostling around, and therefore *coupling energy out of the system* in the form of photons.

The exact spectrum of energy loss will of course depend on the geometry, material and other details of the system under consideration.

The result of all these charges wiggling around results in a spectrum that is predominantly in the infrared, and therefore they show up as ***'heat'*** as we commonly understand it. When you touch the body, the random kinetic energy or vibrations of all its ions is what you feel as heat (transferred to you) and if you direct an infrared detector at it, you will detect the energy of the photons in the infrared band.

I thank Dr. Dipankar, Visiting Professor, IIT Mumbai, for providing the above details

Let us try to understand all that has happened when the *charged capacitor* C_1 was connected to the *uncharged capacitor* C_2.

In the act of connecting a capacitor charged initially with an amount of charge Q with another capacitor, we have ***created a system*** whose steady-state (final) total energy has to necessarily be *less than* the original energy we started with in the singly charged capacitor.

The formula cited earlier proved this. Final stored energy (in the two capacitors) = initial stored energy in C_1 multiplied by $\left(\frac{C_1}{C_1+C_2}\right)$ which took into account the constraints: in the steady-state (after a long time) the potential difference of both the capacitors should be same and, the total charge available for sharing is just the initial charge of the capacitor C_1.

When you connect the two capacitors the **final** potential difference can only ***go down*** and ***not up*** since there is no transformer nor inductor or any other voltage source other than the initial potential difference across the capacitor C_1.

The system will try to get *rid of* the *excess energy* by whatever means available at its disposal and therefore, the *dissipation* of the excess energy (one may call it the difference between the initial and the final energy) in the wires.

You may try as hard as you can to conserve the energy and prevent the loss.....but there is a *poor chance that you will succeed.*

In the act of creating the system of the uncharged capacitor connected to the charged capacitor the *total final energy* possible to store in the capacitors, given a fixed amount of initial charge on one of them, *is fixed.*

The system will "shake away" the excess energy either in the form of *heat* or in the form of *EM radiation* .....depending on the phase of the impedance.

This will continue until charge migration ***stops*** and the dissipation stops and ***steady-state is reached***.

In the rare and impossible event that you prevent both, heat dissipation and *EM* radiation by some means from occurring then, you will *never get* the assumed *rest state.*

The table below shows the **final p.d. and the final energies** for different sets of capacitor values C_1 and C_2.

CapC_1	*Cap*C_2	*Voltage*		*Energy*	
		Initial	*Final*	*Initial*	*Final*
10 pF	10 pF	10 V	5 V	500 pJ	250pJ
10 pF	10 nF	10 V	0.009 V	500 pJ	0.5 pJ
10 pF	10 μF	10 V	0.01 mV	500 pJ	0.0005 pJ

Notice that as the (uncharged) capacitor $\mathbf{C_2}$ is *made larger* and larger in **comparison** with the capacitor $\mathbf{C_1}$ the *final p.d. becomes smaller and smaller* and **most** of the **energy** is **dissipated** away.

When *both* the capacitors are 10 picofarad, the *final p.d. is* **5 volts** or *half* the initial voltage on C_1 and the energy dissipated is half the initial stored energy.

When the capacitors are **10 pf** (C_1) and **10 μF** (C_2), however, the *final p.d is a tiny* **0.1 mV** and the energy dissipated is **99.99 per cent.**

As we shall see, this is the concept behind providing **safety grounds in electrical and electronic** equipment.

2.24 STEADY-STATE AND DYNAMIC EQUILIBRIUM AND TRANSIENT STATE

The situation when the circuit was *shaking away* the excess energy in the form of heat or *EM* waves is a *transient* state.

In many systems, *steady-state is not achieved* until some time has elapsed after the system is started or initiated. This initial situation is often identified as a transient state, start-up or warm-up period.

In electrical and electronic circuits these *sudden changes* are mostly found as the result of the operation of switching devices. Circuit designers use voltage regulators and surge protectors to prevent transicnts in electricity from affecting delicate equipment.

Steady-state is a more *general* situation than Dynamic equilibrium.

While a **dynamic equilibrium** occurs when two or more **reversible processes** occur at the **same rate**, and such a system can be said to be in steady-state.

(There are many physical processes that attain a state of dynamic equilibrium. For example, when a substance boils, the vaporization process and the condensation process are occurring at equal rates. This occurs at a specific fixed temperature called the boiling point of the substance.)

A system that is in steady-state **may not necessarily** be in a state of dynamic equilibrium, because some of the processes involved are not reversible.

Example: The flow of material through a system, or electricity through a network, could be in a **steady-state** because there is a constant flow of material, or electricity.

Conversely, the filling or draining of a tank with material would be an example of a **transient** state.

2.25 WHAT IS GROUNDING?

If you consider that a piece of electrical or electronic equipment with a capacitance of a few picofarad has some excess charge and its potential is say, 415 V with respect to some point at infinity and if the equipment were to be grounded by making a connection to the *earth* whose capacitance we saw is a robust *700 odd microfarad*, the system will immediately reach a *safe value of 6 microvolts.*

The process? Simple mutual repulsion of like charges and a desire to reach static equilibrium with minimum energy at rest caused by charge spreading over a large area using electric fields $E_{NET} = 0$ after equilibrium is reached, all in a flash!

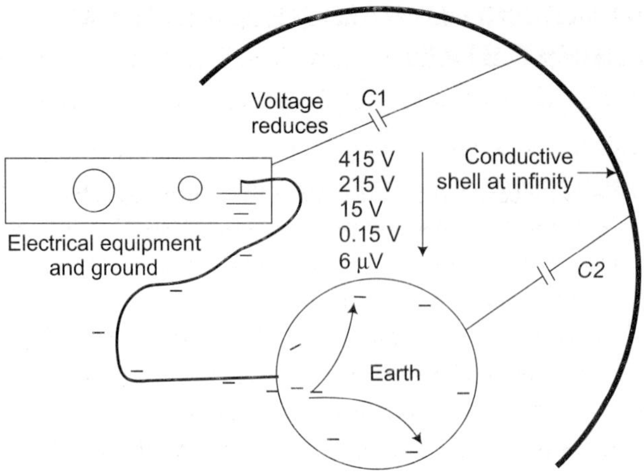

The charges could be of either sign....

2.26 APPLYING A PULSE TO A CR CIRCUIT—ANOTHER CLOSE LOOK

Shown in Fig. 2.65 is the CR circuit of Fig. 2.15 with a "signal ground" added which will be explained shortly.

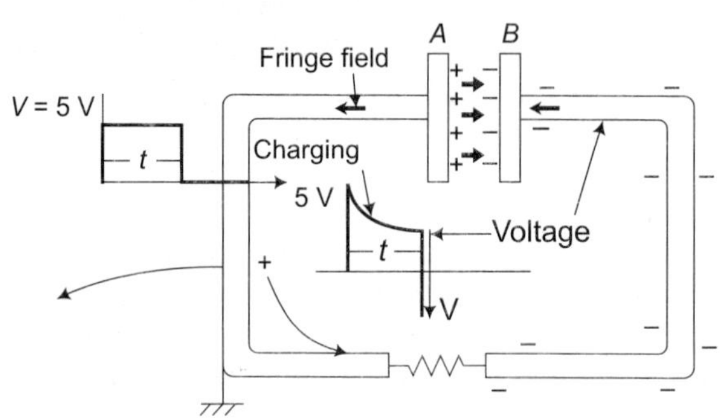

Fig. 2.65 Pulse input – voltage input is made zero.

At the trailing edge of the pulse, shorting the input voltage source (stopping the charge separator machine) makes the few positive (surface) charges on the wire from plate *A* to neutralize the few negative (surface) charges on the resistor end zero volt or close to zero volt.

Note carefully that the electric field flux lines on this section of the wire are *thinly* spaced. This means that the field strength is very tiny here. And at one end the positive charges on the inner surface of plate *A* are tightly bound with the negative charges on plate *B*, shown by tightly spaced thick arrows indicating a powerful field strength. But that does not change the voltage of plate *A* at the instant of the trailing edge transition, with respect to plate B.

How is Zero Volt Possible with Positive Charges on the Plate A?

The truth of the matter is, there was never really "zero" volt, nor a true zero potential either.

All that we began with, when we constructed the circuit was a reference voltage that we can label in the diagram with this symbol ⏚ or ⏊ and as '0' for the origin point in voltage vs. time plots. Physically it is a piece of wire, of conductive material whose bulk is neutral and as per convention we should label it "signal ground" or "circuit zero". *The symbol is not meant to indicate that the point where it is shown is connected physically to the earth.* If it is earthed, then this should be explicitly stated in the diagram.

It is meaningless to imagine that we can, achieve a 'true' 'zero' volt in the circuits that we build and to make voltage measurements with reference to 'true' 'zero'.

Why? Because, recollect that 'zero' potential (see Section 2.21 "Potential of a point charge relative to infinity") is defined only at a point far away from all charges where there would be no electric field and the nearest point that would qualify to be at zero potential would be at "infinity". I have not seen any engineer or scientist with one positive probe of a voltmeter fixed at a point to a circuit and a negative probe of the voltmeter with a one-way ticket boarding an aeroplane to "infinity" to fix the negative probe; and I, for one, have no intention to become the first one!

Those were in a lighter vein. But do not scoff at those who set the zero potential at a point at infinity, because scientists have devised the convention to make it easier to compute the energy stored in various charge configurations.

So, our neutral piece of wire with its millions and millions of neutral atoms is good enough to be labelled as a 'reference' and also as '0' volt. All 'other' voltages in the circuit, are measured with reference to this. And if the wire were connected to the earth (see figure below), the earth being the best charge "spreader" to all its far corners (!) as compared to a simple wire, the potential (compared to a point at infinity) of the wire when connected to it will be even closer to zero volt potential. If allowed by a large enough pulse duration time '*t*' to become fully charged, the plate A of the capacitor will be at the *potential* of the positive terminal of the source i.e +V volts and, the physical centre between the plates of the capacitor will be at +V/2 volts (*potential*).

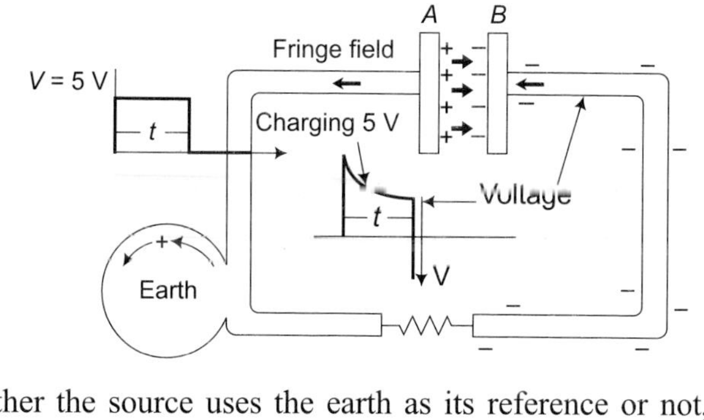

Either way, whether the source uses the earth as its reference or not, the charges on plate B haven't been neutralized instantly at the trailing edge of the pulse input, prevented by the resistance, and therefore, the negative value for the *voltage* at B (with respect to the signal ground or circuit zero).

Why does the wire not qualify to be labelled as 'true' zero potential, since after all, all its atoms are neutral?

There are several reasons for the disqualification of the piece of wire connecting plate A to the resistor from being labelled 'true' zero volts; the foremost being the presence of tiny electric fields that are present to move the electrons when discharging. See Fig. 2.65 and notice the presence of the fringe field even as the capacitor begins to discharge. This is necessary and there we have a field in the "neutral" conductor and the attendant surface charges to enable the conduction process and the neutrality is certainly disturbed on the surfaces and, the field in the wire and the surfaces though tiny, are there, all lined up against qualifying the wire to be at 'true' zero volt.

We can make the wire thicker and reduce the potential difference across its ends but that is the best we can do. In fact, in printed circuit boards the reference lines are made using broad traces for the very same purpose.

Among other reasons, there could be stray charges produced by interfering electric and magnetic fields coupling into the circuit. This is shown by a lone stray positive charge, it could have been negative as well, in the "reference" wire.

The potential of any charged conductor is uniform throughout its interior and on its surface. If it is at 100 volt, the potential of any point on its surface and interior will be 100 volts. If it is at 0 volt, the potential at any point on its surface and interior will be 0 volt.

We now turn our attention to how the voltage of the source is made zero when the pulse input reduces to zero and the sudden transitions in voltage that occur in such circuits. Students find it difficult to visualize the rapid voltage transitions occurring at different points in the circuit especially when circuits are subjected to suddenly transiting voltage inputs like the pulse input, square wave input and so on.

Consider the circuit of Fig. 2.65. When the pulse height was V, the voltage of plate A was at V, say, $V = 5$ volts and when the input was shorted, plate A is at 0 volt, the reference terminal of the voltage source. This is a sudden transition. Before the transition, there were positive surface charges (positive) on the wires between the pulse generator terminal and the plate A and, on the wire between the resistor and the plate B of the capacitor. See Fig. 2.66.

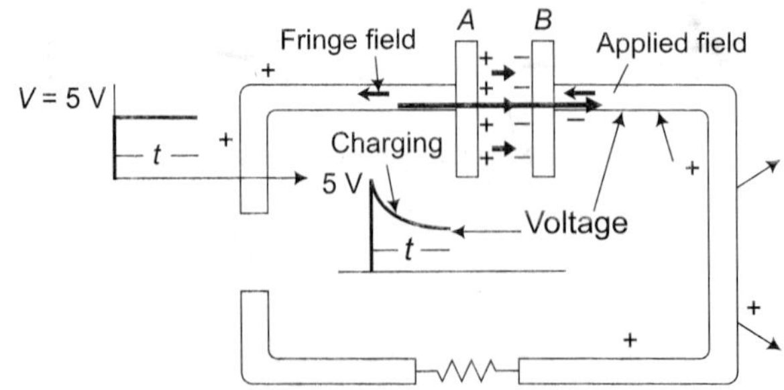

Fig. 2.66 At the instant before the input voltage is made zero.

During polarization, there were negative charges on the plate B and these were tightly bound to the positive charges on plate *A* and the negative charges did not allow the latter (positive charges) to move away. The field vectors will be pointing inward at plate *B*; the work done to bring in a positive charge from plate *A* would have been negative, but the charge would have landed on a potential that is still positive at a time '*t*' much less than RC.

When the pulse generator voltage became zero, the surface charges that had kept the (source) mountain flat top at voltage *V* simply disappeared and, with them the electric field (the field represents the effort the generator used in driving electrons onto plate *B* by overcoming the fringe field of the capacitor and the obstacles in the resistor), is turned off. This is another characteristic of electricity; it is not possible to switch off a gravitational field easily.

As the pulse generator voltage became zero, it left behind in the circuit a mini flat top mountain which it had constructed by charge separation, of the size determined by the voltage between plates *A* and *B* with its flat top at voltage *V*. While the source mountain crumbled when its voltage became zero, it brought down the flat top of the mini mountain to zero volts (input short) with its "stiff" wedge (the resistor in the circuit) between the separated charges causing the voltage of plate *B* to go down below the reference zero volts by an amount equal to *V* (the driving field cut-off everywhere) from whatever value the plate *B* voltage was at as if a spade was used to dig out a chasm at the source mountain's base! see Fig. 2.65.

There is a reduction in the strength of the field emanating from this section of the circuit (the wire from the pulse generator to the plate *A* of the capacitor) because the driving field is off.

The driving field did not become zero merely in the wire from the generator to the plate *A* of the capacitor; it became zero everywhere in the circuit including the wire from the generator to the resistor and the wire between the plate *B* and the resistor, with which it was driving and holding up the charges on plate *B* as well, with a strength of *V*.

So, when the pulse generator voltage became zero, what the thin electric field flux lines mean (see Fig. 2.65; like the long gentle slopes of a mountain) is that the work done by a test positive charge to move from far away towards this section of the wire is "zero" or close to zero, as it would have been if there were no pulse generator at all.

And the tightly spaced flux lines between plate *A* and plate *B* of the capacitor (on the same mountain with *steep* slopes) imply that a test positive charge moving from plate *B* to *A* will rise a short distance, but against a very powerful field,....as if from a chasm at the mountain base, denoting the negative potential at plate *B*, and the test charge will reach zero volt at plate *A*.

If one visualizes a bunch of positive charges on one end of the wire, that is, plate *A* in close proximity to a bunch of negative charges (plate *B*), and computes the potential around the plate *A*, then it will be seen that the potentials will be overall much lowered at fixed points due to a thinning of electric field flux lines elsewhere and a thickening at points on field flux lines directly joining the unlike charges in comparison to the potentials that would occur at these points without the presence of the negative charges. And, when the voltage of plate *A* became zero, the voltage of plate *B* will be *negative, with reference to this "zero" of plate A* which is at the other end of a resistor, that maintains the difference between the two below say, a voltage V = 2.5 volts as depicted in Fig. 2.66. If a piece of wire were in place of the resistor, there would have been a dead short across the capacitor and you would have zero volt everywhere.

Therefore, it would seem all that the pulse generator did was to separate charges (electrons from lattice ions) that were held together as if by tiny springs, by injecting energy into them against the restriction of the resistor and when the generator packed up its bags and left, the capacitor had to rely on its fringe field to begin reuniting with the opposite charges against the resistor.

Let us assume, for a moment that the capacitor was 0.1μF and was charged to 2.5 V. Now that would mean it had a charge of $Q = 0.1 \times 10^{-6} \times 2.5$ coulombs on it. If we had the same amount of charge kept on a sphere with a radius of 1 metre, its potential (with respect to a point at infinity) will be:

$(9.0 \times 10^9 \times 0.1 \times 10^{-6} \times 2.5)/1 =$ **2.25 kV.** It is the presence of the charges of opposite sign on plate *B* that kept the voltage between the plates at 2.5 V.

Many of us attempt to visualize point charges as single charges and dipoles from electrostatics and the electric fields and potentials around them. There is nothing wrong with that. The potentials of single charges and dipoles is of the order of kilovolts at points very close to the charges (for example, potential is $q/4\pi\varepsilon_0 r$ for point charges). A plot of potential lines is shown in Fig. 2.67 of a dipole in isolation. But note that there is a line of zero potential that means the work done to move a test positive charge from infinity along the line will be zero and, work needs to be done to move the test charge to the positive charge, and if moving the test charge towards the negative charge, the electric field does the work (you can simply let it go).

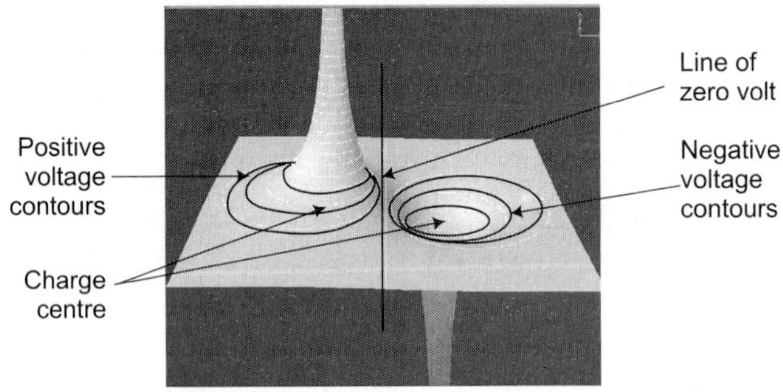

Fig. 2.67 Potential and equipotential lines of a dipole (positive charge on the left).

If the dipole charges come nearer as in Fig. 2.68, the potentials decrease at points surrounding the charges (not in between) due to cancellation of fields of unlike charges. And the voltages are at levels of tenths or multiples of volts at the same locations where they were 100's or 1000's of volts when the charges were farther apart.

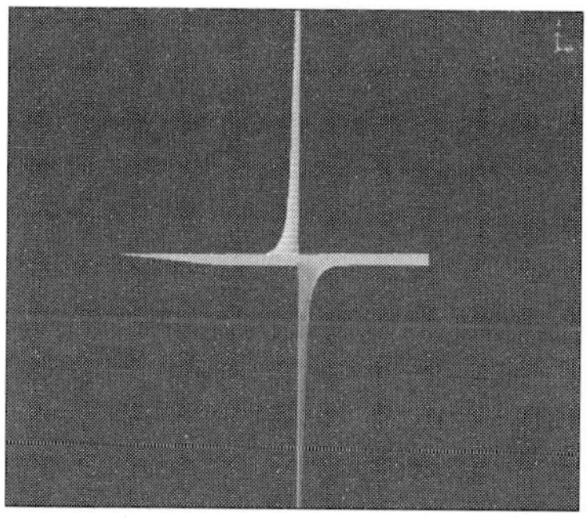

Fig. 2.68 Potential and equipotential lines of a dipole (charges close to each other).

The zero volt level midway between the dipole exists despite the presence of an electric field and does not qualify for a "true" zero potential definition, which is best defined at infinity, where we would believe there is no electric field at *all,* being *far away* from *all* charges.

Similarly, there is a line of zero volt midway between the plates of a parallel plate capacitor. And that is why, the capacitor may be seen as a dipole from far away.

We digress here to consider an average field inside matter, which consists of millions and millions of atoms and molecules.

The Average Field Inside Matter

According to David Griffiths'[8] descriptions in "Introduction to electrodynamics" of the fields inside matter: The electric field inside matter is quite complex at the microscopic level. Very near an electron, the electric field is gigantic, whereas a short distance away it may be small or point in a totally different direction even, due to the interactions with neighbouring atoms and molecules. Moreover, an instant later, as the atoms move about, the field will have altered entirely, due to the rocking ions and darting electrons.

This true microscopic field would be impossible to calculate; nor would it be of interest to calculate. Just as, for macroscopic purposes, we regard water as a continuous fluid, ignoring its molecular structure, so also we can ignore the microscopic bumps and wrinkles in the electric field inside matter, and concentrate on the macroscopic field. This is defined as the *average field* over regions large enough to contain many thousands of atoms so that the uninteresting microscopic fluctuations near the lattice ions and close to the conduction band electrons are smoothed over. The *macroscopic* field is what people mean when they speak of *the* field inside matter, and is also sometimes called an "equipotential" region. In an aluminium metal crystal the regions of the macroscopic field would refer to the volume shown grey in Fig. 2.69.

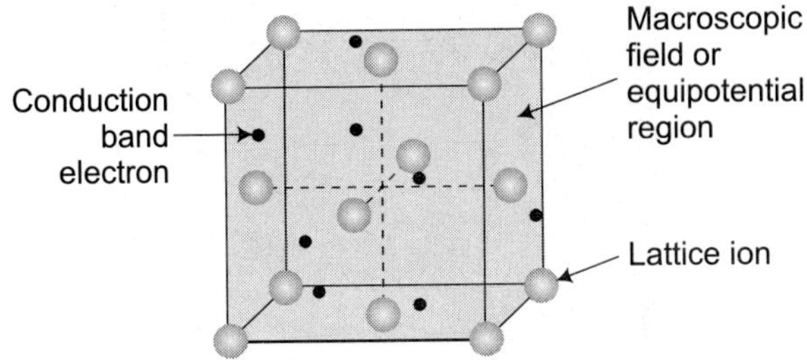

Fig. 2.69 Aluminium metal crystal structure.

While one is tempted to quickly apply the potential profiles of Figs 2.67 and 2.68 to excess charges on the plates of charged capacitors, one should remember that *capacitor plates and wires are conductors and excess charges cannot exist in them as they do in space and*, the close spacings and interactions with neighbouring atoms of the excess charges cause the potentials of the conductors (on the surfaces of which we measure potentials and potential differences) to assume more moderate values.

All the ions and charges within conductors (seemingly) define a *no-man's land* that is out of bounds for the voltages and accompanying fields that are applied in common electric and electronic circuits; one cannot get these voltages to rip off the proton or cause the electrons (or protons) to fuse into each other. In case of metallic conductors one can visualize the conduction band electrons with their incessant movement smoothening out the wrinkles in the electric field and causing the average electric field to be zero or near zero inside, (in static equilibrium) whether *electrified or neutral* and make the interior to be equipotential with the surface.

A simple calculation will show that the potential (relative to infinity) at a location 10^{-10} metre from the centre of a singly charged ion treated as a point particle is a mere 14.4 volts and much less as one moves farther away. The potentials still rise sharply very near the charges. Besides, one is interested in the potential differences of groups of charges in conductors at far greater distances than those between say, electrons and protons, evened out by the properties of conductors that maintain a net *nearly* zero e-field in static equilibrium (near perfect field cancellation between the orbiting electrons and rocking protons noting that the electric fields of the protons and electrons exist all the time) and therefore, we speak of a "net" field that is very tiny with no significant dipoles, as though the two fields are engaged in a tug-of-war competition (both teams alive and existing though the rope barely moves!) and that cause potentials in a conductor to be the same throughout their volumes, and not the individual potentials of the electrons and lattice ions themselves.

How is it that charged conductors (isolated) and current carrying wires and resistors have a potential on the surface as well as the interior if there are no excess charges (being neutral) in the interior but only on their surfaces?

Imagine you wish to lift a rubber ball from the ground. When you grip the ball between your fingers and thumb, a force that is normal to the surface is applied, derived from the fundamental electromagnetic force of repulsion between the electrons of the surface atoms of the skin and the electrons of the surface atoms of the ball. This force squeezes the ball penetrating through the rigidity of all the atoms of the ball. The squeezing force cannot lift the ball because it acts perpendicular to the force of gravity. However, *frictional* forces (again derived from the fundamental electromagnetic force—see Appendix D for the origin of such forces) between the surfaces, then arise and, oppose the weight of the ball, the gravitational force that tends to pull it to the ground. The gravitational force is balanced by the frictional force which is a fraction (called the coefficient of friction) times the normal force due to squeezing. Finally, this balance of forces permits you to lift the object above the ground.

The absence of friction and accompanying frictional forces is what makes it difficult to pick up smooth and hard objects such as ice cubes.

The ball you wish to lift from the ground could have been made from metal and none of the processes would have changed excepting that the forces acting and balancing would be more powerful. After lifting the ball, you would not say, you have lifted only the surface contact atoms of the ball above the ground and that only these have gained in potential energy of the system (of ball and ground). The (entire) ball (its atoms held together by its rigidity) and ground system gained potential energy.

Analogously, when a conductor is electrified by surface charges, it is the equipotential regions of the conductor with its millions of electrons that gives the whole conductor (interior *and* on the surface) the quality of potential. You would not say "only the surface" of the conductor has attained a potential.

This potential is manifested in the electric field surrounding the conductor, while the *resultant* electric field in the interior is *zero* due to the geometry of the arrangement of the surface charges that moved, which is a *natural process over which we can exercise little control.* In the interior the potential is the same as on the surface because the region is equipotential with zero resultant field; a test charge that may be pushed from the outside on the surface of the conductor needs no force to move around in the interior because there is no resultant field there, maintained by the millions of free electrons moving inside.

The electric field around the conductor extends its reach to the entire universe, seeking to induce and *pull opposite* charges towards it and *push like* charges away from it so that it may reach levels of potential energy tending towards zero, whenever possible.

If the charged conductor is isolated, the entire volume of the conductor is equipotential and if it is a current-carrying wire or a resistor in a circuit with a potential difference (see Figs. 2.7, 2.53(b) and 4.16(b)), it has *planes* (*or slices*) of equipotential regions with the potential varying along its length. The slices of equipotential regions are maintained by the millions and millions of electrons within the conductor where they are in abundant supply. The uniform applied field the surface charges impose (which are few in number in comparison with the number of free electrons in the interior), produces the varying potential because voltage (along the length) is E times distance. Significant dipoles are not formed in the *interior* of a conductor; whether it is isolated or steady-state current-carrying, due to an applied electric field. In current-carrying conductors, the maintenance of the equipotential regions is due to the continuous feedback mechanism between the current and the surface charges, as we saw in Section 1.15, Section 1.18 "Neutrality in current carrying conductors" and in Section 1.23 "Field strength inside the wires and the conservation of current".

The fields that are applied at voltages in most low voltage and electronic circuits **shift the outermost electrons and hardly influence the inner ones.** This behaviour makes capacitors store energy*; packing all displaced electrons on one of its plates.*

Returning to the RC circuit to which a voltage pulse is applied, if the input is held at *V* volts for a long time, then, treating the circuit as a closed system (see Section 1.8 "Conservation of charge"), plate *B* attains zero volt with respect to the circuit ground (assumed kept *unearthed*) *and* potential –V/2 volts with reference to a point at infinity (called "potential" by convention when referred to points at infinity) because it is holding a huge bunch of negative charges. See Fig. 2.14. If the circuit ground were perfectly earthed, the plate *B* will attain "zero" potential if the input is held at *V* volts for a long time.

There will be very few or no surface charges on the wire between the resistor and the capacitor plate *B*, because the presence of charges on plate *B* has effectively cancelled the driving field of the pulse generator; the net field is zero. Compare the charges and field on this section of the wire with Fig. 2.66 (during charging) and Fig. 2.15 (after charging). Electric field flux lines that terminate on the negative charges on plate *B* originate on the positive charges on plate *A*.

The charge configurations (excess surface and the bound) everywhere in the circuit determine the potentials near and on conductors and the electric field orientations near the conductors. In conductors and in wires, the potentials have to be same everywhere within; the potential will be different across low conductivity regions (resistive portions) when the charges are moving; the electric field flux lines emanating from conductive portions may be inward, outward, thin or thick at different points in one conductor depending on the charge configurations at a particular instant of time.

Putting all the above fragments together, I find the following a suitable way to visualize the potentials, their formation and destruction for the circuit of Fig. 2.15: When the pulse voltage is high, treating the (physical) centre of the pulse generator (with the circuit ground *unearthed*) and as a "very close" to zero potential, imagine two mountains of concrete (representing charges separated by the pulse generator) stacked one above the other are constructed; *V*/2 volts above zero and *V*/2 volts below zero.This is shown in Fig. 2.70(a). Since the negative terminal of the pulse generator is considered to be the reference or circuit zero volt; the generator mountain is shifted above circuit zero as shown in Fig. 2.70(b).

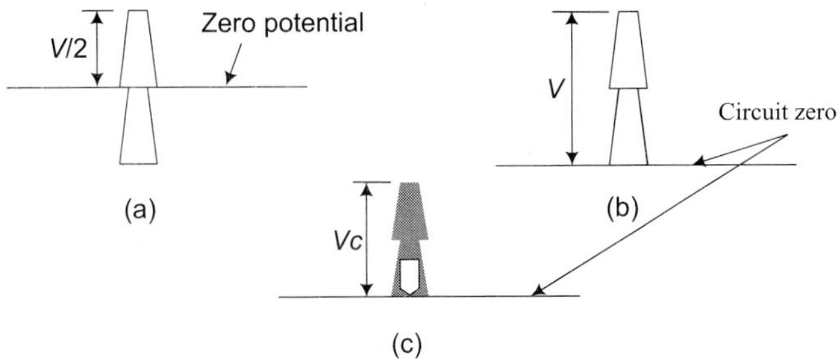

Fig. 2.70 Pulse and capacitor voltage representations for the circuit (Fig. 2.15).

While charging the capacitor, the pulse generator has created two mini-mountains stacked one above the other; one of height $Vc/2$ above zero potential (if the circuit is a neutral object as in Fig. 2.59) and one of height $Vc/2$ below zero potential upto the time t (see Fig. 2.66) to represent the voltage across the capacitor. If one were referring to the figures below Fig. 2.20 with $RC << t$, then $Vc = V$.

The mini mountain representing the voltage across the capacitor may be visualized as made from a mixture of sand and water with a spade-like base; this is indicative of the fragility in its construction that can be easily dissolved by a small wave, as on a beach (because it has no internal source of energy that would maintain its rigidity and an ever-present fringe field ready to discharge it, should it find a path) and the ability to use the spade to sink below zero volts as shown in Fig. 2.70(c). Again, since the negative terminal of the generator is considered a circuit zero volt the base of the lower mountain of the pair is at circuit zero as shown in Fig. 2.70(c).

The visualization using mountains is useful, though not perfect and the reader should refrain from developing analogies in other areas because negative particles (and negative potential energies of electrons near positive ions) are observed only in electricity. See Section 1.13 "Structure of electrical conductors" and Section 2.22 "Potential energy of electric fields".

Circuit references if earthed in addtition to metallic chassis to meet safety regulations, could become critical in industrial environments where several electronic equipment are in operation with long signal lines carrying analog and digital data.

Earthing can result in shifting voltage reference levels caused when currents flow in the finite resistance earth wires when there is a fault (such as insulation failure between live and neutral that is earthed at the power-substation transformer) or, due to other unknown causes.

These shifting reference levels not only provide inaccurate representations of measured parameters in analog/digital signal forms, also, if operations are controlled using these signal forms, unintended behaviour of equipment may result in threats to safety of personnel.

Sometimes, the control equipment earth and power earth are separated to prevent strong earth fault currents from interfering with signal integrity. It is always advised to follow equipment manufacturers' recommendations regarding earthing of equipment.

Inserting a Dielectric Slab Between the Plates of an Air-Filled Capacitor

This topic was deliberately deferred to the end of the chapter because to understand what happens when a dielectric or insulator slab is inserted between the plates of an air-filled capacitor requires some notion about the average field inside matter that was described previously. To increase the quantity of charge and therefore energy stored in a capacitor without increasing its size unduly, it is common practice to introduce a dielectric slab between its plates. (Chabay & Sherwood, 1995, used by permission).[2]

We consider that a plastic insulating slab is inserted between the metallic plates of a charged air-filled capacitor and this is connected to a voltage source. The field applied by the external charges polarizes the molecules in the plastic. The resulting molecular dipoles produce an average field inside the insulator that is in the opposite direction to the applied field as shown in the figure below. The net field inside the plastic (insulator) is weakened, as described next.

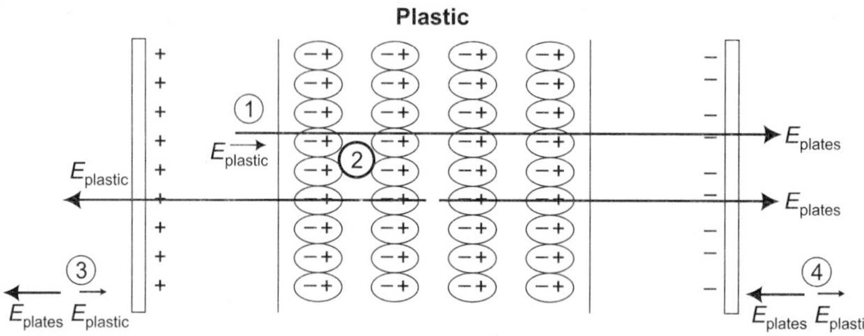

An initial glance would seem to indicate that the field at the location 1 should be more powerful than without the insulator and this is rightly so, due to the presence of the field of the polarized plastic as indicated in the figure. It is not increased much though because, the field due to the plastic is small when compared to E_{plates}. At points 3 and 4, the net field is significantly weakened by the plastic, as can be seen from the field vectors drawn at these points due to the plates and due to the plastic.

At point 2, at the molecular level, the electric field inside the plastic varies a lot in magnitude and direction, because the electric field of a dipole is complicated. So what is shown is an average field, averaged over many (millions) molecules, which is the appropriate field to use in calculating the potential difference from one side of the plastic to the other.

The direction of the field at point 2 which is *inside* the plastic, is shown pointing to the left. How do we know this? The electric field pattern of each molecular dipole is quite complicated, and it is not obvious how to take the right average over a large number of dipoles. To determine this, we use a method of counterargument.

We do know the field at some points due to the polarized plastic as shown in the figure below.

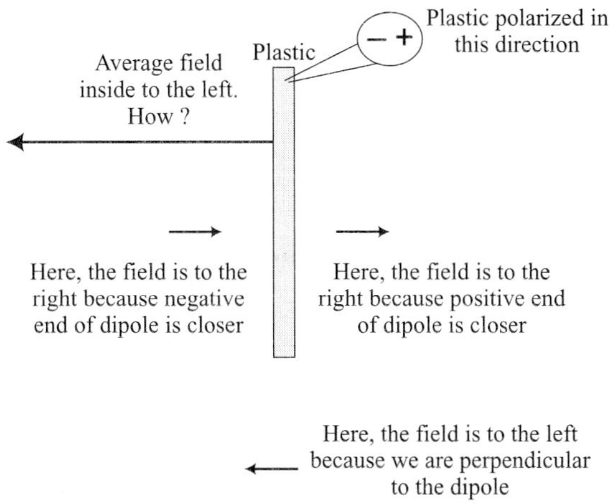

Consider the electric field *just due to the polarized plastic* along this oval path.

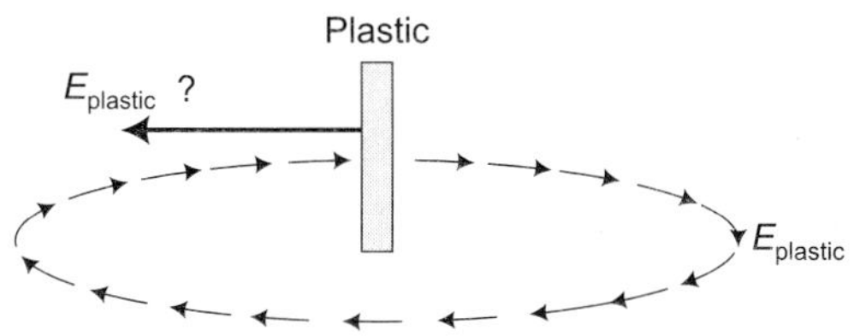

Since the molecular dipoles can be considered to be point charges, the path integral of the electric field due to the molecular dipoles *must* sum to zero. If the average electric field E pointed to the right inside the plastic, then $-\int E.dl \neq 0$ along the round trip path since there are no negative contributions. This is because point charges cannot produce this pattern of electric field.

Therefore, E must have a component opposite to dl somewhere along the path for the integral to evaluate to zero. This is possible only if the average field direction inside the plastic is to the left.

If a plastic slab were to be introduced into a charged air-filled capacitor that has been disconnected from the voltage source, the energy after the plastic slab is introduced is less by a factor $1/k$ where k is the dielectric constant of the plastic. The voltage across the capacitor goes down. The *missing* energy would be felt by the person who inserted the slab, who would feel a *tug* and would have had to restrain it if he had to insert it without acceleration.

The force of attraction arises due to the free charge on the plates and the induced surface charges that appear on the slab. The charged air-filled capacitor expended some energy, the *missing* energy, to produce the charges on the slab by charge separation or induction.

Evidently, if an uncharged air-filled capacitor had a plastic or dielectric slab introduced and were to be charged by a voltage source, the source will need to pump up more charge with the dielectric in place to attain the same voltage. In effect, the capacitance of the capacitor has been increased and will be k (dielectric constant) times the capacitance of the air-filled capacitor if the dielectric fills the capacitor.

I end the section with a picture of a charged capacitor showing its fields between the plates and the fringe field (courtesy www.falstad.com) and the centre zero potential line. The top plate is positively charged with respect to the bottom plate (see picture below).

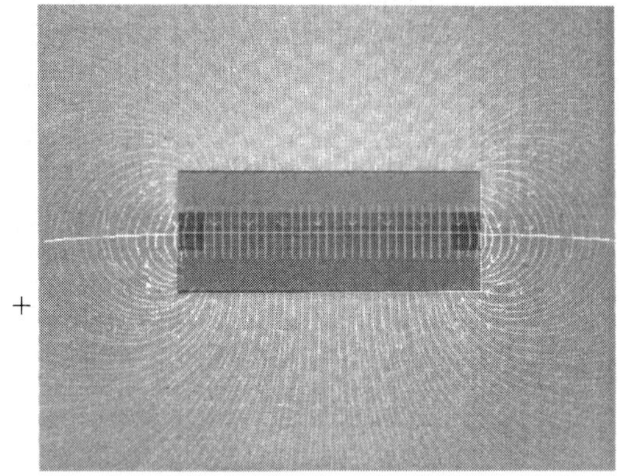

In the next section, we introduce magnetism. This is with a view to explain the concept of electromagnetic energy and how it is transported that will be dealt with in Chapter 4.

2.27 INTRODUCTION TO MAGNETISM

Careful experimentation shows that a moving point charge not only makes an electric field E but also makes a magnetic field B.

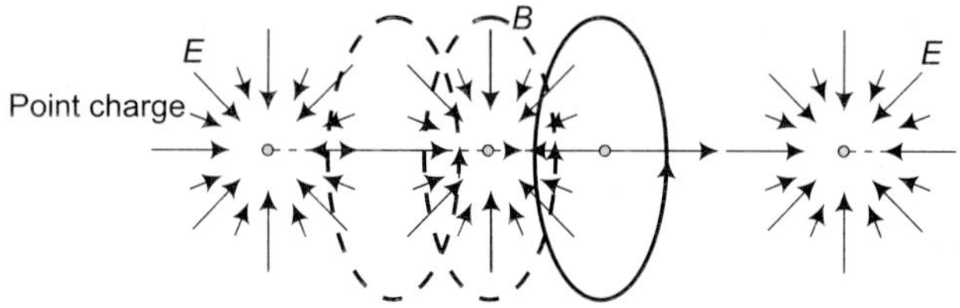

The magnetic field of a single moving point charge is not static in space.

We see that a region of space is *altered* invisibly when an electric charge is introduced, which we call an electric field.

If the charge is *moving*, the region of space surrounding the charge is *further altered invisibly*, and we call this a *magnetic field*.

Magnetic Interactions

Magnetic effects are associated solely with *moving charges* and, *currents in circuits*

...whereas *electric effects* are evident with both *stationary and moving* charges

Magnetic fields...

penetrate through matter.

Magnetic fields ***affect*** moving charges.

The magnetic force in a charge $\boldsymbol{q}$, moving with velocity $\boldsymbol{v}$ in a magnetic field $\boldsymbol{B}$, is

$$\vec{F}_{mag} = q(\vec{v} \times \vec{B})$$

This is known as the "*Lorentz force*" law.

In the presence of both electric and magnetic fields, the net force on q would be

$$\vec{F}_{mag} = q[\vec{E} + (\vec{v} \times \vec{B})]$$

Questions

1. How does a capacitor discharge? Describe with the help of neat sketches.
2. A capacitor is discharged using a wire with a bend in it. Describe the processes that enable the capacitor to discharge and attain static equilibrium with the help of neat sketches.
3. What is meant by "charging" a capacitor? A capacitor is charged with the help of a battery and a resistor and a switch that is closed at a time $t = 0$. Describe the processes that enable charging until it is fully charged and sketch the voltage across the resistor at suitable intervals. Explain the reason for the voltage spike at the time $t = 0$.
4. What is meant by "quasi steady-state" when charging a capacitor? Explain with the help of a simple capacitor charging circuit.
5. What is the fringe field of a capacitor? How does it develop? Derive an expression for the fringe field outside a capacitor (near one of its connecting leads) constructed of two disks of radius R separated by a distance 's' with respect to the field between its plates.

   Indicate typical values of fringe field common in electrical and electronic circuits and compare this with the applied voltage fields again in common electrical and electronic circuits.
6. The expression $J = \sigma E_{NET}$ where J is the current density, σ is the conductivity and E_{NET} is the electric field is a very fundamental relation. Derive the expression from fundamental principles.

7. When a capacitor of capacitance C is charged using a battery of voltage V and a switch, the battery expends an amount of energy QV while the energy stored in the capacitor (after charging) is $QV/2$. Explain where the missing energy is.
8. Assume we apply a voltage pulse input to a capacitor–resistor circuit and that the capacitor is reasonably large. With the help of neat sketches of the input and output voltage (across the resistor), describe the processes with emphasis on the role of the fringe field that gave the output voltage waveform the shape you have drawn.
9. The current in a capacitor circuit leads the input by 90 degrees. Describe with the help of neat sketches the processes that maintain this lead angle in the steady-state. Is it possible to achieve 90 degrees lead angle between the current and the voltage accurately in a typical capacitor circuit. Explain giving reasons.
10. Describe the effects of varying the frequency of a sinusoidal voltage to a capacitor with respect to the current.
11. A capacitor resistor combination is subject to a sinusoidal voltage. With neat sketches explain how the current leads the voltage by an angle that is less than 90 degrees. If the frequency is varied, what effect does this have on the current and the phase with respect to the voltage?
12. What are surface charges really? How do these enable a circuit to split the current in parallel branches? Explain with the help of neat sketches.
13. What are conservative fields and non-conservative fields? Explain with reference to their curl and divergence. Can a non-conservative field be produced using charged objects?
14. With the help of a simple circuit consisting of a battery, resistor and wires, explain what the voltage is due to and state where it exists. What is the nature and strength of the electric field inside the wires and in the space surrounding the wires? Are they uniform?
15. How much force is normally applied by a hammer and chisel? How much force is normally applied when we lean against a wall? Do the atoms of voltmeter probes touch the atoms of wires on which they are pressed to measure the voltage? Explain mentioning the values of nominal force/stress involved at the atomic level.
16. Can a capacitance be associated with a single conductor? Explain with sketches and obtain the capacitance of the earth assuming that the radius of the earth is 6400 km.
17. Can there be a voltage or potential for a single wire connected to a battery? Explain assuming 3 cases; when the battery has no net charge and when the battery has a net charge and when the excess charge on the battery has been neutralized.
18. Explain with a neat sketch how *large regions* within a metal crystal which the conduction band electrons occupy are maintained equipotential despite the huge potentials near individual protons and electrons. What is the "average field" within conductors and where is this field?
19. Explain how the potential is the same as the surface in the interiors of charged isolated conductors and current-carrying wires and resistors with a potential difference.

20. Explain what is meant by "grounding" and the difference between "signal ground" or circuit zero and a circuit ground that has been earthed.
21. Assume we apply a voltage pulse input to a capacitor –resistor circuit (of a small RC time constant) and that the resistor connected to the generator ground is circuit ground. The output voltage is picked across the resistor. Explain how a negative spike voltage is observed at the output corresponding to the trailing edge of the pulse input.
22. Explain how the electric field inside the matter may be considered to be an average field. How does the average field of insulating slabs inserted between the plates of an air filled capacitor increase its energy storage capacity of capacitors? Explain with the help of neat sketches.
23. Will a capacitor whose leads are brought out from the inner surface of its plates, charge when connected with a current limiting resistor to a dc voltage, despite the fact that a strong field that penetrates through matter will come into existence as the capacitor charges? Describe the process and compare with the process of charging a capacitor whose leads are brought out from the external surfaces of its plates.

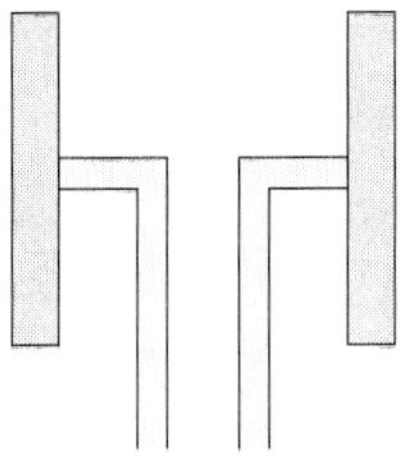

Hint: The round-trip path integral of a conservative electric field (due to a distribution of point charges) should always be "zero"

3

Superposition of ac Signals on dc Voltages; Coupling of ac Signal Sources to Amplifier Inputs

Chapter 3 was developed with a view to describe how superposition of ac signals on dc voltages is achieved. In this I have brought in the role that surface charges and fields play in developing the output voltage of a capacitive coupling network. Beginning with a description of superposition of dc voltages in simple capacitive circuits, the chapter progresses to describe how ac signals are coupled to transistor amplifier inputs.

3.1 SINUSOIDAL VOLTAGE APPLIED TO A CHARGED CAPACITOR AND RESISTOR

When studying amplifier circuits, we use circuits that require the superposition of a small sinusoidal voltage on a steady dc voltage that biases transistors in the active region.

How is this achieved?

Consider a capacitor-resistor circuit in which the capacitor was previously **charged** to a voltage of say, **2 V dc, by an external agent.**

A sinusoidal voltage $v(t) = 0.5 \sin(\omega t)$ with a peak $Vm = 0.5$ volts is applied to the CR combination as shown in Fig. 3.1.

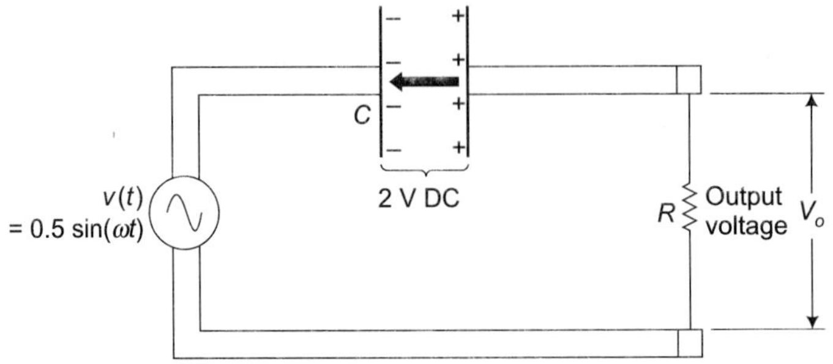

Fig. 3.1 Sinusoidal voltage applied to a charged capacitor and resistor.

Before the sinusoid voltage generator is connected, surface charges appear—positive charges on the wire connected to the resistor *R* and negative charges on the wire connected to the negative plate of the capacitor. The charges form a tight net all over the circuit and there is static equilibrium and *no current* because there is no signal generator as shown in Fig. 3.2.

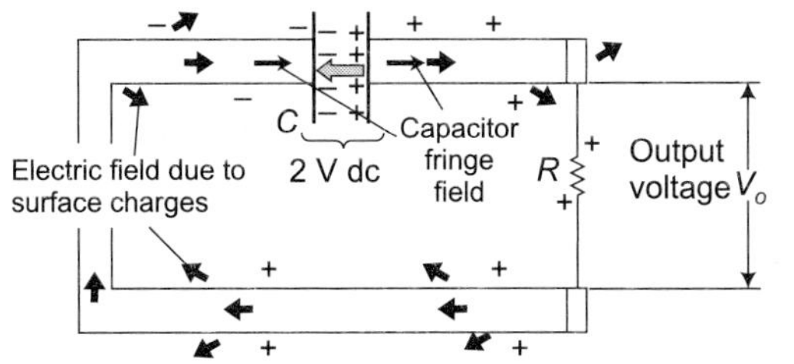

Fig. 3.2 A charged capacitor and resistor (before a sinusoidal voltage is applied).

Now, when the positive half of the sinusoid is applied, positive surface charges move on the wire to the negative plate of the capacitor and negative surface charges move to the wire connected to the resistor and a new configuration of the electric field results as shown in Fig. 3.3.

The electric field (that can penetrate matter) due to the signal penetrates through the capacitor field and adds to the fringe field (remember this is *s*/2R times the field between the plates as was described in Section 2.2).

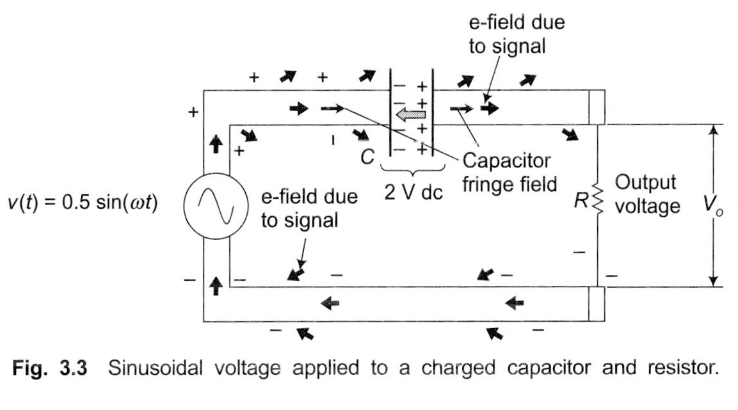

Fig. 3.3 Sinusoidal voltage applied to a charged capacitor and resistor.

Note that the net field in the wire from the capacitor to the resistor *R* is now signal e-field + fringe field.

The resulting large e-field causes surface charges to increase in the wire connected to *R*. The output voltage that results is shown on the right-hand side in the Fig. 3.4.

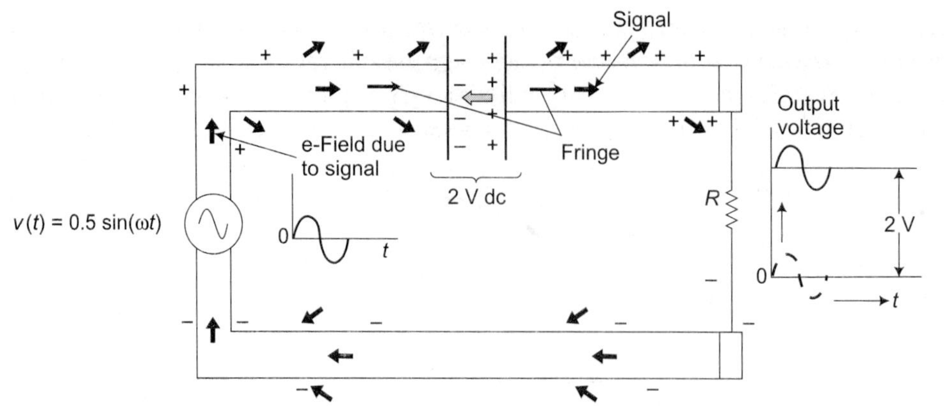

Fig. 3.4 Output voltage when a sinusoidal voltage is applied to a charged capacitor and resistor.

What is the Circuit attempting?

To understand the behaviour of the circuit to a complete sinusoid, let us for a moment assume that the signal is very tiny, or better close to zero or even better at zero voltage but before we do that let us remove the source for a moment and assume the capacitor charged to 2V with polarity as in Fig. 3.2.

Charges will **move** from the positive plate onto the wire and negative charges from the negative plate onto the wire **before** the voltage is made zero.

Now, the voltage is made zero...See Fig. 3.5a.

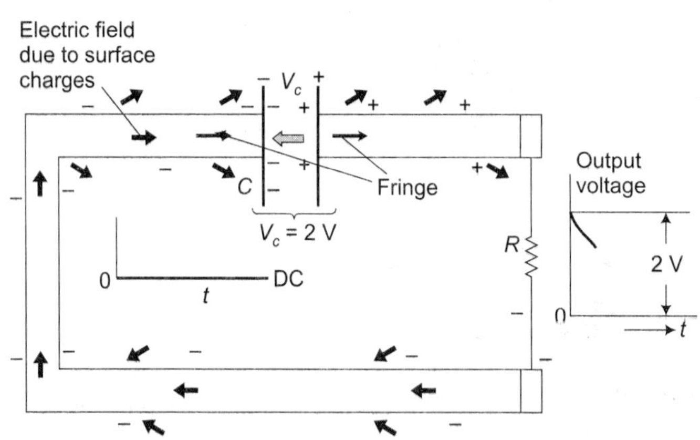

Fig. 3.5a The discharge of a charged capacitor.

Observe that the correct way to depict zero voltage is NOT to open-circuit but to short-circuit the signal source.

The surface charges and electric field they set up cause a discharging current **inside the wires and resistance.** The output voltage decreases exponentially towards zero, as the capacitor discharges and the resistance heats up. Finally, all the charges in the capacitor will be neutralized and the output voltage reaches zero and the fields reduced to zero everywhere as shown in Fig. 3.5b.

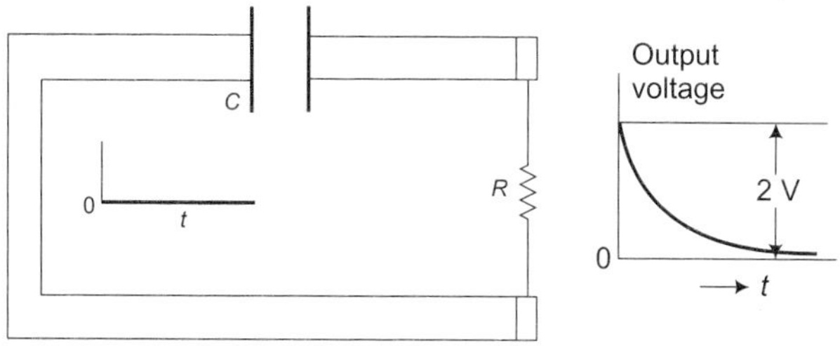

Fig. 3.5b The discharge of a charged capacitor—after a long elapsed time.

Next, we consider the application of a tiny 0.2 V to the capacitor charged to 2V and let us make it steady, say, a dc voltage.

Positive surface charges move out of the positive terminal of the voltage source and negative surface charges move out of the negative terminal, according to rule. (See Section 1.15)

The net electric field zooms up everywhere in the wires and the resistor and the voltage across the resistor rises not to 2 volts but to 2.2 volts—the capacitor has not begun discharging yet. See Fig. 3.6.

The effect of the addition of the source voltage (0.2 V) is felt around the circuit by a **zip** in the electric field (not forgetting that an electric field is an invisible *alteration* of space). The capacitor will have to discharge first before it assumes the voltage of the source and reach static equilibrium. This is what the circuit is attempting.

A dc voltage applied to a charged capacitor and a resistor

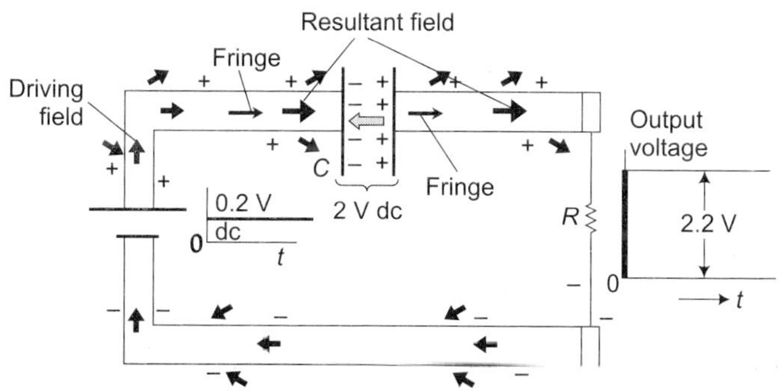

Fig. 3.6 The discharge of a charged capacitor.

What is the Discharge path of the capacitor?

The discharge path is *through the resistance and through the voltage source*!

Don't forget that the voltage source maybe a battery, which is two conducting plates separated in a container containing an electrolyte—and the chemical action had separated charges—negative charges on the negative plate and positive charges on the positive plate.

How can a voltage source close the circuit?

When positive charges move from the wire from the capacitor to the negative terminal of the battery, then charge depletion takes place on the negative terminal. The *duty of the battery* is that when such a charge depletion takes place, a *chemical action kicks in* inside the battery to restore the lost charges on its plate. Let us see the mechanism closely, because it is important how a voltage *source closes the circuit,* for that matter, any circuit.

Let us see how the battery operates to close the circuit for discharge.

For each and every charge that leaves the capacitor, a corresponding one enters the battery terminal...that is the real movement...but, that does not diminish in any way our depiction of the movement. The Fig. 3.7 exaggerates the movement of positive charge from the capacitor plate towards the negative terminal. In reality it is not the same positive charge that left the capacitor plate but another positive charge in the wire close to the battery negative terminal.

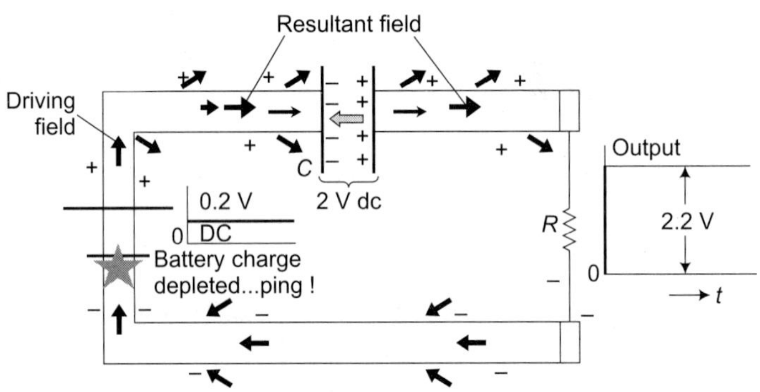

Fig. 3.7 How the battery closes the circuit.

As positive charges move from the capacitor, **simultaneously,** negative charges will have moved to the positive terminal of the battery—**depleting the charge** there.

The battery senses the change because its internal electric field is down, and operates now.

Its chemical action produces a charge pair. Remember that is what a battery is *capable* of—charge separation.

The individual charges move towards the respective plates. This is shown in Fig. 3.8.

The battery maintains the voltage across its terminals by this chemical action that replenishes depleted charge.

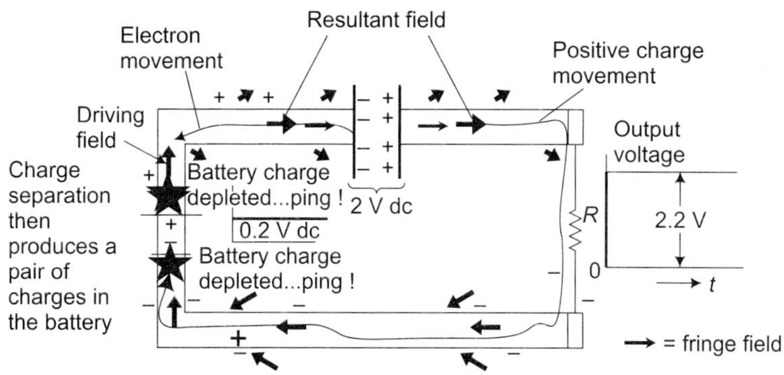

Fig. 3.8 How the battery closes the circuit (contd.).

Let us look at the movement closely, with all the frills removed—the surface charges, the fields. We show the resultant field (that is a combination of the driving field and the fringe field) only.

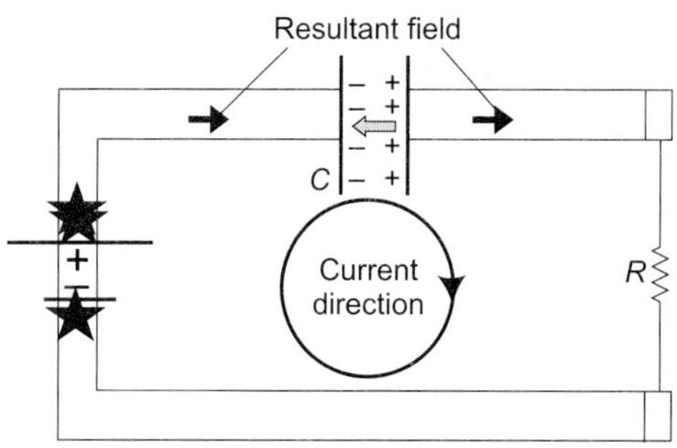

Looking closely, what we see is that there is an upward movement of positive charge, and a corresponding downward movement of negative charge inside the battery—both of which correspond to a *conventional current that is upward* and that is how the *circuit is completed.*

Of course, there cannot be any movement of charges in between the plates of the capacitor, but for the sake of completeness, we say that there is a *displacement current*, all fictitious, but good for bookkeeping.

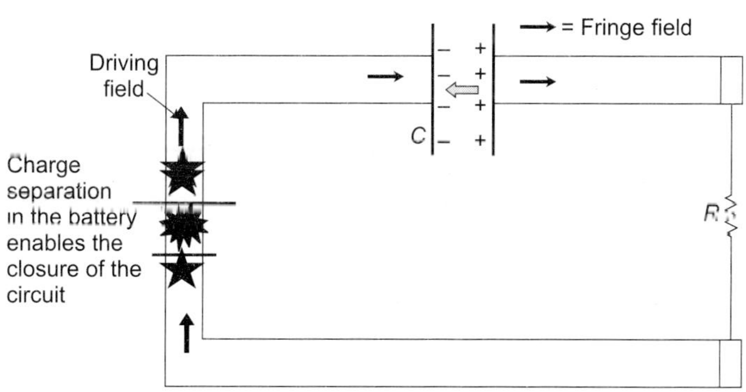

Do not forget that no charge movement is possible without the enabling surface charges with the accompanying electric fields inside the conductor and resistor.

The capacitor will discharge until the voltage across it, namely, V_c becomes 0 volt.

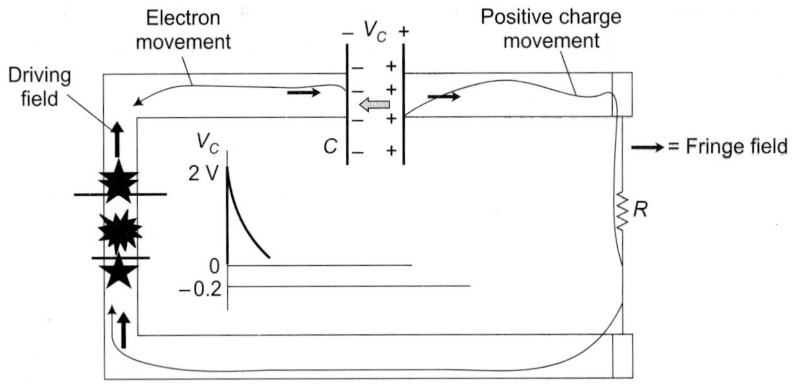

It will then *charge* until the voltage across it reaches 0.2 volt—the voltage of the source.

We briefly digress to describe the superposition of potential principle at work behind the scenes that enables us determine the voltage variation with time, at different points around the circuit.

Superposition of Potential

(Chabay & Sherwood, 1995, used by permission)

It is not difficult to extend the principle of superposition of electric field to prove the superposition of potential. There are two charges $+q_1$ and $-q_2$ located a distance apart and we need to compute the potential difference between two points A and B.

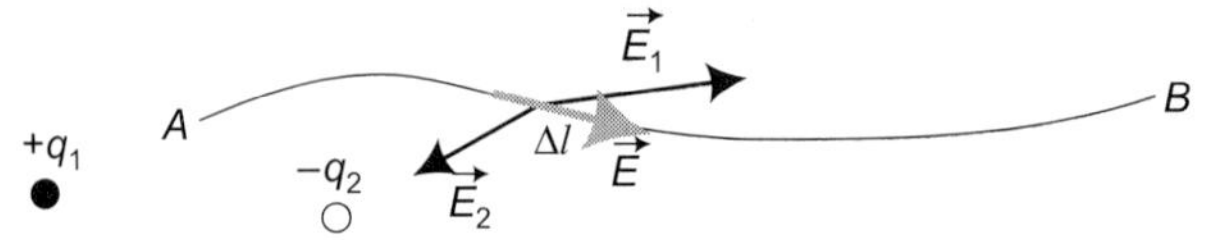

The key point is that the net electric field $\vec{E} = \vec{E}_1 + \vec{E}_2$, so potential difference for each small step of length Δl along the path from A to B is $-\vec{E} \bullet \overrightarrow{\Delta l} = -(\vec{E}_1 + \vec{E}_2) \bullet \overrightarrow{\Delta l}$.

Adding up the contributions

$$\Delta V = -\int_A^B E \bullet dl = -\int_A^B (\vec{E}_1 + \vec{E}_2) \bullet dl = -\int_A^B \vec{E}_1 \bullet dl - \int_A^B \vec{E}_2 \bullet dl.$$

The last two terms are the potential difference due to q_1 alone and the potential difference due to q_2 alone. Since electric field is the superposition of the electric field contributions of all the charges (Appendix A "Superposition of electric fields"), we see that potential difference is also the superposition of the potential difference contributions of all the charges. It is easier to add potentials than it is to add fields, because potential is a scalar, and so is potential difference, while electric field is a vector and so requires vector addition with the use of vector components.

In the circuit comprising the battery, capacitor and the resistor, we have been using the fields due to the charges on the battery, surface charges and the fields due to the charges on the capacitor plates, implicitly using the superposition of potential principle to graph the voltage change with time across the capacitor and resistor.

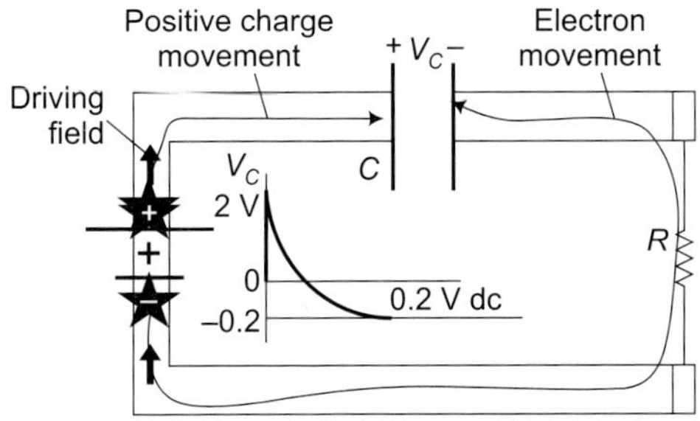

Returning to the circuit, the voltage of the capacitor 0.2 V is shown below the zero line in the graph of the capacitor voltage because we have defined V_c that way (see the polarity assumed for V_c in Fig. 3.5a). Also note the capacitor initial voltage is 2 V not 2.2 V which is the initial voltage of the output. The **output voltage** decreased **exponentially** from its initial 2.2 volts. The action is shown again in the figures below with the output voltage also plotted alongside for the sake of clarity.

As the voltage across the capacitor reaches zero volt, the voltage across the resistor will be 0.2 volt under the influence of the electric field due to the voltage source.

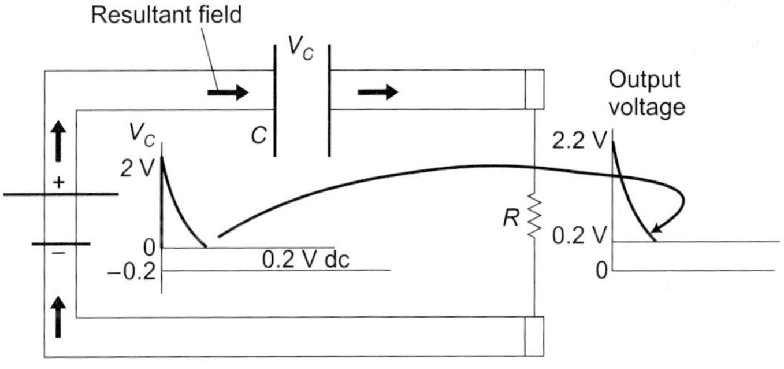

The voltage source now begins to charge the capacitor and V_c therefore, becomes negative.

The voltage source will charge it up until no further movement of charge is possible a state of **static equilibrium.**

The voltage across the resistance *R* will become zero...static equilibrium.

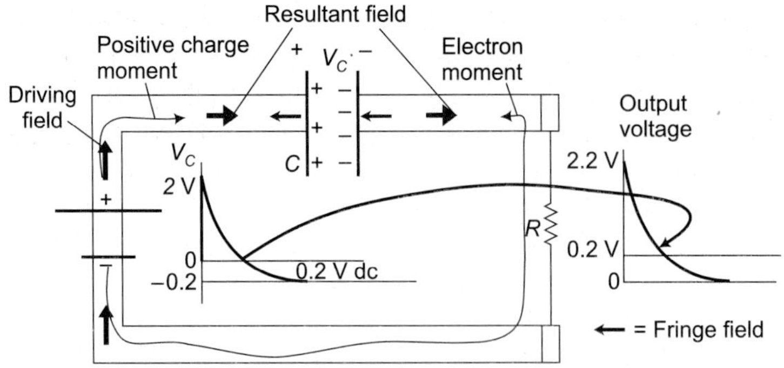

Did you notice that the **current never reversed** during the entire cycle. Yet discharge and charge were built into the unidirection movement.

In all the processes that we saw, only the change in the net field in the wires occurs at nearly the **speed of light** and, the charge transport (drifting velocity) and **t**he ***voltage change*** across the capacitor and the output voltage occur in measurable times ranging from nanoseconds to seconds.

That is to be expected, considering the ***huge*** number of charges, millions and millions that have to be ***moved against the rising opposing fringe fields due to charge accumulation.***

A Reminder

Do note that the capacitor does not have an internal mechanism to separate charges, it is just that simply, a passive device that may store energy but cannot produce it.

This is unlike a battery or a voltage source that have internal mechanisms to effect charge separation and maintain the voltage across them.

This also means that the *sizes* of the capacitor and the resistance will influence the voltage *decreasing times* across the capacitor and the output voltage. Because $Q = CV$, therefore for a fixed V, a small capacitor will have less charge than a large capacitor.

If RC is large then, it may be expected that huge amounts of charge have to be removed—so the time is prolonged—we say the time constant is large. Even if C were small, then a large R will again *slow things down*. This is depicted in Fig. 3.9.

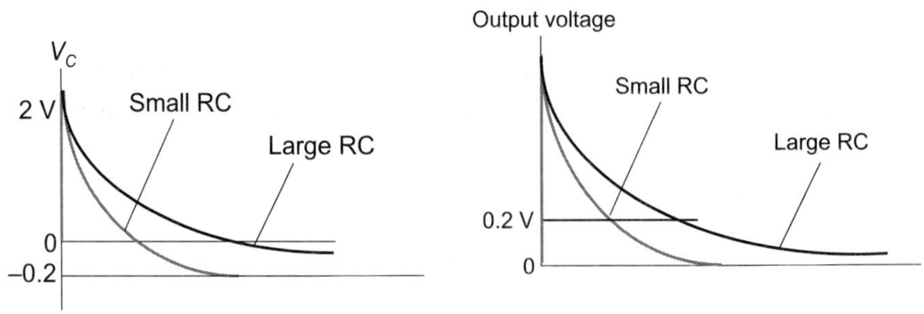

Fig. 3.9 How the sizes of the capacitor and the resistor influence the charging and discharging times.

Notice that the voltage source is depicted in the circuits with a shorting line across the supply terminals. This is merely to emphasize the fact that the circuit has a closed path.

In case the source is the winding of a transformer the winding may be considered a closed path as well and the induction process causes surface charges to appear to develop the so called induced emf or voltage.

In either case, the *depletion* of charges at their terminals *causes* the charge separation mechanism to kick in; chemical action in the case of batteries and induced emf action in the case of transformer windings—and the circuit is completed.

3.2 APPLICATION OF A SMALL SINUSOIDAL VOLTAGE TO A CHARGED CAPACITOR

In the light of what we have seen with the application of a small dc voltage to the CR combination, let us take another look at what happens when we apply a sinusoidal voltage with the capacitor charged externally (not shown to 2 V DC). This is shown in Fig. 3.10.

To understand intuitively the operation of the circuit, we depart from convention and show positive excursions of the voltage with a battery whose voltage varies sinusoidally and, negative excursions with the battery whose voltage varies sinusoidally, connected in reverse. The e-fields and the surface charges are not shown for the sake of clarity now, but they are present in the circuit to enable the processes.

Positive half of the input sinusoidal voltage

The movement of charges (discharging cycle) is clockwise as per convention during the positive cycle of the sinusoid. Also refer to Fig. 3.6 for a confirmation of the direction of charge movement where a small dc voltage of 0.2 V dc was applied to the circuit.

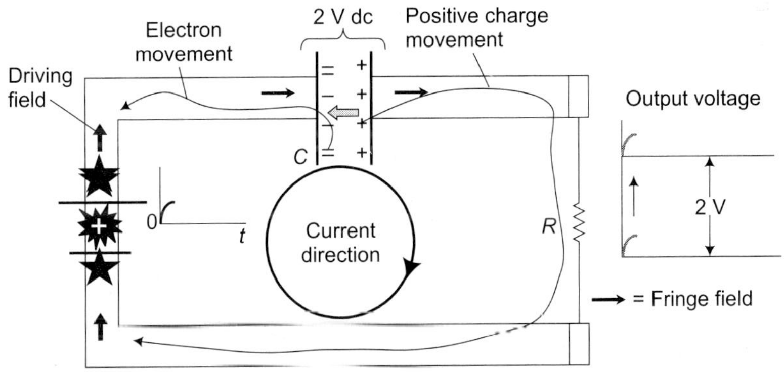

Fig. 3.10 The discharging cycle with a charged capacitor

We make one assumption though here, that the CR product is large, specifically, we make the capacitor large.

We could have as well made the resistance also large ....and that will not affect the subsequent operations as we shall show now.

The meaning of making CR large is that *not enough charges* have moved away from the capacitor plates during each cycle of the sinusoid that would cause the voltage across the capacitor to reduce to zero. At the end of the positive cycle, therefore, there are still charges, quite large, in fact, left on the capacitor plates.

Notice that the capacitor is still charged with a positive voltage. It lost only *some* of its charge while discharging during the positive cycle of the input. It is therefore at a voltage less than 2 V dc but still a little close to 2 V dc at the end of the first positive cycle. The large circle in the middle of the diagram is drawn to emphasize the fact that conventional current is clockwise in the circuit.

Accordingly, the output voltage is a sinusoid riding on the 2 V dc (nearly) as shown in Figs 3.10 and 3.11.

This will happen if the sinusoid is fast enough, or technically speaking RC>>T, the period of the input and further, the amplitude of the signal 0.5 volt is small in comparison with the initial voltage of the capacitor 2V dc so, its influence is small.

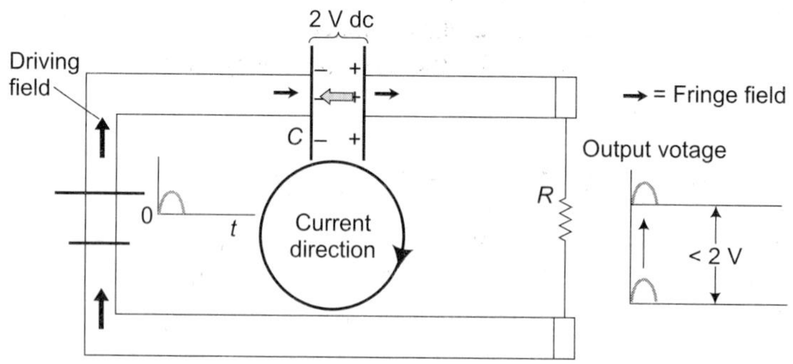

Fig. 3.11 The discharging cycle with the positive half sinusoid applied.

One can visualize the process of making CR large, by *imagining a tall and large diameter tank* connected from a tiny outlet near the bottom to a small pipeline, wherein the pipeline may have *small* flow fluctuations but the liquid in the large tank will always drain into the pipeline slowly.

Negative half of the input sinusoidal voltage

We reverse the battery connections, to simulate the negative cycle of the input as shown in Fig. 3.12 and remember this is the first cycle and that the input peak voltage is a mere 0.5 volt.

The driving electric field reverses direction. Notice that there is ***no change*** in the **directions** of the movement of the charges from the capacitor plates; the positive charges on the capacitor (which is less than 2 V and for simplicity we assume 1.8 V) but close to 2 V move towards the weakly positive battery terminal. And similarly, the negative charges (electrons) move towards the negative terminal.

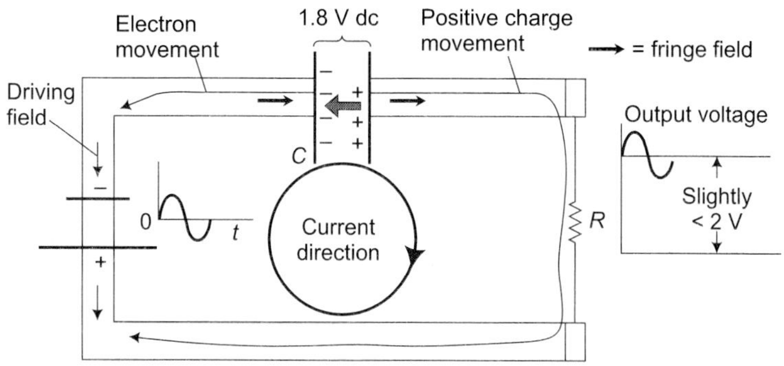

Fig. 3.12a The nagative input cycle.

We now enter into one of the *most crucial phases* in our understanding of the operation of the circuit.

During the negative cycle, notice that the current in the battery continues in the *same direction*. One way to visualize without cluttering the mind with visions of the chemical processes involved is to open a separate line inside the battery that keeps the charges in motion inside.

The charges continue through the battery. The battery acts as a *policeman* keeping a watch on its terminals and *separating or combining charges* as and when necessary to *maintain its emf* according to its nature.

During the negative cycle, there *will hardly* be any replenishment by the battery because the charges moving in from the capacitor plates have a sign that is the same as the sign of the charges on its (battery) terminals.

Most of the charges will use the bypass route and the *replenishing mechanism* is not called upon to do much.

But should there be movement at all from the capacitor plates to the battery terminals? Because it *appears* as if negative charges are moving from the negative plate of the capacitor towards the negative terminal of the battery, and similarly positive charges are moving from the positive plate of the capacitor towards the positive terminal of the battery.

And, the answer is

There will be a movement of charges and, these are moving no doubt from a positive to a positive potential and from a negative to a negative potential. But you have to remember that positive charges are moving from a higher positive potential (capacitor) towards a *smaller positive potential* (battery) and negative charges (electrons) from a more negative (capacitor) potential towards a *lesser negative potential* (battery). See the resultant field and its direction in the wires, as shown in Fig. 3.12b

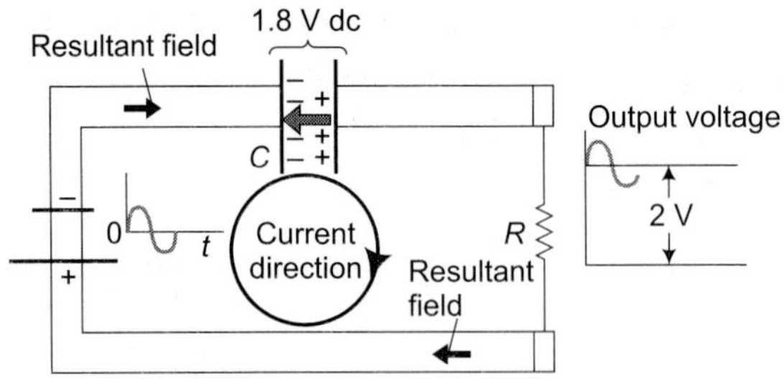

Fig. 3.12b Negative input cycle – resultant field.

Let us study the action one more time during the negative cycle of the input sinusoid. Notice how the *opposite charges cancel each other inside the battery, and register as a current.*

The capacitor voltage is shown reduced, say, 1.8 V dc, that reflects the small amount of charge that was lost during the positive cycle. This is shown in Fig. 3.13.

When necessary, a few charges on the battery terminals may cancel each other as well to provide the sinusoidal input voltage condition. This is a valid visualization after all, because we are using a battery to simulate the sinusoidal voltage.

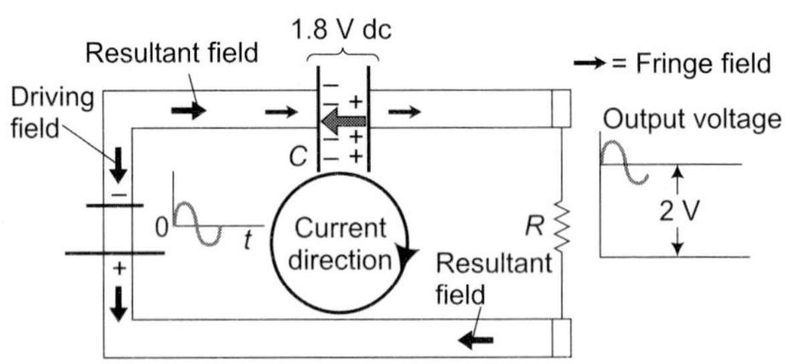

Fig. 3.13 A second look at the negative input cycle.

Notice that the current did not change direction during the negative cycle. This was on account of two reasons.

First, the small signal input voltage that was assumed, a valid one because, in most amplifiers where such circuits are employed, the sinusoidal input voltage is rarely greater than the biasing voltage and second, the capability of the battery or sinusoidal voltage source to permit current in both directions, there would be an open circuit and the circuit will come to a halt.

There is yet one more **nagging question** that needs to be answered.

How does the voltage vary at the output? **How is the signal voltage conveyed across the capacitor?** Is not the capacitor voltage influencing the voltage on the resistor?

Yes, and as the input voltage varies, the *electric field varies* in the wires connected to the capacitor plates.

These in turn cause a change in the surface charge conditions on the wires and the *potential varies* accordingly.

We will **combine these actions** to understand the operation—first few cycles.

Notice that it is the electric fields and surface charges combination that enabled the voltage at the output (across the resistor) to vary.

The actions are shown in Figs. 3.14.

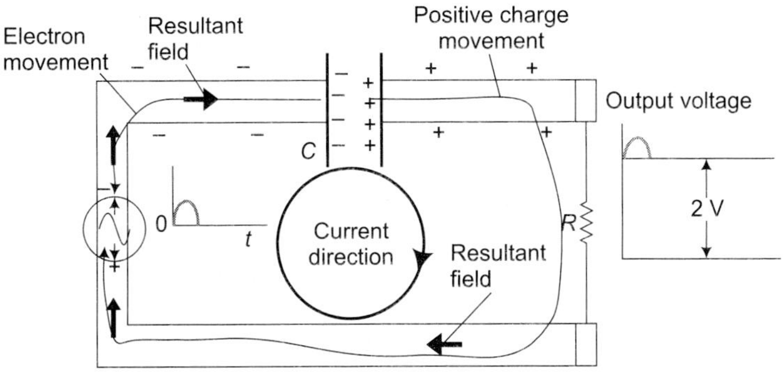

Fig. 3.14a The positive input cycle.

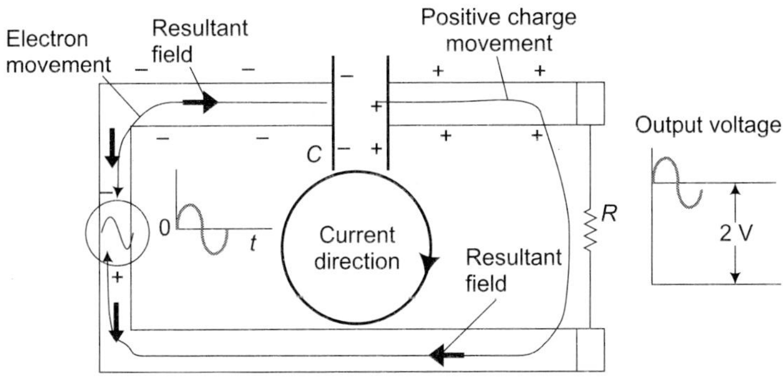

Fig. 3.14b The negative input cycle.

The output is a sinusoid *superimposed* on the 2 V dc steady voltage that *progressively decays* as more number of cycles pass.

The capacitor will be *discharged* in a *few cycles* and the few charges that may be present on it in every cycle cancel out each other during the *alternating* positive and negative cycles.

This is shown in the Figs. 3.15 and 3.16.

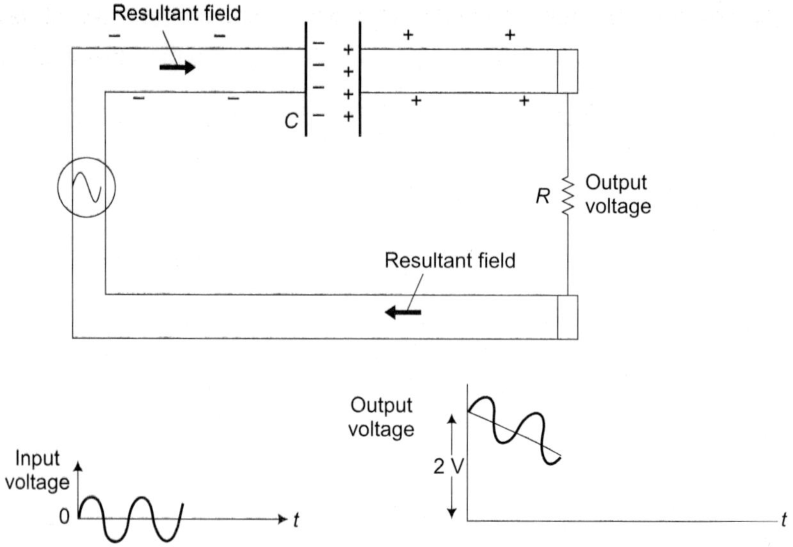

Fig. 3.15 The first few cycles.

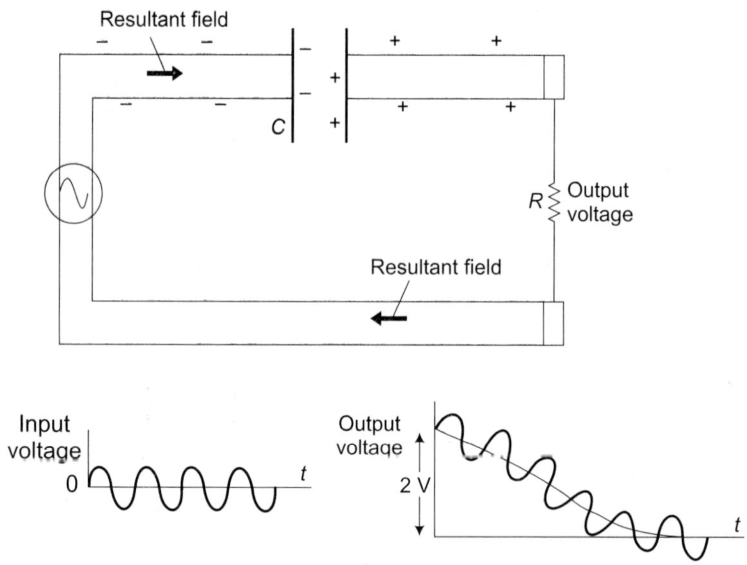

Fig. 3.16 The first few cycles. (Contd.)

After several cycles have elapsed, the capacitor is *discharged* and the current will be clockwise during the positive and counterclockwise during the negative half of every cycle.

The output voltage is now a *faithful reproduction* of the input under the conditions that the RC time constant is *large* compared with the period T of the sinusoidal input. This is shown in Fig. 3.17.

The first few cycles are shown compressed in the figure for the sake of clarity.

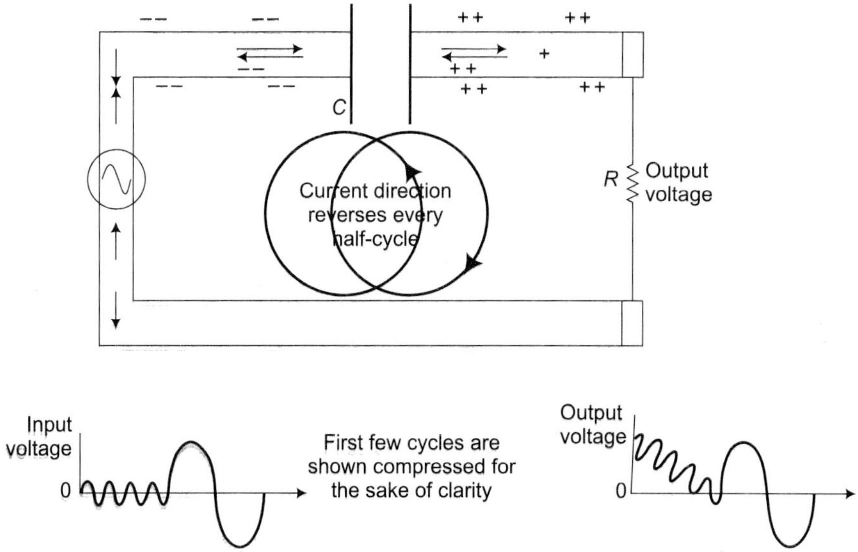

Fig. 3.17 After several cycles have elapsed.

It must be borne in mind that the current is now (after several cycles have elapsed and the capacitor is discharged) *forward* during the positive cycle and *reverse* during the negative cycle. A screenshot of the results generated by a simulated run of a CR circuit with a charged capacitor and a sinusoidal generator with TINA TI V9 software (special complementary basic edition) is included in the power point slides of Chapter 3 in the CD with this book.

3.3 SINUSOIDAL VOLTAGE SUPERIMPOSED ON A DC VOLTAGE

We are now ready to develop a circuit that can superimpose a sinusoid on a dc voltage.

Such a circuit will *maintain the operating point* of a transistor in the *active region* permitting the input which is a sinusoid to be *superimposed* on a dc voltage and be amplified and provide power gain with the ability to drive loads such as loudspeakers.

The matter rests on being *able to somehow maintain the voltage across the resistor R* of the CR combination.

We simply incorporate the resistor in a *divider* arrangement from an available regulated power supply or a battery using one more resistor.

We maintain the requirements of a $0.5 \sin \omega t$ sinusoid to be superimposed on a 2 V dc steady voltage.

The circuit comprising two resistors R1 and R2 and a voltage source (a battery) are used as shown in Fig. 3.18.

The resistors R1 and R2 form a voltage divider and their values are computed so that the *voltage across R2 is 2 V dc.*

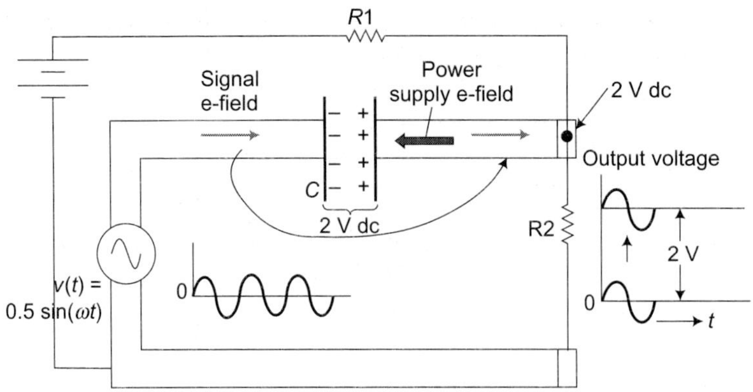

Fig. 3.18 Circuit that will maintain the voltage across the capacitor.

During the *first few cycles*, the capacitor is forced to charge up to 2 V dc by the voltage on R2 carrying the input sinusoidal voltage along with it. This is shown in Fig. 3.19.

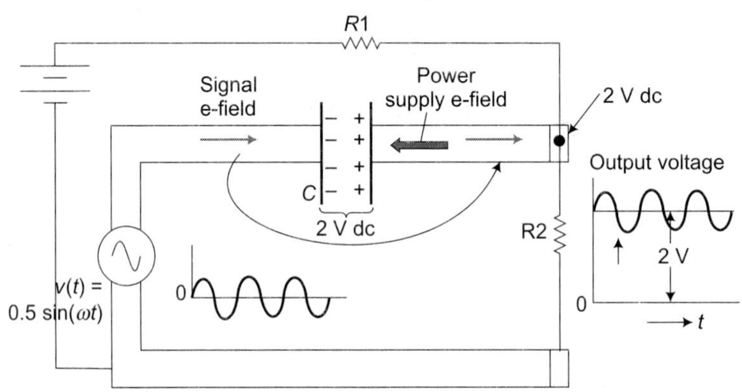

Fig. 3.19 Circuit that will maintain the voltage across the capacitor.

Supposing a *deficit of charge* occurs on the capacitor (should be equal to 2 V dc) reflected by a *failing fringe field*. This is shown in Fig. 3.20.

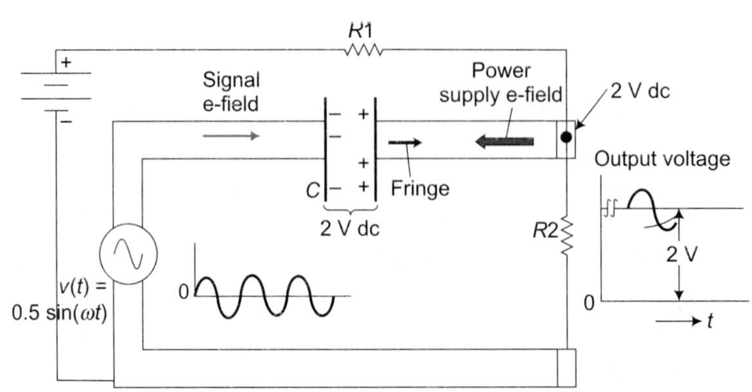

Fig. 3.20 Deficit of charge on the capacitor and falling fringe field.

This will be *immediately replenished* by charges pushed by the *power supply e-field,* and the dc voltage restored. This is shown in Fig. 3.21.

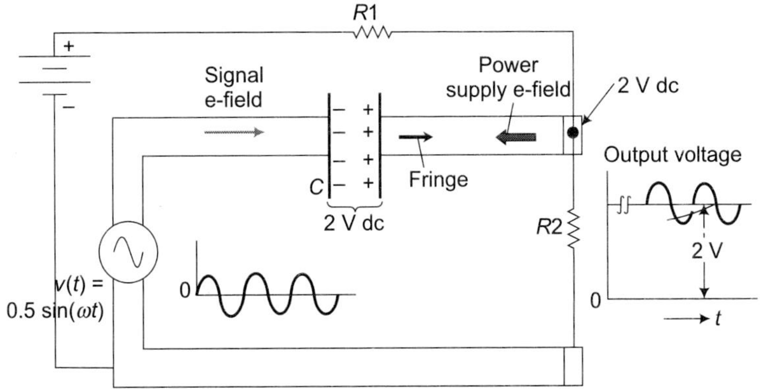

Fig. 3.21 Deficit charge on the capacitor being replenished.

This is how the voltage divider circuit with a capacitor provides a *means to superimpose* a continuously varying voltage on a steady dc voltage.

Using this circuit technique, ac signal voltages are coupled to transistor amplifiers as shown in Fig. 3.22 for the self-biasing arrangement of a common emitter amplifier.

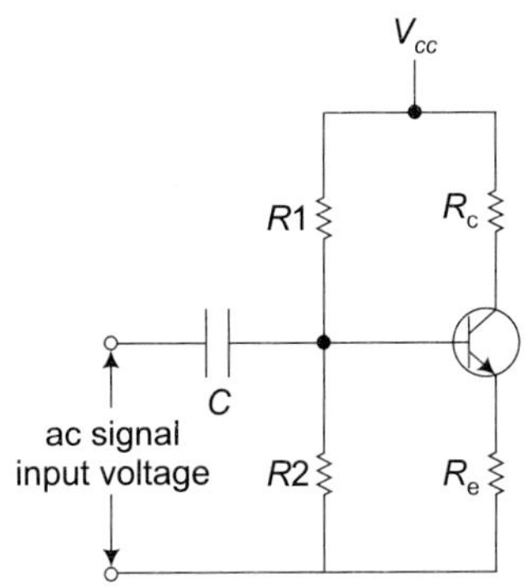

Fig. 3.22 A transistor self-biasing circuit with ac signal input voltage.

A screenshot of the results generated by a simulated run of a circuit with a capacitor, a resistor divider network, a DC source and a sinusoidal generator with TINA TI V9 software (special complementary basic edition) is included in the power point slides of Chapter 3 in the CD with this book.

Sinusoidal voltage superimposed on a dc voltage

The following questions are with reference to Fig. 3.21.

What will the output voltage be if the capacitor is replaced with a short circuit?

(a) Zero

(b) Two volts dc

(c) The input signal voltage

The answer is 'c'.

What will the output voltage be if you replace the capacitor with another resistor?

(a) The signal voltage superimposed on a dc voltage less than 2 volts

(b) An attenuated form of the signal voltage superimposed on a dc voltage less than 2 volts

(c) The signal voltage superimposed on 2 volts dc

The answer is 'b'.

What will the ouput voltage be if you **remove** the potentiometric divider and simply **replace** the capacitor with a 2 volts battery?

The output will be satisfactory for a short period of time, i.e., the ac signal voltage will be superimposed over 2 volts dc.

But, who is to replace the battery if it were to discharge!

Questions

1. How does a charged capacitor discharge and then charge to the voltage of a small dc voltage when connected with a resistor as shown in the figure below?

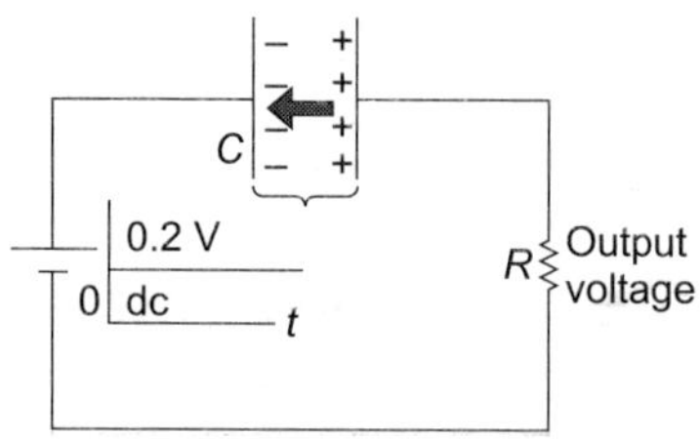

2. For the circuit of question 1, describe with the help of neat sketches the output voltage and the capacitor voltage with respect to time and what will be the voltage after a long time has elapsed.
3. A resistor R and charged capacitor C series combination is made to discharge by short circuiting their open ends. Describe with the help of neat sketches how the size of the capacitor and the resistor affect the time taken to discharge.
4. What is the effect of using a small value of capacitance for the capacitor C in the circuit of question 1 if the input is a small amplitude sinusoid? Explain with the help of neat sketches.
5. Describe with the help of neat sketches the output voltage with respect to time of the circuit of question 1 when subjected to a small amplitude sinusoidal voltage for a few cycles and after several cycles have elapsed.
6. Explain how the circuit of question 1 can be used with suitable modifications to couple ac signals to common-emitter transistor amplifiers where the requirement is to superimpose the ac signal on a dc voltage that biases the transistor in the active region.
7. Describe how the principle of superposition of electric fields may be used to validate the principle of superposition of potential. Is the principle valid to use in circuits? Explain.

4

Electrical Energy, Electromagnetic Energy and Power

In this chapter, we qualitatively understand the presence of and the transport of electrical energy in simple circuits.

Beginning with a very simple and yet powerful analogy of the energy in a rotating bicycle wheel to explain the concept of electric and magnetic energy, the next sections describe how the two fundamental forms of energy—electric and magnetic—combine to enable its transport in circuits.

When studying the transport of energy in circuits, most among us think that the electrical energy is carried by individual electrons. But this is wrong. An energy audit of a simple electric circuit that reveals the energy storage, transport and dissipation components is carried out and is presented with the aid of the Poynting theorem. In later sections, the energy dissipated component represented by the Poynting vector 'S' which is the cross-product of the electric and magnetic fields and its equivalence to the wattage given by voltage times current is presented.

4.1 WHERE IS THE ELECTRICAL ENERGY?[11]

Is Electric Energy carried by individual electrons? Wrong

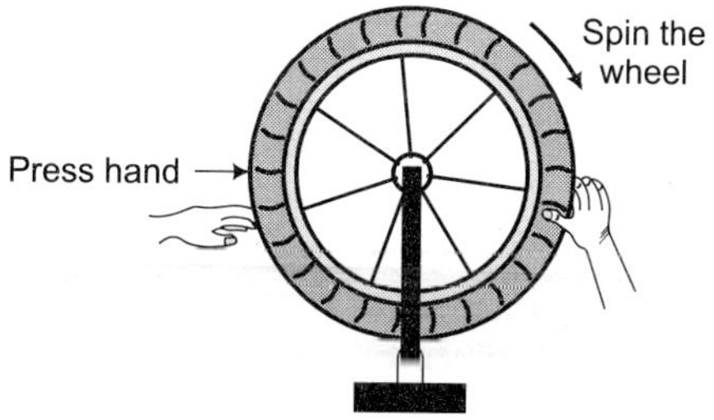

Consider a bicycle wheel mounted on a stand. Spin the wheel once using your ***right hand***. When you spin the wheel, your (right) hand ***injects energy*** into the entire wheel ***all at once***.

Remove your right hand.

Now put your ***left hand*** lightly against some part of the tire so the spinning wheel is slowed and stopped by friction.

Your *left hand gets hot*.

Your hand extracts energy from the entire wheel, all at once, and the whole wheel slows down.

Website: http://amasci.com/miscon/eleca.htm#batt2

Would it be right to say that the "power" hand fills each rubber molecule with energy, that the molecules travel to the "friction" hand and dump their energy? Then the empty molecules return to the "power" hand and get refilled? No, of course not! If this were true then the energy would be forced to travel only as fast as the rubber. Your "friction" hand wouldn't experience any friction until those *magically energized* rubber molecules made their way around the rim.

Part of the wheel would be *spinning while part* would be de-energized and *unmoving* and this would be really a *strange sight to see*!!

Instead, one hand spins the wheel and fills the *whole thing* with kinetic energy, and the rubbing or pressed left hand *sucks* the energy back out at the *same* time. The wheel rotates, and *energy flows almost instantly* across the wheel, going from the "spin" hand to the "friction" hand.

Is electric energy carried by individual electrons? Wrong

A flashlight circuit is like the bicycle wheel. The *electrons* in the copper wire circuit are *like the rim of the wheel.*

Electrons are like an *invisible drive-belt* hidden inside the wires. *The battery* causes *all* the electrons in the loop of wire to begin moving *and* in this way it *injects energy into the* WHOLE CIRCUIT all at once, just like a hand that spins a bicycle wheel.

As soon as the *battery moves the circuit's electrons*, the distant lightbulb lights up. The electrons moving into the bulb's filament are exactly the same as the ones moving out; the bulb doesn't change them or extracts energy from them.

(Did your left hand do anything to *change* the rubber molecules as it rubbed on the bicycle wheel? It just slowed down the entire wheel.) A hand can extract energy from an entire bicycle wheel instantly, and the hand heats up by friction.

Same thing with the ***bulb*, it slows down ALL** the electrons throughout the entire circuit, and in this way ***the bulb extracts energy from the whole circuit as it lights up.***

This is depicted in the circuit comprising a battery, connecting wires and a bulb as shown in Fig 4.1.

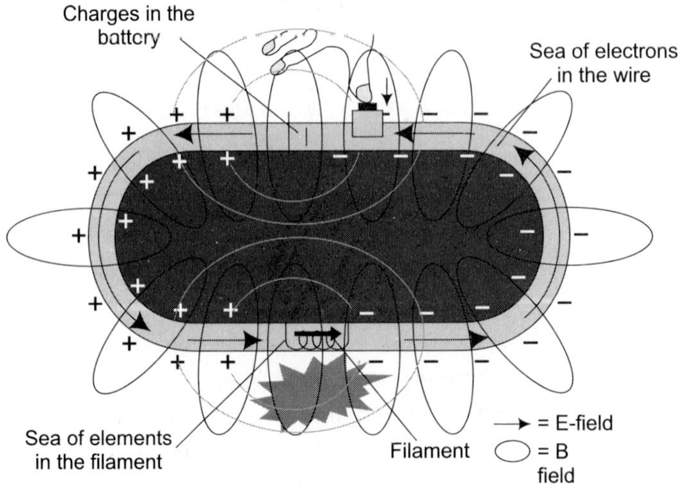

Fig. 4.1 The surface charges, electric and magnetic fields in the circuit.

When the circuit is closed there is first an initial transient phase. During this phase, the surface charges are set up all around the circuit.

Positive charges from the positive terminal and negative charges from the negative terminal enter the wires near the terminals.

The electric field set up by the charges gradually becomes uniform all around the circuit in the wires and more powerful in the filament of the bulb.

Shown in Fig. 4.1 are the surface charges, the electric field and the magnetic field.

After the initial transient is over, lasting a few picoseconds at most, all the electrons, the millions in the wires and the millions in the filament all begin moving with a start-stop motion.

The bulb lights up immediately as heat and light is generated by the collision of the electrons with the lattice ions. See also Section 1.18 and the descriptions related to Fig. 1.25 about how the electrons transfer their energy to the lattice ions.

This is shown in Fig. 4.2.

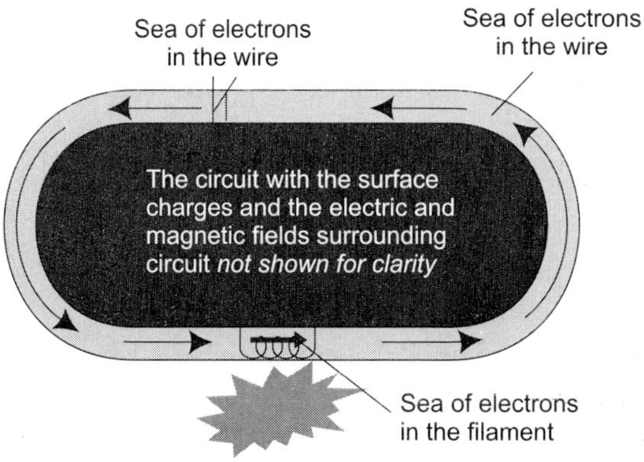

Fig. 4.2 All the electrons are pushed when the circuit is closed.

The bicycle wheel analogy has no problem explaining ac. Just wiggle the bicycle wheel back and forth instead of spinning it continuously. The wiggling wheel will rub upon the distant unmoving "friction" hand, and it will heat up that hand.

Energy can essentially travel instantly across the bicycle wheel, even though the wheel itself rotates slowly. Energy can travel instantly between the two hands even if the wheel moves back and forth instead of spinning.

What determines the direction of this energy travel? It's simple. If one hand spins the wheel, *it throws energy out* into the wheel, and if another hand rubs on the tire, *it extracts energy* from the wheel.

Also notice that the *energy doesn't care about the wheel's direction of rotation.* The energy flows one way, from one hand to the other, even if the wheel reverses direction, and even if the wheel vibrates back and forth rather than continuously turning.

Another analogy using sound waves !

Would you say that your vocal chords place energy into air molecules, then each air molecule *zooms* out of your mouth at *720 miles per hour*, and eventually crashes into the ears of distant listeners?

Is Electric Energy carried by individual electrons? Wrong

That is silly. There is no supersonic wind coming out of our mouths.

Sound is carried by waves in the air, **not by air molecules** launched at immense velocities out of your mouth.

The air molecules remain at their place most of the time, but pass along the energy by wiggling and jiggling!

Electric energy is wave energy; electric energy moves along columns of electrons like sound moves through the air, and when electrical energy flows across a circuit, the electrons **DON'T** flow along with it.

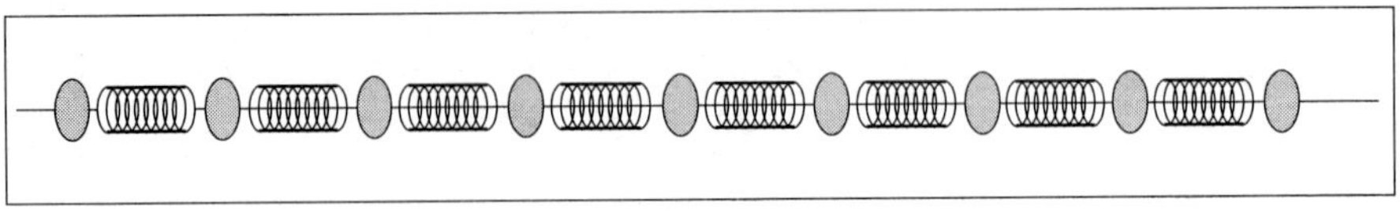

See animation in power point slide in the CD Chapter 4.

The above animation[14] portrays a medium as a series of particles connected by springs. As one individual particle is disturbed, it transmits the disturbance to the next interconnected particle. This disturbance continues to be passed on to the next particle.

The result is that energy is transported from one end of the medium to the other end of the medium *without* the actual transport of matter.

A wave can be described as a disturbance that travels through a medium, transporting energy from one location to another location.

The medium is simply the material through which the disturbance is moving; it can be thought of as a series of interacting particles.

Electrons do not push each other to carry the electric energy *unlike* in the case of air molecules that push each other and carry the sound energy.

In electric circuits, an electric field is set up by surface charges and this electric field after the very initial transient, **jolts all** the electrons into motion—**all around the circuit at the same time.**

4.2 THE NATURE OF ELECTRICAL ENERGY

Many things in nature wiggle and jiggle
We call a wiggle in *time* a *vibration* and...
a wiggle in *space* a *wave*...

A wave *cannot exist in one place* but must extend from one place to another.
Similarly, a vibration *cannot exist in one instant* but needs time to move to and fro.
Light and sound are *both vibrations* that propagate through space as *waves.*
But of two very different kinds...

Is Electric Energy carried by individual electrons? Wrong

Sound is propagation of vibrations through a material medium.
If there is no medium to vibrate, then no sound is possible.
On the other hand, light is a vibration of *nonmaterial* electric and magnetic fields, a...

...vibration of pure energy.

The *propagation* through space of these vibrating fields make up light waves
Although light waves *can* pass through many material media, *they need none*; light can propagate through a vacuum.

Sound and light are fundamentally different.

But the source of all waves, whether sound, light or whatever, is something that is *vibrating.*

For sound it can be the vibrating prongs of a tuning fork and for **light**, **the vibrations of electrons in an atom.**

Electrons are a wave medium,
and *electrical energy is wave energy.*

Remember, **electric current** is actually a flowing motion of charged particles. Electric current is a very ***slow*** flow of charges.

On the other hand, **electric energy** is different. It is made of ***waves in electromagnetic fields*** and it moves VERY rapidly.

Electric energy moves at a ***different speed*** than electric current, so obviously they are two different things flowing in ***wires at the same time***.

In an electric circuit, the **path of** the **electric charges** is **circular**, while the **path of** the **energy is not**.

If electric current is **like wind**, then electrical energy is **like** some **sound waves**, and **electrons are like the molecules of the air.** For example, sound (energy) can travel through a pipe if the pipe is full of air molecules, and electrical energy can flow along a wire because the wire is full of movable charges.

Sound moves much **faster** than wind, correct? And electrical energy moves **much faster** than electric current for much the same reason.

Air in a pipe can flow fast or slow, while sound waves always move at the same very high speed.

Charges in a wire can flow fast or slow, while **electrical energy always flows along the wire at the same incredibly high speed.**

Whenever sound (energy) is flowing through a pipe, the air molecules in that pipe are vibrating back and forth.

Is Electrical Energy carried by individual electrons? Wrong

When **waves of ac electrical energy** are flowing along a wire, the electrons in that wire are vibrating back and forth at 50 times per second; (60 times per second in some countries)

The description of the sound waves and electrical energy travelling in pipes and wires are just an analogy, and sound waves ***are not exactly*** like electrical energy.

For example, *sound energy can flow inside an air-filled tube, while...*

...electrical energy ***is coupled*** with compression waves in the electrons of the wire

Electron-waves travel inside the wires coupled with the ***energy they carry in the invisible fields surrounding*** the wires.

We cannot see gravitational fields though they exist.

We can "see" electric fields and magnetic fields using grass seeds and magnetic compasses.

Electric currents in copper wires are a flow of electrons, but these electrons are not supplied by batteries. Generators do not 'generate' them. Instead the electrons come from the wire. In a copper wire, copper atoms supply the flowing electrons.

The *millions of electrons* in a circuit were already there *before* the battery was connected.

They (*millions of electrons*) were even there *before* the copper was mined and made into wires!

Batteries and generators do not create electrons, they merely *pump* them and the electrons act like a pre-existing fluid which is always found within all wires.

Source: http://amasci.com/miscon/eleca.html#batt2

Most textbooks teach that the energy is trapped inside the electrons, and these electrons flow inside the wires. Doesn't this mean that electrical energy flows INSIDE the metal wires? No.

Because electrical energy is not trapped inside the electrons.

Instead the energy is coupled with invisible *magnetic fields and electric fields* which *surround* the electrons, and these fields surround the wires.

Electrons don't flow fast like the energy does, instead they *ooze along slowly* (millions of them everywhere in the conductors... with a very high speed start-stop motion with a drift superimposed) to produce an electric current.

Let us consider an example of how much *kinetic energy*[27] is really there in the electrons in a copper wire of 1 mm^2 carrying a current of 1 amp powering a medium sized domestic appliance, say a refrigerator.

Each cubic centimetre of copper contains $8{\cdot}4830 \times 10^{22}$ free electrons per cc.
The mean velocity (drift) is 0.075 mm/sec or 0.00027 km/hr.

Taking the mass of the electron as 9×10^{-31}kg, the *power delivered* by the moving electrons as kinetic energy per second is....
a *teeny weeny* 1.5×10^{-20} watts (or joules/second).

Now, if this current (1 amp) is being delivered with a potential difference of 230 V then the power consumed is nearly a whopping 230 watts!

Or 10^{22} *times* the power of the moving electrons.*

We therefore have to *conclude* that *the real burden of the signal and energy transfer is being carried elsewhere, and* ***not*** *simply by the electrons in the wires.*

*http://www.st-andrews.ac.uk/~www_pa/Scots_Guide/audio/part6/page2.html

It is tempting to assume that the energy flows through the wires of the circuit, like water in a pipe. But if we look carefully we find that the energy density in the wires is *much too small* to account for the amount of energy in transit.

In a sense, the wires guide the energy but do not carry it. For electromagnetic fields, the total energy density (in empty space) is u_{tot}

$$u_{tot} = u_{electric} + u_{magnetic} = \frac{1}{2}\left[\varepsilon_o E^2 + \frac{B^2}{\mu_o}\right] \qquad ...(1)$$

and the speed of flow is c, the speed of light. We can also compute a momentum called the *electromagnetic momentum* that the fields themselves carry! (Refer to Chapter "Momentum" in "Introduction to electrodynamics" by David J. Griffiths).

The calculation of the total energy is extremely complex involving changes in the configuration of the fields and the electromagnetic field density.

The *rate* of flow of energy (power) in uniform wires in circuits involving the fields is a more direct calculation and is given by the Poynting vector to be described later.

We ignored the energy in the fields in the worked example in Section 1.7 because the velocities of the charged balls are quite small, the charge movement is sluggish because of the mass of the balls they are riding on. As a matter of fact, we can compute the energy by writing $E = E_1 + E_2$ (the total electric field is the sum of the field due to particle 1 and the field due to particle 2), and get for the energy $\frac{1}{2}\varepsilon_0(E_1 + E_2)^2$. Squaring out the binomial, we get three terms: one involving E_1 alone (the self-energy of charge 1), which we ignored because it is the same before and after, and hence is irrelevant for the calculation; one involving E_2 alone (to which the same remark applies); and a "cross-term" involving E_1E_2. This last, when integrated over all space, gives the interaction potential energy kqQ/r that we used in the example in Section

1.7. (For details, see section "Work and Energy in Electrostatics" in Introduction to Electrodynamics by David J. Griffiths).

There is a tiny amount of electromagnetic energy radiated when the balls accelerate from rest; this is negligibly small in comparison with the kinetic energy of the balls.

However, when electrons are moved by an applied electric field in the *wires* (good conductors) in a circuit, they accelerate freely between collisions with the atomic cores, and so can attain large velocities that are yet non-relativistic, during those short intervals between collisions. Their total kinetic energy is quite negligible. But, *they* contribute to the *huge* amounts of electromagnetic energy transported. It may be said that the energy is entirely "potential" in much the same way as we did in the example of Section 1.7 because their drift velocity, due to the start-stop motion, is tiny, for example, 1 metre/hour with a current of 115 amps, as we saw in Section 1.18.

But, wait a minute, does the expression $\frac{1}{2}\varepsilon_o(E_1 + E_2)^2$ represent the energy being transported?

We must be careful to distinguish the energy stored in the fields from the energy that is imparted to the charges in circuits and also from the energy transported that is dissipated or lost, never to return.

Suppose we have some charge and current configuration which, at time t, produces fields E(electric) and B(magnetic). In the next instant, dt, the charges move around a bit. This is the kind of movement that occurs in a wire carrying a current.

How much work, dW, is done by the electromagnetic forces acting on these charges in the interval dt?

According to the Lorentz force law (see Section 2.27), the work done on a charge q to move it a distance dl is

$$\vec{F}.dl = q(\vec{E} + \vec{v} \times \vec{B}) \bullet \vec{v}dt = q\vec{E} \bullet \vec{v}dt \quad \text{...(2)}$$

where v is the velocity of the charges. The second term in (2) is zero because the dot product of two vectors perpendicular to each other is zero.

Now,

$$q = \rho d\tau \quad \text{...(3)}$$

and,

$$\rho\vec{v} = \vec{J} \quad \text{...(4)}$$

where ρ is the charge density in the wire, $d\tau$ is an infinitesimal volume element and, J the current density. Substituting ρ from (4) in (3), we get, $q = \frac{\vec{J}}{\vec{v}}\, d\tau$. Substituting this expression for q in (2), we get, the rate at which work is done on all the charges in a volume V is

$\frac{dW}{dt} = \int_V (\vec{E}.\vec{J})d\tau$ where $\vec{J}$ is the current density established in a conductor.

Evidently, $\vec{E}.\vec{J}$ is the work done per unit time, per unit volume, which is to say, the *power* delivered per unit volume.

The quantity $\vec{E}.\vec{J}$ can be expressed in terms of the fields alone (for details see section Poynting's Theorem in "Introduction to Electrodynamics" by David J. Griffiths) as

$$\vec{E}.\vec{J} = -\frac{1}{2}\frac{\partial}{\partial t}\left(\varepsilon_o E^2 + \frac{1}{\mu_o}B^2\right) - \frac{1}{\mu_o}\nabla \bullet (\vec{E}\times\vec{B})$$

Putting this into the equation for $\frac{dW}{dt}$ above, and applying the divergence theorem to the second term, we have

$$\frac{dW}{dt} = -\frac{d}{dt}\int_V \frac{1}{2}\left(\varepsilon_o E^2 + \frac{1}{\mu_o}B^2\right)d\tau - \frac{1}{\mu_o}\oint_S (\vec{E}\times\vec{B})\bullet d\vec{a} \qquad ...(5)$$

where S is the surface bounding the volume V and da an infinitesimal area element on this surface. This is Poynting's theorem where the first integral on the right is the total energy *stored* in the fields $u_{tot} = u_{electromagnetic}$ and the second term represents the rate at which energy is (dissipated in the circuit) *carried out* of the volume V (representing the volume of a wire say, in a circuit) across its boundary surface, by the electromagnetic fields.

The work done 'W' on the charges by the battery will *increase their* mechanical energy (kinetic, potential) as the atoms (millions of them !) stretch and the electrons (millions) are accelerated. Many among us miss this component of the energy in circuits.

Poynting's theorem (5) says then, that the work done on the charges by the electromagnetic force is equal to the decrease in energy stored in the field, less the energy that flowed out through the surface as for example, in heating a resistive element.

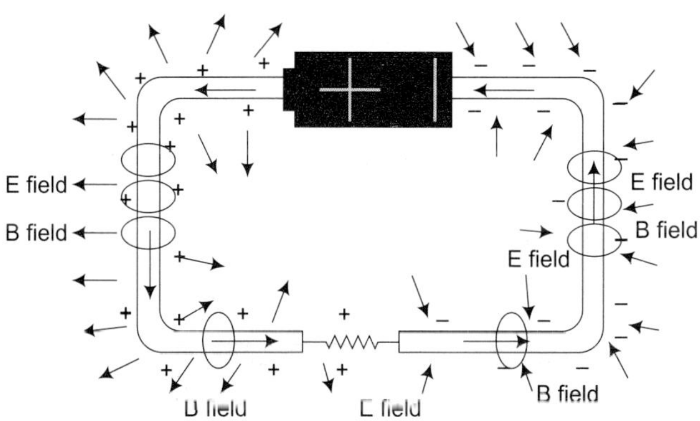

The energy *per unit time*, *per unit area*, transported (and, dissipated in resistive elements in a circuit) by the fields is called the Poynting vector S:

$$S = \frac{1}{\mu_o}(\vec{E} \times \vec{B}).$$

In Section 4.4, we will see how the Poynting vector 'S' integrated over the surface of a wire $\frac{1}{\mu_o}\oint_S (\vec{E} \times \vec{B}).da$ gives us the familiar expression $V{\times}I$ for the power delivered to a load resistor. The transport of energy in circuits may be likened to a water delivery system with a storage tank, an inlet pipe near its bottom, connected to a source of water supply with a pump and, an outlet pipe also near the bottom that supplies water to a faucet, as shown in the figure below. The work done by the pump to sustain the flow of water in the pipes represents the dW/dt term, the water storage represents the electromagnetic energy storage term $\frac{1}{2}\left[\varepsilon_o E^2 + \frac{B^2}{\mu_o}\right]$ (electric fields due to charges on the battery and surface charges and, magnetic fields due to moving charges), the flow of water represents the energy transport by the moving charges and, the frictional losses in the pipes represents the Poynting vector 'S' term, the energy dissipated in the conductor and resistor.

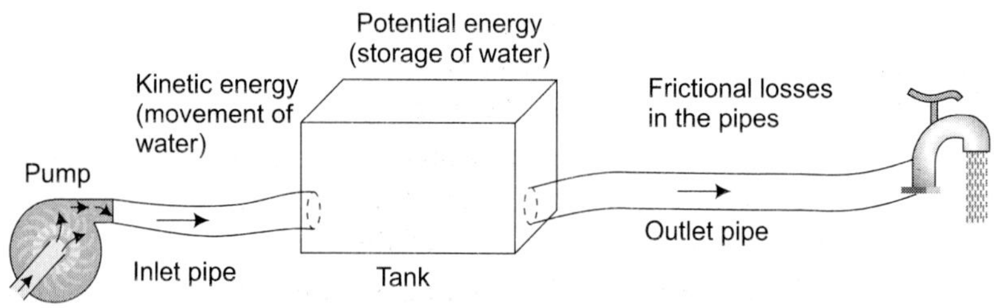

When delivering energy to resistive loads, the electromagnetic energy causes the electrons in the *resistors* (poor conductors) to gain large kinetic energies (and higher drift velocities in thin than in thick wires; see Fig 1.34) between collisions, that is *directly* transferred to the lattice during collisions, resulting in its dissipation as heat.

Whenever a pair of wires is connected to a battery or generator, the two wires *become oppositely charged*, and they become surrounded with an invisible *electrostatic field.*

This field **is** associated with the *"electrical energy"*.

During an electric current, the wires become *surrounded* with *magnetic field.*

This field **is** associated with the *"electrical energy"*.

Magnetic and electric fields exist in the empty space surrounding your *lamp cord*, and these *fields* are associated with the *'wattage'* that powers the light bulb.

4.3 IN A CIRCUIT WHY ARE TWO WIRES NEEDED?[11]

Let us answer this question with a question. When a circular belt is passing over two pulleys, why are two belts needed?

The answer: THERE ARE NOT TWO BELTS !

There is only one belt, and the belt is circular. It looks like there are two belts, with one of them flowing leftwards and the other one flowing right. But in truth, there is just one belt, and it is rotating.

So, why are two wires needed?

The answer: THERE ARE NOT TWO WIRES! Instead there is only one wire, but it is connected in a circle.

All metals are full of movable electrons, so when we connect a wire in a circle, we are forming a kind of "electric drive-belt" which can move inside the wire.

In the Fig 4.3, is seen a simple circuit with a battery connected to a resistor such as a light bulb. The battery converts its chemical fuel into waste products, and the resistor or the light bulb gets hot.

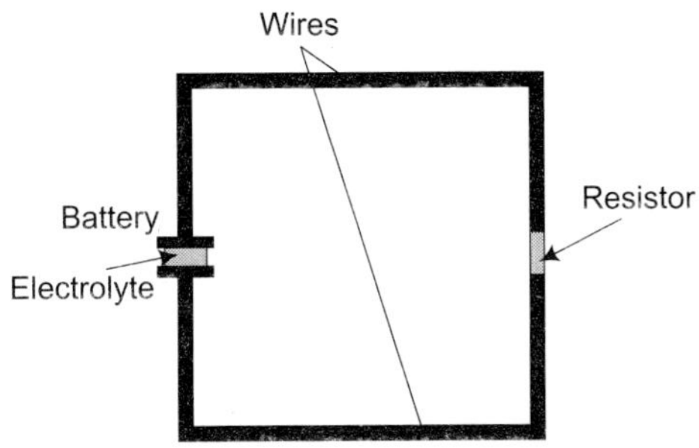

Fig. 4.3 A simple circuit

In the circuit of Fig 4.3 the battery converts its chemical fuel into waste products, and the resistor gets hot, from the energy it receives in the form of electric and magnetic fields from the battery.

Refer : www.amasci.com

Electronics students commonly assume that electrical energy flows inside metal wires. Physics students know differently. Electrical energy only flows extremely slowly through metals. In fact, the energy sent out by batteries and generators is located in empty space: it takes the form of electromagnetic fields surrounding the wires.

Inductors store energy as a magnetic field outside the windings, capacitors store energy as an electric field in the empty space between the plates, and an electric circuit as a whole does both at once.

The energy which flows across a circuit is not moving through the interior of the metal wires.

Whenever a battery powers a light bulb, the battery spews electrical energy into space! The electrical energy is then guided by the wires, and eventually it dives into the light bulb, where it drives the charges against the resisting force of electrical "friction." An electric circuit is a transmission line for electromagnetic energy.

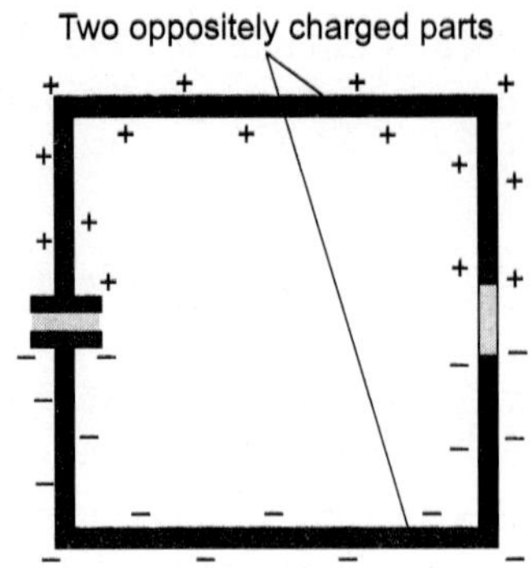

Fig. 4.4 Two charged conductors: *Voltage*

Everything connected to one battery terminal acquires the same electrical potential (voltage.) The circuit acts like two separate conductors, one with a positive charge imbalance and one with negative, as shown in Fig. 4.4.

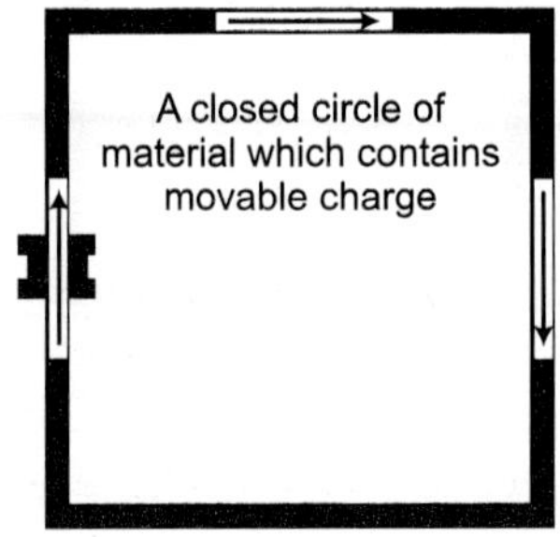

Fig. 4.5 The conductive path: *Current*

All conductive materials contain movable charges. The resistor and the battery's electrolyte both are conductive. When we include them with the wires, we can see that an electric circuit is a complete circle which is full of "fluid" charge. It acts like a liquid flywheel; a flywheel hidden inside a closed ring of pipe, as shown in Fig. 4.5.

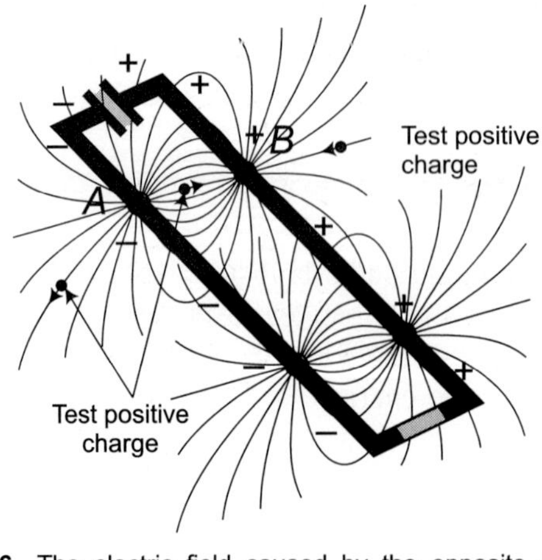

Fig. 4.6 The electric field caused by the opposite charges.

Again, in Fig. 4.6 is a 3D oblique view. The two halves of the circuit act as opposite-charged wires with e-field flux connecting them. We need to draw a third pattern between the two above, then draw more between those until the whole wire is covered with bent sheets of electrostatic flux which arcs between the wires.

External work has to be done to push the test **positive** charge towards the wire connected to the + terminal and the voltage or potential difference from point A to B is

$$V = -\int_{A}^{B} E.dx$$

The magnetic (B) field caused by the current loop is shown in Fig. 4.7.

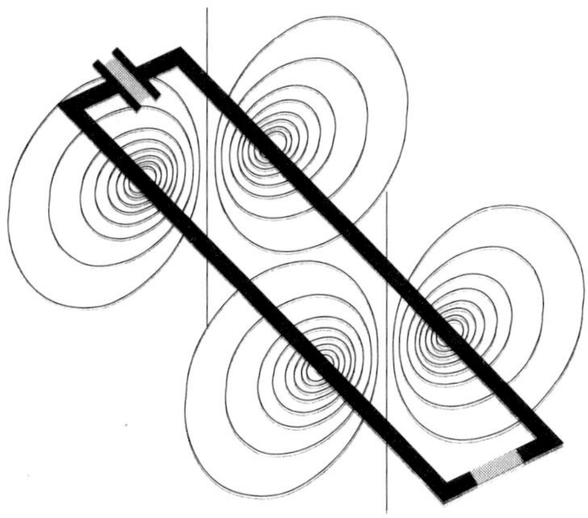

Fig. 4.7 The magnetic field caused by the current loop.

To be more accurate, we need to draw more than just two patterns. Between the two patterns above, draw a third. Then between each of those draw more and more. The end result looks like "tubes" of magnetic flux surrounding the wires. Figure 4.8 depicts a 3D oblique view of the two fields, e-field and b-field.

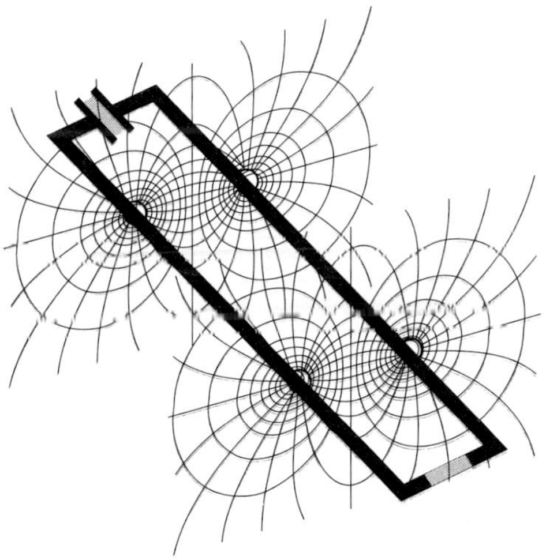

Fig. 4.8 A e-field and b-field together.

Add more and more patterns between two e-field and b-field patterns, until the empty space is packed full of "hair." Note that most of the flowing energy lies between the two wires... but quite a bit also surrounds the "cable pair" as a whole.

Electromagnetic energy flows out of the battery and into the empty space around the circuit. It flows parallel to the connecting wires, then it dives into the resistor, as shown in Fig. 4.9.

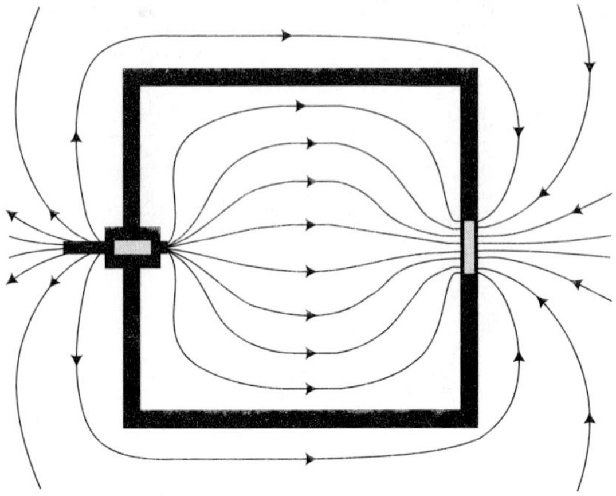

Fig. 4.9 The energy flow (Poynting field).

The field of energy flow is found by multiplying the e-field by the b-field (E × B vector cross-product) as seen in Fig. 4.9. Figure 4.10 shows the Poynting energy flow field and the e-field together.

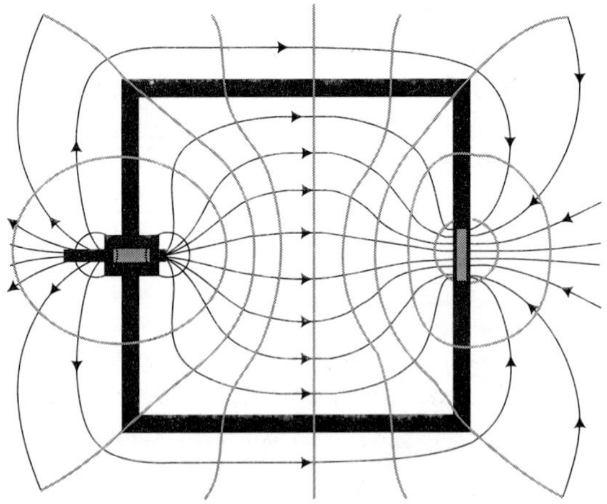

Fig. 4.10 Energy flow field with e-field in grey.

Note that the energy always flows perpendicular to the flux lines of e-field.

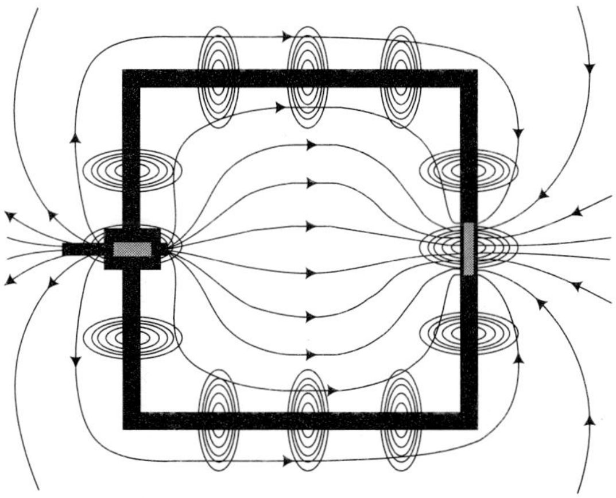

Fig. 4.11 Energy flow with b-field in gray.

Note that the energy always flows perpendicular to the lines of b-field too, as shown in Fig. 4.11.

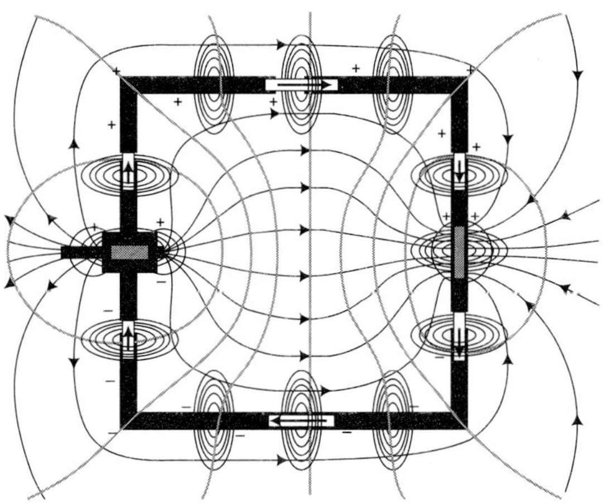

Fig. 4.12 A simple circuit?

When all the separate invisible phenomena are displayed together, you can see why "electricity" might be a bit hard to understand. And Fig. 4.12 only shows a two-dimensional slice; a sort of side view of the fields. The real fields are 3D and volume-filling, so an accurate drawing would look like a black glob of hairs.

4.4 THE POYNTING VECTOR

The *Poynting vector* describes the energy flux Joules m^{-2} s^{-1} or Watts m^{-2} of an electromagnetic (em) field. It is named after its inventor John Henry Poynting.

It *points in the direction of energy flow* and its magnitude is the power per unit area crossing a surface which is normal to it.

It is given the symbol $\vec{S}$ (it is a vector) and, in SI units, is given by:

$$\vec{S} = \vec{E} \times \vec{H} = \frac{1}{\mu}(\vec{E} \times \vec{B})$$

where E is the electric field, H and B are the magnetic field strength and magnetic flux density respectively, and μ is the permeability of the surrounding medium.

For example, the Poynting vector near an ideally conducting wire is along the wire axis—so electric energy is flowing in space outside of the wire, as shown in Fig. 4.13(a).

The Poynting vector *tilts towards the wire for a resistive wire*, indicating that *energy flows from the em field into the wire,* producing resistive Joule heating in the wire, as shown in Fig. 4.13(b).

Application of the Poynting Vector for Energy Transported in a Transmission line

Consider a long resistanceless transmission line of length 'L' delivering power from a battery to a resistive load. A steady current i exists as shown in Fig. 4.13(a).

The magnetic field strength H will be in the ϕ direction and at the surface of the wire it will have a value $H_\phi = \dfrac{i}{2\pi r}$ where r is the radius of the line. See Section 5.2 for details of this expression.

Let the voltage drop be V, then, $E = \dfrac{V}{L}$.

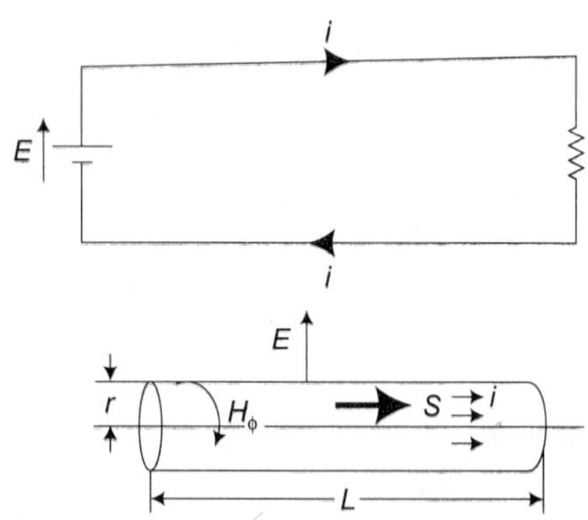

Fig. 4.13(a) Poynting vector to calculate the energy of a resistanceless transmission line.

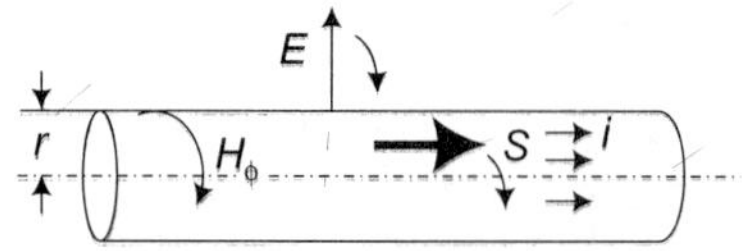

Fig. 4.13(b) Tilt of electric field on resistive wire towards the wire.

and
$$H_\phi(r) = \frac{i}{2\pi r}.$$

The Poynting vector is integrated over the surface of the cylinder. ***S*** is directed radially inward, the caps at the ends do not make any contribution to

$$\oint S.da$$

where, $da = 2\pi r dz$ is the surface area for an element length dz of the line.

Now,
$$\oint S.da = \int (E_Z \times H_\phi).da$$

Observe that E_z and H_ϕ are mutually perpendicular to each other.

So,

$$\oint S.da = \int_0^L E_Z H_\phi 2\pi r dz$$

Using the relations $E = \frac{V}{L}$ and, $H_\phi(r) = \frac{i}{2\pi r}$ in the above expression, we obtain,

$$\oint S.da = \frac{Vi}{L}\int_0^L dz = Vi$$

Note that the quantity Vi represents *ohmic loss* in the resistor.

Therefore, we may conclude that the *loss* is supplied by energy *entering through* the surface of the resistive load.

Within the resistive load, the Poynting vector is directed radially inward (see Fig. 4.14) so that the ohmic loss equal to i^2R is supplied by the electric and magnetic fields within the resistor. However, in the free space surrounding the wires between them, the Poynting vector is directed towards the load.

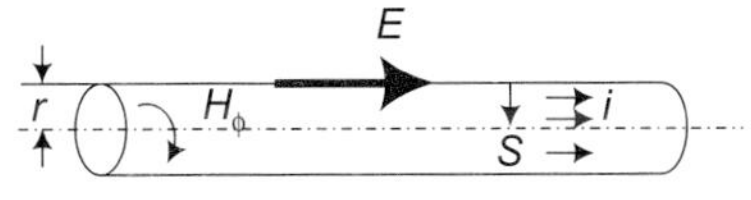

Fig. 4.14 In a resistive conductor the Poynting vector is directed radially inward.

This is a result of the electric field component E_l of E due to surface charges pointing mostly outwards (with a very tiny field component E_2 of E along the axial direction in the wire) so that the energy represented by $S = E \times H$ is mostly directed toward the load (resistor) and the energy flows through the dielectric (free space) *around the* wires coupled with the moving electrons inside and *not through the wires itself.*

See Figures 4.9 to 4.12 and verify the direction of energy flow between the wires applying the right-hand rule for cross products as described in Section 5.2.

4.5 POWER FLOW WHEN A CAPACITOR DISCHARGES INTO A RESISTOR

We consider a simple RC circuit, where a charged capacitor is made to discharge into a resistor to which it is connected with a wire. This wire offers a high resistance formed by narrowing the wire. The electric field must be larger in the high-resistive region (and smaller in the low-resistance conductor portion) to have uniform current.

The plots of the field vectors and equipotential lines were made from computer-based simulation studies by Norris W. Preyer, Department of Physics and Astronomy, College of Charleston, Charleston, South Carolina, USA and reproduced with permission in the Figs. 4.15 and 4.16.

The Fig. 4.15(a) illustrates the orientation of the electric fields in the space surrounding the circuit when quasi steady-state is reached that will ensure that the energy placed in the capacitor is dissipated in the resistor. Notice the intricate arrangement of positive and negative surface charges that will achieve the necessary electric field magnitudes and orientations to drive the current in the wires and the resistor.

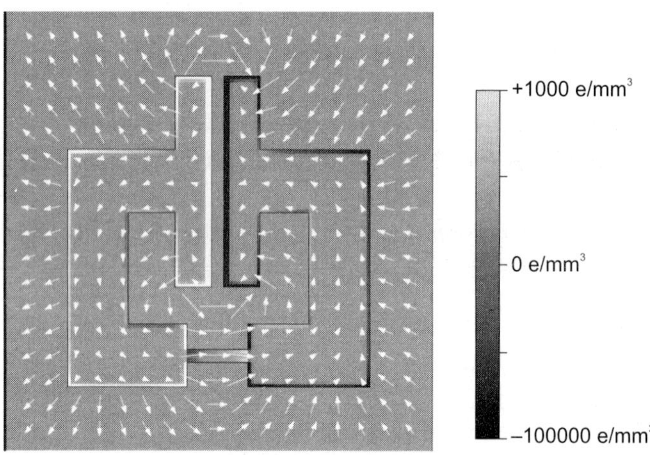

Fig. 4.15a The electric field vectors of a capacitor discharging in an RC.

The various gray shades represent the excess charge density, ranging between +/–5000 e (charge of an electron)/mm^3. The charge density on the capacitor plates exceeds this range, which was chosen to prominently display the surface charges on the wires. The tiny arrows plot the square root of the electric field. The large field vectors between the plates have been omitted.

The electric field outside the conductors is important for ensuring that the Poynting vectors point in the correct directions, and that high-resistance regions have a greater energy flow than low-resistance regions. This model illustrates capacitor-like interface charges piling up on either side of the narrow wire resistor to create a powerful electric field inside the resistor as shown in the magnified view of the resistor and the wire in left of Fig. 4.15(b), which will produce uniform current in the entire circuit. The right in Fig. 4.15(b) shows only the surface charge with polarities without the outline of the wires and resistors to better illustrate the intricacy in the geometry of the surface charge and interface charge arrangement.

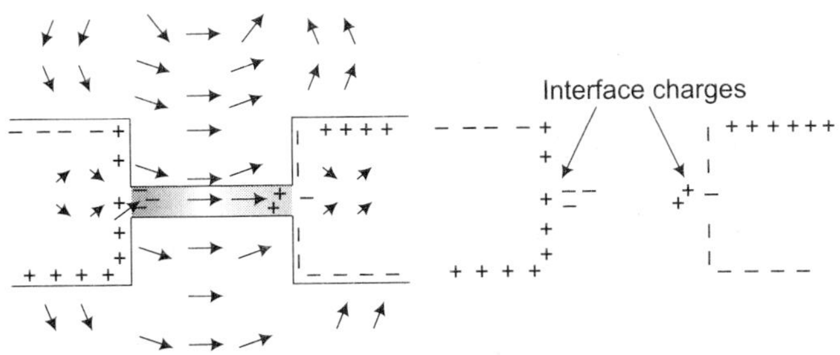

Fig. 4.15b A magnified view of the resistor and the wire showing the electric field vectors, surface and interface charges.

Notice how these interface charges at the ends of the resistor increase the electric field and hence current density through the narrow region and, simultaneously decrease the field and current density in the wider wires. These and the surface charges on the wires will also vary at every instant of the quasi steady-state to ensure that the conservation of current law is upheld.

The Poynting vector will be directed into the resistor because a large amount of energy is absorbed here and to a smaller extent into the conductor.

While the equipotential lines are interesting by themselves, these lines also show '*S*', the Poynting vector. Recollect from Section 2.26 that charged insulated conductors are equipotential on the surface as well as in the interior. Similarly, in the electric circuit, there will be equipotential surfaces of varying magnitude along different cross-sections of the wire and resistor. An equipotential line (surface) is one for which no work is required for a test charge to move along (on) it. Since, *E* is the negative *gradient* of potential ($-\nabla V$), then, we interpret the *direction* of the gradient (or, in other words of *E* that set up the potential) as follows:

If one stands on the "surface" of potential, one would see the surface rising in some direction and sloping downward in the opposite direction. There *is* one direction in which a short step will take you *lower* than a step of the same length in any other direction. *That* is the *direction* of *E* or rather the *direction* of the gradient (represented mathematically by the symbol '∇') of potential.

A field flux line through any point and the equipotential surface through that point are perpendicular to one another, just as, on a contour map of hilly terrain, the slope is steepest at right angles to a contour of constant elevation.

Due to symmetry, *B*, the magnetic field, is perpendicular to the figure, so *S* is perpendicular to *E* and *B*. Hence, the equipotential lines are *parallel* to the Poynting vector (which is $E \times B$) and you can see most of the power going to the resistive regions.

The equipotential lines for the circuit with a discrete resistor are shown in Fig 4.16(a).

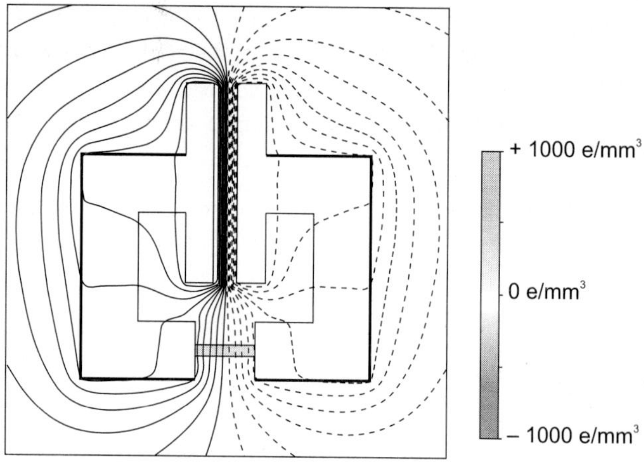

Fig. 4.16a The equipotential lines (parallel to the Poynting vector).

In this figure, note how the equipotential lines are thickly crowded in the resistive region, indicating a steep voltage change from one end to the other (see Section 2.6 for a description of variation of potential *across* these lines) and, a large flow of power into the narrow-wire resistor. For the entire circuit, these lines are parallel to the Poynting vector and are considered equivalent in terms of orientation and density to the Poynting vector.

To better appreciate the representation using equipotential surfaces (lines) of the voltage, refer to Section 2.20 "Visualizing voltage", especially Fig. 2.53(b) for points in the interior of the conductor, wherein there is no work done penalty to be paid for a test charge to move towards the surface along this line.

As one moves *across* these lines along a conductor or resistor, the potential changes and, this change is what is commonly called "voltage drop".

In the conductor regions, the lines are spread far apart indicating a very small voltage change from one end to the other and, a smaller flow of power into these regions.

For the sake of completeness, shown in Fig. 4.16(b) in the left is a magnified view of the wire and resistor showing the equipotential lines, surface and interface charges, and in the right picture the same view but without the wire, resistor and equipotential lines that shows more clearly the intricate arrangement of surface charges that will produce the electric fields in the resistor. The full lines indicate a positive potential and the dashed lines indicate a negative potential.

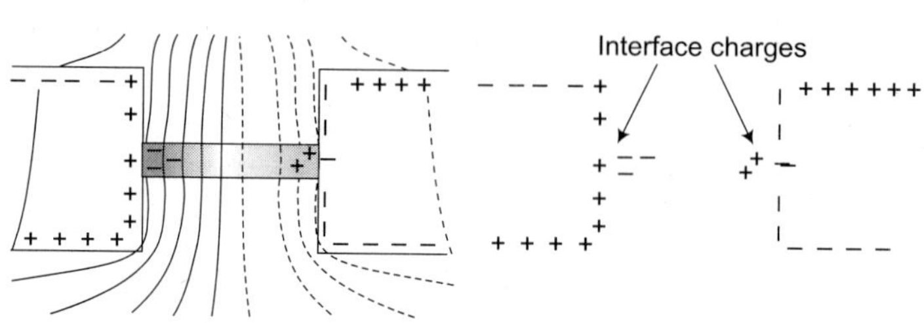

Fig. 4.16b A magnified view of the wire and the resistor showing the equipotential lines, surface and interface charges.

I urge the reader to visit the site http://galaxy.cofc.edu for more examples and animations of these circuits.

4.6 AC ENERGY

When you plug a lump into a wall socket, you *shouldn't imagine* that the ac energy is a mysterious invisible entity traveling back and forth inside the wires.

Instead you should think of ac energy as a mysterious invisible flow that comes out of the outlet, runs along the outside of BOTH wires, then dives into the filament of the light bulb.

Your *electric company* is sending out long "sausages" or tubes of electrical energy, the wires are guiding them, and your appliances (such as lamps, fans and TVs) are *absorbing* them.

4.7 HOW DOES THE LIGHT IN A DISTANT BULB TURN ON IMMEDIATELY WHEN YOU TURN ON THE SWITCH?

It is not necessary for charges to move physically to a distant place to cause a change at that place: a rearrangement of charges due to tiny movements produces a change at the speed of light in the electric field at a distance.

It is important to keep in mind that the effects do take some time to propagate (nanoseconds). The propagation is not instantaneous, but it is very fast.

When you turn on the light switch in your room (making the final connection in a circuit), it could take hours on the average for electrons in the switch to drift to the overhead light.

But the light comes on right away, because the rearrangement of surface charges in the circuit takes place at about the speed of light.

The final steady state of the circuit is established in a few nanoseconds, after which the electron sea....

...circulates slowly and majestically around and around everywhere in the circuit, including through the switch and the wires and the light bulb.

Most lighting actually uses "alternating current," in which case the electron sea doesn't drift continuously but, merely sloshes back and forth very short distances, everywhere in the circuit.

We conclude the chapter with an answer to an oft-asked question about the nature of light which is electromagnetic (em) radiation (electric and magnetic fields at right angles to each other) produced by an accelerated charge. Descriptions of the nature of em radiation and the manner of its production is beyond the scope of this book and the reader is referred to "Electric and Magnetic Interactions" by Chabay and Sherwood, 2010, Wiley for an excellent treatment of this topic.

How does an opaque body (such as a cardboard) block light when we had stated that electric fields and magnetic fields penetrate through matter? (Chabay & Sherwood, 1995, used by permission).[2]

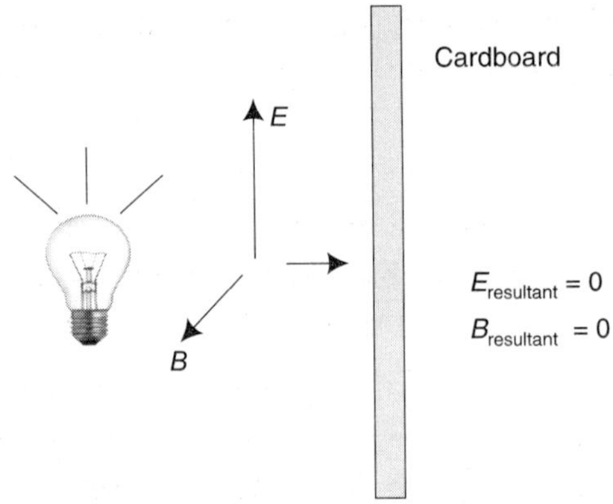

The electric and magnetic fields radiated by the accelerated charges in the filament of the bulb are certainly present to the right of the bulb and yet, there is no light on the other side of the cardboard, so the net field must be zero there.

There must be an additional source of electric and magnetic field somewhere, producing a field that adds to the radiative field resulting in the zero net field.

The electrons in the atoms of the cardboard molecules are accelerated by the electric field of the light radiation in such a way that the re-radiated field (because this new field is from the cardboard) is equal and opposite to the incident time-varying electric and magnetic fields from the light bulb. The net field on the other side of the cardboard is the superposition of the two fields, and is zero.

Based on topics from "Introduction to electrodynamics" by David. J. Griffiths the following pragraphs provide a qualitative view of the magnitudes of electric and magnetic forces, a sort of comparison, though it must be understood that the forces act differently on charged particles.

Typically, electric forces are enormously larger than "magnetic ones". This has to do with the sizes of the fundamental constants $\in_0$ and μ_0.
How is it that we "notice" magnetic forces at all?

The answer is that both in the production of a magnetic field (Biot-Savart law) *and* in its detection (Lorentz) it is the current (charge times velocity) that enters, and we can compensate for a smallish velocity (remember it took an hour for charges to move 1 metre in a wire carrying 115 amps) by pouring huge quantities of charge down the wire.

Ordinarily, this charge would simultaneously generate so large an electric force as to "swamp" or mask the magnetic one.

But, if we arrange to keep the wire "neutral", by embedding in it an equal amount of opposite charge at rest, "the positive atomic cores", the electric field cancels out (other than the one due to surface charges), leaving the magnetic field to stand alone.

This is precisely what happens in an ordinary current-carrying wire.

Questions

1. Describe the various types of potential energy and how it is stored.
2. Derive the expression for the Poynting theorem $\frac{dW}{dt} = -\frac{d}{dt}\int_V \frac{1}{2}\left(\varepsilon_o E^2 + \frac{1}{\mu_o} B^2\right) d\tau - \frac{1}{\mu_o}\oint_S (E \times B).da$ with a description of all the energy processes applicable in a circuit making sure to identify the imparted energy, stored energy, and the dissipated energy denoted by the theorem.
3. Draw the electric field flux lines and the magnetic field lines for a simple series circuit comprising a battery, a resistor (dissipative element) and wires. Derive an expression for the energy transported by applying the Poynting vector

$$S = \vec{E} \times \vec{H} = \frac{1}{\mu} (\vec{E} \times \vec{B})$$

and show that $\oint S.da = VI$, where V is the voltage of the battery and I the current in the circuit. Describe the direction of the energy transport between the wires and in the load resistor.
4. Describe all the processes that occur when a distant electric bulb is turned on by switching on a voltage supply to it.
5. Typically, electric forces are enormously larger than "magnetic ones". Then, how is it that we notice and can measure magnetic forces in circuit experiments? Explain.

5

Magnetic Field and Faraday's Law; Coulomb and Non-Coulomb Electric Fields; Self-Induction; Seeing Atoms

In Chapter 5, the association of electric fields to time-varying magnetic fields called Faraday's law is presented. Self-induction is next treated and the subtleties in the processes involved that produce "induced emfs" are presented.

The current in an inductor circuit lags an applied sinusoidal voltage and the processes that produce this steady state operation are then described. The behaviour of inductive circuits when the frequency of a sinusoidal voltage they are subject to is varied, is explained.

The production of high voltages using inductors is important because, while this phenomenon is exploited in producing sparks for automobiles, they prove to be an irritant when switching inductive loads using electronic devices. This aspect is treated in later sections.

5.1 MAGNETIC INTERACTIONS

Magnetic interactions are ***similar to electric interactions***:

1. There is both repulsion (as between like charges so between like poles) and attraction (as between unlike charges so between unlike poles of a magnet);
2. The superposition principle is valid. (E1 + E2 + E3...) and net field of two or more magnets is the vector sum of the individual magnetic fields (B1 + B2 + B3...) and,

...Magnetic fields... invisibly alter the space where they exist and they penetrate through matter.

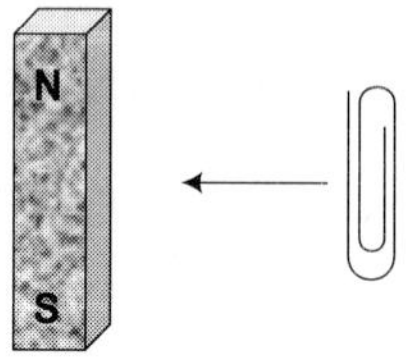

The space between the magnet and the metallic paper clip is altered invisibly when the clip is attracted to the magnet. It is the same magnetic field that surrounds current carrying wires as shown in the diagrams below.

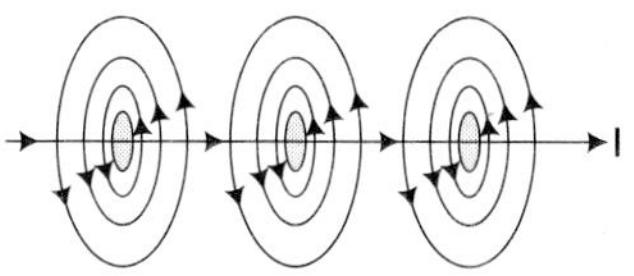

Diag A
(Wire carrying a current)

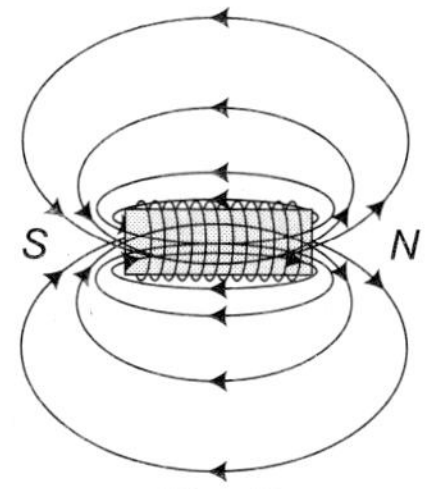

Diag B
(Solenoid carrying a current)

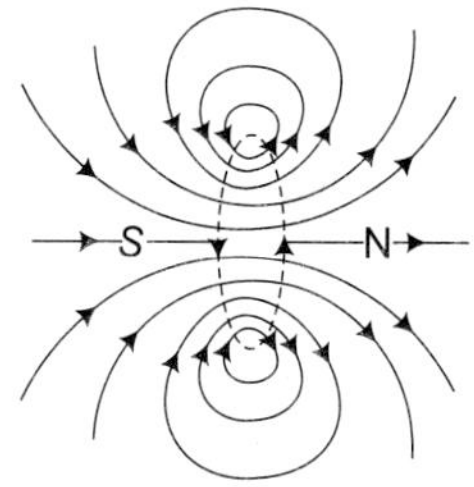

Diag C
(Circular current loop)

Magnetic field lines always form closed loops and are never unterminated. In Diag. C some lines are shown unterminated to fit into the figure.

There are some differences, however, between electric and magnetic interactions:

1. A magnet is *permanently magnetized* but, a charged plastic pen loses its charge within a relatively short time (see Section 1.25)...
2. A magnetic field around a current-carrying wire has a *curly* pattern that we don't see with electrostatic fields.

Magnetic effects are associated solely with ***moving charges*** whereas electric effects are evident with ***stationary and moving charges***.

5.2 MAGNETIC FIELD

We now know that a moving charge produces an invisible magnetic field and to determine the direction of the magnetic field we apply the right-hand rule as shown in Fig. 5.1. We speak of a magnetic field due to a steady current and not of a single point charge as we did in Section 2.27. Steady currents (millions and millions of electrons everywhere in the wires moving with a start-stop motion with a drift superimposed, not piling up anywhere) produce magnetic fields that are constant in time just as stationary charges produce constant electric fields in space.

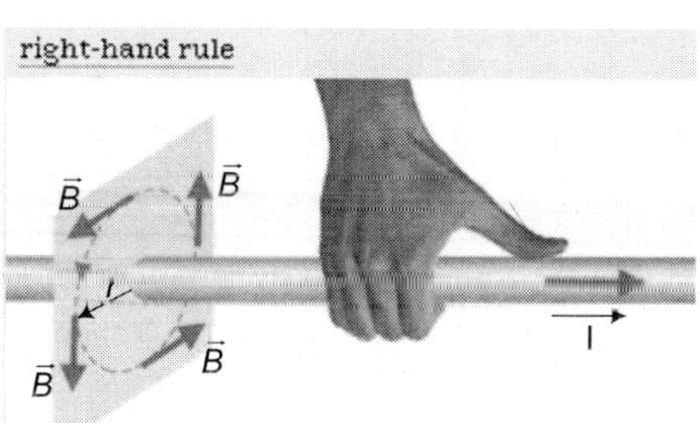

Fig. 5.1 The right-hand rule used to determine the magnetic field.

To apply the right-hand rule for determining the direction of the magnetic field produced by a steady current, imagine gripping the wire with your right hand so that your thumb points in the direction of current flow. Your fingers naturally curl in the direction in which the magnetic field circulates.

What forces produce the magnetic flux? Is it the current only? Or is its direction important?

Magnetomotive forces (mmf) and Magnetic Field Strength, H

Experiments show that for a homogeneous medium, for a long wire carrying current I (Fig. 5.1), **B**, the magnetic flux per unit area is proportional to $\frac{\mu I}{r}$, where, r is the distance from the wire and μ is a constant that depends on the medium. When the experiment was performed in vacuum, it was found that the proportionality factor is $\frac{1}{2\pi}$ giving, $B = \frac{\mu I}{2\pi r}$. This may be written as $B = \mu H$ where $H = \frac{I}{2\pi r}$ Amps/metre.

The magnetic field strength H is thus defined by this relation in terms of the current which produces it and the geometry of the system. H, the magnitude of the magnetic field strength may be visualized as a "stress" or a force per unit length that produces a flux density through its medium. Many students confuse magnetic field strength H with the magnetic field or flux density B. The magnetic field strength H is the cause and B, the flux density is its effect. The magnetomotive force (mmf) is a quantity that is defined as the line integral: $mmf = \int_a^b H.ds$ along the path between two points a and b. If the path is circular of radius r about the current- carrying wire, with the wire at the centre, H has the constant value $\frac{I}{2\pi r}$ and is directed along the path, so that $mmf = \int H.ds = I$ because the distance moved along the circle is $2\pi r$.

It is only a coincidence that mmf is I in this particular case and therefore cannot be generalized.

In an electric circuit, the current is due to the existence of an electromotive force. By analogy, we may say that in a magnetic circuit the magnetic flux is due to the existence of a magnetomotive force (mmf) caused by a current flowing through one or more turns. The value of the mmf is proportional to the current and the number of turns, and is expressed in ampere turns; but for the purpose of dimensional analysis, it is expressed in amperes, since the number of turns is dimensionless. Hence, the unit of magnetomotive force is the ampere.

If there is a current of I amperes in a coil of N turns, as shown in the figure below of a toroid, the magnetomotive force F is the total current linked with the magnetic circuit, namely IN amperes.

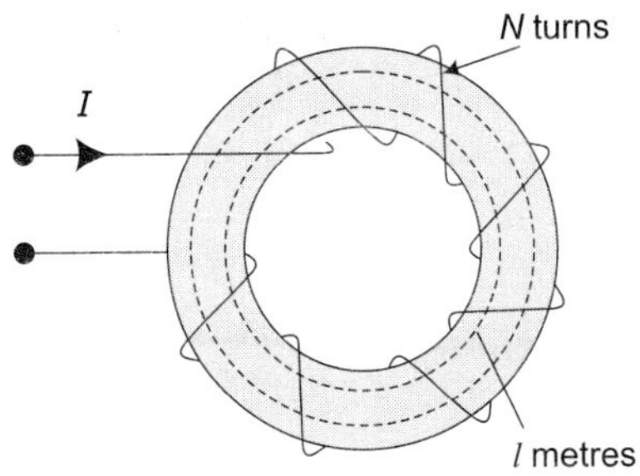

A toroid

If the magnetic circuit is homogeneous and of uniform cross-sectional area, the magnetomotive force per metre length of the magnetic circuit is also the magnetic field strength *H* and if the mean length of the magnetic circuit is *l* metres,

$$H = \frac{N \times I}{l} \quad \text{amperes per metre and,}$$

$$H = \frac{F}{l} \quad \text{where the magnetomotive force } F = NI \text{ amperes.}$$

Biot-Savart Law

Careful experimentation shows that a moving point charge not only makes an electric field but also makes a magnetic field that is given by the Biot-Savart law.

The Biot-Savart law for a single moving source charge gives the magnetic field B at the observation location as shown in Fig. 5.2 by

$$\vec{B} = \frac{\mu_o}{4\pi} \frac{\vec{q}v \times \vec{r}}{r^2}$$

in *tesla* where, *v* is the velocity of the source point charge *q* and *r* is a unit vector that points from the source charge to the observation location.

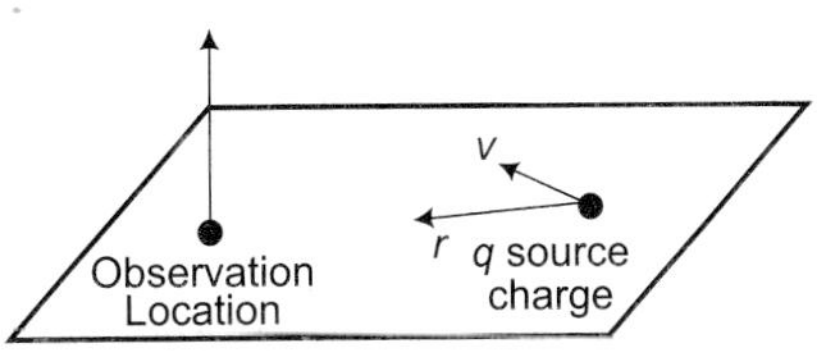

Fig. 5.2 Biot-Savart law to determine the magnetic field due to a moving charge.

Right-hand rule for vector cross-product (Chabay & Sherwood, 1995, used by permission)[2]

A simple method for determining the direction and magnitude of a vector cross-product is described below with reference to Fig. 5.3.

Keep your right hand open and flat (fingers and thumb in the same plane), with your thumb perpendicular to your fingers, point your fingers in the direction of the first vector (*v*).

Now, bend your fingers toward the second vector (r), noting the angle of bend θ, which is always less than 180°.

You may have to rotate your wrist to get your hand into a position that lets you bend your fingers this way.

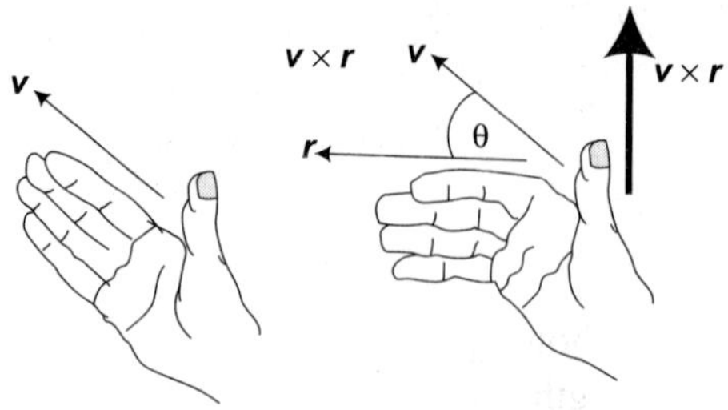

Fig. 5.3 The right-hand rule for the magnitude and direction of $\mathbf{v} \times \mathbf{r}$.

The thumb points in the direction of cross product $\mathbf{v} \times \mathbf{r}$.

The magnetic field calculated by applying the Biot-Savart law (Chabay & Sherwood, 1995, used by permission)[2] for a circular loop of radius a and conventional current I at a distance x from the centre, along the axis, is as shown in Fig. 5.4:

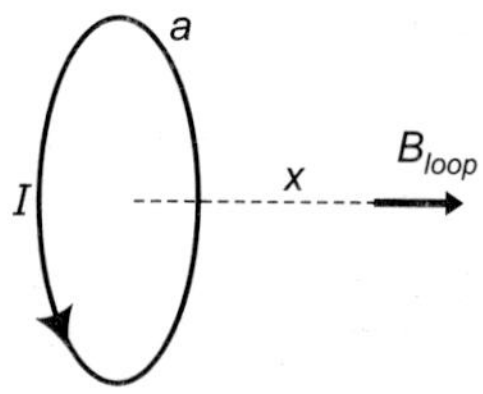

Fig. 5.4 The Biot-Savart law used to find the magnetic field B.

and is given by,

$$\vec{B} = \frac{\mu_o}{4\pi} \frac{2\pi a^2 I}{(x^2 + a^2)^{3/2}}.$$

The magnetic field for a circular loop of radius a and conventional current I at a distance x from the centre, *far away* along the axis, is as seen in Fig. 5.5 and,

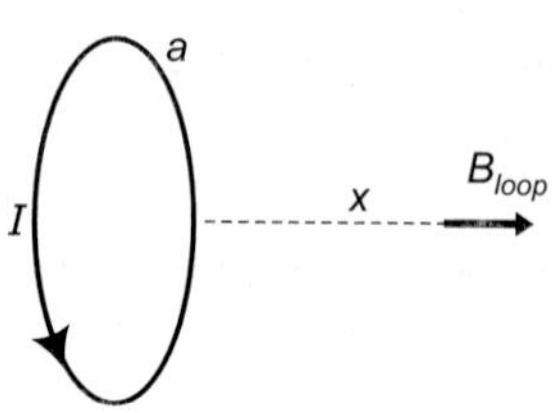

Fig. 5.5 The Biot-Savart law used to find B far away from the loop.

is given by

$$\vec{B} = \frac{\mu_o}{4\pi} \frac{2\pi a^2 I}{x^3}.$$

In other words, the magnetic field falls away fairly rapidly by $1/(\text{distance})^3$ as we move away from a current carrying loop.

Magnetic Flux vs Magnetic Field (Chabay & Sherwood, 1995, used by permission).[2]

The magnetic flux on a small area ΔA is $\vec{B}.\vec{n}\,\Delta A$ where, n is a dimensionless unit vector perpendicular to the area ΔA (n is called the normal in the surface). A diagram that illustrates the flux of a magnetic field B is shown below.

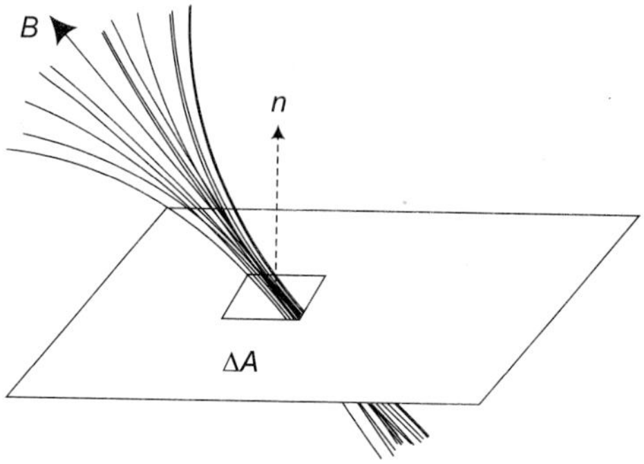

We add up all magnetic field contributions over an extended surface to get the magnetic flux

$$\phi_{magnetic} = \int_{Surface} \vec{B}\cdot\vec{n}\,dA$$

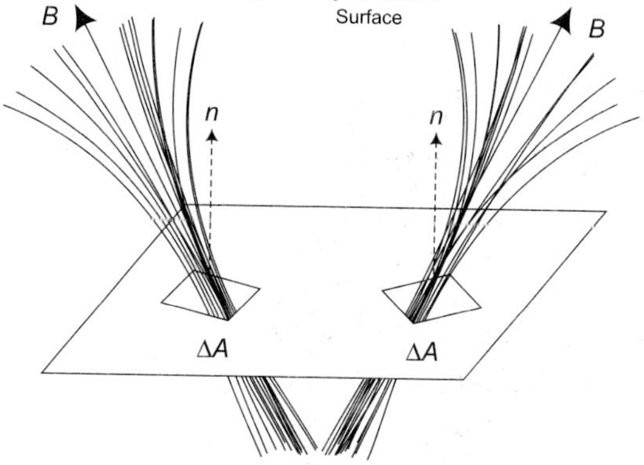

5.3 CHARGES AND ELECTRIC FIELDS AND MAGNETIC FIELDS

We have learnt electric and magnetic interactions as separate phenomena. Charges ***make*** electric fields, and electric fields ***affect*** charges.

Moving charges ***make*** magnetic fields, and magnetic fields ***affect moving*** charges

Yet, there is a deeper relationship between the two kinds of fields.

A ***stationary*** charge makes ***no*** magnetic field.

A ***time-varying*** magnetic field can ***produce*** an electric field.

The unit of magnetic field is "tesla".

5.4 CURRENTS AND MMF

The effect of the current in a coil of wire is to provide an mmf (magnetomotive force) just as the chemical reactions within a car battery provide an emf.

The electric circuit and the magnetic circuit are analogous and just as low resistance copper wire connects the battery to the high resistance of the headlight bulb filament, so the ferrite core connects the coil with the air gap.

To produce a larger emf, we connect batteries or sources of emf ...

... in series.

To produce a larger mmf, we connect coils with currents ...

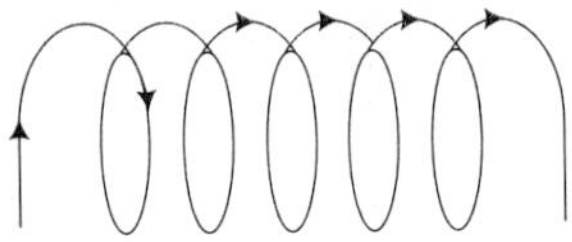

... in series.

mmf = No. of turns × current

The electric circuit with its increasing emf produces more current.

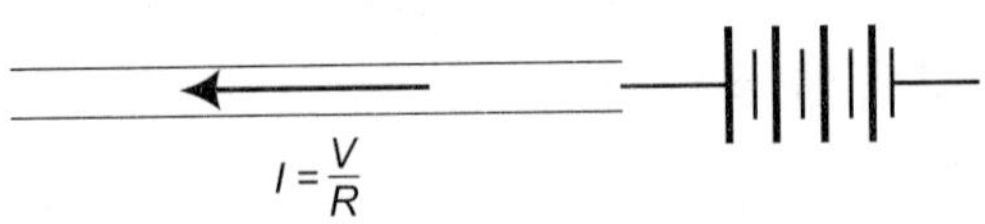

The magnetic circuit with its increasing mmf produces more magnetic field lines and flux.

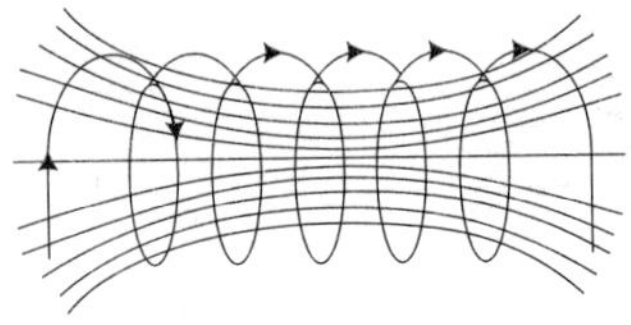

$$\textbf{Flux } \phi = \frac{MMF}{Reluctance}$$

Reluctance is a concept used in the analysis of magnetic circuits and is analogous to resistance in electric circuits. But, while a resistance dissipates energy, reluctance stores magnetic energy in the magnetic field. Reluctance is defined as the ratio of the magnetomotive force (mmf) that produces a magnetic flux ϕ in a magnetic circuit. It is expressed in ampere-turns per weber. Air and vacuum have high reluctance, while easily magnetizable materials such as soft iron have low reluctance. In later chapters, we shall see that the air gaps between the rotating structure and the stationary magnetic field sourcing coils in motors and generators are kept small to make the gap reluctance small.

5.5 PRODUCING ELECTRIC FIELDS

There are two different ways by which an electric field can be produced:

1. By charges according to ***Coulomb's law***.

When charges are separated, a coulomb electric field is produced around them, for example, charges on the end plates of a battery and dipoles of atoms and molecules.

2. By ***time-varying magnetic fields***.

An electric field co-occurs with a time varying magnetic field, and the phenomenon is referred to as magnetic induction.

The historical term "magnetic induction" is often used to describe the phenomenon, and one says that the time-varying magnetic field "induces" the electric field (that we shall see has a different pattern in space from that produced by point charges). This is somewhat misleading, as the source of both the time-varying magnetic field and the curly electric field is varying (nonconstant) currents, which we will simply refer to as Faraday's law. ("Electric and Magnetic Interactions", Chabay and Sherwood, Wiley, 2010).

5.6 EFFECT OF ELECTRIC FIELDS ASSOCIATED WITH TIME-VARYING MAGNETIC FIELDS ON CHARGES

No matter how an electric field is produced, it has the same effect on a charge q; Force = qE.

But the non-Coulomb electric field that occurs with a time-varying magnetic field has a different pattern in space than the Coulomb electric field $E = \dfrac{q}{4\pi\varepsilon_0 d^2}$.

In a "Coulomb" electric field the round-trip path integral is 'zero'.

But, a round-trip path integral of the "non-Coulomb" electric field associated with a time-varying magnetic field is *not zero*.

5.7 SOLENOIDS AND INDUCTORS

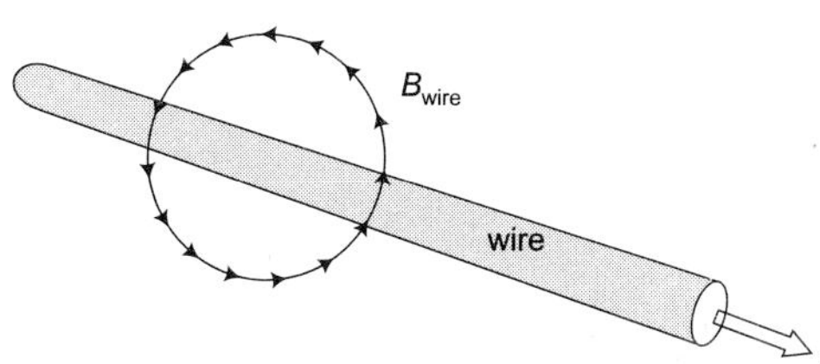

A long straight coil of wire can be used to generate a nearly uniform magnetic field similar to that of a bar magnet. Such coils, called solenoids (used as electromagnets and for actuation of mechanical valves) and inductors (used to smoothen out current variations in electronics and electrical circuits), have an enormous number of practical applications. The field can be greatly strengthened by the addition of an iron core. The figure below shows the schematic of an inductor and inductors of different sizes, shapes and packaging.

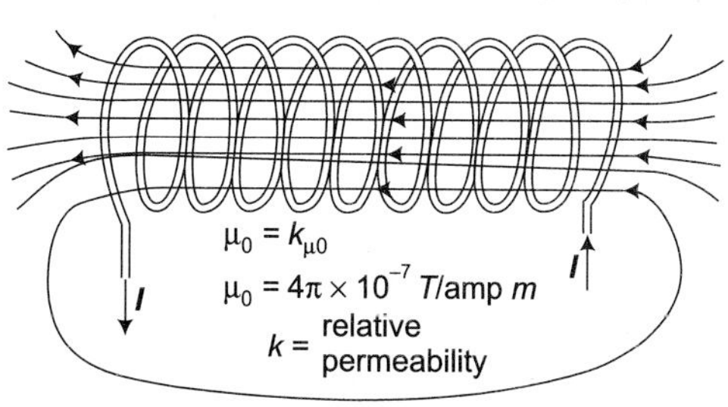

Consider a very long solenoid, a long hollow coil of current-carrying wire as shown in Fig. 5.6.

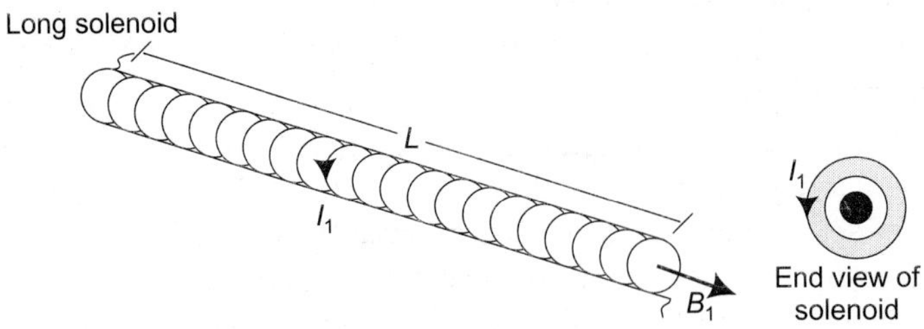

Fig. 5.6 Magnetic field inside a current carrying solenoid.

With a current I_1 in the solenoid of N turns and length L, the magnetic field B_1 inside the solenoid (not too near the ends) is

$$B_1 = \frac{\mu_0 N I_1}{L}$$

The magnetic field outside the solenoid is (not too near the ends) very small and for a very long solenoid can be taken to be "zero". The derivation of this expression can be found in "Physics" by Resnick and Haliday[13].

5.8 CURLY ELECTRIC FIELDS

A charge moving outside the central region of this solenoid will not experience an electric or magnetic force because the electric and magnetic fields outside the solenoid are essentially zero *if B_1 is constant in time* as shown in Fig. 5.7 (Chabay & Sherwood, 1995, used by permission)[2].

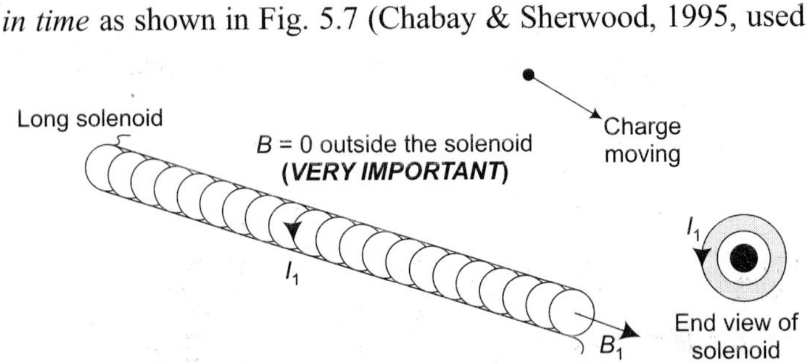

Fig. 5.7 Charge moving outside does not experience a force.

But something remarkable happens if the magnetic field B_1 inside the solenoid varies with time. There is still no magnetic field outside the solenoid, but, we observe an electric field both inside and outside the solenoid as shown in Fig. 5.8(a) and 5.8(b).

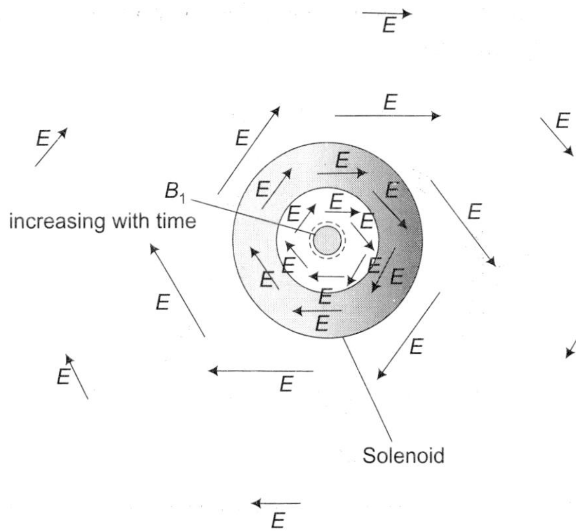

Fig. 5.8(a) Electric field co-occurring with a time-varying magnetic field.

Outside the solenoid, the electric field is proportional to $1/r$. Inside the solenoid, the electric field is proportional to r, the distance from the axis (smaller near the axis). These are indicated by the varying size of the electric field vectors inside and outside the solenoid.

This electric field *curls* around the axis of the solenoid.

This peculiar electric field is proportional to dB/dt, the *rate of change* of the magnetic field. This odd-looking electric field acts in the usual way:

a charge q experiences a force $F = qE$

A proton placed above the solenoid will be initially pushed to the right as shown in Fig. 5.8(b). The curly electric field does not originate or terminate on point charges and that is the reason why we call such an electric field a **"non-Coulomb"** field.

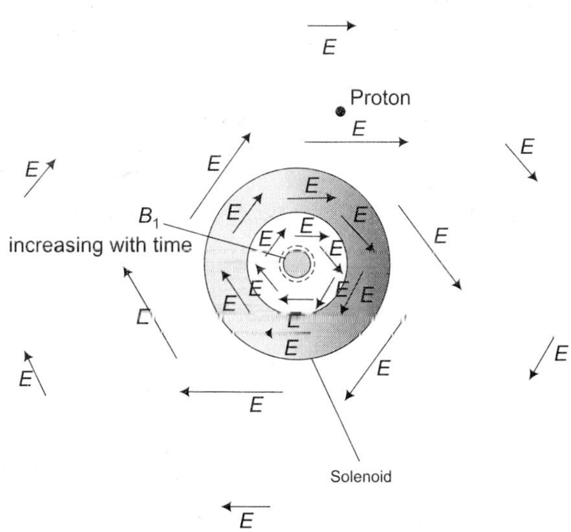

Fig. 5.8(b) Electric field co-occurring with a time-varying magnetic field.

Suppose we place a circular metal ring around the solenoid as shown in Fig. 5.9. (Chabay & Sherwood, 1995, used by permission).[2]

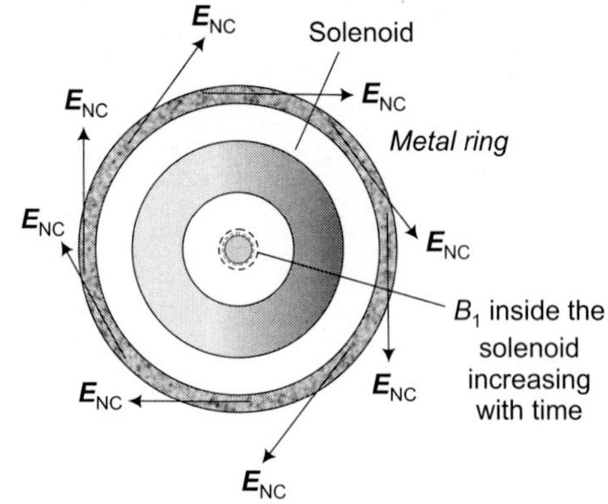

Fig. 5.9 Non-Coulomb electric field co-occurring with a time-varying magnetic field.

With the magnetic field pointing out and *increasing with time* the *non-Coulomb electric field* inside the metal will drive conventional current clockwise around the ring (Chabay & Sherwood, 1995, used by permission)[2].

The practical importance of the occurrence of a non-Coulomb electric field is that it can make current run in a wire just as though a battery were present.

Recollect what was said about the nature of the electric field in Sections 1.4 and 1.11: an electric field invisibly alters space and exerts a continuous and relentless force on charged particles causing them to accelerate and is the cause of the current (counterclockwise start-stop movement of millions and millions of electrons everywhere in the wires with a drift superimposed) in the ring of Fig. 5.9. The electric field fills the ring.

In Chapter 2, we had asked a question: Is it possible to set up a configuration of charged objects in such a way that the electric field along a circular path looks like this? **The answer was "No". But this curly non-Coulomb field E_{NC} field can be associated with a time-varying magnetic field.**

Note: Some researchers of the phenomenon of the curly electric field appearing with a time-varying magnetic field, state that the rate of change of *B* is the effect, with the source of the effect being the curl of *E*! Because the distinction is not so important, Sherwood and Chabay in the third edition of their book "Matter and Interactions", John Wiley & Sons, 2010, state that the changing magnetic field and the curly electric field, co-occur.

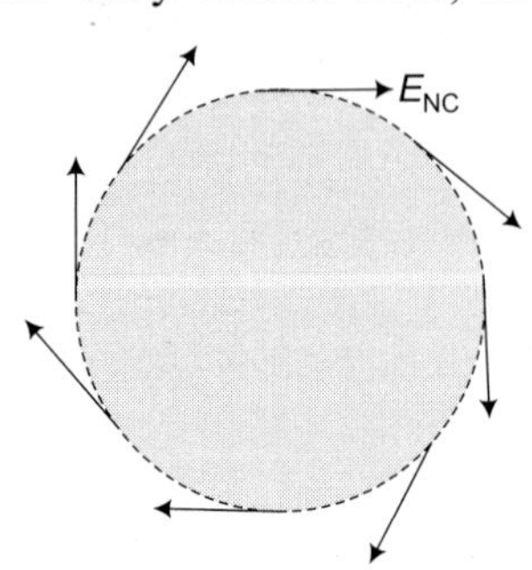

Most Physics texts show figures similar to the one below to describe Faraday's law which associates a non-Coulomb electric field with a time-varying magnetic field.

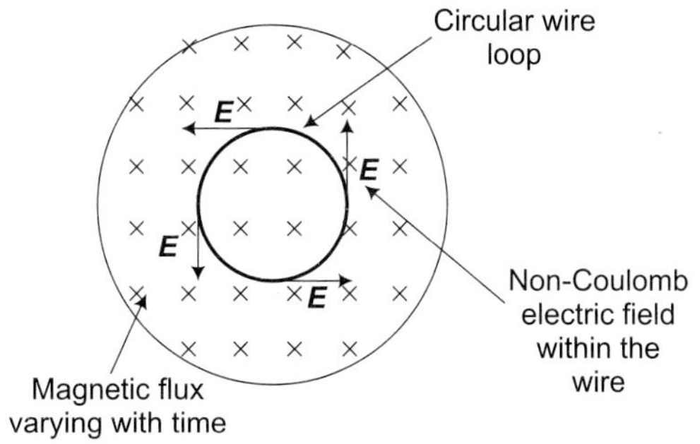

Notice how the circular wire loop is totally embraced and is cut by the changing magnetic flux.

But, see Fig. 5.10 in the next section and you will notice that what counts is the magnetic flux encircled by the circuit, and not the magnetic flux on a closed surface such as a box or sphere as we will see in section 5.12, Fig. 5.11b.

5.9 CURLY ELECTRIC FIELDS AND POTENTIAL DIFFERENCE

An electric field co-occurs in an area where there is no magnetic field with a time-varying magnetic field.

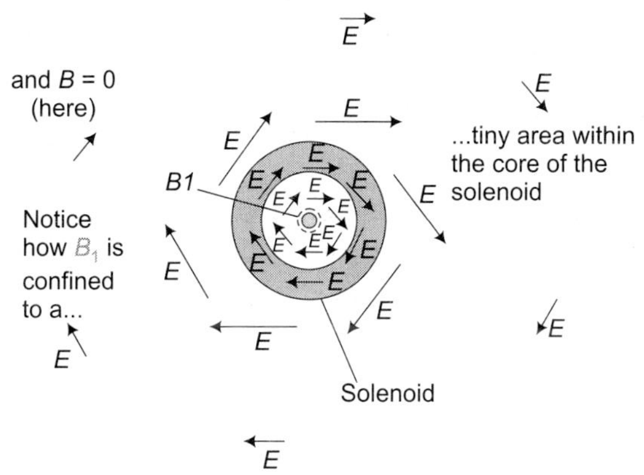

Fig. 5.10 Non-Coulomb electric field (in an area where there is no magnetic field) co-occurring with a time-varying magnetic field.

There is no magnetic field outside the solenoid.

Yet, an electric field that has a curly pattern co-occurs with the time-varying magnetic field (due to the varying current in the solenoid wire) outside the solenoid where, $B = 0$ or nearly zero. An electric field also occurs inside the solenoid.

Non-conservative fields

From: Massachusetts Institute of Technology lecture notes[12].

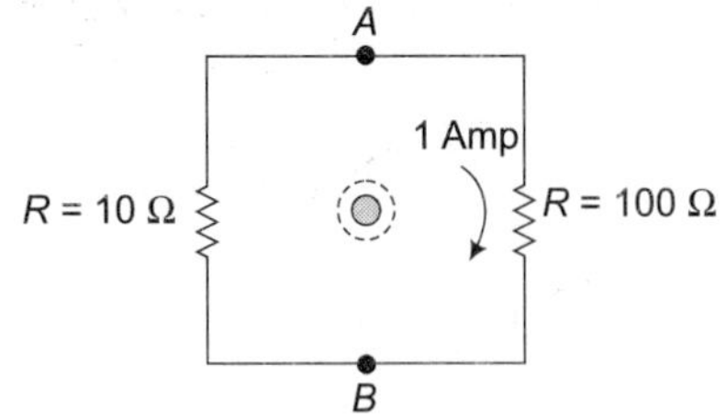

A ***magnetic field penetrates*** this circuit ***outwards***, and is increasing at a rate such that a current of 1 A is induced in the circuit (which direction?)

The ***potential difference*** $V_A - V_B$ is:

1. +10 V	2. –10 V	3. +100 V	4. –100 V
5. +110 V	6. –110 V	7. +90 V	8. –90 V

9. None of the above

The Answer is 9. None of the above.

The question is meaningless. There is no such thing as potential difference when a changing magnetic flux is present. This is because by Faraday's law, a non-conservative ε is induced (that is, its integral around a closed loop is non-zero).

A live demonstration of non-conservative fields was given by Professor Walter Lewin of Massachusetts Institute of Technology, Boston, USA and is available at http://ocw.mit.edu/courses/physics/8-02-electricity-and-magnetism-spring-2002/video-lectures/lecture-16-electromagnetic-induction and http://www.youtube.com/watch?v=G3e14SVDyME for a circuit as shown below.

The two voltmeters connected across the same points A and B show different values for a short time when the magnetic field is increasing in the loop set up with the two resistors R_1 and R_2. The circuit was arranged to develop an induced emf of about 1 volt. The voltage across the resistor R_2 was recorded at 0.9 V and across R_1 at –0.IV, and are not equal, as they should have been according to Kirchchoff's law. The reason for this behaviour is because the loop is in a non-conservative field and the results are not as one would have expected.

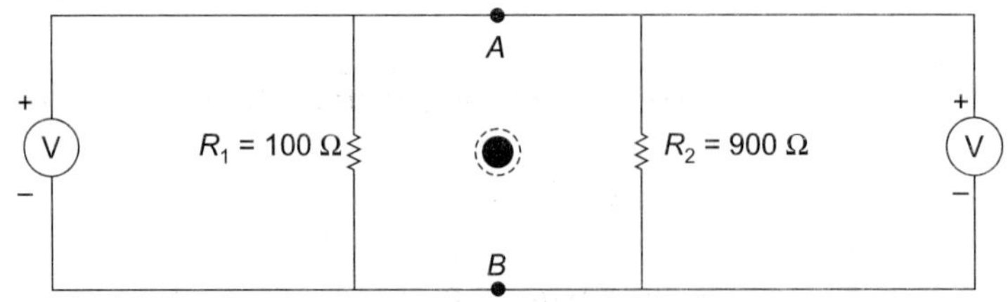

A step-by-step derivation of the above astounding result is given by Prof. Walter Lewin and is available at http:/ocw.mit.edu/courses/physics/8-02-electicity-and-magnetism-spring-2002/lecture-notes/lecsup315.pdf. In the article, it is demonstrated that Kirchhoff's law is applicable only to conservative fields and Faraday's law is applicable to non-conservative fields.

5.10 CONSERVATIVE AND NON-CONSERVATIVE ELECTRIC FIELDS

Let us for a moment assume that potential had a meaning in Faraday's law. If so, then imagine how you would place the surface charge on the ring.

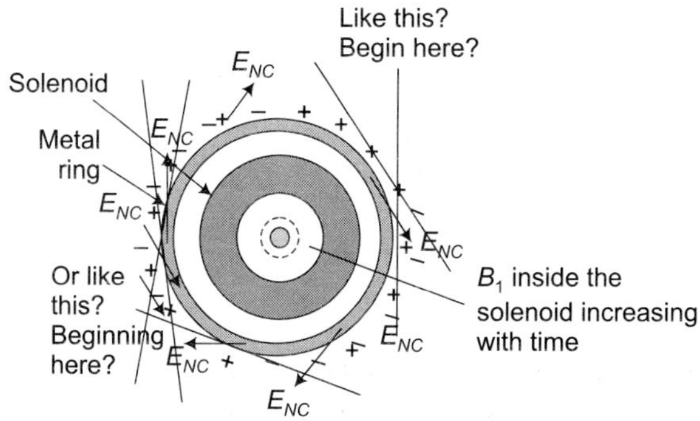

The symmetry of the ring makes it **impossible** to have a surface charge gradient along the ring.

The electric fields that occur in association with a changing magnetic flux are not associated with point charges. Although both kinds of electric fields exert forces on charges, there is a *difference between them*. The simplest manifestation of this difference is that flux lines of E associated with a *changing magnetic flux* can form *closed loops*.

Flux lines of E associated with *charges* always can be drawn to start on a positive charge and end on a negative charge.

5.11 CONSERVATIVE FIELDS (CHARGE SEPARATION-COULOMB FORCES)

We know that the potential difference between the two points is

$$V_b - V_a = \frac{W_{ab}}{q_o} = -\int_a^b E.dl$$

and we know that if potential is to have a real meaning then this integral (and W_{ab}) must have the same meaning for every path connecting a with b and is true for every case as explained in physics textbooks.

An interesting special case is when the points a and b are the same; the starting point and the ending point are the same. Then the path connecting them forms a closed loop and V_a must be identical with V_b and this equation reduces to

$$\oint_{a,b} E.dl = 0$$

This happens when the fields are "*conservative*". See Section 2.19 "Conservative and non-conservative electric fields" and the topic "Allowable patterns of electric fields using charged objects".

Recollect from Section 2.17 "What is voltage?" that the round-trip path integral of the electric field due to a point charge (includes any static charge distribution and surface charges of circuits) is "zero" no matter which path is taken.

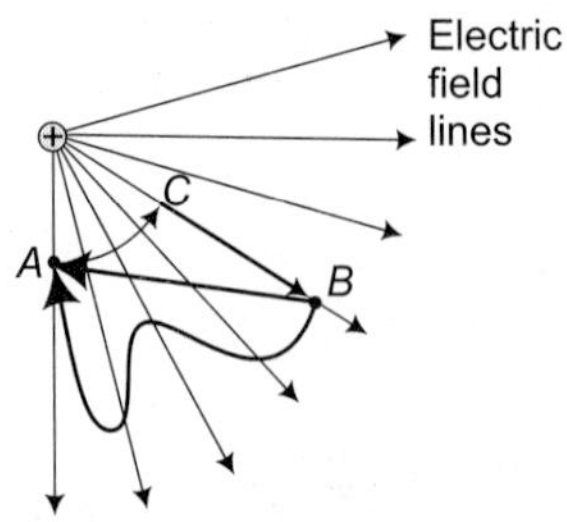

The work done to move a test charge from A to C to B and back to A is "zero".

Such distributions can have a *divergence* (the electric field flux lines diverge and curve around from the distribution originating and terminating on point charges) but never produce a curl that will close the field flux lines on themselves.

5.12 FARADAY'S LAW AND NON-CONSERVATIVE FIELDS

However, when changing magnetic flux is present,

$$\oint_{a,b} E.dl$$

is precisely *not zero* and is according to Faraday's law:

$$-d\phi_B/dt$$

where ϕ_B is the total magnetic flux encircling the circuit as described in Section 5.2. Faraday's law is a major physical law and cannot be derived from other known physical principles.

Electric fields associated with stationary charges are *conservative*, but those associated with changing magnetic fields are *non-conservative*.

Electric potential, which can be defined only for a conservative force, has *no meaning* for electric fields associated with *time-varying magnetic fields*.

Non-conservative fields can't have potentials associated with them.

As long as the curly non-conservative Faraday electric field occurs with a changing magnetic field B, the divergence of the field is zero. Such a field may have a non-zero curl and its field flux lines will close on itself; and, if it is pure, that is, not influenced by a coulombic field due to point charges around, it will have a curl (due to the curly pattern), but its divergence will be "zero".

Compare the field as in Fig. 5.8 with the field due to point charges and static charge distributions of Fig. 2.63 and the figures in Section 1.6. Also refer Section 2.19 "Conservative and non-conservative electric fields".

What would be the reading of a voltmeter whose probes are connected together around a solenoid whose current is made to vary thereby causing the magnetic field B_1 to vary as shown in the Fig. 5.11(a)? (Chabay & Sherwood, 1995, used by permission)[2].

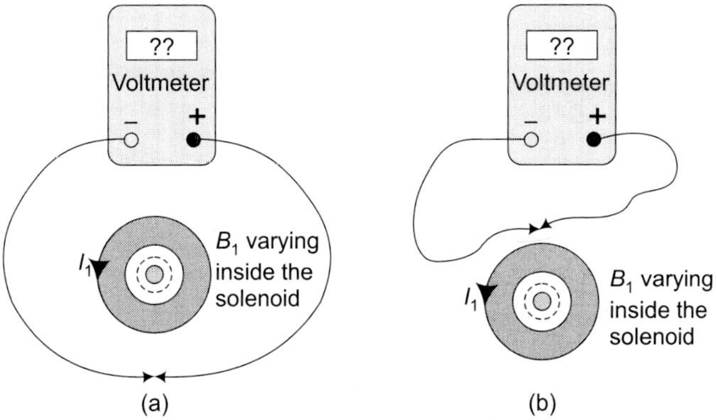

Fig. 5.11 emf due to a time-varying magnetic field.

What would be the reading of a voltmeter whose probes are connected together around a solenoid as shown in the Fig. 5.11(b)?

A voltmeter is an ammeter with a large resistance in series; a voltmeter may give a reading in the presence of a time-varying magnetic flux.

Since the electric circuit encircles the magnetic field inside the solenoid, the voltmeter will show a nonzero emf in Fig. 5.11(a) and is shown in Fig. 5.11(c).

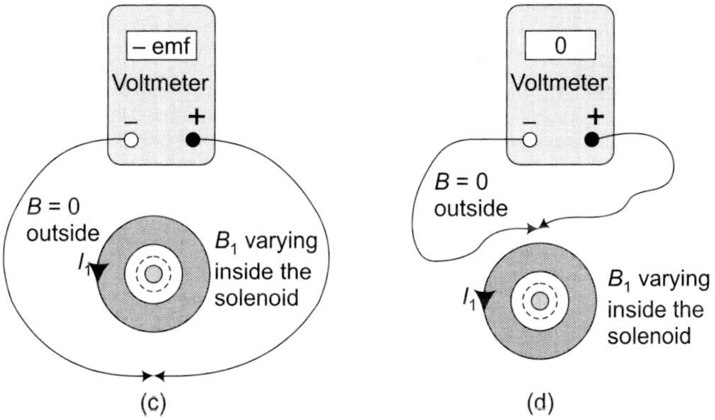

Fig. 5.11 emf due to a time varying magnetic field.

Due to the changing flux through the loop, a tiny current runs through the voltmeter from – to + and the reading is negative.

The voltmeter will show 0 volt when connected outside the solenoid and does not encircle it as in Fig. 5.11(b). Because there is no changing flux through the loop, there is no emf; therefore, the reading is zero as shown in Fig. 5.11(d).

Notice that what counts is the magnetic flux *encircled* by the circuit, **not** the magnetic flux on a *closed surface* such as a box or sphere.

Being quantitative about the emf, experimentally it is found that the magnitude of the induced emf is numerically equal to this:

$|emf| = \left|\frac{d}{dt}\left(B_1 \pi r_1^2\right)\right|$ where r_1 is the radius of the solenoid and B_1 the magnetic field in the solenoid.

The quantity $(B_1 \pi r_1^2)$ is called the "magnetic flux" Φ_{mag} (as we described in Section 5.2) on the area encircled by the circuit (the ammeter and its wires).

5.13 DIRECTION OF INDUCED EMF AND LENZ'S RULE

Consider an inductor with a current that is increasing such that the magnetic field due to it is pointing out and increasing as shown in the Fig. 5.12(a) (after Chabay & Sherwood, 1995).

We have seen the direction of the non-Coulomb electric field earlier. It is clockwise.

If the magnetic field points out of the page and is decreasing, the non-Coulomb electric field curls around in the opposite direction as shown in the Fig. 5.12(b). It is not the direction of $\vec{B}_1$ that determines the direction of $\vec{E}_{NC}$, but rather the direction of the *rate of change* of $\vec{B}_1$.

In the Figs. 5.12 a,b,c and d are shown all four cases, with the magnetic field pointing out or into the page, and increasing or decreasing with time (Chabay & Sherwood, 1995, used by permission)[2].

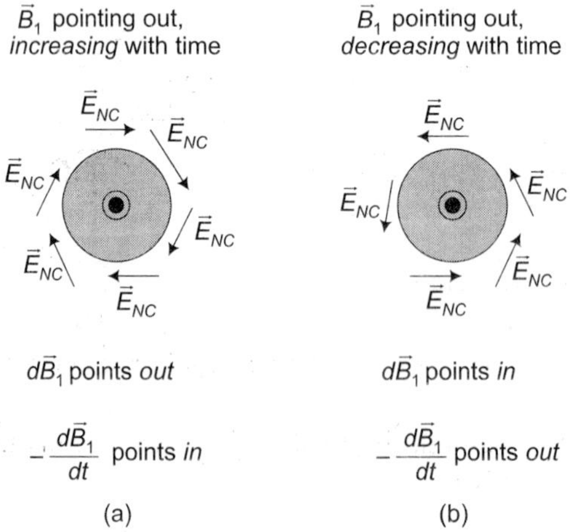

Fig. 5.12 Non-Coulomb electric field with a time-varying magnetic field.

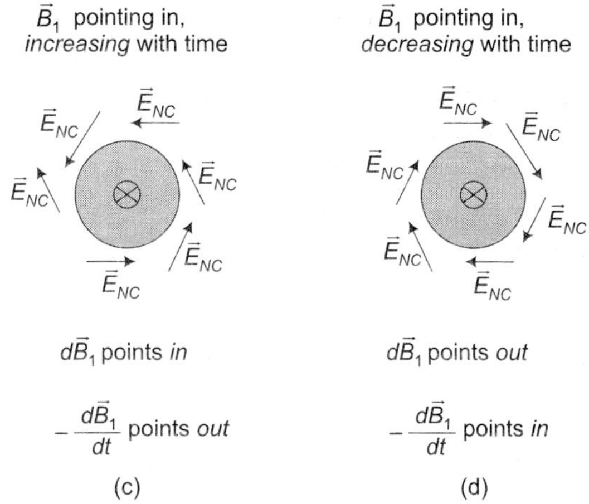

Fig. 5.12 Non-Coulomb electric field with a time-varying magnetic field.

A right-hand rule covers all of these cases. With the thumb of your right hand pointing in the direction of $-\dfrac{d\vec{B}_1}{dt}$, your fingers curl around in the direction of $\vec{E}_{NC}$.

Note: A good way to find the direction of $-\dfrac{d\vec{B}_1}{dt}$ is to draw the magnetic field $\vec{B}_1(t)$ at a time t and the magnetic field $\vec{B}_1\,(t + \Delta t)$ at a slightly later time $(t + \Delta t)$, and oberve the change $\Delta\vec{B}_1$. Then $-\ \Delta\vec{B}_1$ is the direction of $-\dfrac{d\vec{B}_1}{dt}$.

Lenz's rule is an easier rule to use in many situations. Imagine that you place a wire around the changing flux, so that the non-Coulomb electric field drives current in the wire, then, according to Lenz's rule, the (non-Coulomb) electric field would drive current in a direction to make a magnetic field that *attempts* to keep the flux *constant*.

For example, if $\vec{B}_1$ points out of the page and is increasing, the non-Coulomb electric field curls *clockwise* around the solenoid as in the Fig. 5.12(a).

If a wire encircles the solenoid, conventional current runs clockwise in the wire, and this induced current produces a magnetic field in the region. The direction of the magnetic field due to the clockwise current will be *into the page* at the centre of the wire, decreasing the net flux

This magnetic field tries to keep the magnetic flux constant despite the increase in B_1. The induced current does not succeed at keeping the flux constant (unless the wire is made of superconducting material). In particular, if B_1 is increasing at a constant rate, the current I_2 in the ring is constant by Faraday's law, so the net flux does increase.

If on the other hand B_1 points out of the page and is decreasing (the Fig. 5.12(b)), there is a counterclockwise non-Coulomb electric field which would drive conventional current counterclockwise in a wire, which would make a magnetic field out of the page, in the direction to try to keep the magnetic flux constant despite the decrease in B_1. (Again, the net flux will nevertheless decrease, unless the ring is superconducting.)

The reader is urged to go through the examples Figs. 5.12(c) and Fig. 5.12(d) and verify that Lenz's rule does correctly summarize the experimental data for the direction of the non-Coulomb electric field.

5.14 DIFFERENCE BETWEEN "EMF" AND "POTENTIAL DIFFERENCE"

It is important to keep in mind that although the units of emf are volts, the emf is not a potential difference (Chabay & Sherwood, 1995, used by permission)[2].

Potential difference is a path integral of the electric field made by real charges. On the other hand, emf is the "non-Coulomb" work per unit charge as is the case with chemical actions and say, a conveyor belt in the mechanical battery that causes a separation of charges by other than coulombic means. An example of separation of charges by coulombic means is by the external charge that caused polarization in a metal resulting in a field due to surface charge as shown in Fig. 1.13.

Although the units of emf are volts, the emf **is not** a potential difference.

For example, the **emf** of a flashlight battery is 1.5 **volts**, and is associated with charge separation due to chemical reactions in which electrons are reactants or products.

In the mechanical battery that was described in Chapter 1, $\mathbf{F}_{NC}$**s** in the energy conservation equation represents the work done in moving one electron through the mechanical battery, or work per unit charge (joules per coulomb, or volts). **This is the emf.**

5.15 BOTH COULOMB AND NON-COULOMB ELECTRIC FIELDS IN A NON-UNIFORM CONDUCTOR

We have seen the co-occurrence of a *non-Coulomb electric field* with a time-varying magnetic field and a current due to the resultant emf in a uniform circular metal ring.

If there is a thin section of ring, however, charge will *pile up* at the ends of the thin section, and there will be surface charge all along the ring (Chabay & Sherwood, 1995, used by permission)[2]. This results in the production of a coulomb electric field.

We had seen earlier that when a uniform metallic ring encircled a solenoid producing a time-varying magnetic field, there was a current driven around the ring and there were no surface charges along the ring.

Consider a non-uniform metallic ring encircling a solenoid as shown below in Fig. 5.13. A $\mathbf{E}_{NC}$ (a non-Coulombic electric field) associates with the changing magnetic flux. Surface charges will pile up at the edges of the non-uniform section. These surface charges will produce a Coulomb electric field $\mathbf{E}_C$ which adds vectorially to the non-Coulomb electric field.

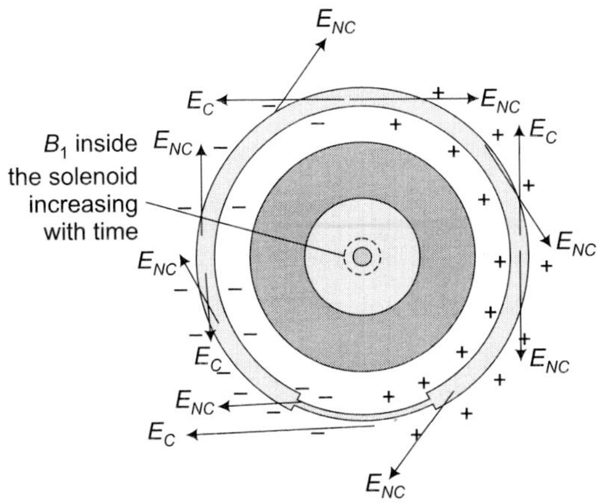

Fig. 5.13 Both Coulomb and non-Coulomb electric fields in a ring.

The net electric field $E_{NC} + E_C$ is what drives the electrons around the non-uniform ring.

5.16 CURLY ELECTRIC FIELDS AND INDUCED EMFS

An attempt to *change the current* (increase or reduce) in a coil induces an emf in *the same coil*. Since this self-induced *emf acts* in a direction to *make a current* that would *oppose* the flux *change*, there is a kind of sluggishness in any coil of wire. This is demonstrated in the circuit in Fig 5.14 (Chabay & Sherwood, 1995, used by permission)[2].

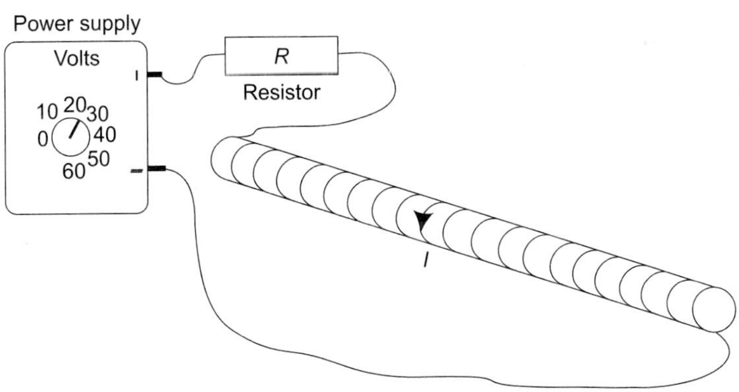

Fig. 5.14 Self-induction in a coil.

It is *hard* to change the current. The attempt to change the current in the coil induces an emf in the same coil. The self-induced emf acts in a direction to make a current that would oppose the flux change which is the attempt to increase or reduce the current, say, by increasing or decreasing the voltage in the circuit of Fig. 5.14. This makes the coil of wire act sluggishly. The current will change slowly in response. If there was no coil, the current would have changed *instantly* in response to a change in input voltage. Energy would simply be radiated away in the resistor, not to *return*. A resistor does not store energy. It takes it away from a circuit.

Now, let us go back to Fig. 5.14.

If the current is increasing, the **non-Coulomb electric field $\mathbf{E}_{NC}$** ***curls clockwise*** as shown in Fig. 5.15.

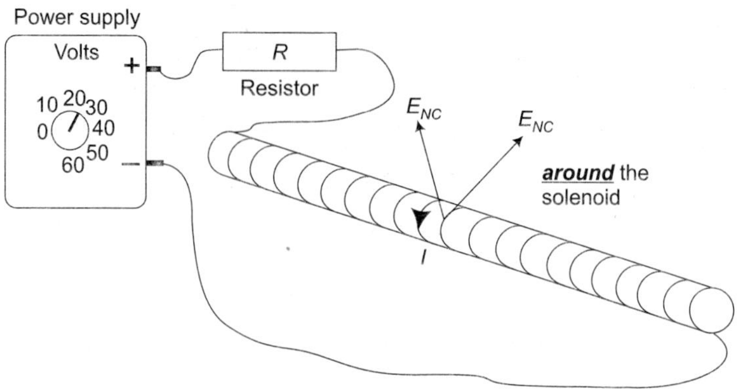

Fig. 5.15 Direction of the self-induced emf.

The electric field E_{NC} polarizes the solenoid and *surface charges* appear that produce a coulomb electric field E_C as shown in Fig. 5.16.

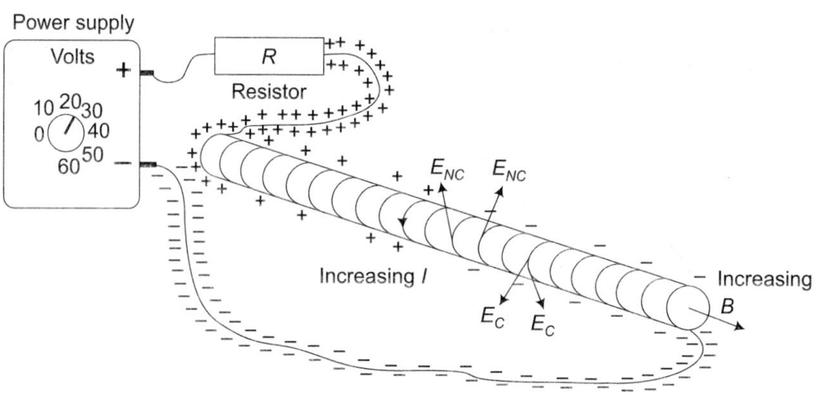

Fig. 5.16 Polarization of the solenoid.

If the current increase is more rapid, the non-Coulomb electric field E_{NC} grows and (remember that this is an electric field that co-occurs due to Faraday's law and not due to point charges) continues to polarize the solenoid and *surface charges* that appear *produce a large Coulomb electric field* E_C. In the final stage of development, a strong electric field is set up in the resistor and the field (not forgetting that an electric field is an invisible *alteration* of space) follows the wire and points opposite to the *non-Coulomb field* (Chabay & Sherwood, 1995, used by permission)[2]. This is shown in Fig. 5.17.

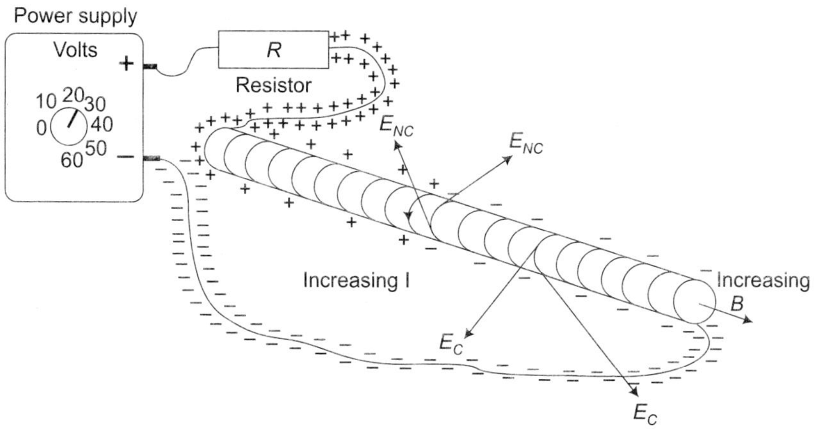

Fig. 5.17 Production of a Coulomb electric field in a solenoid.

If the resistance of the solenoid is very small, the *magnitude of the two electric fields is nearly equal* ($E_C \sim E_{NC}$). Note the similarity to a battery: a *non-Coulomb* $emf_{induced}$ maintains a charge separation and, there is a potential drop ΔV along the solenoid that is numerically equal to $emf_{induced}$. If the wire resistance $r_{solenoid}$ is significant, we have $\Delta V = emf_{induced} - r_{solenoid} \times I$, just like a battery with internal resistance. This similarity to a battery is shown in Fig. 5.18.

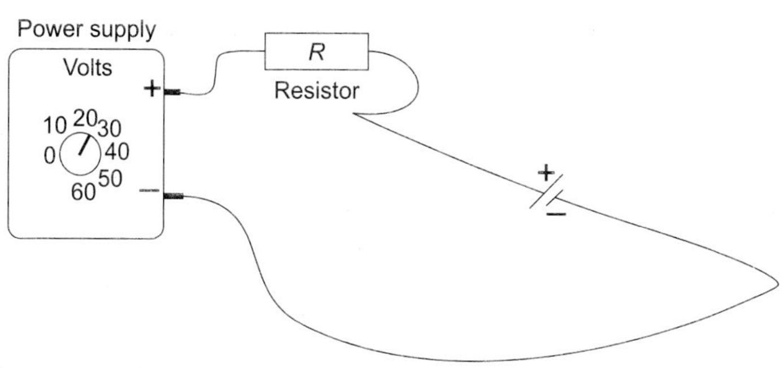

Fig. 5.18 Production of a Coulomb electric field in a solenoid.

It might not have occurred to you that a curly electric field cannot directly oppose the applied e-field due to the increasing emf that is attempting to change the current in the solenoid.

The curly field polarizes the solenoid and shifts the electron sea resulting in surface charge production.

These charges in turn produce a **Coulomb** electric field inside the solenoid coil wire that follows the coil and opposes the **applied** emf. The electric field fills the coil wire. As the applied voltage grows, it **overcomes** the self-induced (**opposing**) emf and **establishes** the current: that is the moment the energy is transferred. This is depicted in Fig. 5.19.

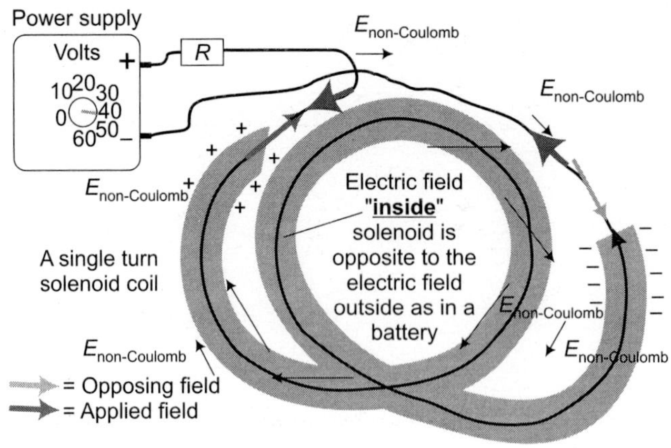

Fig. 5.19 Coulomb field of polarized coil opposes the applied field.

The **emf induced** in a coil when its current is changed is given by Faraday's law:

$$v = L\ \frac{di}{dt}$$

where L = the inductance of the coil, and for a coil of N turns uniformly wound on a non-magnetic ring of uniform section and A its cross-sectional area, as in the toroid shown in Section 5.2, is given by, $4\pi \times 10^{-7} \times \dfrac{AN^2}{l}$ henrys, where l is the length of the magnetic circuit in metres.

Does a long straight wire have an inductance? Yes indeed ! Remember, a magnetic field penetrates through matter. For one derivation of the inductance from the energy stored in the magnetic field and another assuming infinite multiply-connected loops which includes all the current in a straight wire, refer to a paper "Self-inductance of a long wire: infinite multiple loop approach"[29].

5.17 CURRENT IN AN INDUCTIVE CIRCUIT – SINUSOIDAL INPUT

Why does the (steady-state) current lag the voltage in an inductor subjected to a sinusoidal voltage?

Since the resistance of the coil is assumed neglibible, we can regard the whole of the induced e.m.f as being equal to the applied voltage. Hence the curve of applied voltage to an inductor is the same as that of the induced e.m.f.; and since the former is sinusoidal, the wave of induced e.m.f must also be a sine curve and the current as well, in the steady-state, a few cycles after switch ON as shown in the Fig 5.20.

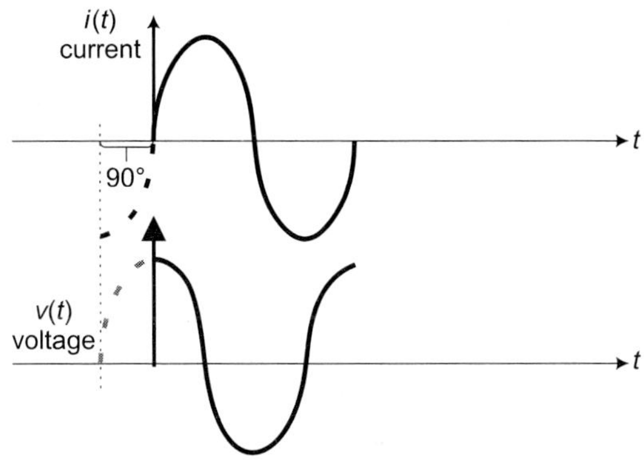

Fig. 5.20 Current lags the voltage in an inductive circuit.

At $t = 0$, the current rises from 0 at a quick rate; the induced emf is a maximum. A little later, the current continues to grow in magnitude, the induced emf is smaller because the rate of change of current is smaller than that at $t = 0$; the current is governed by $J = \sigma E_{net}$ where E_{net} is the resultant of the applied field and the Coulomb field due to polarization by the non-Coulomb electric field. As the applied voltage reduces in magnitude, the current is increasing because the induced emf is smaller.

When the applied voltage crosses zero and reversing note that the current is a maximum and has ceased growing because the applied voltage is nearer zero but, has not yet begun to decrease; consequently, the rate of change of current is then zero. The induced emf will therefore have decreased from a maximum to zero. The energy stored in the magnetic field of the inductor will also be a maximum. As the current decreases, the energy in the magnetic field of the inductor reduces; the induced emf will be in the same direction as the current and the energy will begin to be *returned* to the source.

It is possible to trace the full voltage waveform following the above line of reasoning and from the figure it is seen that the induced emf which is nearly equal to the applied voltage attains its maximum positive value a quarter of a cycle earlier than the current; in other words, the current in a purely inductive circuit lags the applied voltage by 90°.

In an inductor, the opposition to the applied voltage that is changing the current is instantaneous (the *current* change triggers the occurrence of the non-Coulomb field that by polarization then produces a Coulomb electric field which is in a direction opposite to the applied electric field) so, the current can only follow *after* the applied field has overcome the opposing emf. An inductor hates the change of high frequencies and thus has large voltages there, and, for sinusoidal currents, the current lags the voltage by 90 degrees if the inductor is pure, and less if a resistance is in series with it; the inductor fights *before* current flows. If an inductor weren't to fight, you will get energy for free!

A capacitor fights *after* the current deposits charge on its plates that develops the opposing fringe field; therefore the current leads the voltage in a capacitor circuit. It does not fight earlier, because it is *not* an inductor that stores energy in a magnetic field due to a current to do that. To visualize the behaviour of inductive circuits, it is helpful to consider a simple mechanical analogy (the most appropriate one, in my opinion) as described in Chapter 10 of "Electrical Technology" by Edward Hughes.[15]

Take a heavy metal cylinder **C**, as shown in the figure below and roll it backwards and forwards using the flat palms of your hands, on a horizontal surface like a strong table between two extreme positions **A** and **B**.

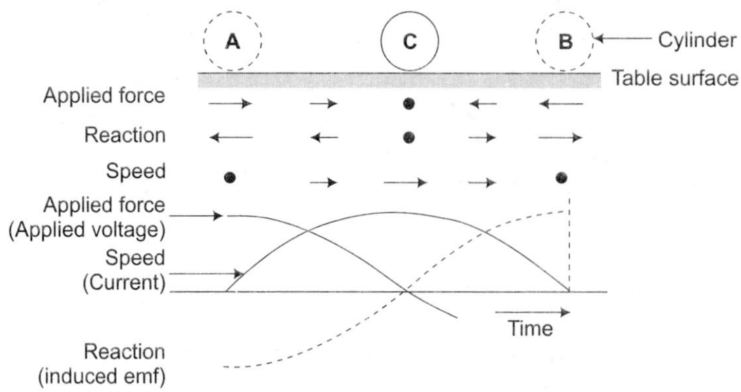

At first the speed is zero, but the force applied to the body is at its maximum, causing **C** to accelerate towards the right. You will need to reduce the applied force – as indicated by the length of the arrows in the Figure—until it is zero when **C** is midway between **A** and **B; C** ceases to accelerate and will therefore have attained its maximum speed from left to right. Immediately after **C** has passed the mid-point, the direction of the applied force is reversed and increased until the body is brought to rest at **B** and then begins its return movement. The reaction of **C**, on the other hand, is equal and opposite to the applied force and corresponds to the emf induced in the inductive circuit.

From an inspection of the arrows in the figure, it is seen that the speed in a given direction is a maximum a quarter of complete oscillation after the applied force has been a maximum in the same direction, but a quarter of an oscillation before the reaction reaches a maximum in that direction.

This is analogous to the current in a purely inductive circuit *lagging* the applied voltage by a quarter of a cycle. Also it is evident that when the speed is maximum the applied force is zero, and that when the applied force is maximum the speed is zero, and that during the second half of the movement indicated in the figure, the direction of motion is opposite to that of the applied force. These relationships correspond exactly to those found for a purely inductive circuit in the steady-state.

Note that when we start the cylinder rolling from rest from the centre position **C**, the speed is zero though the applied force may be large; a transient period of a few cycles when the roller movement will not be perfectly sinusoidal is necessary before steady-state conditions result. Similarly, the current in an inductive circuit is zero at switch ON assuming that the sinusoidal voltage starts from zero and the current and opposing emf will make adjustments for a few cycles (a transient phase) before attaining steady-state.

An inductor has a non-Coulomb field on application of the input sinusoidal voltage. It then polarizes and develops the opposing coulombic electric field almost instantaneously to closely match the e-field of the applied input voltage because there is no resistance (which we assume), as shown in the Fig 5.21 ...of course due to surface charges, that are not shown in the figure.

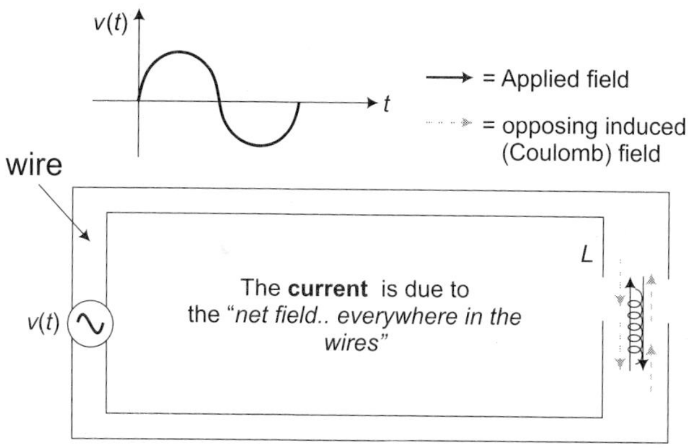

Fig. 5.21 Inductor develops a Coulomb opposing field instantly.

The existence of a *sinusoidal* current resulting from the application of a sinusoidal voltage to an inductor is a characteristic of the non-Coulomb electric field that is proportional to the rate of change in current causing a changing magnetic field. The relation between the "RMS" (square root of the mean of the squares) current and voltage of an inductor is given by:

$$\frac{V}{I} = 2\pi f L = \text{inductive reactance}$$

where, f = frequency of the sinusoidal voltage and L is the inductance of the inductor.

The inductive reactance is expressed in ohms and, as we saw in section 1.2 "Current", is useful to determine steady-state current values to sinusoidal voltage inputs. It is to be noted that though the value of the reactance is expressed in ohms, it was not obtained by computing a resistance, as of resistors. The current is a result of an opposing Coulomb electric field that is a result of polarization by the non-Coulomb electric field associated with the changing magnetic field, and the current resulting thereof by applying the relation $J = \sigma E_{NET}$ where E_{NET} is the resultant field of the applied field and the Coulomb electric field.

What is meant by steady-state in an inductive circuit subject to a sinusoidal voltage input?

Inductive circuits when subject to a sinusoidal voltage input will exhibit a transient behaviour for a few cycles initially before settling down into a steady-state. The remarks that we made regarding steady-state in capacitive circuits are also applicable to inductive circuits. Refer to Section 2.10 "Current in a Capacitor circuit-Sinusoidal input".

If we increase the "*frequency*" of the **input voltage**, the "rate of change" of the input voltage and the applied field is "greater than" the rates obtained with applied voltages at lower *frequencies*.

At low frequencies, this causes a smaller induced opposing electric field and emf, therefore, large currents will flow within small intervals of time in the coils of the inductor. This is shown in Fig. 5.22. Further reduction in frequency will cause the current to increase and the induced electric field and the resulting polarization of the inductor will be very tiny. There is barely an induced emf to oppose the input field.

In the limit, if the input is a dc voltage, the current will become so large that the inductor will burn out (unless the inductor coil was wound using a high resistance wire, as in dc coil operated relays and contactors used for switching ON and OFF motors).

At high frequencies, even as the current begins to grow and sets up a magnetic field, a powerful non-Coulomb electric field associated with the high growth rate of the current, polarizes the conductor (see Section 5.16) and produces an electric field *within* the inductor that opposes the applied field.

This process occurs at nearly the speed of light. The opposition to the applied field therefore, is also almost instantaneous and results in a net field E_{NET} in the inductor which we use in the relation $J = \sigma E_{NET}$ to compute the current.

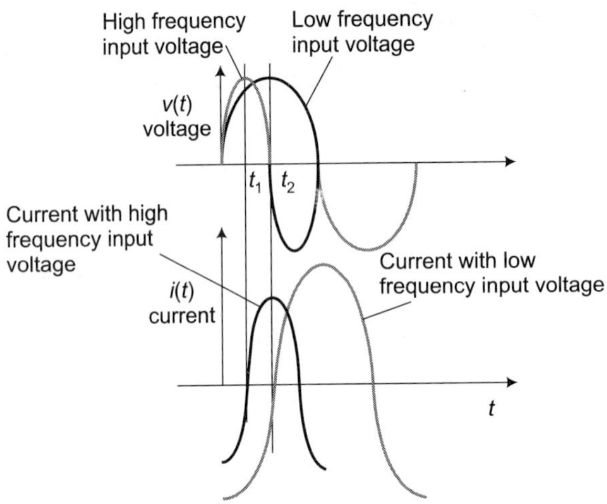

If we increase the frequency of the input voltage, the rate of change of the input voltage and the applied field is greater than that with smaller frequencies of applied sinusoidal voltages and is fast enough for the resulting powerful opposing induced emf to restrict the current.

Fig. 5.22 Current when the sinusoidal input voltage frequency is increased.

The result: The current is "smaller than" that at lower frequencies of input voltage. Many students find it difficult to visualize the production of the non-Coulomb electric field as an instantaneous phenomenon and seemingly *wait* until the current is in the inductor, not realizing that it is the speed of the rise of the applied voltage that gets the current up affecting the rate of its growth (and consequently the *rate of* growth of the magnetic flux in the inductor). With high magnetic flux growth rates, the induced emf is stronger to oppose the applied field and is given by $|emf| = \left|\frac{d\phi_{mag}}{dt}\right|$ (as we saw in Section 5.12). Some among them also make the mistake of matching the applied and the induced emf exactly. This is inconsistent with the observed presence of a current in the inductor. See Section 5.16. On the other hand, if the frequency is reduced, the non-Coulomb electric field and the resulting polarization of the inductor is smaller.

With smaller frequencies, the rate of change of the applied field is smaller and a smaller induced opposing emf is produced. The current is larger, determined by the relation $J = \sigma E_{NET}$ where E_{NET} is the resultant of the applied field and the opposing small induced field due to the induced emf. Refer Section 5.16.

As we saw in the case of capacitors, it may appear that there are two relations (see "inductive reactance" in a previous paragraph) that the current seems to be required to satisfy: $J = \sigma E_{NET}$ and $\frac{V}{I} = 2\pi fL$; here again, both relations are satisfied. Between two circuits whose resistances are insignificantly small (one cannot realize "zero resistance" in any practical circuit) yet different, the E_{NET} change in the wires of the two circuits is so *tiny* that the inductive reactance relation will appear not to have been violated, and becomes an *ideal approximation* (Refer "Ideal Approximation" in "Electronic Principles" by Paul Malvino[20]) while noting that E_{NET} regulates the total charge forward and reverse movement in the inductor wire, and is neither easily measured nor computed in a practical circuit. If the resistance were to assume larger values then, the inductive reactance relation does not apply to the circuit as a whole and the circuit becomes an Inductor-Resistor circuit.

How does the current increase in the inductor even as the applied sinusoidal voltage is reducing? between times t_1 and t_2 in the high frequency input voltage waveform of Fig. 5.22?

We begin by visualizing the current in the time interval 0 to t_1. The current is negative and decreasing (towards zero) as the applied voltage and so the applied electric field grows in the interval. This is due to the braking effect the applied field exerts on the charges that form the negative current.

The rising applied voltage at $t = 0$, pulls the charges back ....like sort of applying a "brake" on the charges and this causes the current to decrease. As a result of the decreasing current, the magnetic flux (see Fig. 5.12(d)) decreases and the Coulomb electric field (effect of polarization by the co-occurring non-Coulomb electric field), is opposite to the applied electric field, but weak, because of the slow rate of change of the current (and therefore of the magnetic flux) as is clear from the slope line at $t = 0$, that is shown in the Fig. 5.23(a). In the figure, on the right is shown the weak non-Coulomb field by thin E_{NC} field vectors.

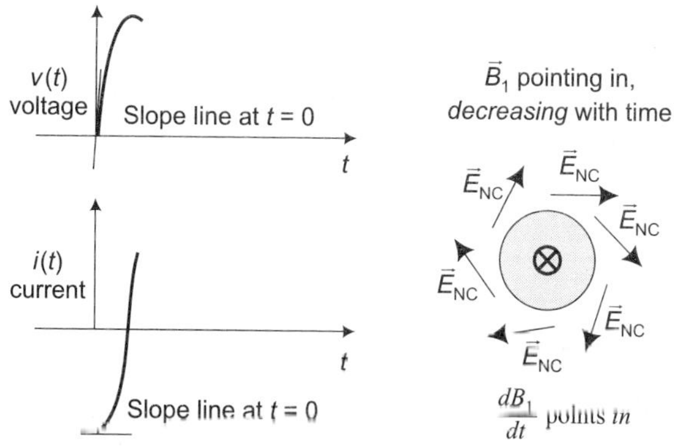

Fig. 5.23(a) The negative current decreases as the input voltage rises at $t = 0$.

As the applied voltage continues to rise, the braking effect is more powerful and this in turn causes the charges to slow down further; the rate of current decline is faster as can be seen if we draw a slope line at $t = t_A$, a time we choose midway between $t = 0$ and t_1 shown in Fig. 5.23(b). The associated electric field is stronger now and is shown by slightly thicker E_{NC} field vectors in the figure on the right in Fig. 5.23(a).

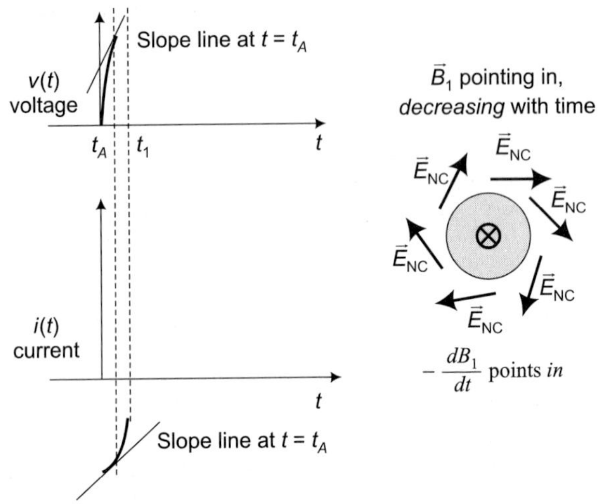

Fig. 5.23(b) Negative current when the input voltage is midway to its peak.

The associated electric field is stronger at $t = t_A$ but because the applied field is stronger (note that the voltage has risen considerably at $t = t_A$), therefore, the current (negative) continues to decrease more rapidly towards zero. The rapid decrease of current is the reason for the strong associated field.

The reversal of the current as it crosses zero at $t = t_1$ is very eventful.

I have described this in section 2.10 and urge the reader to once again visualize the rapid movement (start-stop) of the charges in one direction, their slowing down, halting and gradual movement in the opposite direction.

The acceleration at $t = t_1$ is huge assuming a reasonably high frequency and one can get a feel for this by walking rapidly in one direction, slowing down, halting and then walking backwards!

The associated electric field at $t = t_1$ is stronger as shown in the figure on the right in Fig. 5.23(c) and totally opposes the applied field and from the relation $J = \sigma E_{NET}$ the magnitude of the current is very small (because E_{NET} is the small result of one powerful slowly varying field and an almost equally powerful but oppositely directed induced field) around the time t_1 though its rate of change is very high.

At $t = t_1+$, the current in the inductor is small and now positive and still has a very steep rate of rise, keeping in mind that it is the end of the process of rapid current reversal. The magnetic field due to it is reversed as shown in Fig. 5.23(c). Checking with the corresponding Lenz rule Fig. 5.12(a), we find the non-Coulomb field does *not reverse*.

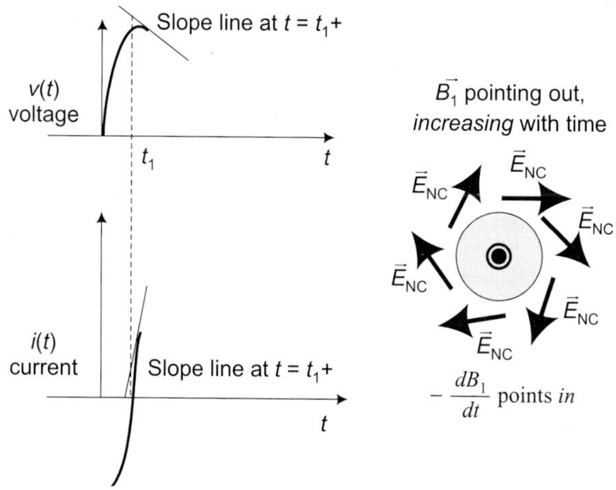

Fig. 5.23(c) A small current when the sinusoidal input voltage is near its peak.

The applied field is also powerful since the applied voltage is close to its peak value.

At $t = t_1+$, the rate of current rise being high, the rate of change of magnetic flux is high and the strength of the non-Coulomb electric field is also high. Consequent polarization of the inductor and the opposition to the applied field is almost total, resulting in the small current at $t = t_1+$.

Further, at $t = t_1+$, the applied field though large has a small rate of decline as is evident from the not-very-steep slope line at $t = t_1+$ in Fig. 5.23(c). The corresponding non-Coulomb electric field is large as evident by the steep slope line for the rate of growth of current (and therefore the high rate of growth of the magnetic flux) and this is shown by the thick E_{NC} arrows in Fig. 5.23(c).

The resultant electric field in the wires given by applied field—E_{NC} will be correspondingly small and so, the current given by $J = \sigma E_{NET}$ will be small.

The crucial point in the visualization is the slow rate of decline of the input voltage (and therefore, the applied electric field) at points near the time t_1. The decline of the applied field itself causes a decrease in the current and the rate of current growth reduces at the time $t = t_B$ which is say, midway between t_1 and t_2 shown in the Fig. 5.23(d & e).

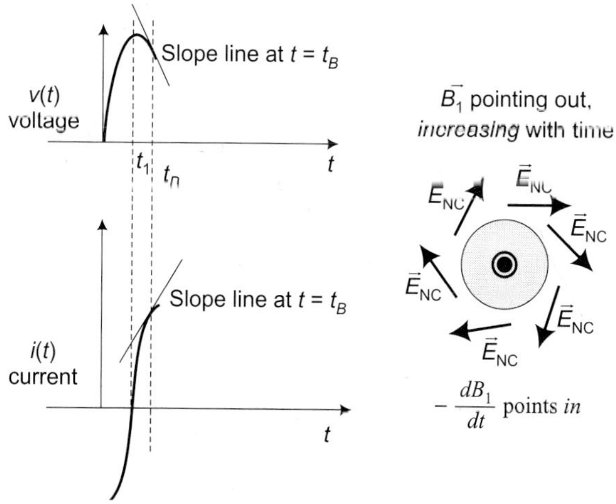

Fig. 5.23(d) The rate of growth of current is diminished at $t = t_B$.

At $t = t_B$, the current is still growing but its rate is diminished because of the reduced applied electric field. The co-occurring weaker non-Coulomb electric field E_{NC} is shown by thinner arrows in Fig. 5.23(d). The *resultant* electric field in the inductor wire is more than that at $t = t_1+$ in spite of the applied field having reduced in magnitude and the current given by the relation $J = \sigma E_{NET}$ will therefore be more than that at $t = t_1+$.

As the input voltage reduces along the downward segment of the positive portion of the sinusoid, the applied field is near zero at $t = t_2-$. This is shown in Fig. 5.23(e). The current nearly "peaks" because $J = \sigma E_{NET}$ and E_{NET} is almost entirely the applied field close to zero.

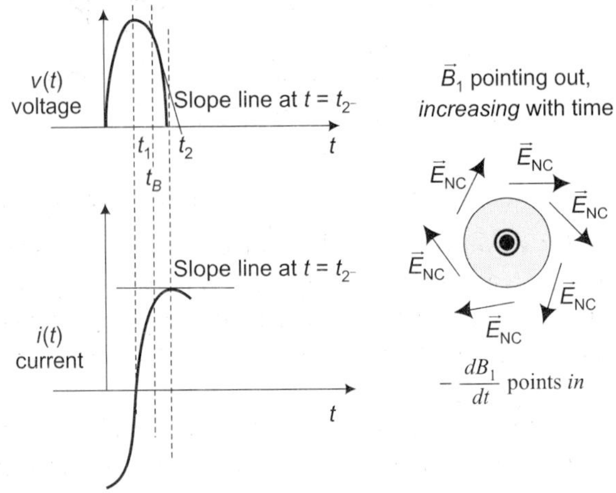

Fig. 5.23(e) The current levels off at t_2 and its rate of growth is almost zero at $t = t_2-$.

There is barely an electric field because the current is not changing (and therefore, the magnetic flux is not changing). The current levels off at t_2 and it remains constant and at this time the rate of change of voltage is also a maximum as seen from the slope line at $t = t_2$. See Fig. 5 23(e).

Thus, we see how the current increases between the times t_1 and t_2, even as the voltage is decreasing.

The current is maximum and is due to the large conductivity σ of the inductor wire, even when the applied field is tiny with almost no opposing non-Coulomb electric field! Currents smaller than the peak value, are due to the resultant of more than one field in the wire.

I urge the reader to analyse the current waveform for the negative half-cycle of the sinusoid and verify that the current indeed lags the applied voltage by 90° as shown in Fig. 5.20 for an inductor applying Lenz's rule of Fig. 5.12 that I have indicated in the figure below for all the portions of the applied sinusoidal votage.

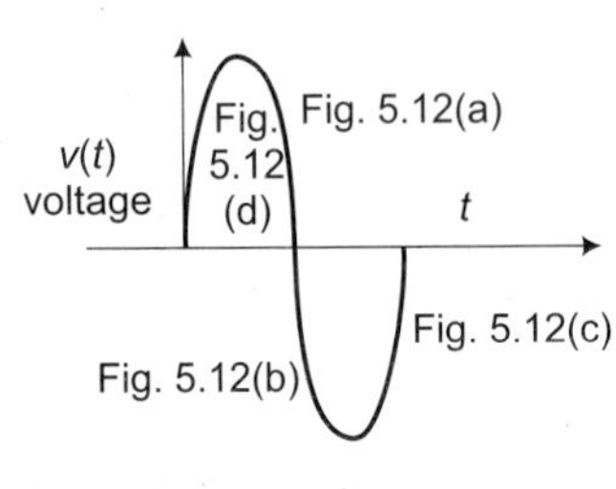

If the inductance of inductor L is made smaller then, the non-Coulomb electric field and thence the Coulomb electric field by polarization, does not build as fast as it would have if L were to be large. With a small L (imagine a fewer number of turns in the coil), the attempt to create a magnetic field is easier and causes the current to build faster with a smaller self-inducing (opposing) emf in a given interval of time as evident from the inductor current-voltage relation $i = \frac{1}{L}\int v.dt$. The (opposing) Coulomb electric field is smaller and the net field consequently is greater. Therefore, at a fixed frequency, the current increases ($J = \sigma\ E_{NET}$ in the wires) as the size of the inductor is decreased and this is encapsulated in the inductive reactance expression, $2\pi fL$.

5.18 APPLICATION OF A SINUSOIDAL VOLTAGE TO AN INDUCTOR AND RESISTOR COMBINATION–LR CIRCUIT

When an inductor is connected in series with a resistor R as shown in Fig. 5.24, and a sinusoidal voltage is applied to the combination, then the current is "*diminished*" in magnitude in comparison to the current that would result if the resistor were not there. A part of the input voltage will be used to push charges through the resistor.

When an inductor is connected in series with a resistor R, then the current is diminished in magnitude and the applied voltage $v(t)$ is split between the inductor and the resistor in a slightly more complex way.

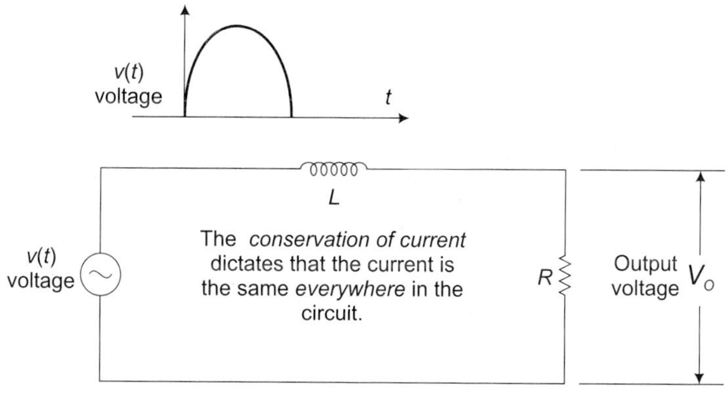

Fig. 5.24 Current when a sinusoidal voltage is applied to an inductor resistor combination.

Surface charges appear and bunch up at the junction of the inductor and resistance.

The initial current which is negative is not as large as that in the inductor circuit (see Fig. 5.23(a)). It is restricted by the resistor in series.

All the processes that lead to the current lagging the voltage across the inductor... mind... despite there being a resistor in series... by 90 degrees... hold.

But, due to the resistor, the voltage across the inductor is not the same as the input voltage. It is smaller in magnitude.

This is shown in Fig. 5.25.

The initial negative current is not as large as that in the inductor circuit. It is restricted by the resistor in series.

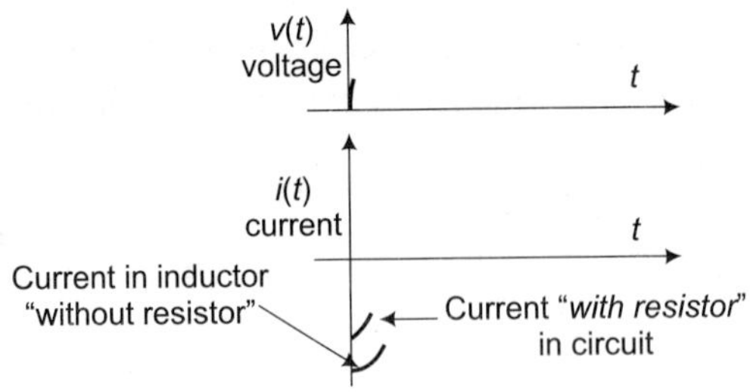

Fig. 5.25 Comparison of current in the inductor circuit (Fig. 5.21) and in an inductor-resistor combination circuit subjected to a sinusoidal voltage.

Consider the situation when steady-state is reached and from a previous cycle the current through the inductor is negative and the input voltage is now rising from zero. See Fig. 5.25.

The initial negative current is reducing in magnitude and approaches zero, even as the voltage across the inductor is rising.

The current *levels off* and remains constant as the voltage across the inductor reaches zero and is just becoming negative.

Remember, the current is a *constant* when the *rate of change* of the voltage is a maximum across an inductor which is when it is reversing. See Fig. 5.23(e) and Fig. 5.29 and note that the instant this occurs is *after* the applied voltage peaks.

The situation is shown in Fig. 5.26.

At this instant, the rate of growth of the input voltage and the rate of growth of the inductor voltage result in a constant "net" electric field in the wires that maintains the current constant "after" the input voltage has reached its peak and also note that at this instant the voltage across the resistor which has the same phase as the current is also constant through for a very brief period (see Fig. 5.29).

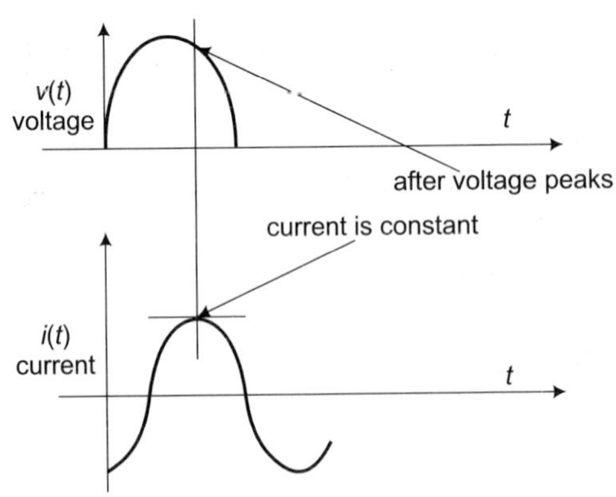

Fig. 5.26 Current peak in an inductor-resistor combination circuit subjected to a sinusoidal voltage.

Why does the current peak and remain constant briefly "after" the applied voltage has peaked? While this was happening in the inductor "only" circuit (see Fig. 5 22) when the applied voltage was crossing zero?

When there is a resistor, the current is reduced in the circuit, therefore, the rate of change of the magnetic flux will also be smaller in comparison with the circuit without a resistor.

Note from the current and voltage waveforms in Fig. 5.29, the negative current at $t = 0$ is smaller and decreases towards zero earlier than it would have when there was no resistor in the circuit; with a resistor in circuit it crosses zero when the applied field (applied voltage) is smaller in magnitude than its peak value.

Also, note that the induced emf of the inductor when the current crosses zero is maximum and is expected since the rate of change of current and therefore of the magnetic flux is maximum.

In the inductor-resistor circuit, as the current becomes positive and even as it continues to grow, the applied voltage peaks. The applied voltage decreases in magnitude after peaking but the rate of *change of the current* is also decreasing, and therefore, the non-Coulomb electric field E_{NC} decreases. Closer observation reveals that the rate of change of the input voltage and the rate of change of the voltage across the inductor is the same at this instant of time. See Fig. 5.29.

In other words, the rate of growth of the applied input field and the rate of growth of the associated non-Coulomb field and resulting emf due to polarization of the inductor wire (at different magnitudes though) is the "*same*" at this instant of time, as we shall soon see, resulting in a constant net field, therefore, the current remains steady since $J = \sigma E_{NET}$ = a constant.

As the input voltage continues to reduce in magnitude, the current begins to reduce too but since its rate of change is small there is a weak non-Coulomb electric field.

However, in the inductor circuit (see Fig. 5.22), the instant when the current remains constant occurs when the input voltage and so the inductor voltage (being the same as the input voltage) is going through "zero" and their rates of change match at every instant.

In the inductor resistor circuit, as the applied field continues to reduce along the sinusoidal curve, it contributes towards establishing a resultant field in the inductor wire that firmly brings the current down.

Even as the applied voltage crosses zero and becomes negative, the current soon follows and also crosses zero becoming negative as seen in Fig. 5.27. Before the current becomes negative, it is the induced emf that sustains the current as can be verified by checking the Figs. 5.23(d) and 5.23(e). With the applied voltage negative and the current still positive, energy will be returned to the source. (The transport of energy and the processes involved are described in Chapter 4.)

Further reduction in the input voltage makes the current cross "zero" and reverse under the influence of the dominant induced emf. The induced emf has assumed control over the current upto some point "after" the input voltage reverses, at 180 degrees. This is shown in Fig. 5.27.

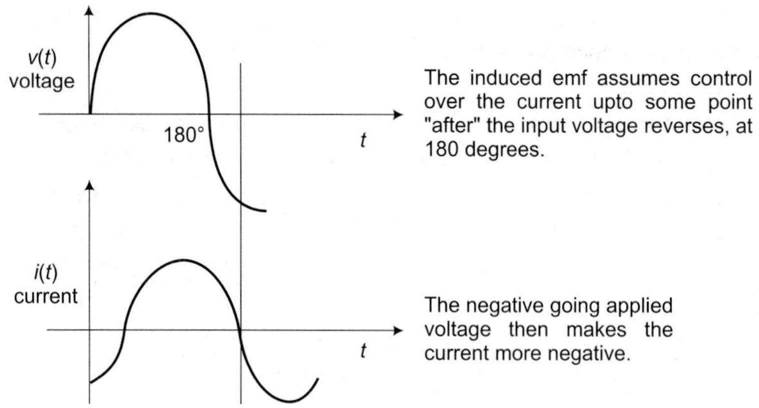

Fig. 5.27 The current in the inductor resistor combination subjected to a sinuoidal voltage.

After the current crosses zero and becomes negative, the negative going applied voltage then makes the current more negative.

The current therefore, we say, "lags" the applied input voltage and by an amount less than 90 degrees.

The current in a circuit comprising an inductor with a resistor is smaller and lags the input voltage by an angle *less than* 90 degrees.

The complete voltage and current cycle is shown in Fig. 5.28.

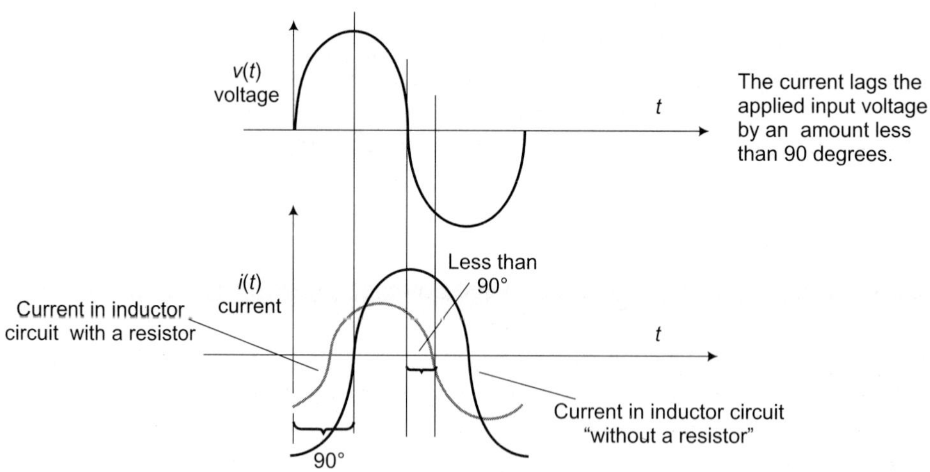

Fig. 5.28 Comparison of the current in the inductor resistor combination circuit with that in the inductor circuit (Fig. 5.20) subjected to a sinusoidal voltage.

If we measure the output voltage across the resistor of an LR circuit as in the figure below, we will observe that the voltage across the resistor lags the input voltage by the same amount as the

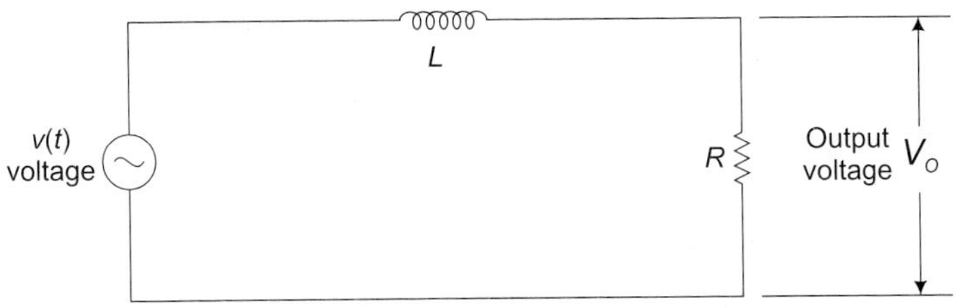

current because they are in phase for a resistor. Therefore, the output voltage (across the resistor) *lags* the input voltage by the same amount as the current.

The current, which has to be the same all around the circuit, *lags* the voltage across the inductor by 90 degrees, a rule that is *never violated.*

Note the *slope* of the inductor voltage change at *the zero crossing point* of the input voltage is the *same* as the slope of the input voltage change after the input peaks, where the current is *constant.*

Observe that the output voltage is *smaller* in magnitude than the input voltage, as it should be.

The waveforms of input and output voltages and the voltage across the inductor are shown in Fig. 5.29.

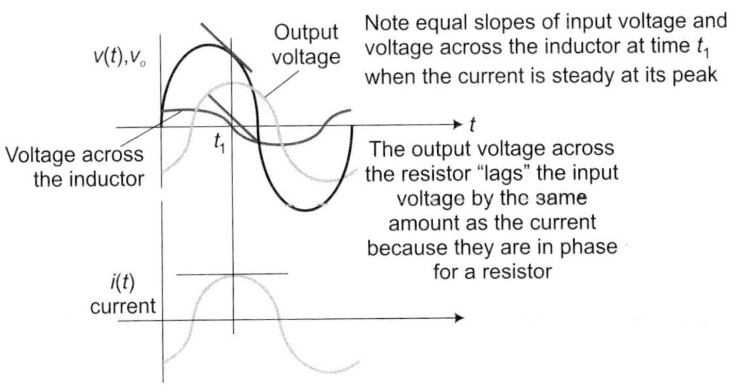

Fig. 5.29 Voltage waveforms of input, output and the current in a L-R combination circuit subjected to a sinusoidal voltage.

I urge the reader to review the comments made on visualizing currents in Section 2.5. One must not focus on the current in one element and then the other of a circuit. One has to think about what happens to every element in the circuit all at the same time.

5.19 APPLICATION OF SINUSOIDAL VOLTAGES AT VARIOUS FREQUENCIES TO AN L-R COMBINATION

What happens when the frequency of the input sinusoidal voltage is decreased?

If the frequency of the applied voltage is decreased, the associated non-Coulomb electric field will decrease because the rates of current growth and decrease (and therefore that of the magnetic flux in the inductor) will be smaller. There will be less opposition to the establishment of current and so the current will be greater.

Together with the resistor limited current, the slowly varying applied field (due to its low frequency) causes the opposing induced emf to be weak and these result in the current lagging the applied voltage by a small amount.

At much lower frequencies, the resistor limited current and the slow growth and fall of the applied voltage causes the rates of growth and fall of the current and the magnetic flux in the inductor to be weaker, and this results in the output voltage being almost in phase with the input voltage.

The waveforms for the circuit subject to a low frequency sinusoidal input voltage are shown in Fig 5.30.

The output voltage picked across the resistor is able to reproduce the input signal with less attenuation.

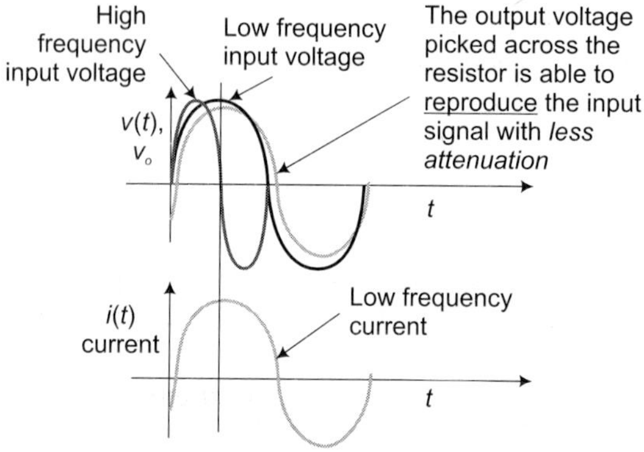

Fig. 5.30 Voltage waveforms of input, output and the current in an L-R combination subject to a low frequency sinusoidal voltage.

The L-R Combination – More on the non-Coulomb Field

Do not forget that the non-Coulomb field is due to the *rate* of growth (or decay) of current and the current is due to the applied field.

The current is limited by the resistor and the slowly varying applied input field. The current cannot grow to large magnitudes and its rate of rise is further limited by the opposing induced emf that builds up due to the *rate* of the changing magnetic flux which is again due to the current. At low frequencies, this *rate* being small, the induced emf is small.

And that is the reason why at low frequencies, the associated non-Coulomb electric field is unable to build enough strength to "oppose" the applied field.

Therefore, we say the inductor acts as a "short circuit" at low frequencies, as if the inductor was not there but a wire.

What happens when the frequency of the input sinusoidal voltage is increased?

If we *increase the frequency* of the input voltage, the strength of the associated non-Coulomb electric field will increase because the rates of current growth and decrease (and therefore that of the magnetic flux in the inductor) are steeper. There is a stronger opposition to the establishment of current and so the current is smaller which is further restricted by the presence of the resistor, and the R-L circuit impedance will be large. The waveforms of output voltage and current, when a high frequency sinusoidal voltage is applied, are shown in Fig. 5.31.

While the resistor limited current causes the rates of growth and decay of the current to be gentler, the more rapid reversals due to the high frequency of the applied voltage (and therefore

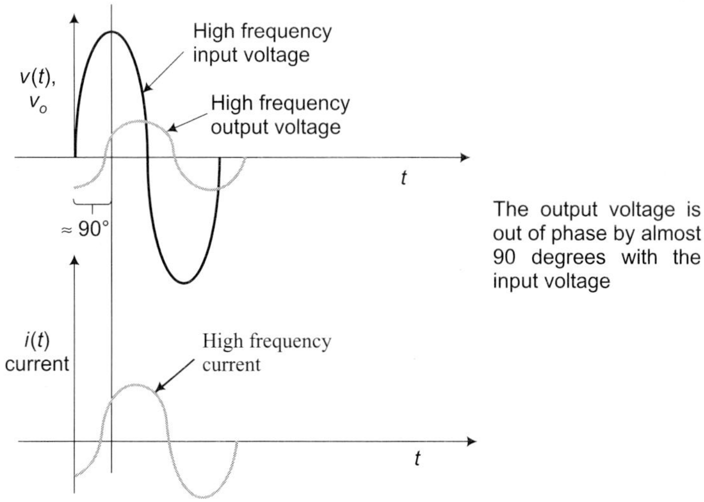

Fig. 5.31 Voltage waveforms of input, output and the current in an L-R combination subject to a high frequency sinusoidal voltage.

of the applied electric field) causes the opposing induced emf to be yet strong enough to cause the current to lag the applied voltage by almost 90 degrees as shown in the Fig. 5.31.

The output voltage therefore, lags the input voltage by almost 90 degrees and its magnitude will be small.

The output voltage is unable to follow the input signal being increasingly blocked by the inductor's induced emf and therefore, the *attenuation* is also high.

The potential difference across the inductor is frequency dependent and current flows after the induced emf/potential difference appears across the inductor coil; therefore, the current "lags" the applied voltage by a phase angle less than 90° but greater than 0°.

So, we say, the inductor acts as an open circuit to high-frequency input voltages.

As the frequency of the applied voltage (e-field) is increased, the applied voltage is opposed by the induced emf of the non-Coulomb electric fields that are stronger (resulting in a powerful opposing induced emf due to polarization) thereby making it difficult to establish large currents and, the R-L circuit impedance (see Section 1.2 "Current") will be "high".

From the qualitative descriptions given above of the behaviour of an inductor to voltages of different frequencies, it may be observed that an inductor blocks high frequency inputs.

It therefore, finds use as a low pass filter in electric and electronic circuits.

A smoothing circuit is used to reduce the ripple of a dc voltage produced by rectifying an ac voltage. Such a smoothing low-pass filter using an inductor L is shown in the Fig 5.32.

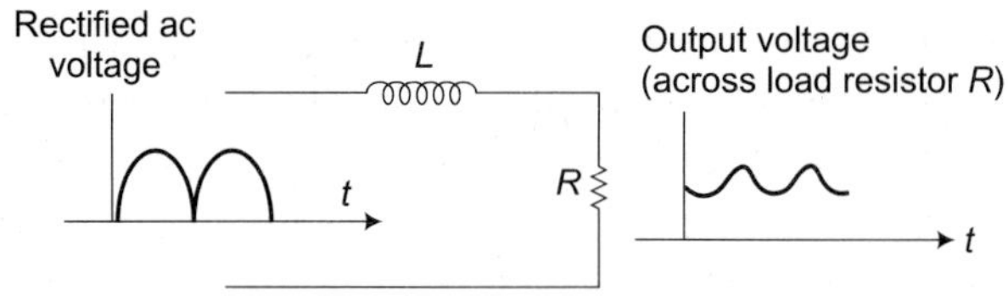

Fig. 5.32 Smoothing circuit that reduces the ripple of a rectified ac voltage.

5.20 CURRENT WITHOUT A VOLTAGE!

Inductors – One Strange Phenomenon

Just as Isaac Newton's first Law of Motion ("an object in motion tends to stay in motion; an object at rest tends to stay at rest") describes the tendency of a mass to oppose changes in velocity, we can state an inductor's tendency to oppose changes in current as such: ***"Electrons moving through an inductor tend to stay in motion; electrons at rest in an inductor tend to stay at rest"***.

Hypothetically, an inductor left short-circuited will maintain a constant rate of current through it with no external assistance.

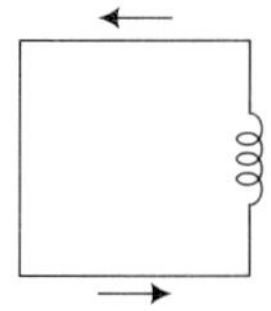

Current sustained with
the inductor short-circuited

The electric field will be "***zero***" everywhere around the inductor and also inside the inductor wire as ***neither*** the current ***nor*** the magnetic field ***is changing***. ***There is no voltage across its ends, either***.

We have a current without a voltage. Practically speaking, however, the ability for an inductor to ***self-sustain current*** is realized only ***with superconductive wire***, as the wire ***resistance*** in any *normal* inductor is enough to ***cause current to decay*** very quickly with no external source of power.

Just as a stationary electron has an electric field surrounding it, we see moving electrons have a magnetic field. The magnetic field sustains the current and the energy stored in the field is given by $\frac{1}{2}(LI^2)$ where L is the inductance and I the current in the inductor.

5.21 INDUCTORS – PRODUCTION OF A HIGH VOLTAGE

We are aware that a current carrying inductor stores energy in the magnetic field in the space surrounding the coil.

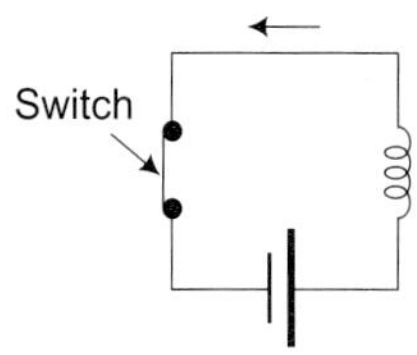

Fig. 5.33 Production of a high voltage using an inductor.

Consider the circuit in Fig. 5.33 which shows an inductor carrying a current with a switch in the circuit. If we break the circuit by opening the switch, we would be interrupting the current which is a manifestation of the magnetic energy in the coil.

Breaking the circuit is an attempt to destroy magnetic energy ... but energy cannot be destroyed, so, the circuit reacts by inducing an emf.

Since the current is stopped abruptly, a large emf will be produced across the contacts of the switch and the air between the contacts of the switch will ionize.

An arc will appear across the contacts and the energy of the inductor will be dissipated as electromagnetic radiation in the form of heat (infrared) and light (visible spectrum) as shown in Fig. 5.34.

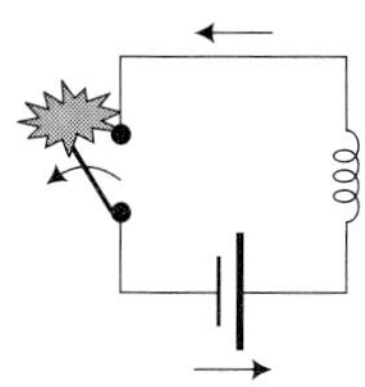

Fig. 5.34 Arcing across the contact resulting in a high voltage.

This method of producing a high voltage is adopted to produce the sparks in spark plugs to ignite the fuel in motor cars, scooters and motorbikes using ignition coils.

The same method is used to initiate the discharge in fluorescent lamps or tubelights. Another way to interpret the production of the high voltage induced.

Since an attempt is made to stop the current abruptly, ***di/dt* is large** and so is the induced emf *v*.

The production of a spark is exploited in the spark plug circuit of petrol engine driven automobiles. However, this could cause catastrophic failures of sensitive electronic driver circuits like transistors when used to switch a relay coil.

See Section 9.13 for a description of how transistors that drive relay coils are protected using a diode.

What is the sequence of events at the microscopic level if we insert a resistor in the inductor circuit made of superconductive wire with a current in it, as shown below in Fig. 5.35?

Fig. 5.35 Inserting a resistor in a current carrying inductor.

5.22 VOLTAGES AND CURRENTS IN A TRANSMISSION LINE

In previous chapters, we studied the nature and effects of electric field, surface charges and how they enable the process of conduction, and capacitance and inductive (even in straight wires) effects.

These and other phenomenon produce a behaviour of voltages and currents in transmission lines of long lengths that are peculiar, because changes in the electric field do not propagate instantly but at the speed of light and due to the inductance and capacitance of the line and the presence of surface charges. At high frequencies, the peculiar behaviour is more pronounced.

To view an interactive animation of the simulation of a transmission line, please visit http://www.astrophysik.uni-kiel.de/~hhaertel/TL_tutorial/TL_html_uk/index_e.htm (reproduced here with permission from Hermann Haertel[6], University Kiel, Germany) and click "voltage and current pulses on a double line" and see the progress of the voltage and current pulses. A description of the pulses is provided in the right panel of the web page.

Note that the current is *not uniform* throughout the line as is observed in simple electric and electronic circuits of small dimensions.

We conclude this chapter with a description in the next section of the fascinating world of neutral atoms and this is followed by a note on the attempts made by physicists to see atoms.

5.23 NEUTRAL ATOMS

What is fascinating about an atom is how it remains "neutral" despite being a collection of sources of powerful electric and magnetic fields. The electrons at very high speeds (the velocity of the electron of a hydrogen atom is about 2.19×10^6 metres/second and is yet only about 1/137th the speed of light) orbit around the central core (in the classical view), so perfectly balanced that the "net" electric field is very close to zero outside the atom. Because of this astounding assembly of sub-atomic particles... for all time.. the e-field never becomes significantly non-zero for the tiniest fraction of time... the cancellation of the field outside is ever so precise.

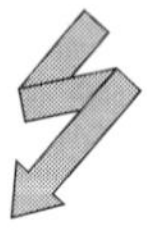

If it were not so, we would be living in a "*shocking*" world! and you would be shocked by anything you touch!

It is this "dance" (that I would like to imagine) that Neils Bohr and Schrodinger and Fermi have tried to encapsulate in their theories, equations and solutions using quantum wells and potential wells.

Little wonder then that seeing atoms has been a fascination for all–laymen, engineers and scientists.

But, alas we can probably never see this dance and probably never ever can, as described in the article "Intro to Nano: Seeing atoms" at the link below:

http://www.nisenet.org/nano-101-seeing-atoms

Scientists doubt that we'd ever be able to observe an atom, or anything else smaller than a few hundred nanometres, because of the limitations of the optical microscope. This *microscope* uses light to make an image, so its performance is constrained by the nature of light.

Light behaves as a wave, and for an optical microscope to form a clear image of a sample, *the sample has to be larger than a wavelength of light* (the distance between successive wave crests or other identical parts of a wave).

The wavelengths of visible light range from about 700 nanometres to about 400 nanometres. *But atoms are only 0.1 nanometres to 0.5 nanometres in diameter.*

Seeing atoms

This intriguing nanoscale image, called the Quantum Corral, was created by Donald Eigler, Crommie, and Christopher Lutz in 1993 at IBM's Almaden Research Center. The team first used a scanning tunneling microscope (STM) to move iron atoms (which you see as turquoise mounds) one by one to form a corral on a copper surface. Inside the corral, trapped copper electrons, which ordinarily would move freely, form standing waves—waves that appear to be standing still.

After the iron atoms were in place, the team used the STM to scan back and forth over the sample to collect information about the contours of the surface, and the computer displayed the information as a black-and-white topographic map. Then, according to Don Eigler, "we used the computer to make it look like a real surface illuminated by light. We like to use colour because it can help differentiate parts of the image and it's more visually interesting."

Visit http://www.org/image/quantum-corral to read the full article and http://www.almaden.ibm.com/vis/stm/corral.html#stm7 for more descriptions of the quantum corral and http://www.almaden.ibm.com/vis/stm/lobby.html for a description of how the image was made.

Questions

1. What is a magnetic field? Describe its nature and using the Biot-Savart law find the magnetic field B at a distance x far away from a current carrying loop in the z-y axis carrying a current I.

2. How is a curly electric field produced? Describe with the help of neat sketches. Why is this field called non-conservative? Explain.
3. Consider an inductor with a current that is increasing such that the magnetic field due to its pointing out and increasing as shown in the figure below.

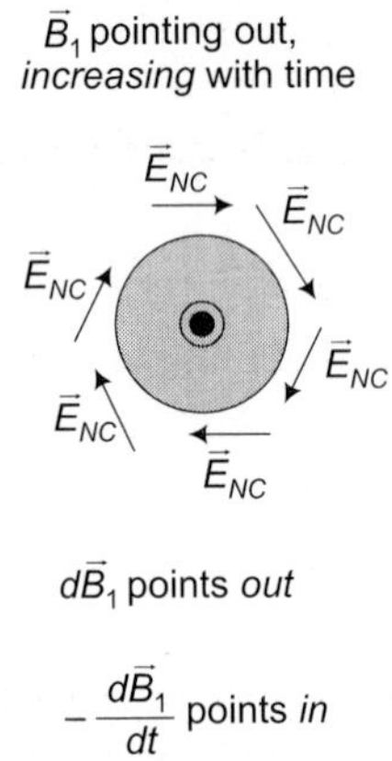

The non-Coulomb electric field that will result is indicated in the figure. Sketch the non-Coulomb electric field for the 3 cases with B_1 pointing out (increasing), pointing in (increasing) and pointing in (decreasing).

4. An otherwise uniform metallic ring with a thin section surrounds an inductor whose current is varied. Sketch the effect this will have on the potential of the ring. Describe the nature of the electric field inside the ring.
5. A varying dc voltage is applied to an inductor. Describe the processes that lead to the production of a coulombic electric field in the inductor.
6. The current in an inductor circuit lags the input by 90 degrees. Describe with the help of neat sketches the processes that maintain this lag angle in the steady-state. Is it possible to achieve 90 degrees lag angle between the current and the voltage accurately in a typical inductor circuit? Explain giving reasons.
7. Describe the effects of varying the frequency of a sinusoidal voltage to an inductor with respect to the current.
8. An inductor-resistor combination is subject to a sinusoidal voltage. With neat sketches explain how the current lags the voltage by an angle that is less than 90 degrees. If the frequency is varied, what effect does this have on the current and the phase with respect to the voltage?
9. How are high voltages produced in the ignition circuit of an automobile? Explain with reference to the conservation of energy principle and the expression for the induced emf.

   Explain why we are not shocked when we touch objects considering the fact that all matter is made of atoms that contain electrons and protons and the potential very close to them can be of the order of thousands of kilovolts.
10. The unit of resistance and inductive reactance is ohm. Describe how the two are different qualitatively.

6

Magnetic Forces; Magnetic Forces Do No Work!; How a Wire Moves

In the first section of this chapter, the phenomenon of magnetic forces is presented. In the following sections the processes that enable a current carrying wire to move when placed in a magnetic field are described. The fact that magnetic forces do no work is explained with the help of an experiment that makes a mass attached to a current carrying loop to hang in mid air when placed in a magnetic field. The description includes calculations that lead to prove this important aspect of magnetic forces.

6.1 MAGNETIC FORCES

Reproduced with permission from Pearson Publications, David J. Griffiths', "Introduction to Electrodynamics".[8]

Magnetic fields ***affect*** moving charges.

The magnetic force in a charge q, moving with velocity v in a magnetic field **B**, is

$$F_{mag} = q(\vec{v} \times \vec{B})$$

This is known as the *Lorentz force* law.

The experiments described below provide evidence of the magnetic force.

Two wires are hanging from the top a few centimetres as seen in Fig. 6.1(a). When the current is turned ON they jump apart evidently *repelling* one another as shown in Fig. 6.1(b).

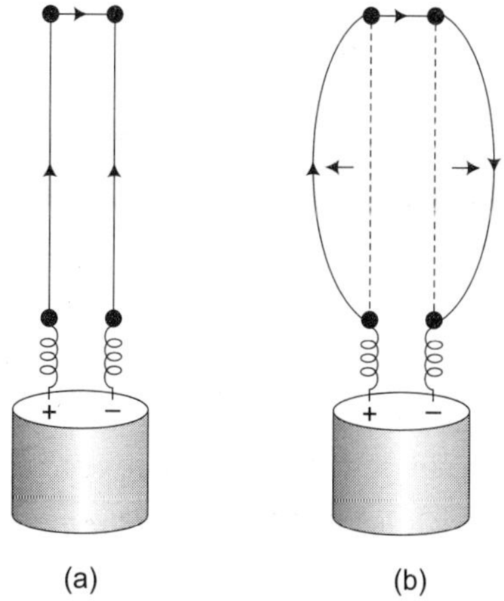

Fig. 6.1 Force in a current carrying wire.

In Fig. 6.2 is shown a system of coils that can carry a current to produce a magnetic field.

An **electron** is made to move by an electric field (not shown) towards a **coil system** with a velocity v.

A current is switched on in the coil(s) and a horizontal magnetic field is produced.

The moving electron is *deflected* vertically by the horizontal *magnetic field.*

$$F_{mag} = q\vec{v} \times \vec{B}$$

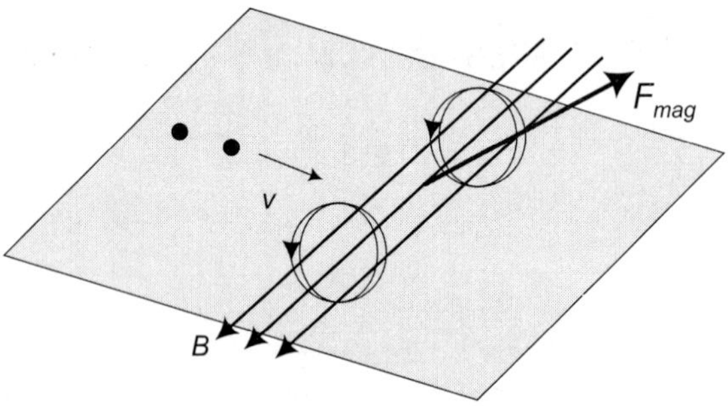

Fig. 6.2 A moving electron deflected by a magnetic field.

Note: If the electron were to be stationary in the vicinity of the magnetic field there would be no movement of the electron affected by the magnetic field.

In Fig. 6.3 are shown 2 wires in close proximity to one another with a current I in each of them. The charges move with a velocity v.

Current in wire 1 produces a magnetic field B around wire 2 and which can be determined using the Biot-Savart law.

The force on the charges having velocity v in wire 2 is given by $F = q\vec{v} \times \vec{B}$.

The direction of the force is determined by applying the right hand rule for vector cross products (see Section 5.2).

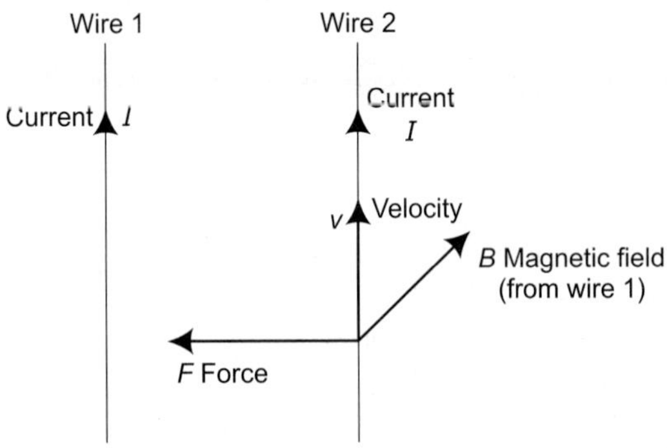

Fig. 6.3 Forces of two current carrying wires in close proximity to one another.

It must be remembered that a single wire with a current in it cannot exert a force on itself!

It is the *external* magnetic field that applies a force on a current carrying wire as shown in Fig. 6.4.

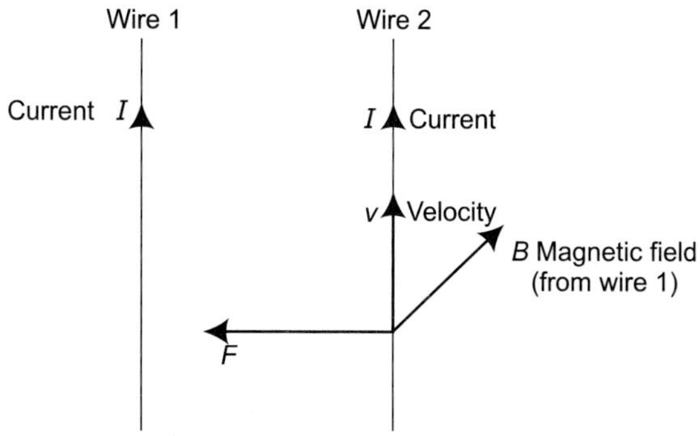

Fig. 6.4 Force is due to external magnetic field.

A similar force will result due to a magnetic field set up by wire 2 on wire 1 and the result is *attraction*. This is shown in Fig. 6.5.

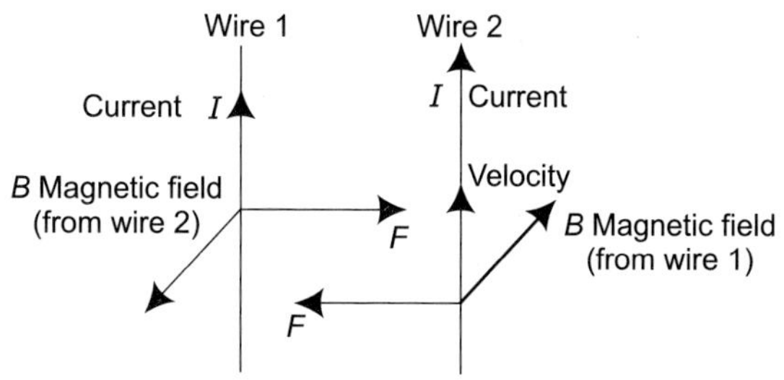

Fig. 6.5 Force of attraction of two current carrying wires in close proximity.

Expressed vectorially, as shown in the Fig. 6.6, the magnetic force F is given by

$$F = q\vec{v} \times \vec{B}$$

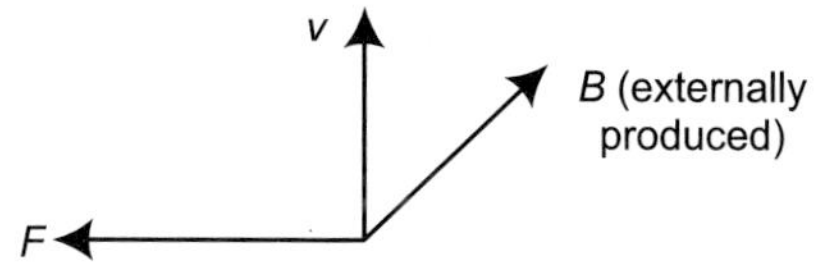

Fig. 6.6 Graphical view of the force vector = $\vec{F} = q\vec{v} \times B$.

If the experiment was conducted using copper wires, it makes the results even more astounding. Because, copper is not a magnetic material.

You can test this out by using the strongest magnet that you can find to pick up copper materials such as copper coins or copper wires. The magnet will *not be able to attract these items*, thus showing that copper is a *non-magnetic* metal.

Then, why does the wire move?

In the next few sections an answer to this question will be provided. We begin by describing the action of a magnetic field on a current carrying conductor.

6.2 A MAGNETIC FIELD ACTING ON A CURRENT CARRYING CONDUCTOR

To understand this, consider a metallic conductor that is carrying a current being subjected to a magnetic field as shown in Fig. 6.7 (after Chabay & Sherwood, 1995).

Under the influence of the magnetic field the electrons are deflected downward as shown (apply $qv \times B$ to determine the direction taking care that the particles are electrons).

Note: The direction of force determined using the right hand rule for vector cross products described in Section 5.2 is straightforward to apply for moving positive charges. When applying the right hand rule for moving electrons in a magnetic field, the direction of $v \times B$ will be 180 degrees (opposite) the direction indicated by the thumb.

As time goes on, more and more electrons moved due to the action of the magnetic field will pile up on the bottom of the wire with a corresponding deficiency of electrons on the top.

However, this piling-up won't continue forever...

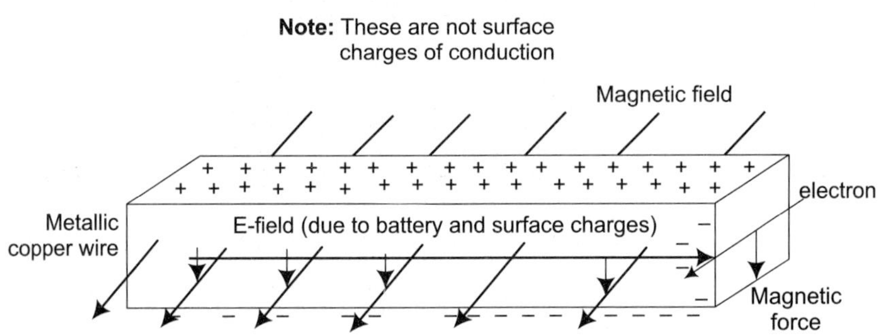

Fig. 6.7 Charges pile on the top and both bottom of the wire.

The extra surface charges produce an *electric field* between the upper and lower surface of the metallic wire.

The extra surface charges exert an electric force qE (shown in grey in Fig. 6.8 and on electrons this will be upward) on the moving electrons which in the steady state just balances the magnetic force that caused their movement to the bottom, qvB and so, $qE = qvB$.

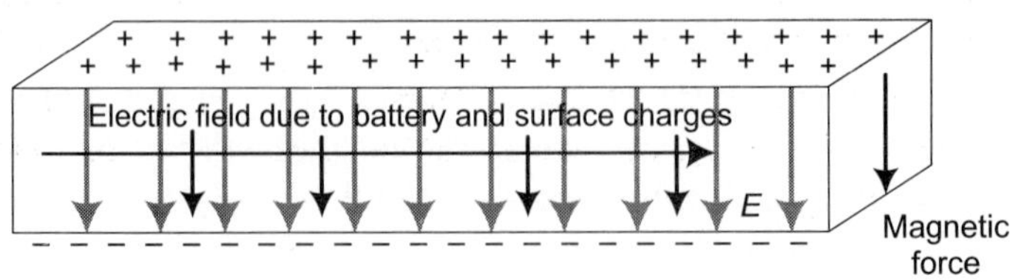

Fig. 6.8 Extra surface charges exert a force qE on the moving electrons.

Since qE becomes equal to qvB, the moving electrons now continue through the wire without being deflected as shown in Fig. 6.9.

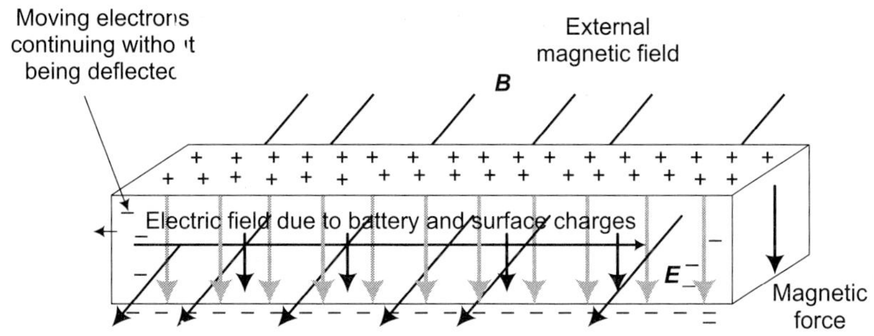

Fig. 6.9 $qE = qvB$ and the moving electrons continue without being deflected.

6.3 BALANCED FORCES WITH MOVEMENT

Another example of balanced forces with movement: the carrom player in the image below pushes the striker, the downward force of gravity is balanced by the upward reaction of the carrom board surface.

The striker moves horizontally between these vertically *balanced opposing forces* on the board.

And that is exactly how the electrons move through the wire (not forgetting that it is a start-stop motion of millions and millions of electrons everywhere in the wire with a drift superimposed) between the electric force and its balanced oppositely directed magnetic force in the wire carrying a current when immersed in a magnetic field.

6.4 HALL EFFECT

The phenomenon of charges piling up on opposite surfaces of a conductor carrying a current in the presence of a **magnetic field** is called the **Hall effect** as shown in Fig. 6.10.

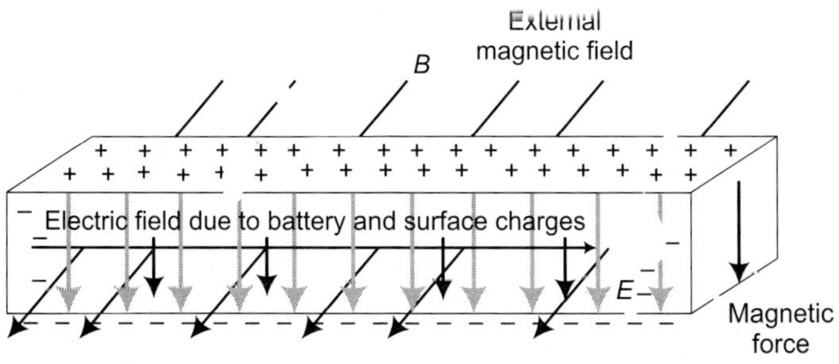

Fig. 6.10 The Hall effect.

There is a subtle point about the magnetic force (Chabay & Sherwood, 1995, used by permission)[2] that a magnetic field exerts on a current-carrying wire, the magnetic force only acts on moving particles which in the case of a metal wire are the drifting electrons.

The stationary positive atomic cores do not experience a magnetic force. So why do we see the entire wire move?

We have seen how the extra surface charges exert an electric force qE on the moving electrons which in the steady state just **balances the magnetic force** $\boldsymbol{qvB}$.

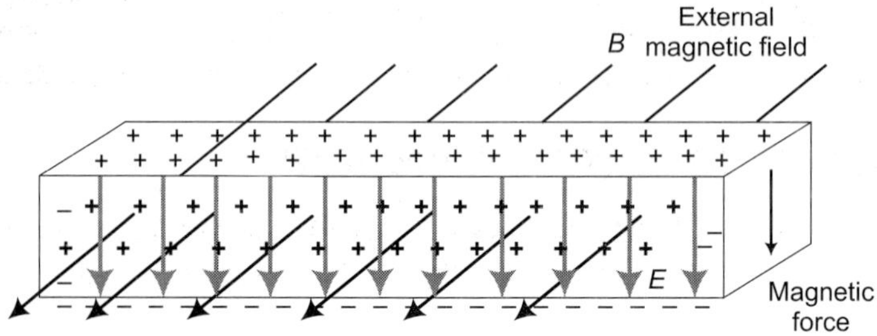

The extra surface charges also exert an electric force qE on the "*stationary*" positive atomic cores.

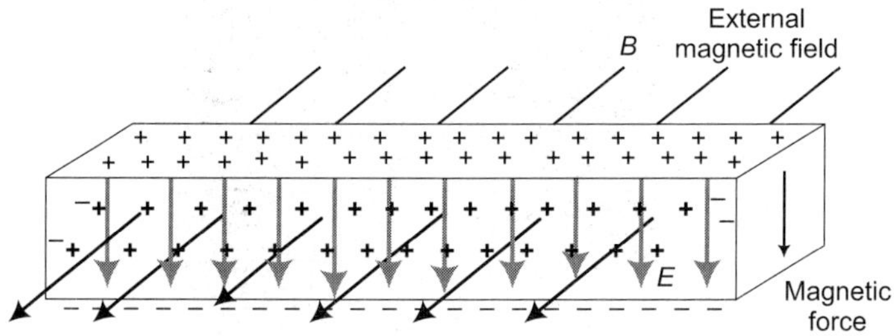

Fig. 6.11 The electric force acts on the positive atomic cores.

And this electric force is an unbalanced force (as shown in Fig. 6.11). There is no counter to this force because the atomic cores initially are not moving and hence are not subject to a magnetic force.

And that is the reason why *the wire moves* and an *electric force* acts on the positive atomic cores as shown in Fig. 6.12.

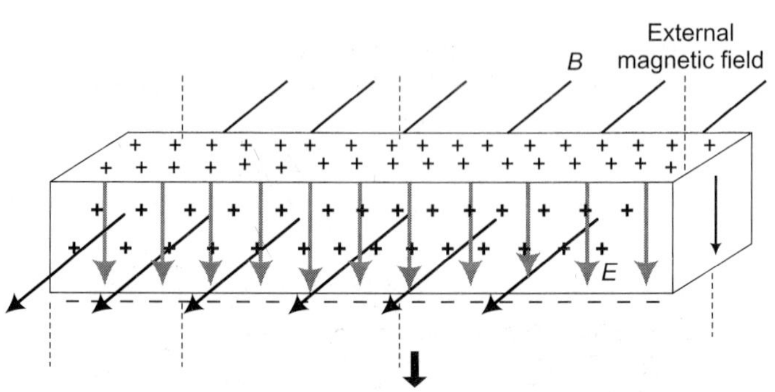

Fig. 6.12 The electric force acting on the positive atomic cores causing it to move.

The electric force due to the Hall-effect surface charges force the wire as a whole to move in the direction of $I(d\vec{L} \times \vec{B})$ (where dL is an elemental length of the conductor of length L carrying a current I) and the magnitude of the force is equal to this force.

6.5 MAGNETIC FORCES DO NO WORK

Reproduced with permission from Pearson Publishers, David J. Griffith's, "Introduction to Electrodynamics".

Wait a minute! It is not over yet!

Magnetic Forces do...

***no* work**

Magnetic forces may *alter* the *direction* in which
a charged particle moves ... but...
they cannot... *repeat*... cannot
speed it up
... nor slow it down.

Consider the magnetic force acting on the moving charges that caused these charges to pile up without accelerating or slowing them down.

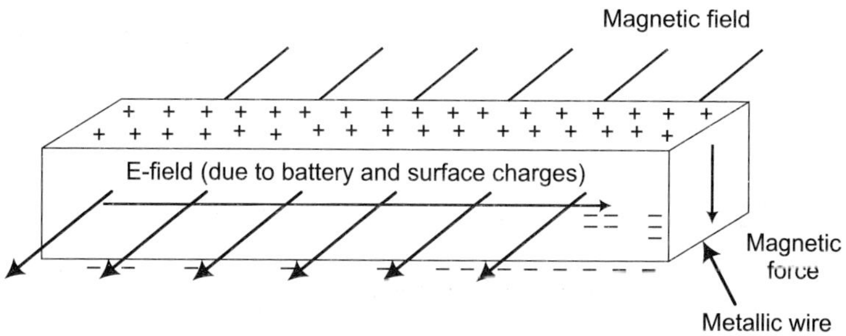

Compute the work done by the magnetic field as follows:

If the charge 'q' moves an amount dL in time dt, then the charge q moves by

$$dL = v.dt$$

where v is the velocity of the charge.

$$\text{Then, } dW_{mag} = F_{mag}.\ dL$$

$$= q\ (v \times B).\ vdt = 0$$

And the work done is...

0!!

Note: $v \times B$ is perpendicular to v, therefore, $(v \times B).\ v = 0$.

If the charge 'q' moves an amount dl, the magnetic field does not do any work! = 0!

Not convinced? Observe carefully

The *force* is always acting in a *direction that is perpendicular* to the motion of the charge or electron as shown in Fig 6.13. As is observed here the magnetic force acts perpendicular to the motion of the charge pushing them down.

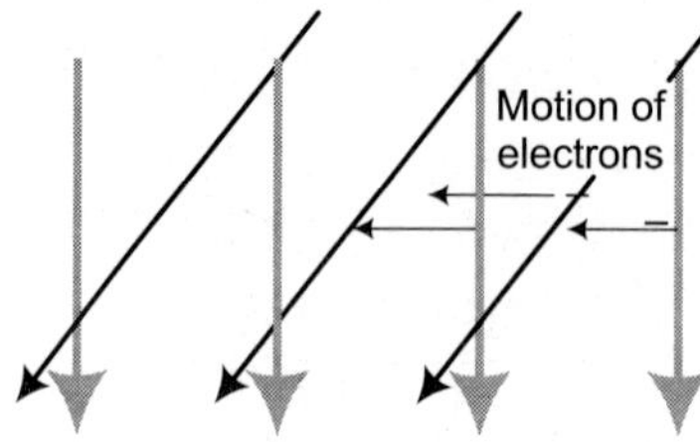

Fig. 6.13 The magnetic force acting perpendicular to the motion of charge.

For work to be done, the movement *along the direction of the force* only counts.

$$W_{mag} = F_{mag} \cdot dL$$

or, the dot product

$$= F \cos\theta dL$$
$$= FdL \cos 90°...$$
$$= 0.$$

Then, when an electromagnet is used to lift a heavy weight, does the magnetic field *not* do the work?

The answer to this is:

Let us study an example and see for ourselves.

We begin by defining a current. The current in a wire is the *charge per unit time* passing a given point.

A line charge λ (mobile charges per metre) travelling down a wire at speed v constitutes a current as shown in Fig 6.14.

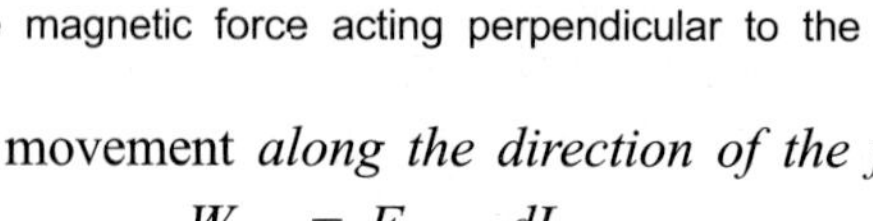

Fig. 6.14 Definition of a current.

The current $I = \lambda v,$

because, a segment of length $v\Delta t$, carrying a charge $\lambda v \Delta t$, passes point P in a time interval Δt.

The magnetic force on a segment of current-carrying wire is evidently,

$$F_{mag} = \int (\vec{v} \times \vec{B}) dq = \int (\vec{v} \times \vec{B}) \lambda dl = \int (I \times \vec{B}) dl$$

Since I and dl both point in the same direction, we can write this as

$$F_{mag} = \int I(d\vec{l} \times \vec{B})$$

Typically, the current is constant along the wire, and therefore, I comes outside the integral,

$$F_{mag} = I\int (d\vec{l} \times \vec{B})$$

Consider an experiment in which a rectangular loop of wire supporting a mass m hangs vertically with one end of the loop in a uniform magnetic field B as shown in Fig. 6.15.

What should be the current I in the loop that will balance the gravitational force downwards?

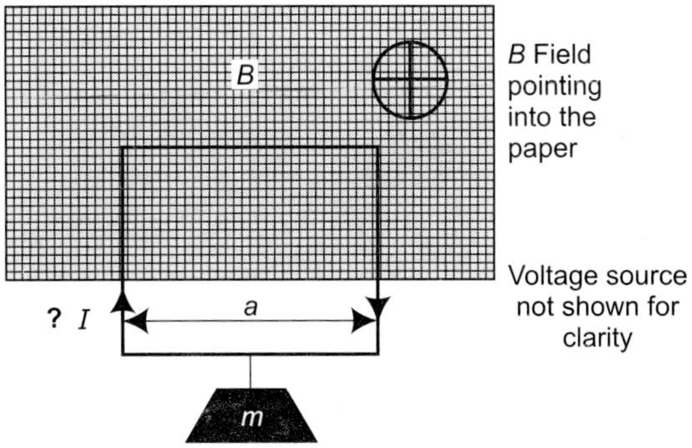

Fig. 6.15 A current carrying loop in a magnetic field suspends a mass.

First, the current must be clockwise so that the magnetic force on the horizontal segment is upwards.

Since $$F_{mag} = I\int (dl \times B)$$

Then, $$\boldsymbol{F_{mag} = I\ B\ a}$$

where, a = width of the loop because the integral on all the small current segments Δl is 'a' as shown in Fig. 6.16.

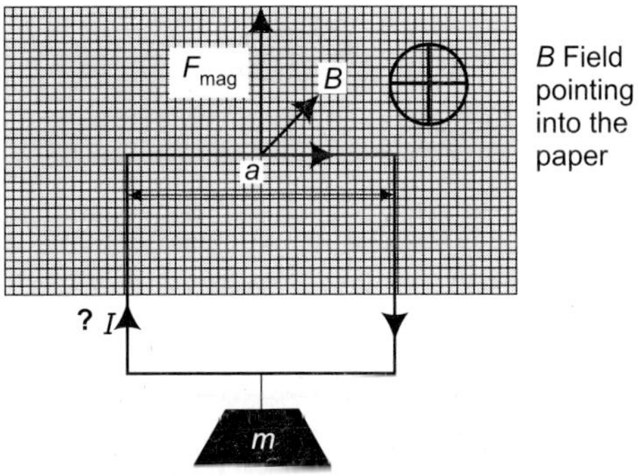

Fig. 6.16 Current must be clockwise so as to apply a force F_{mag} upwards.

The magnetic forces on the two vertical segments cancel as shown in the figure below.

For F_{mag} to balance out the weight mg:

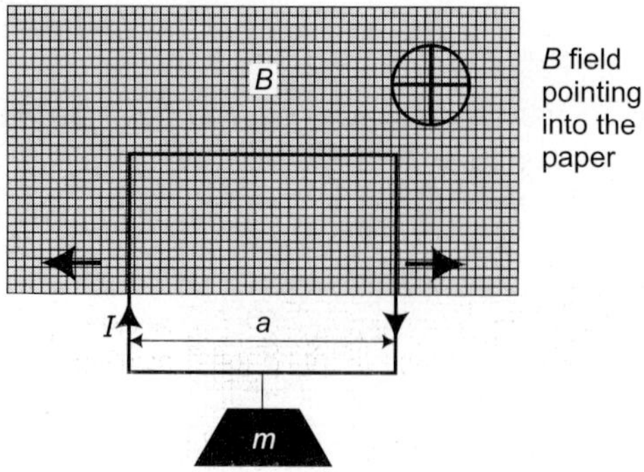

$$F_{mag} = I\ B\ a = mg$$

So, we have, $I = mg/Ba$

For the current I, then, both the forces are in perfect balance and the weight is hanging in mid air! This is shown in Fig. 6.17 (see side view).

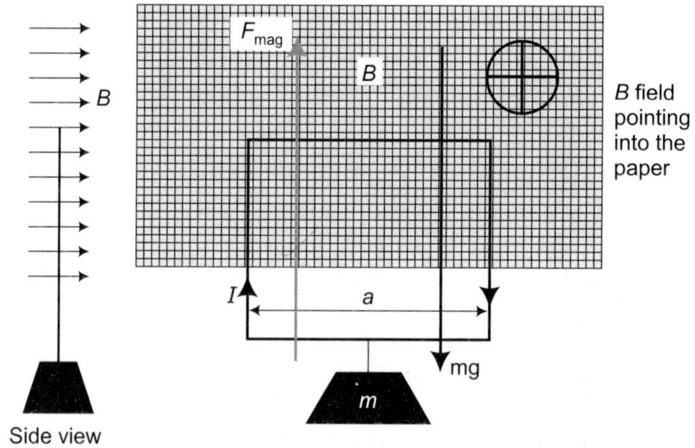

Fig. 6.17 The mass is hanging in mid air!

At a microscopic level, the force holding the weight up is due to the *electric field* force on the positive atomic cores in the wire that is developed as a result of Hall effect phenomenon of separation of charges.

What happens when we *increase* the current?

The upward magnetic force increases and there is a corresponding increase in Hall effect surface charges and the electric field strength increases its force on the positive atomic cores.

This force now exceeds the downward force of gravity, and the loop rises.

And that is the reason why the loop rises. An electric force acts on the positive atomic cores as shown in Fig. 6.18.

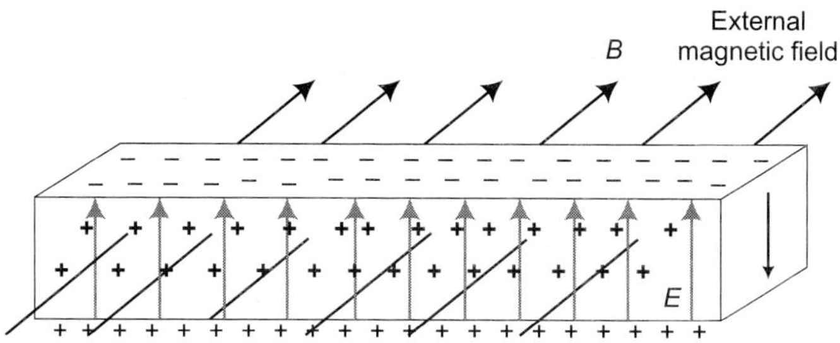

Fig. 6.18 An electric force acts on the positive cores raising the loop.

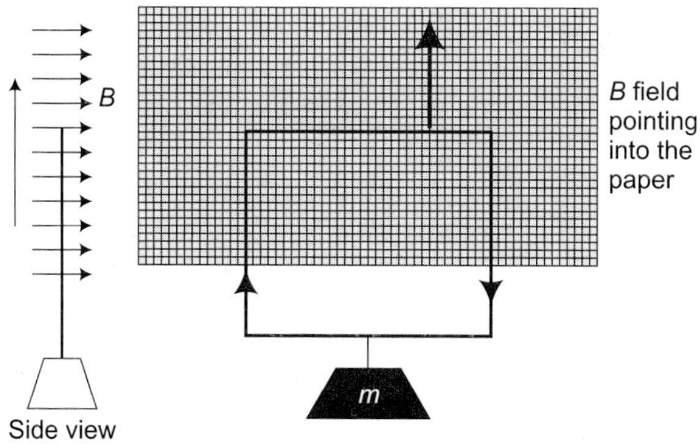

Somebody's doing the work and it looks as if the magnetic force is responsible. We are tempted to write

$$W_{mag} = F_{mag}h = IBah$$

where h = height, the loop rises.

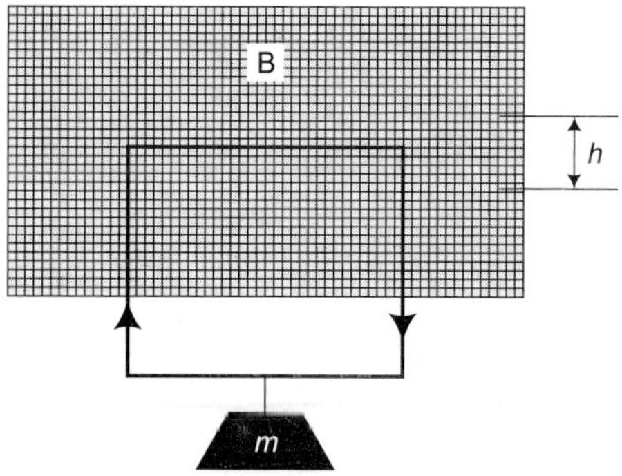

But we know that magnetic forces never do work.

So what is going on here?

When the loop starts to rise, the charges in the wire **are no longer** moving horizontally (with speed w) only, as shown in the figure below.

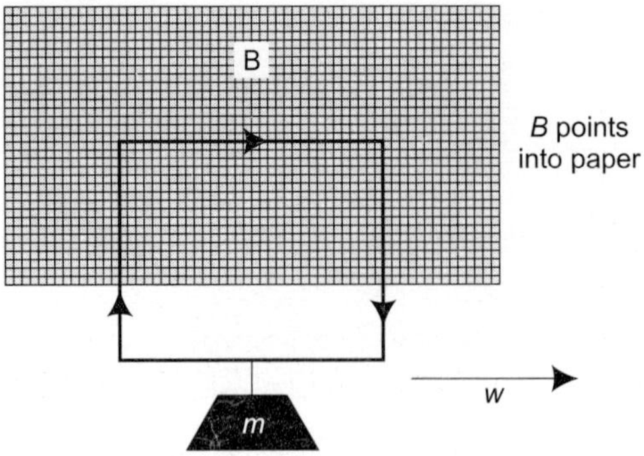

Their (charges) velocity now acquires an upward component u, which is the upward speed of the loop in addition to the horizontal component associated with the current I as shown in Fig. 6.19.

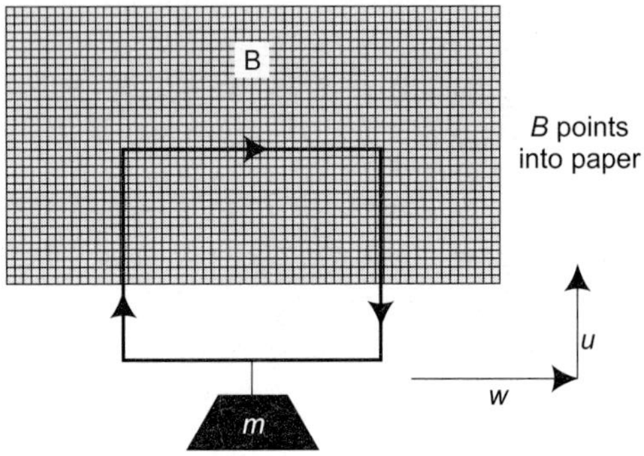

Fig. 6.19 Upward component 'u' of velocity of charges.

But the magnetic force which is always perpendicular to the velocity (resultant), check this, no longer points straight up, but *tilts* back.

Figure 6.20 shows the velocity components and the resultant velocity 'v' of the moving charges and Fig. 6.21 shows all the components of the velocity plotted inside the wire.

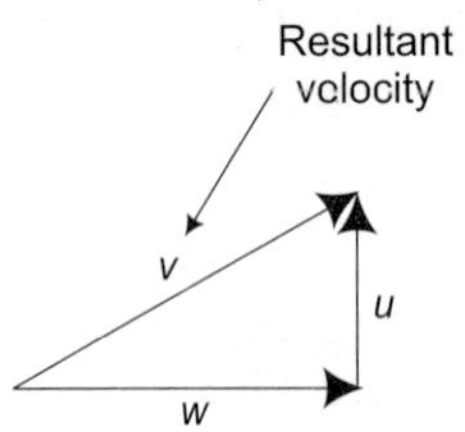

Fig. 6.20 Resultant velocity of the charges.

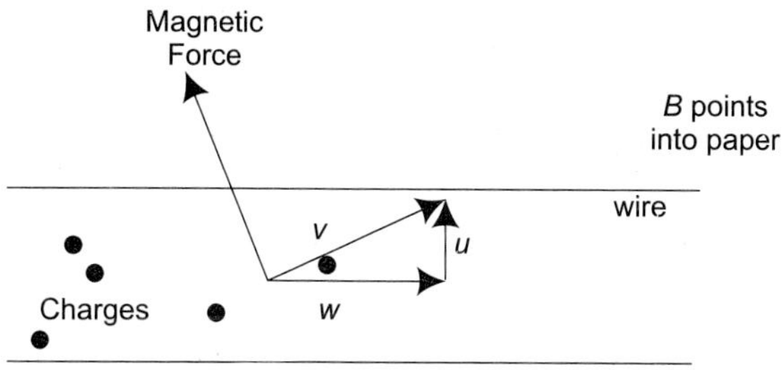

Fig. 6.21 The velocity components of the charges plotted inside the wire for clarity.

Do you want to know how it feels to be an electron—subjected to such a force? Imagine that you are walking on a long narrow carpet and the carpet ahead is lifted all of a sudden! as shown in Fig. 6.22.

Fig. 6.22 How the charges "feel" the magnetic deflecting force and their resultant velocity.

The magnetic force is perpendicular to the net displacement of the charge (which is in the direction of v) and therefore, it does no work on q.

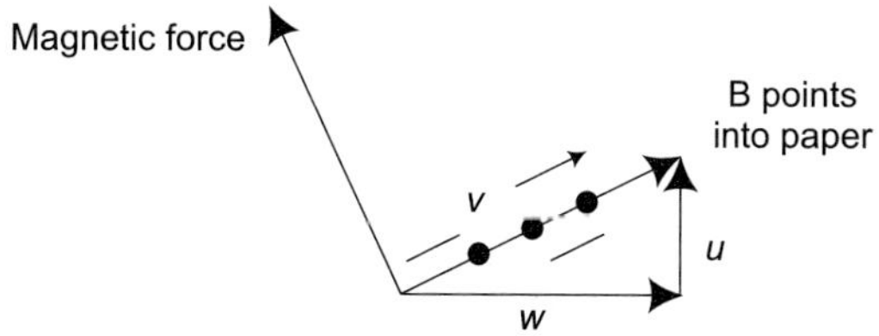

It does have a vertical component (qwB), indeed. The net vertical force on all the charges (λa) in the upper segment of the loop is (see Fig. 6.23) given by,

$$F_{vert} = \lambda\ a\ w\ B = IBa$$

where, λ is the charge per unit length, I = current, B = magnetic field, a = width of the loop and w = horizontal speed of the charges.

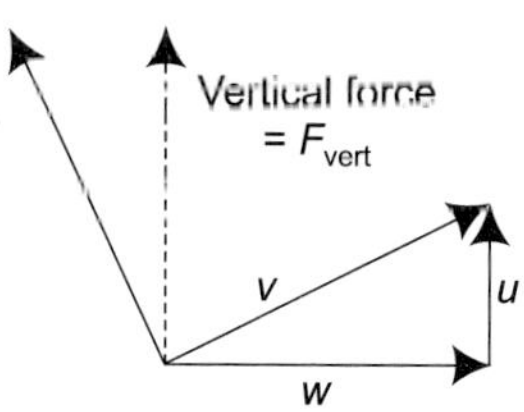

Fig. 6.23 Net vertical force on all the charges.

But now it also has a horizontal component quB, which *opposes* the flow of current and imposing a *braking* effect on the charges causing them to *pile up* at one end and produce an electric force against the applied voltage (electric force) that set the charges in motion in the loop of wire—in short a *back emf.*

This is shown in Fig. 6.24.

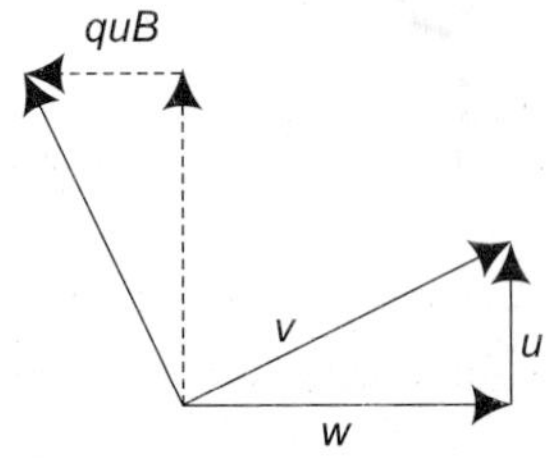

Fig. 6.24 Horizontal component of net force on all the charges.

Whoever is in charge of *maintaining* that current, therefore, must now push those piled up charges along—against the backward (braking) component of the magnetic force—against the back emf as shown in Fig. 6.25.

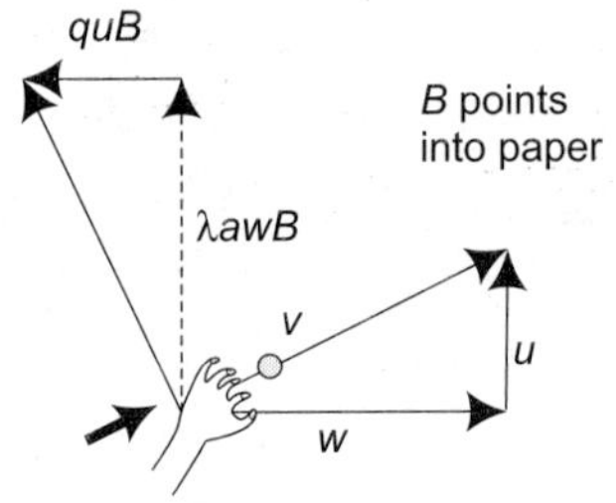

Fig. 6.25 Pushing the charges forward.

The total horizontal force on the top segment of the loop is

$$F_{horiz} = \lambda auB$$

where, λ is the charge per unit length.

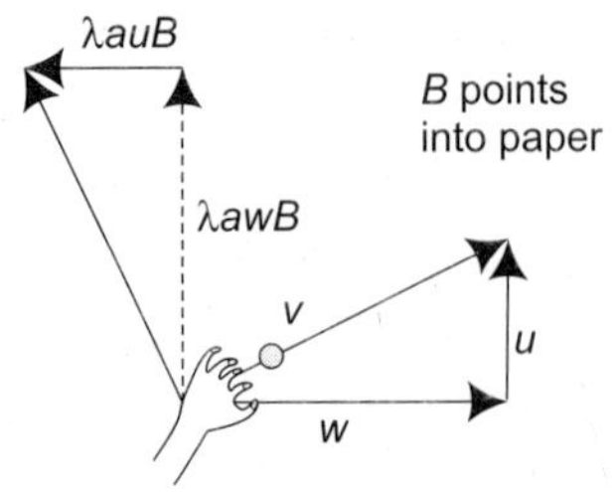

In a time dt the charges move a (horizontal) distance wdt. So, the work done by this agency such as a battery or a generator is:

$$W_{battery} = \int F_{horiz} w dt = \lambda a B \int u w dt = IBah$$

where h is the height the loop rises and because $I = \lambda w$ and integral $udt = h$.

This is what we attributed (wrongly!) to the magnetic force and wrote $W_{mag} = IBah$.

Was work done in the process?

Absolutely!

Who did it?

Not me, said the "magnetic field"...

Then, who did it?

It was the Battery!

Questions

1. Explain how magnetic fields affect moving charges and state and explain "Lorentz law".
2. Describe the processes that cause two current-carrying wires in close proximity to each other to attract.
3. With the help of neat sketches explain how a current-carrying wire moves when placed in a magnetic field.
4. Explain why magnetic forces do no work using "Lorentz law"?
5. A current-carrying rectangular loop (of width 'a') of wire carrying a current I is placed in a perpendicular magnetic field B. A mass m is suspended from the loop. Describe all the processes that cause the loop to rise when the current is increased and to prove that magnetic forces do no work show that the work done to raise the weight is due to the voltage source.

7

Faraday's Law; Motional emf; How Generators and Microphones Operate

The application of Faraday's law which is a major physical law is the theme of Chapter 7. Beginning with an introduction to the phenomenon of the co-occurrence of a non-Coulomb electric field with a changing magnetic field and its use in determining the induced emf when a current carrying coil is placed near another, the important concept of motional emf is explained and gradually developed to describe how generators and microphones operate.

7.1 MAGNETIC FIELD: A REVIEW

Magnetic interactions have some similarity to electric interactions:

1. There is *both* attraction and repulsion.
2. The *superposition principle* is valid.
3. The interactions pass *through matter*.

The *differences* between magnetic and electric interactions are:

1. A magnet is *permanently* magnetised while a charged plastic pen loses its charge within a relatively short time (see Section 1.25).
2. The magnetic field around a current carrying wire has a *curly* pattern, which is *not* seen with electric fields of point charges.

7.2 THE PHENOMENON OF ELECTRIC FIELD ASSOCIATED WITH A TIME-VARYING MAGNETIC FIELD

Consider two coils of wire *near each other*, positioned on a common axis, and one coil is connected to a *power supply* with an ammeter in series as shown in Fig 7.1 (after Chabay & Sherwood, 1995).

The second coil is *short-circuited at both its ends* with an ammeter in series. There is *no voltage* source in the second coil.

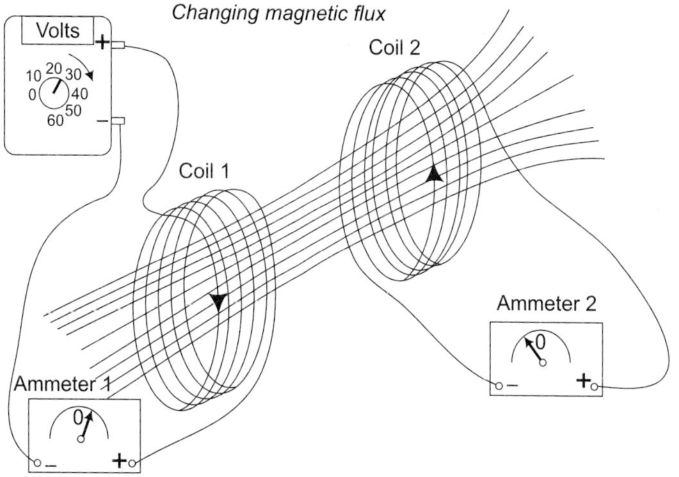

Fig. 7.1 The phenomenon of magnetic induction.

What we see is that when we increase the current in the coil 1 at a steady rate, there is a steady negative current measured by ammeter 2 as shown in Fig 7.1.

If we keep the current constant in coil 1, there is no current in the second coil (not shown in Fig 7.1).

And if we decrease the current 1 at a constant rate, there is a large steady positive current in ammeter 2.

7.3 LENZ'S RULE FOR THE DIRECTION OF THE INDUCED EMF

Recollect our observations in Chapter 5, that a changing magnetic field of a solenoid or coil is associated with a non-Coulomb electric field as shown in Fig 7.2.

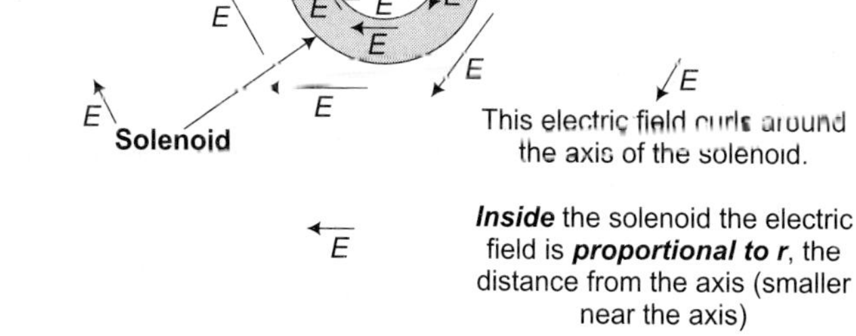

Fig. 7.2 The co-occurrence of an electric field with a time-varying magnetic field.

Lenz's rule, which we studied in Section 5.13, is used to determine the direction of the induced emf in many situations. Imagine that you place a wire around the changing flux, so that the non-Coulomb electric field drives current in the wire.

The non-Coulomb electric field would drive current in a direction to make a magnetic field that attempts to keep the flux constant.

For example, if B_1 points out the page and is increasing, as shown in the figure above, the non-Coulomb electric field curls clockwise around the solenoid. If a wire encircles the solenoid, conventional current runs clockwise in the wire, and this induced current produces a magnetic field in the region. See Fig 7.3.

In what direction is the magnetic field inside the solenoid, due to the clockwise current?

Current flowing clockwise through the wire makes a magnetic field into the page at *centre* of wire, decreasing the net flux.

7.4 CORRECT APPLICATION OF LENZ'S RULE

If a metallic ring encircles the solenoid, then the "induced" current driven by the non-Coulomb electric field produces a magnetic field into the page at the centre of the wire, *decreasing* the net flux

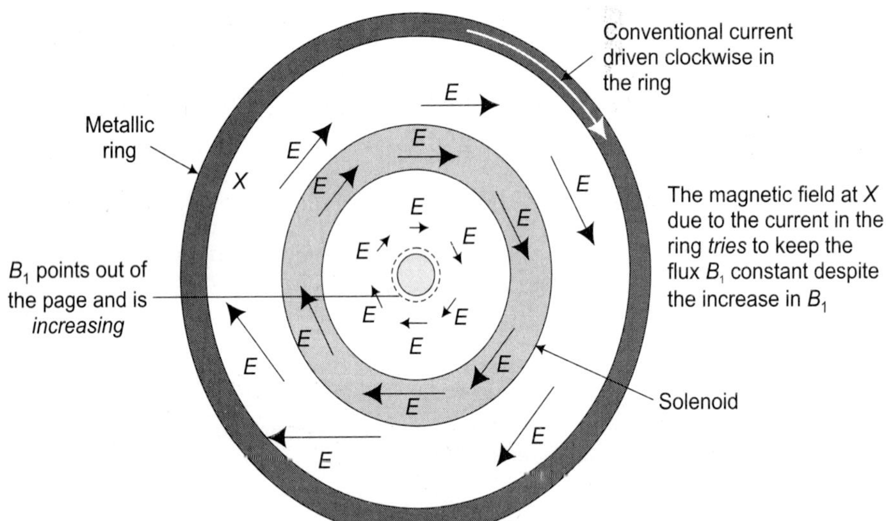

Fig. 7.3 Correct application of Lenz's rule.

Applying this rule, we can see that if we increase the current at a steady rate in the coil 1 in the circuit of Fig 7.1, there is a steady negative current measured by ammeter 2, and this current makes a steady magnetic field B_2 to the left inside the coil and so the emf and current are constant (Chabay & Sherwood, 1995, used by permission).[2] This is shown in the Fig 7.4.

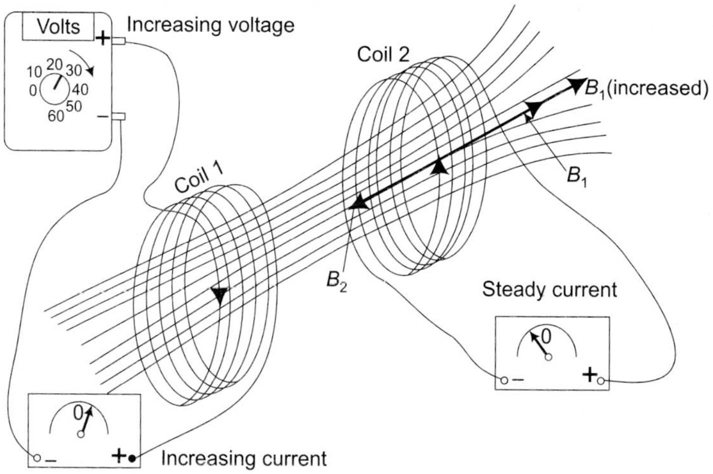

Fig. 7.4 Steady current in coil 2 when the current in coil 1 is increasing at a constant rate.

7.5 INDUCED CURRENT WHEN THE SECOND COIL IS TILTED

Is the *induced emf* and therefore, the current in the second coil affected when the second coil is *tilted at an angle* θ? (after Chabay & Sherwood, 1995).

When the second coil is tilted by an angle θ the **flux** *involves* (that induces the emf) the *perpendicular* (to the coil face) component of $\boldsymbol{B}_1$ and, therefore, the induced current in the second coil is reduced by the factor **cos** θ (Chabay & Sherwood, 1995, used by permission).[2] This is shown in Fig 7.5.

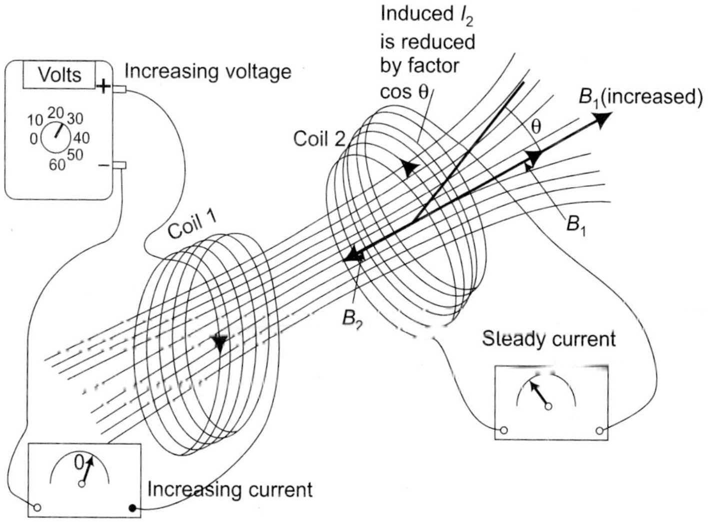

Fig. 7.5 Coil 2 tilted at an angle θ, causes induced current to reduce by factor cos θ.

7.6 FARADAY'S LAW

The principle of inducing an emf in the second coil due to a changing current in the first coil on the common axis is used in transformers to *transform* voltage.

The phenomenon discovered by Michael Faraday, and known as Faraday's law of *induced emf* is a major physical law which *cannot be derived* from any of the other fundamental principles we have studied. The induced emf according to the law, is given by

$$|emf| = \left|\frac{d\phi_{mag}}{dt}\right|.$$

The induced emf along a round-trip path is equal to the rate of change of the magnetic flux on the area encircled by the path, with ϕ_{mag} defined:

$$\phi_{mag} = \int_{surface} \vec{B}.\vec{n}\, dA$$

where, n is a unit vector perpendicular to the area ΔA, a tiny area of the surface on which the flux is present. See Section 5.2 also.

Faraday's Law Applied to Moving Coils

We have seen a non-Coulomb electric field associated with a *time-varying magnetic flux*.

But wait, this is *not the only way* to produce a varying flux.

With a *steady* current in the first coil, you can *move* the first coil closer to the second coil. This is shown in Fig 7.6.

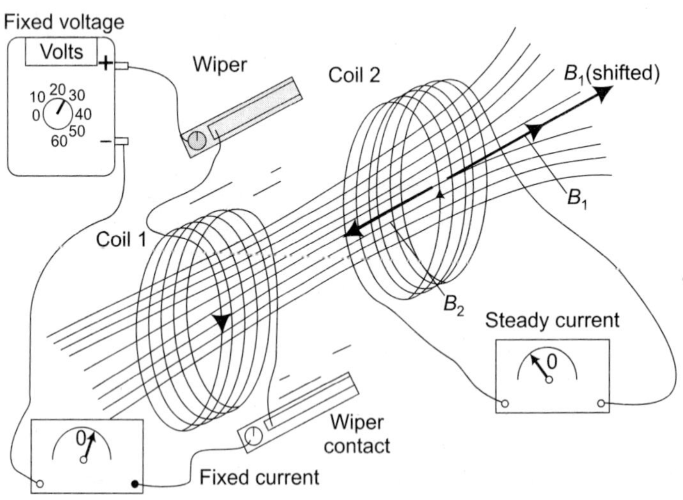

Fig. 7.6 Induced emf by moving the 1st coil with a steady current towards the 2nd coil.

The magnetic flux inside the second coil increases, and while the flux is increasing there is an emf (and a current in the second coil). It does not matter which coil moves: in both cases there is a change of magnetic flux through coil 2 and an induced current.

We next study the very important phenomenon of motional emf. It is fundamental to the understanding of how generators and microphones operate.

7.7 MOTIONAL EMF

We have seen that a *changing* magnetic field is associated with a curly (non-Coulomb) electric field.

However, there is one more way that magnetic flux can *change—or does it really?* We shall study this.

And, this is the way that many types of electric *generators, galvanometers, loudspeakers, microphones, motors and several devices operate.*

While, in this method Faraday's law still correctly predicts the induced emf, there is a *subtlety involved in the process.*

We shall demonstrate one more phenomenon of charge by observing how charges can be moved—simply moved.

Can you *move* a bunch of charges by holding them in your fist?

It is really very *simple.*

Simply grab *any* piece of wire, even *without a current in it,* you have a bunch of jumpy energetic conduction band electrons. Move the wire and, lo and behold, you have successfully moved a bunch of charges! See Fig 7.7.

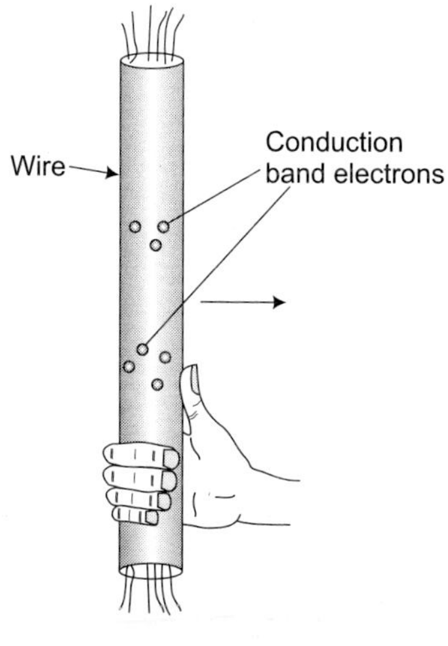

Fig. 7.7 Moving a fistful of charges.

Now, stretch your imagination and don't you think *whatever you move contains some charges* or not ? Simple, was it not?

Consider the case of pulling a *metal bar* at a constant speed v along frictionless *metal rails* through a uniform constant magnetic field B pointing into the paper as shown in Fig 7.8. A resistor with *resistance R* is connected *between* the *rails.* As the bar is pulled along the rails the resistor will heat up (Chabay & Sherwood, 1995 used by permission).

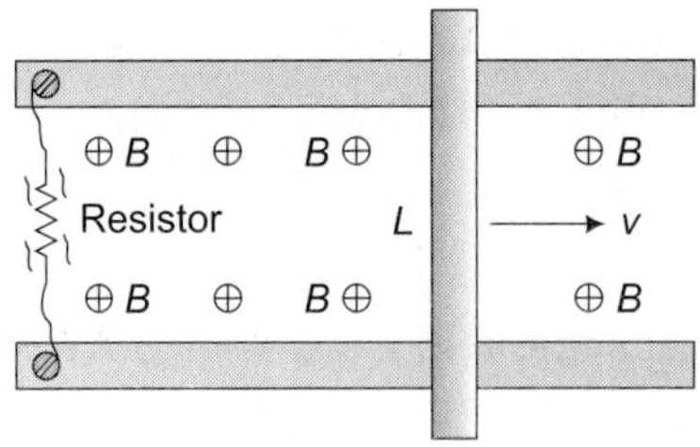

Fig. 7.8 Moving a bar on rails immersed in a constant magnetic field.

The *additional flux* enclosed by the circuit loop consisting of the moving bar, rails, and resistor is

$$\text{additional flux} = \Delta\varphi_{\text{magnetic}} = B\Delta A = BL\Delta x = BL(v\Delta t)$$

where L = length of the bar and Δx is a small incremental change in the distance the bar moves when it is pulled with a velocity v.

The area, the loop comprising the resistor R, the rails and the bar encloses, is A, and ΔA is a small incremental change in the area when the bar moves a distance Δx.

And, the rate of change of flux

$$= \frac{d\varphi_{magnetic}}{dt} = \frac{(BLv)\Delta t}{\Delta t} = BLv$$

This is also equal to the emf generated in the circuit.

The amount of current through the resistor is

$$I = \frac{emf}{R} = \frac{BLv}{R}$$

Faraday's law is very general and any *change* in magnetic flux, whether *due to* a change in magnetic field or, a *change in area*, produces an emf equal to the rate of change of flux.

In fact, both can be present simultaneously, as when you drag the metal bar along the rails, *and* the magnitude of the uniform magnetic field *is increasing* as shown in Fig. 7.9.

The flux is ***BA***, and the rate of change of the flux is due *in part to the change* in the magnetic field, and *in part to the change* in *the area* enclosed by the changing path.

The *resistor* will heat up!

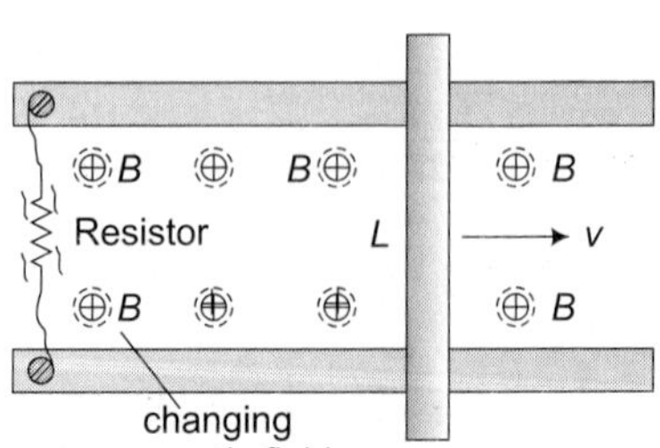

Fig. 7.9 Moving a bar on rails immersed in a magnetic field that is changing.

The equation of the rate of change of flux

$$\frac{d\varphi}{dt} = \frac{d}{dt}(BA) = \frac{dB}{dt}A + B\frac{dA}{dt} = \frac{dB}{dt}A + B\frac{Ldx}{dt}$$

Change of B, fixed path (first term) — Change of path (second term)

$$= \frac{dB}{dt}A + BLv$$

The calculation is a bit mysterious. Physically, *where* did the emf represented by BLv come from if it is *not* associated with a *changing* magnetic field?

We will solve this *puzzle* now.

Let us take another look at how emf was produced when the *magnetic field* was constant in our experiment and the *area* of the loop enclosed by the circuit *was changing*. See Fig. 7.10.

We have a conducting metal bar and we are moving it.

Let us not forget that we are moving the jumpy *energetic conduction band electrons* in it as well. See Fig. 7.11.

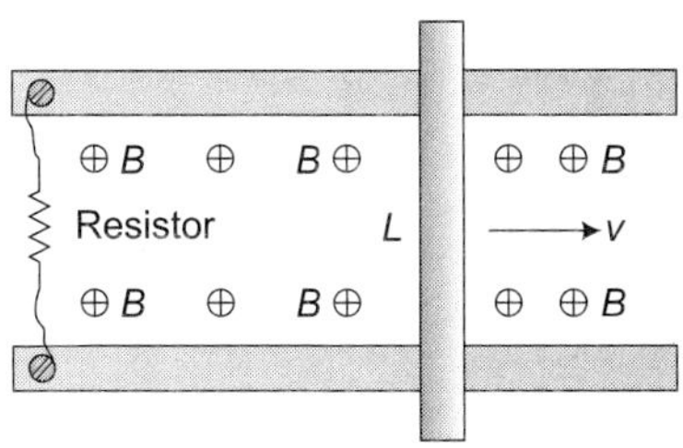

Fig. 7.10 Moving a bar on rails immersed in a magnetic field.

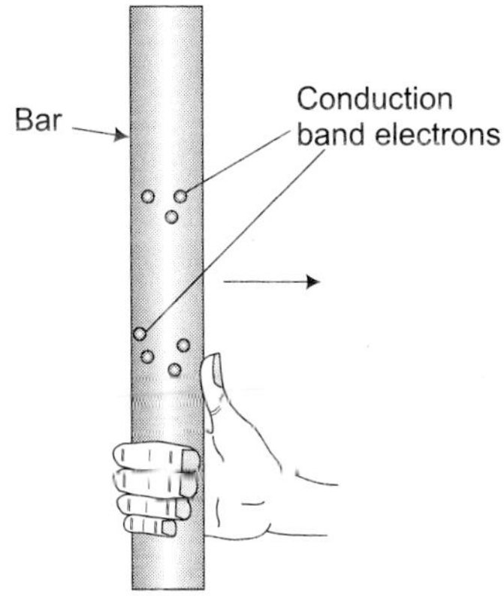

Fig. 7.11 Moving charges in the bar.

Now let us stick in a *uniform* ***magnetic field*** pointing ***into*** the area where the bar is moving and move the bar as shown in Fig. 7.12.

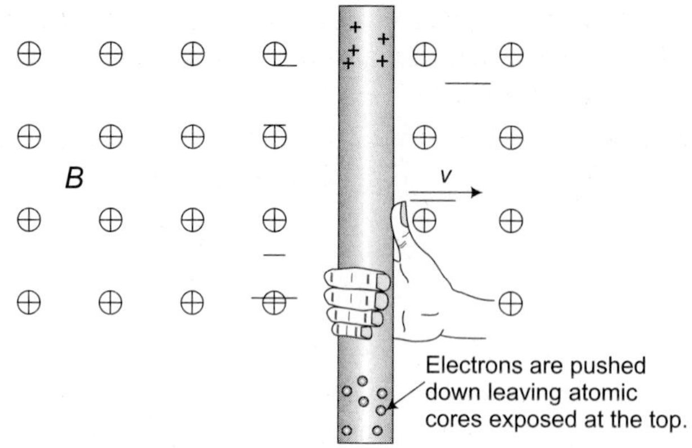

Fig. 7.12 Polarizing the bar by moving it on rails immersed in a uniform magnetic field.

As a result of the magnetic force (due to the presence of the uniform magnetic field), the free electron sea shifted, the bar polarized and a Coulomb electric field develops as in Fig. 7.13.

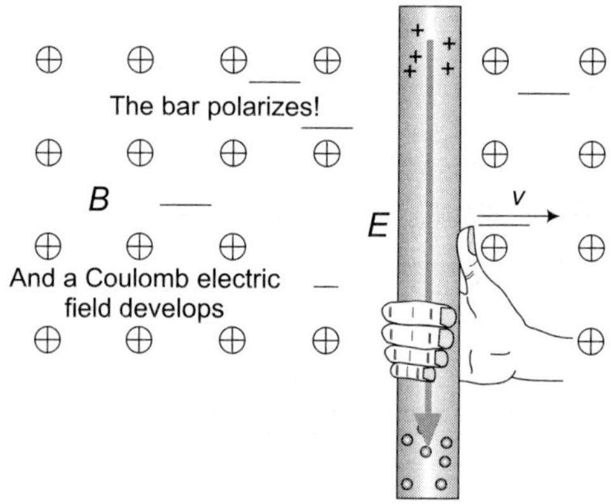

Fig. 7.13 Development of a Coulomb electric field inside the bar.

Move the bar the ***other way*** and it is oppositely polarized as shown in Fig. 7.14.

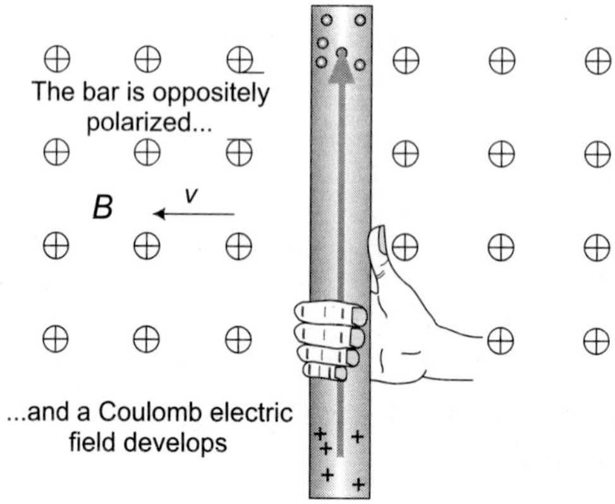

Fig. 7.14 Move the bar the other way and the bar is oppositely polarized.

Therefore, the *induced emf* is due to *separation of charges* in the bar as it moved. This we call motional emf, the *BdA/dt* term in the emf equation, which can result even when the magnetic field is *not changing* or is constant.

When does the *electron sea* **stop shifting?** Precisely when, the $qv \times B$ magnetic force is balanced by the Coulomb electric force as shown in Fig. 7.15.

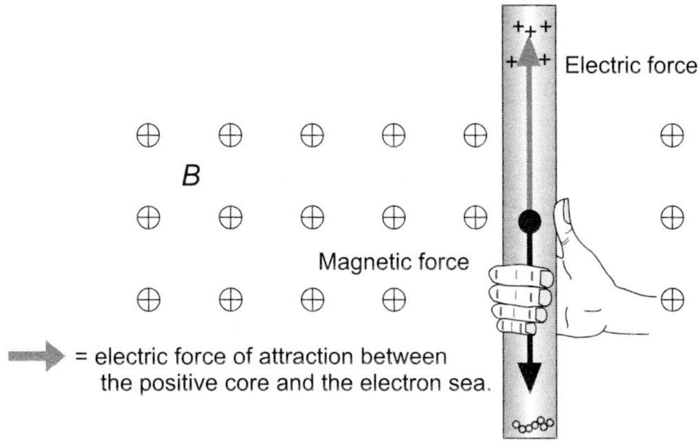

Fig 7.15 The sea stops shifting when the magnetic force equals the Coulomb electric force.

If the **bar** is kept moving at a constant rate $\boldsymbol{v}$ it is possible to calculate the Coulomb *electric* field produced by the trick of polarization or what we call separation of charges.

The electron sea will stop shifting when

$$F_{electric} = F_{magnetic},$$

or,

$$qE = qvB \sin 90°$$

using the rule $v \times B$ to determine $F_{magnetic}$.

Therefore,

$$E = vB$$

Notice that we have a nonzero electric field inside a metal in static equilibrium (bar moving at constant speed, no acceleration). There is no current, because the electric force is balanced by a magnetic force: the net force on an electron inside the metal is zero.

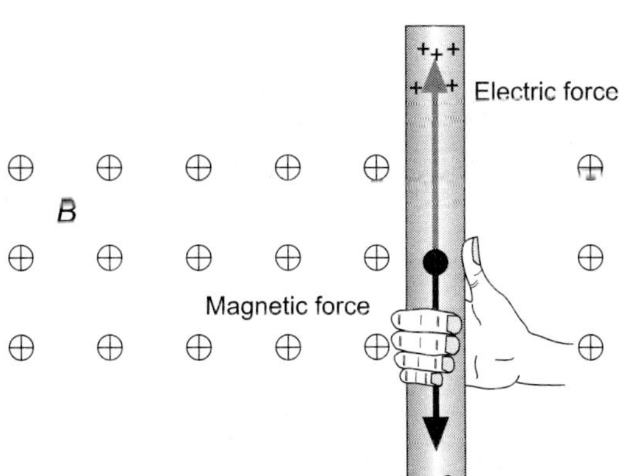

Once the bar is completely polarized, how much force must we apply to the bar to keep it moving at constant speed?

None! Since the bar is moving at constant speed, the *net* force on it must be zero. No force is needed to keep it moving, since no other forces act on the bar.

And the *polarized* bar acts like a battery.

In a battery, charges are *separated* by chemical reactions and an *emf* develops between its plates and an *electric field as well* in between and in the space surrounding.

Similarly, charges are separated *between the two ends* of the bar and an *emf* develops therefore *between* its two ends and an electric field as well in between and in the space surrounding.

If we let this metal bar slide on rails as before, with a resistor connecting the rails, a current will run in the circuit and we will have to exert a force to keep the bar moving.

Also notice that electrons run the *wrong way* through the moving bar, moving towards the *negative* end of the bar as if *inside* a battery.

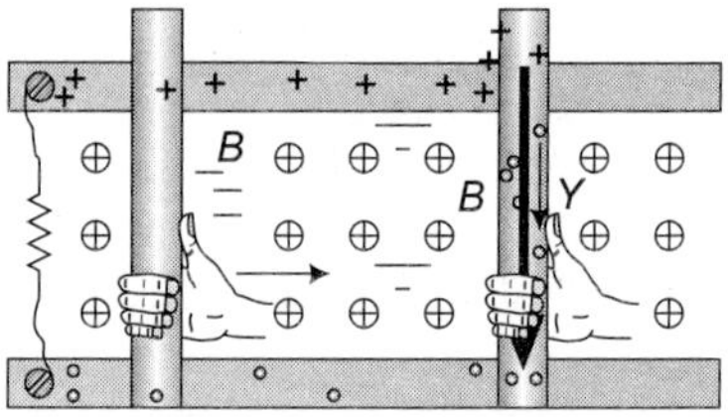

Fig. 7.16 The electrons move the wrong way inside the bar.

Why do the electrons move the wrong way inside the bar (shown by the arrow marked *Y* in the Fig. 7.16)?

The current through the *resistor* is *continually depleting* the charges on the ends of the moving bar, with the result that the electric field inside the bar is always *slightly less* than needed to balance the magnetic force.

And since the upward electric force qE is *slightly less* than the downward magnetic force qvB, so the electrons move towards the negative end of the bar.

A *reduction* of internal field due to *depletion* of terminal charges and replenishment by the *separation of charges inside the bar....* electrons stripped from immobile *positive atomic cores*...motional emf process in generators as shown in Fig. 7.17.

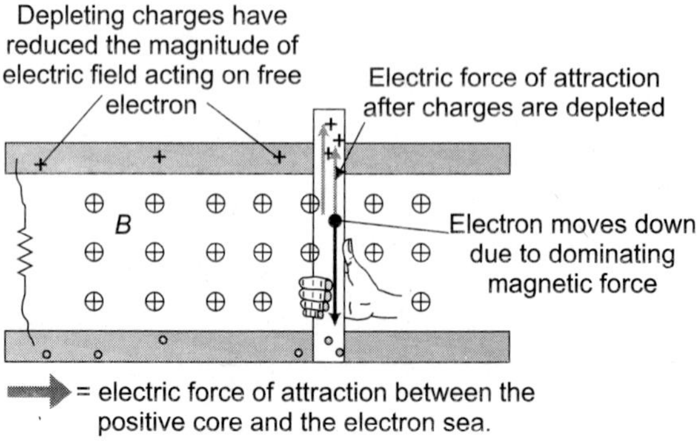

Fig. 7.17 Reduction in Coulomb electric field due to depletion of positive charges.

So, what we see is that the emf due to magnetic forces (in the presence of a uniform *constant* magnetic field) is equal to vBL, which is (*by coincidence*?) numerically equal to $d\varphi_{magnetic}/dt$ for this situation. This emf is called *motional emf.* The motional emf is also termed back emf when current carrying loops of conductors immersed in a magnetic field develop a torque as in motors which we shall study in Chapter 8.

A moving element in a circuit implies motional emf due to magnetic forces, and it can be shown that this motional emf is always numerically equal to the $d\varphi_{magnetic}/dt$ associated with the change in path surrounding the flux.

7.8 MOTIONAL EMF AND SURFACE CHARGES

In Fig. 7.18 is shown the motional emf with surface charges in the external circuit. It is the surface charges that cause a current to flow in the circuit, the rails and the bar.

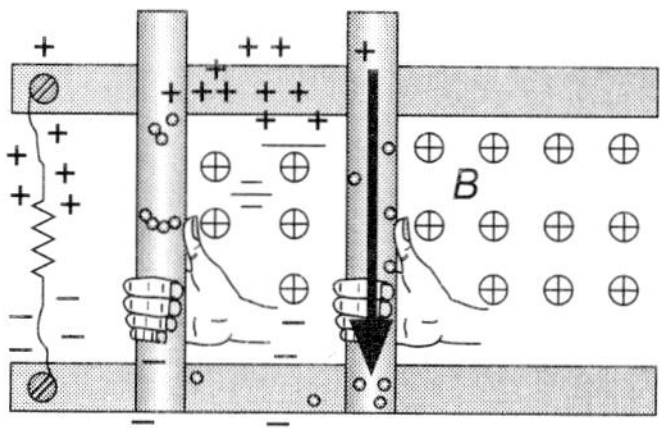

Fig. 7.18 Motional emf with surface charges in the external circuit.

The above principle of producing an emf is used in the design of *generators* and *microphones*.

To produce an emf, all you have to do is simply wave an object containing free electrons and immobile positive cores in a magnetic field.

The electrons of the electron sea move and an electric field is established between regions where they bunch up and regions where the lattice ions are exposed. Therefore, an emf is created between the two regions.

But, is it *physically possible* to drag a metal bar on metal rails to generate an *emf*? How long should the rails extend? How long a *magnet* will you use to generate the magnetic field *B?*

This seemingly difficult problem is easily overcome by *folding* over the rails and locating the sources of magnetic field around a cylinder.

That is how electric *generators* are constructed.

We will now observe how an emf is produced in such a machine by *rotating* a rectangular wire loop fixed on an axle in a magnetic field.

7.9 GENERATING AN EMF IN ROTATING MACHINES

Suppose we turn a crank to rotate a loop of dimensions $w \times h$ at a constant angular velocity $\omega = d\theta/dt$ inside a uniform magnetic field B. Refer to Fig. 7.19.

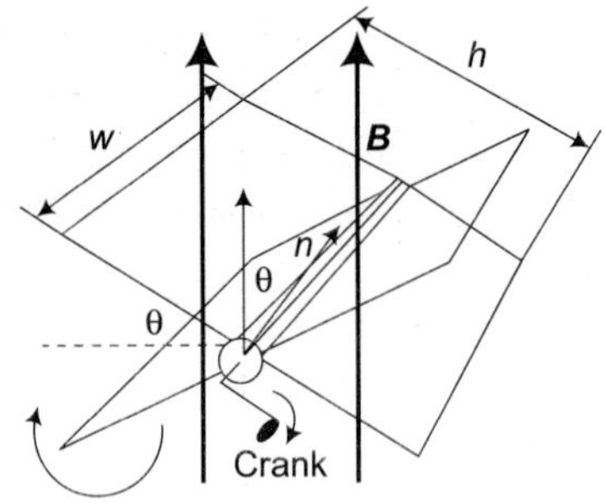

Fig. 7.19 Rotating a rectangular current carrying loop in a magnetic field.

When the loop is at an angle θ from the horizontal, the normal n to the loop is at an angle θ to the magnetic field B.

The area of the loop is $A = wh$.

From the expression derived in Section 5.2, the total flux Φ = sum of $B.n\ \Delta A = B\cos\theta$ $(wh) = B\ wh\ \cos\theta$, where $B\ \cos\theta$ is the *projected* area on the horizontal.

Since,
$$\text{emf} = \frac{d\Phi}{dt} = \frac{d}{dt}\,Bwh\,\cos\theta$$
$$= Bwh\,\frac{d}{dt}\,\cos\theta = Bwh(-\sin\theta)\,\frac{d\theta}{dt}$$

or,
$$\frac{d\Phi}{dt} = Bwh(-\sin\theta)\,\frac{d\theta}{dt}$$

The rate of change of the angle is the angular velocity in radians per second,

$$\omega = \frac{d\theta}{dt}.$$

Therefore, the emf $= Bwh(-\sin\theta)\,\dfrac{d\theta}{dt} = \omega Bwh(\sin\theta)$

7.10 POLARITY OF EMF IN ROTATING MACHINES

When a loop is driven with a force of rotation F_{rot} it rotates and reaches the position X, the emf developed is determined using the relation emf = $\omega Bwh(\sin\theta)$ that was derived in Section 7.9.

The force F_{rot} can be resolved into two components $F1$ and $F2$ perpendicular to each other to determine the *direction* of velocity of the charges. This is shown in Fig 7.20a.

The component F1 does *not* contribute to the development of the emf since it is *in line* with the magnetic field.

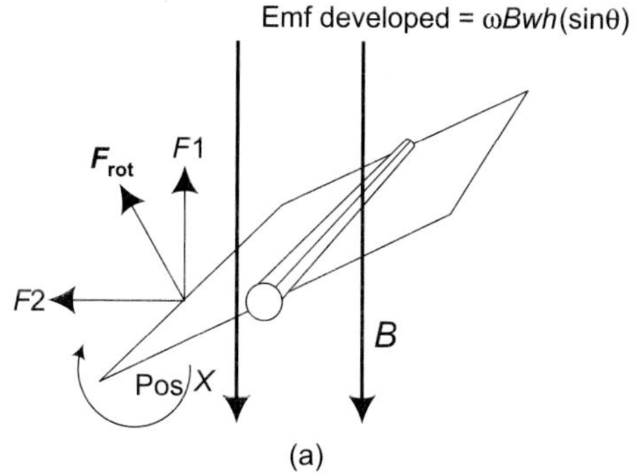

Fig. 7.20a Resolving the force causing rotation into components.

The emf developed by the component F2 will be as shown in Fig. 7.20(b) using $v \times B$ applied to the movable charges (which are electrons) inside the wire which have a *velocity component* v in the direction of F2 and thus, we determine the *direction* of the magnetic force and the resulting separation of charges.

Note: When applying the hand rule for vector cross products described in Section 5.2 to moving electrons in a magnetic field, the direction of $v \times B$ will be 180 degrees (opposite) the direction indicated by the thumb.

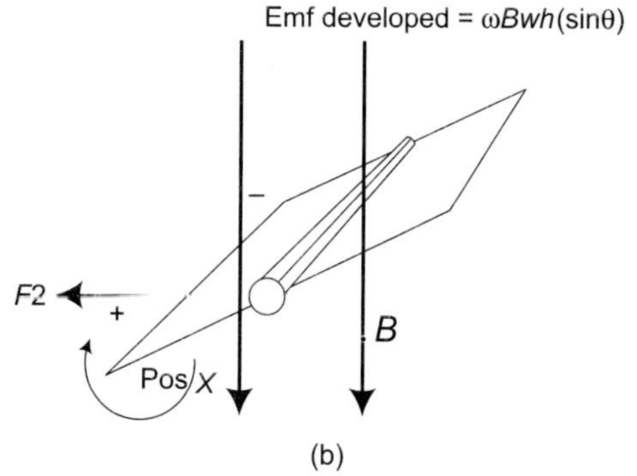

Fig 7.20b Determining the direction of movement of electrons and resulting emf.

A similar analysis when carried out on the loop when it has rotated to position Y as shown in Fig. 7.21 shows that the *emf* developed is due entirely to the *velocity of charges* (*electrons*) in the direction of F3.

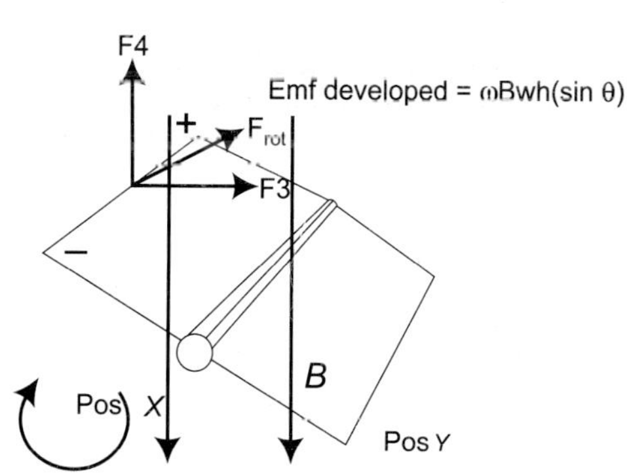

Fig. 7.21 Rotating the loop to position *Y*.

Notice that the *polarity* of the emf has reversed in this orientation in comparison to the previous orientation of Pos *X* as shown in the figure below with the *X* and *Y* positions combined.

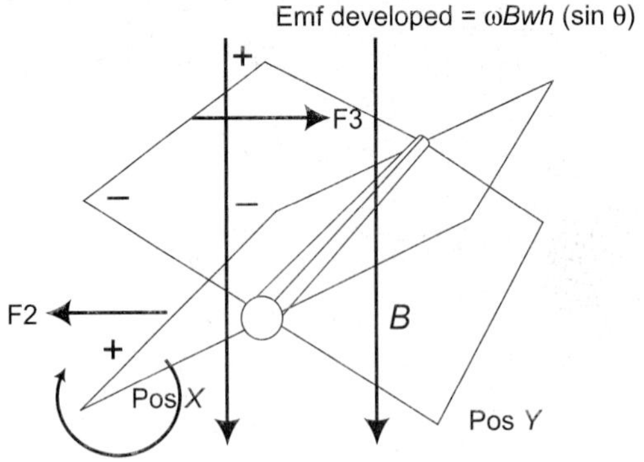

If we were to construct an emf generator using such a loop then it would generate an emf that *changes polarity* after every half cycle of rotation as seen from the Fig. 7.22.

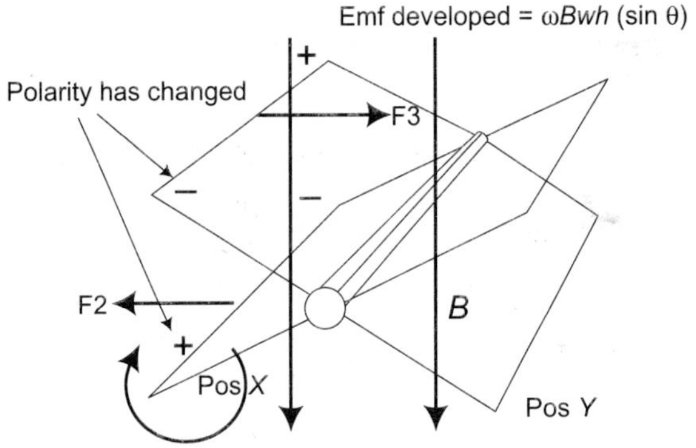

Fig. 7.22 The polarity change at the ends of the conductor when moved from position *X* to position *Y*.

7.11 UNIDIRECTIONAL EMF

The key to develop an emf with an *unchanging or unidirectional* polarity is to make electrical connections to the coil in such a way that just as it crosses the position where its polarity is changing we *reverse* the connection to the coil.

This is achieved by using a *split-ring commutator*, a device that is connected to the shaft of the generator and that rotates along with it which collects the emf, automatically *reversing* the connections with the coil.

Armature and Commutator

Figure 7.23 is a photograph of an armature and commutator of a dc machine disassembled from the machine (reproduced with permission Chuck Morrison www.morrisonprairie.com) Generator operation will be discussed in the next section.

Fig. 7.23 Armature and commutator of a dc machine.

7.12 GENERATOR OPERATION

Figure 7.24 shows the production of emf in a generator using the essential components.

The generator consists of a yoke with field poles in which coils produce a magnetic field directed towards a cylindrical structure called an armature carrying the main conductors. The conductors in which the generator's emf will be induced, are placed in slots of the armature. The armature is capable of being rotated on an axle by a drive such as a *turbine.*

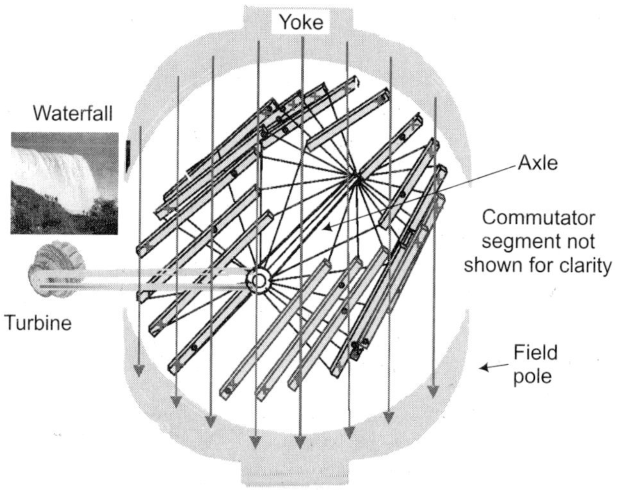

Fig. 7.24 Generator operation.

Figure 7.24 is an attempt to capture the essence of a dc generator operation. The armature and the commutator rings are not shown for the sake of clarity.

Shown are the yoke and field poles with the magnetic field and the armature conductors.

The generator is shown being driven by a turbine via a belt and pulleys arrangement. The turbine is driven by water falling in a waterfall.

In the conductors the charges are shown being separated (electrons and exposed positive atomic cores) as the conductors rotate in the magnetic field driven by the axle on which the armature is mounted. The separated charges are responsible for the production of the emf in the conductor.

What constitutes a *load* for a generator?

It is the electrical energy that is produced as a result of the *conversion* from mechanical energy of the drive to the generator.

The electrical energy is the *combination* of the *emf produced with the current* in a *load* such as a *resistor* or an electrical network or a motor that constitutes the *load.*

Figure 7.25 shows the action of a generator with a split-ring commutator.

The armature is made from copper conductors that consist of positive atomic cores and free electrons.

You may note how the emfs produced on opposite sides of the coil (due to separation of charges) add up like *batteries connected in series* and the action of the commutator that reverses the coil connections so that the *emf remains unidirectional* at the terminals of the generator...the *load is a lamp.* The commutator is made from a conductive ring split in two halves and fixed on a insulated cylinder coupled to the axle and which rotates alongwith the axle of the generator. The generated emf is picked off the commutator using two leaf springs pressing against it. In large generators this is achieved by pressing carbon brushes against the commutator *segments.*

In the series of figures from Figs. 7.25 to 7.28 is shown the rotation of the armature and the process of commutation. The separation of charges is shown in the conductors in the Figs. 7.25-7.28.

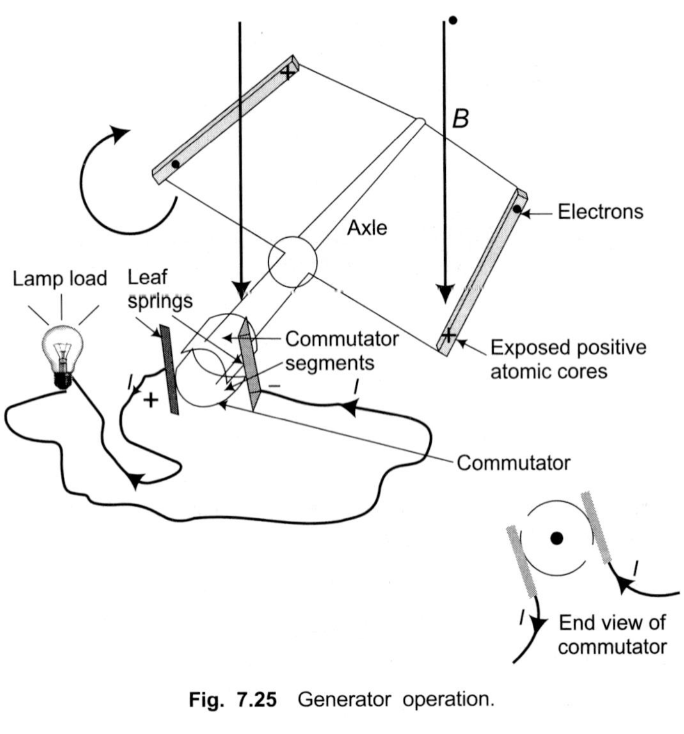

Fig. 7.25 Generator operation.

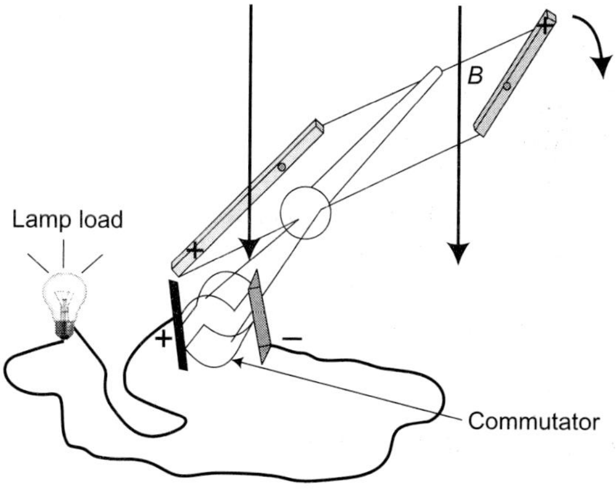

Fig. 7.26 About half rotation later.

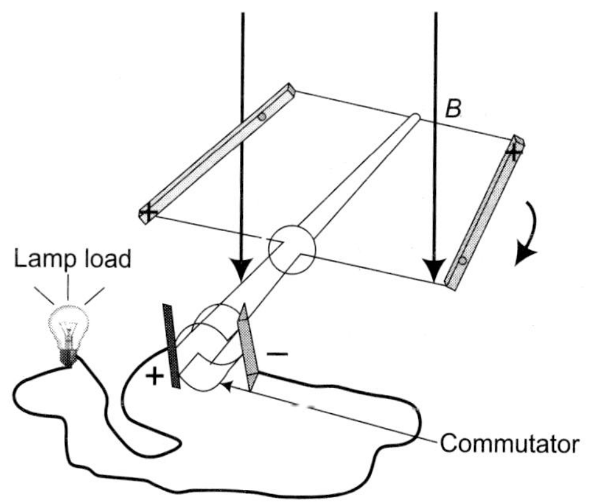

Fig. 7.27 Just before commutation occurs.

As the emf in the conductor reverses, the commutator segment in contact with the leaf spring slips over to make contact with the other leaf spring and this maintains a unidirectional emf at the generator terminals.

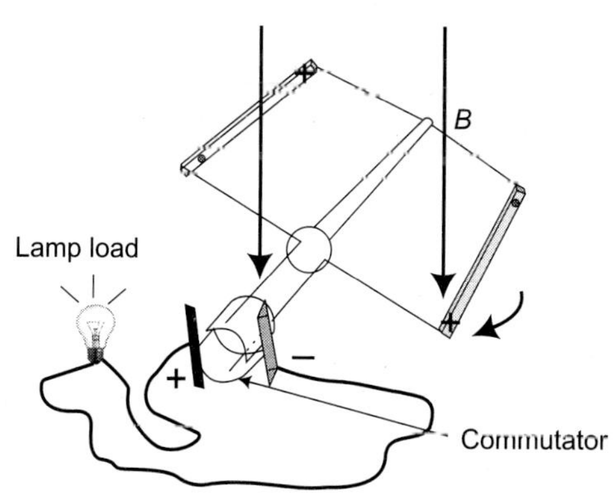

Fig. 7.28 Commutation has completed.

Commutation is now complete as shown in Fig. 7.28 and, though the emf in the conductors has reversed (compare the polarity of the emf in the conductors in Fig. 7.28 with Fig. 7.27), the commutator has ensured that the voltage applied to the lamp load is unidirectional.

Most commutators in DC machines for industrial applications are multi-segment in construction as shown in Fig. 7.23 and not the two-sement one as in Fig. 7.25. There is no difference in construction between a D.C. motor and a D.C. generator. The armature (see Fig. 8.46 of a D.C. motor) has slots on its periphery to provide mechanical security for several coil windings and to enable a shorter air gap for the flux to cross between the pole face and the armature teeth. The armature slots accommodate several coils, made from a few hundreds of conductors.[15] The ends of the coils are soldered to the segments against which carbon brushes are pressed (the brush is about a segment wide) to collect the generated emf from all the conductors which may be connected in series to increase the terminal voltage or in parallel to increase the current capacity though with a smaller terminal voltage.

7.13 MAGNETIC FIELD LINES

Notice that we have shown the magnetic flux in the air gap between the poles and the armature *unterminated* for lack of space. Remember magnetic field lines are never unterminated and must always form *closed loops.* The air gap is kept small to provide a low reluctance (Section 5.4) so that the magnetic field generated in the poles find a path through to the armature conductors.

In a real generator how do the magnetic field lines form loops?

Magnetic field lines follow the path of *least reluctance* similar to the way in which current takes the path of least resistance.

In dc generators and motors the path of *least reluctance* lies in the frame or yoke of the motor and the field lines form *closed loops* taking the *following path:* the field poles, the air gap, inside the armature structure and through the air gap again and into the frame or yoke and return to the field poles as seen in the Fig. 7.29.

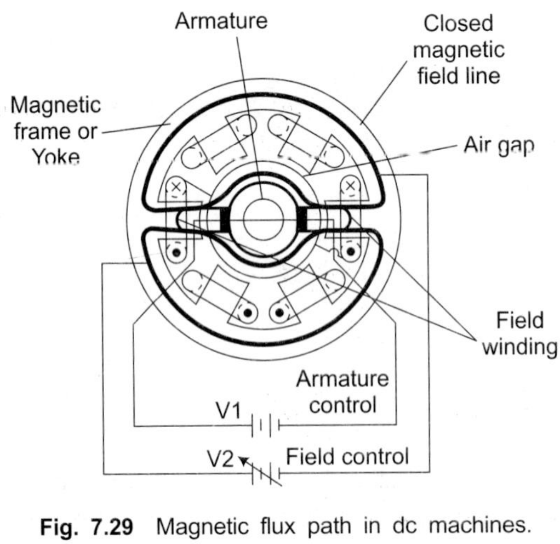

Fig. 7.29 Magnetic flux path in dc machines.

7.14 MICROPHONES

A microphone is an electromechanical device that uses vibration to create an electrical signal or emf proportional to the vibration, which is usually an air pressure wave such as that produced by voice or sounds.

Sound is produced by a rapid variation in the average density or pressure of air molecules above and below the current atmospheric pressure. These minute changes in atmospheric pressure are referred to as **sound pressure** and the fluctuations in pressure as **sound waves**. A vibrating sound source such as a loudspeaker cone causes the surrounding air molecules to bounce off each other with a force proportional to the disturbance. This creates ripples of more dense (higher pressure, called compression) to less dense (lower pressure, called rarefaction) air molecules. The back and forth oscillation of pressure produces the sound wave that impinges on our eardrums and which we perceive as sound.

The softest perceptible sound, corresponds to a pressure variation of 0.00002 psi (pounds per square inch) which is less than a billionth variation around normal atmospheric pressure of 14.7 psi and the loudest sound that can cause ear damage is 0.002 psi variation. For the sake of comparison, a bicycle tire tube is inflated to a pressure usually in the range of 25-100 psi.

Microphones operate based on the same principle as generators.

In *generators* notice that all the conductor loops are arranged such that the direction of motion of the conductors is a *twist*, so the direction of the uniform and constant magnetic field on opposite sides of a *loop* was the same.

There is one subtle difference between a generator and the *microphone.*

In a *generator* the loops of coils can twist so that when these are moving in a *constant magnetic field* their opposite ends produce an *emf* that are additive, like connecting batteries *together in series to produce a larger voltage.*

Microphones consist of a *diaphragm* that is attached to a *coil of wire* as shown in Fig. 7.30.

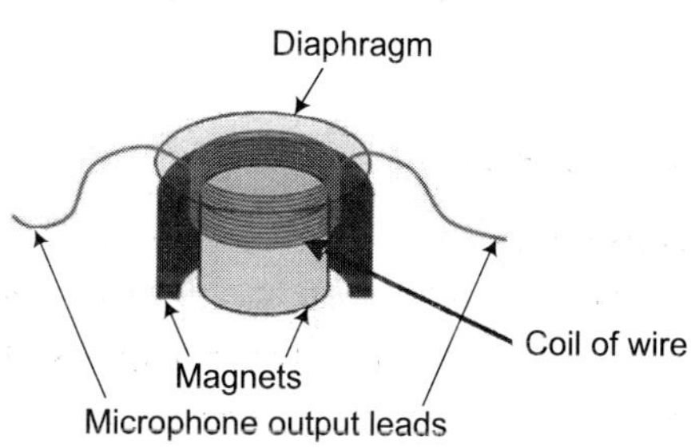

Fig. 7.30 Construction of a microphone.

When sound waves hit the diaphragm, it vibrates and vibrates the entire coil along with it.

As there is no *twisting* action of loops of wire, microphones require that the magnetic field be oriented uniformly all around using *circular magnets* or other such devices such that the emfs produced by opposite sides of *all the loops* are additive thus *maximizing* the emf generated.

In Fig. 7.31 it is seen that the magnetic field is directed radially inward around the microphone through the coils of wire. Therefore, the emfs generated in every segment of the coil when it moves is additive.

Microphone Operation

Figure 7.31 shows a view of the generation process of emfs in a microphone.

Shown in the figure are the coils (grossly exaggerated using rectangular cross-section wires) and the diaphragm that vibrates when sound waves strike it. The diaphragm is attached to the coil and so causes the coil to vibrate along with it leading to the generation of an induced emf at the ends of the coil, the audio frequency electric signal.

The circular magnet is of special construction such that its magnetic field is directed radially inwards.

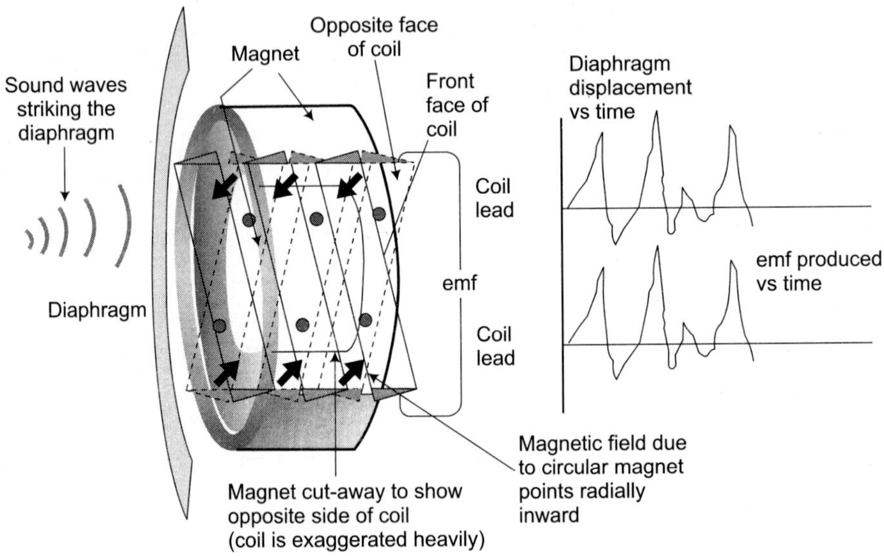

Fig. 7.31 Generation of emf by the microphone.

Summarizing the effects of Motional emf

In the presence of a uniform *constant* magnetic field the motional emf is due to magnetic forces on the moving electrons, that enable a separation of charges to *opposite* ends of a conductor.

Of course, the *electrons* that are *shifted* provide a negative potential while, the *immobile positive cores* are simply "uncovered" to provide a positive potential.

And calculations from Faraday's law predict this *induced emf* correctly—even though it is not associated with a changing magnetic field.

7.15 ENERGY CONSERVATION AND MOTIONAL EMF

We next consider whether energy has been conserved in the experiments in which a metal bar was moved along rails immersed in a magnetic field.

If the metal bar in Fig. 7.32 slides *effortlessly* along at a constant speed v, and the *resistor* heats up the surroundings, we would be getting *something for nothing*!

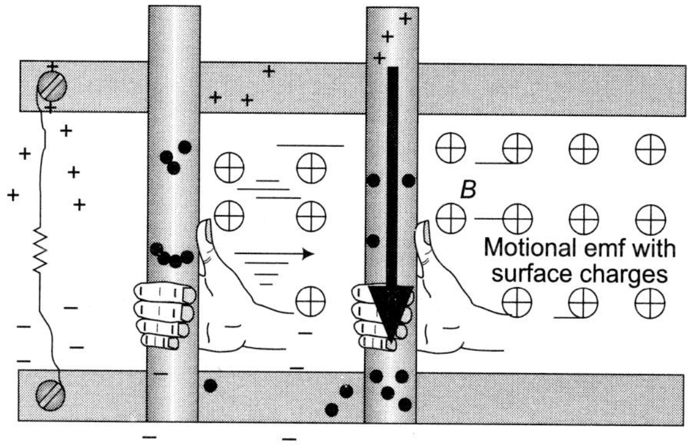

Fig. 7.32 Metal bar sliding on rails immersed in a magnetic field.

There must be an energy input into this system. It turns out that we must exert a force on the bar to keep it moving at constant speed, because there is a magnetic force on the moving bar, due to the current I in the bar.

The bar now has a current flowing through it, and feels a magnetic force to the left, in the direction of $dl \times B$, since conventional current runs upward through the bar. This is shown in Fig. 7.33.

Of course, the force is a result of the Hall effect charges producing an electric field (not shown in the figure) that exerts a force on the positive atomic cores in the bar.

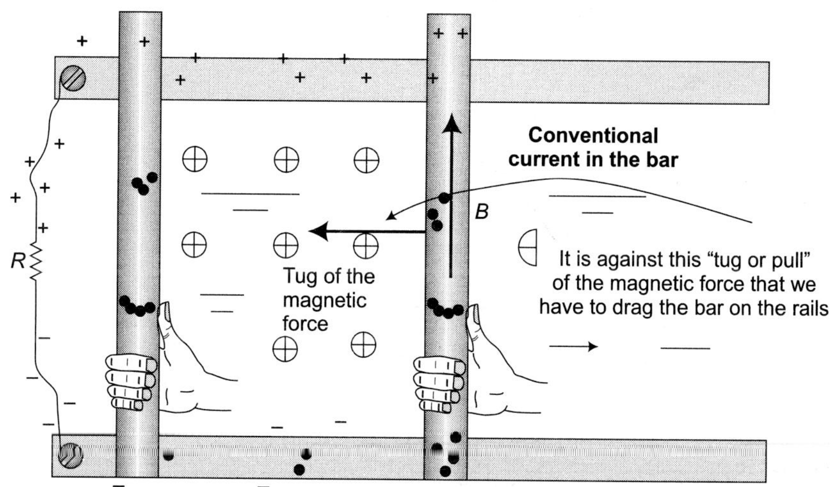

Fig. 7.33 Tug of the magnetic force.

It is *against* this *tug or pull* of the magnetic force that we have to *drag* the bar on the rails!

Now you know why electricity is not *cheap!*

It comes at a price!

The price we have to pay to "overcome" the magnetic tug.

Now, we can understand the role of the *turbine* in our generator.

It *overcomes* the tug of the generator conductors when current in them is delivering power to the electrical load.

A little maths now! To calculate the amount of magnetic *tug* on the bar on rails (see Section 6.5)

$$F_{magnetic} = \left|\int_{bar} I d\vec{l} \times \vec{B}\right| = \int_{bar} IBdl \sin 90^\circ$$

where I is the direction of conventional current upwards in the bar, dl is an elemental length of the bar of length L.

$$= IBL = \left(\frac{vBL}{R}\right) BL = \frac{v(BL)^2}{R}$$

where R is the load resistance and $I = vBL/R$ is from the expression derived in Section 7.7.

So, we must apply a force of magnitude $v(BL)^2/R$ to the right to keep the bar moving.

In a short time Δt we move the bar a distance Δx. We do an amount of work $F\Delta x$ in a time Δt, so we have to supply energy at the rate of $F\Delta x/\Delta t = Fv$.

So, power input Fv is numerically equal to the power output of the resistor.

$$Power\_input = F\frac{\Delta x}{\Delta t}$$

$$= Fv = \left(\frac{v(BL)^2}{R}\right) v = \frac{(vBL)^2}{R}$$

Power dissipated in resistor $= I\ \Delta\text{Voltage} = I^2R$

Since $$I = \frac{emf}{R} = \frac{BLv}{R}$$

we get, power dissipated $= I^2 R = \left(\frac{v(BL)}{R}\right)^2 R = \frac{(vBL)^2}{R}$

Since, power input = power dissipated, therefore, the energy **is conserved!**

Questions

1. Describe the phenomenon of magnetic induction with reference to two separate coils placed along the same axis a little distance separating them and when the current in one of the coils is changed with the other coil short-circuited.
2. What is motional emf? With the help of neat sketches describe the processes that produce this.
3. Derive an expression for the emf produced in rotating machines using the example of a loop of dimensions $w \times h$ rotating at a constant angular velocity $\omega = d\theta/dt$ inside a uniform magnetic field B.
4. With the help of neat sketches showing a rectangular loop rotating in a magnetic field that causes an emf whose polarity changes, explain how a unidirectional emf can be produced.
5. Using appropriate expressions, prove that the energy is conserved when a bar of length L is pulled at a uniform speed v over two metal rails, two ends of which are connected together using a resistor of resistance R and the rails are immersed in a magnetic field B.

8

Magnetic Forces; Magnetic Moment; Magnetic Forces on Magnetic Moment; Torque Production in Motors; Motor and Load Torque; Back emf; How Motors and Loudspeakers Operate; Magnetic Fields in Matter

The first section of this chapter deals with the action of magnetic forces on moving charges and then is developed the concept of magnetic moments and their behaviour when left to rotate freely in magnetic fields. This is followed by a description of the magnetic force that results when a magnetic moment, a current carrying loop of wire, is placed in a magnetic field. Then, is shown how it is this phenomenon that leads to the generation of torque in motors. One section is devoted to a description of the concept of load torque and the effect of its imposition on motor torque and how this causes the motor to draw power from the power supply. The phenomenon of how it is the 'back emf' in motors that regulates the motor torque to match load torque is described in some detail.

8.1 MAGNETIC FORCE AND MAGNETIC MOMENT

When an electric charge is placed in a magnetic field it experiences a force as we saw in Section 6.1.

The magnetic force in a charge q, moving with velocity v in a magnetic field B, is, from "Introduction to Electrodynamics" by David J. Griffiths[8], reproduced with permission,

$$F_{mag} = q\ (v \times B)$$

This is known as the *Lorentz force* law.

In the presence of both electric and magnetic fields, the net force on q would be

$$F_{mag} = q[E + (v \times B)]$$

The magnetic field for a circular loop of *radius a* and conventional *current I,* at a distance

x from the centre, **far away** *along the axis*, is (Chabay & Sherwood, 1995, used by permission)[2]

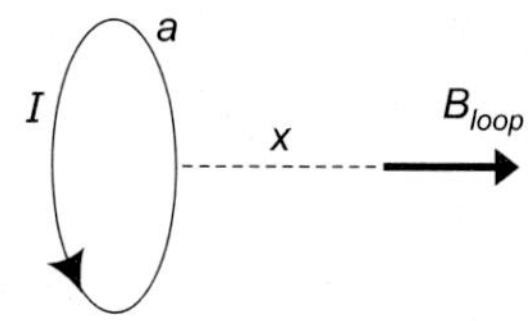

Fig. 8.1 The magnetic field for a circular loop.

$$\vec{B} = \frac{\mu_o}{4\pi} \frac{2\pi a^2 I}{x^3}$$

In this formula the quantity $\pi a^2 I$ in the formula is called the **magnetic moment.**

It is usually referred to by the letter μ.

More generally, $\mu = \mathbf{IA}$

where, A is the area of the loop and is equal to πa^2 for circular loops.

With this definition of the magnetic moment, we write B_{axis} as

$$B_{axis} = \frac{\mu_o}{4\pi} \frac{2\mu}{x^3}$$

8.2 MAGNETIC MOMENT AS A VECTOR

The ***magnetic moment*** of a current carrying loop is considered to be a *vector* μ whose **direction** is the direction of the **magnetic field** at the centre of the coil. This is shown in Fig. 8.2.

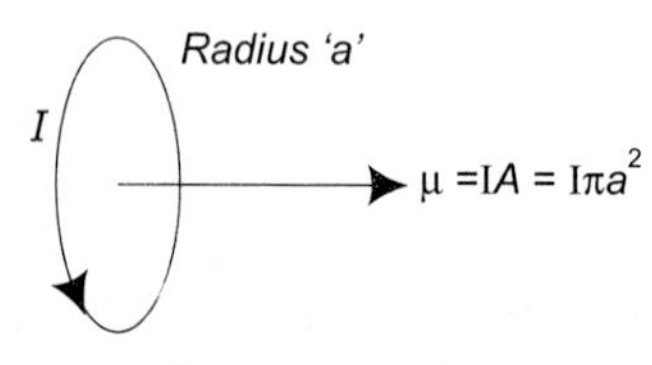

Fig. 8.2 The magnetic moment of a current carrying loop.

8.3 MAGNETIC FORCES ON A MAGNETIC MOMENT

A magnetic field can cause a current-carrying coil of wire to *twist.* (Chabay & Sherwood, 1995, used by permmission).[2]

Consider a rectangular current-carrying loop of wire in a uniform magnetic field (that is, the magnetic field has the same direction and magnitude throughout this region).

The loop is pivoted on a horizontal axle. If the conventional current I is as shown in Fig. 8.3, then, the magnetic forces on each of the four sides of the loop will be as shown.

The forces are determined by $F = qv \times B$ applied to a wire.

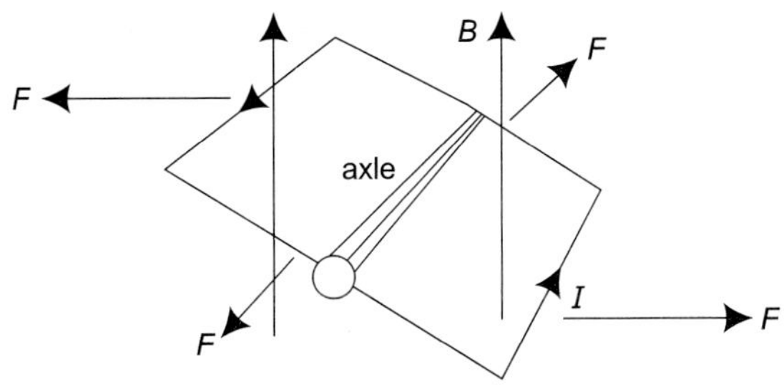

Fig. 8.3 The magnetic forces on a magnetic moment.

A simple method for determining the direction and magnitude of a *vector cross product* $v \times B$.

Keep your right hand open and flat (fingers and thumb in the same plane), with your thumb perpendicular to your fingers, point your fingers in the direction of the first vector (v) as shown in Fig. 8.4.

Now, *bend* your fingers toward the second vector (B), noting the angle of bend θ, which is always less than 180°. You may have to "rotate" your wrist to get your hand into a position that lets you bend your fingers this way.

The thumb points in the direction of cross product $v \times B$.

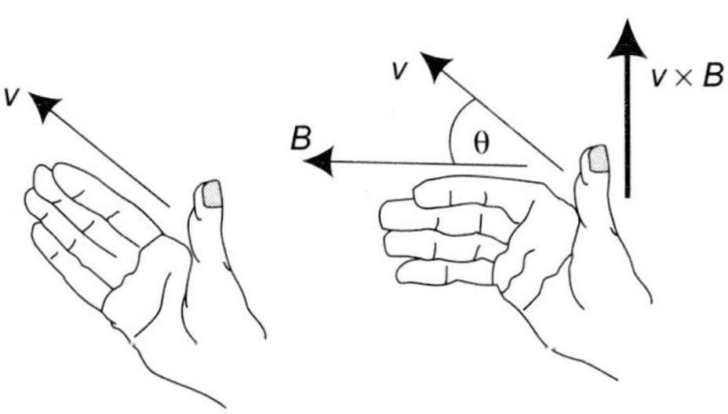

Fig. 8.4 Determining the magnitude and direction of $v \times B$.

This rule may be applied to verify the magnitude and direction of the forces given by $qv \times B$ in the previous Fig. 8.3.

When the plane of the loop is perpendicular to the magnetic field, the magnetic forces ***don't*** exert any twist as shown in Fig. 8.5. Why? Using the vector product rule, check how the forces only tend to pull the loop apart.

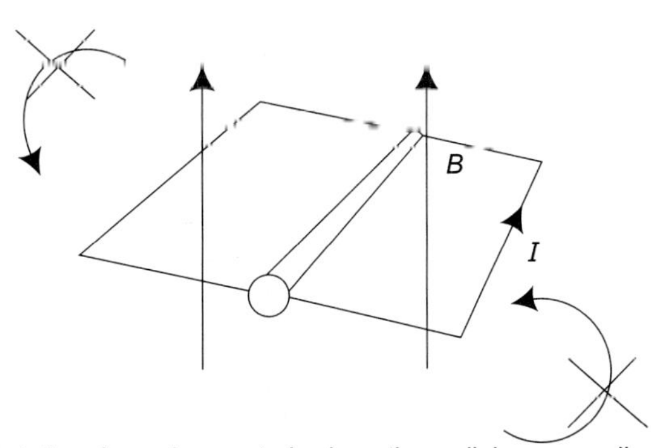

Fig. 8.5 No twisting force is exerted when the coil is perpendicular to the field.

Let us take a closer look at how the magnetic forces cause the loop to twist. The forces at the top and bottom of the axle cancel each other. The left and right side forces will act as shown in Fig. 8.6.

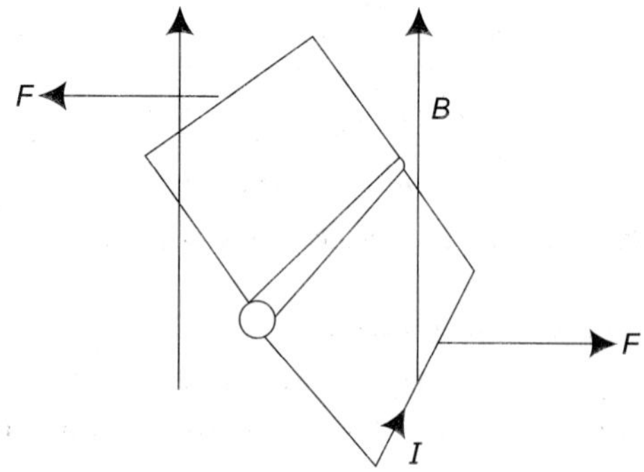

Fig. 8.6 The forces acting on a current carrying rectangular loop in a magnetic field.

The force acting on the left side loop can be resolved into 2 component force vectors $F1$ and $F2$ *perpendicular* to each other as shown in Fig. 8.7.

Component $F1$ is cancelled out by the reaction forces $2 \times (F1)/2$ along the arms of the loop connected with the axle.

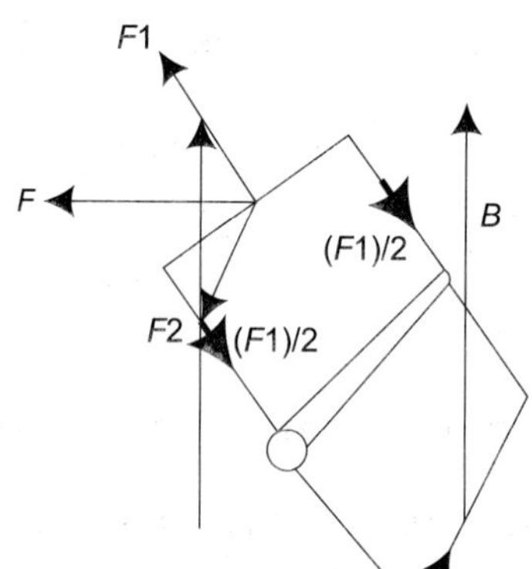

Fig. 8.7 Resolving the (left) force F into its components.

This leaves $F2$ capable of imparting a movement to the left edge of the loop as shown in Fig. 8.8.

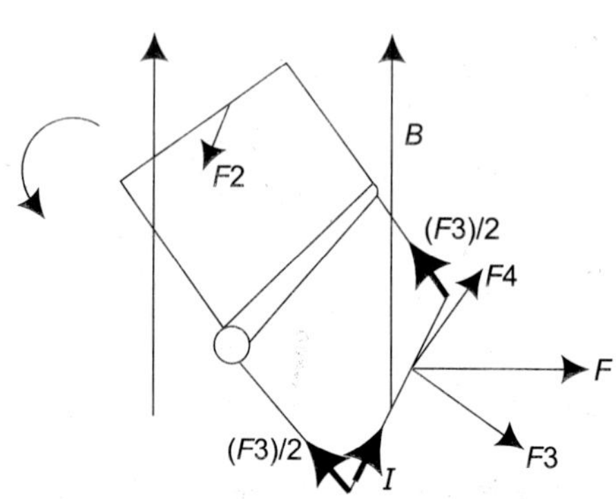

Fig. 8.8 Resolving the (right) force F into its components.

A similar analysis of forces carried out on the right side of the loop shows component *F*4 capable of imparting a movement to the right edge of the loop as shown in Figs. 8.8 and 8.9.

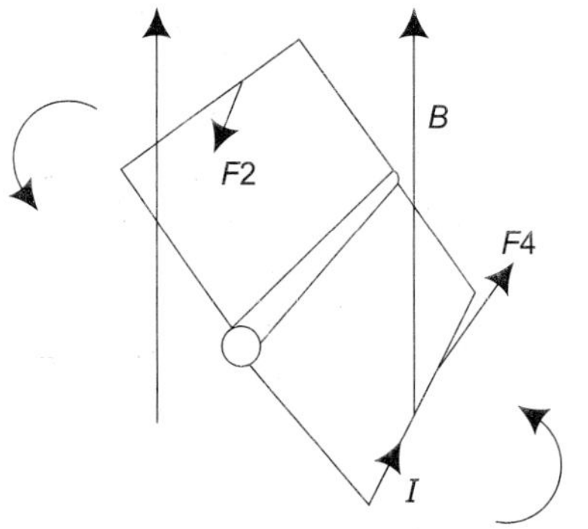

Fig. 8.9 The forces *F*2 and *F*4 capable of rotating the loop.

We see in Fig. 8.10 that the two components *F*2 and *F*4 rotate the loop *counterclockwise.*

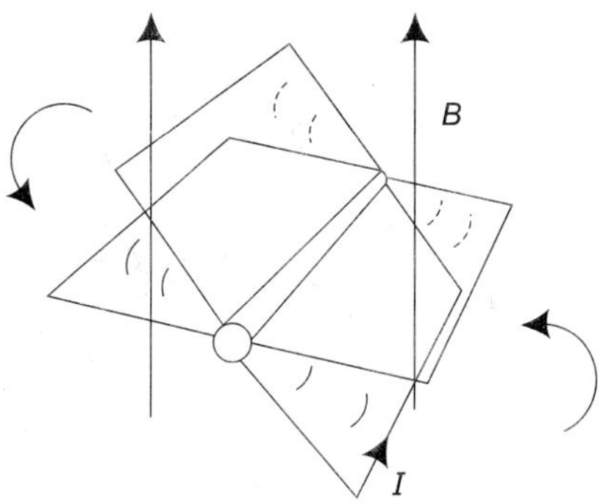

Fig. 8.10 The forces *F*2 and *F*4 rotate the loop counterclockwise.

8.4 STABILITY OF THE MAGNETIC MOMENT

Given the orientation of the loop and the current *I* as shown in Fig. 8.11, if you *nudge* the loop slightly, will the loop flip?

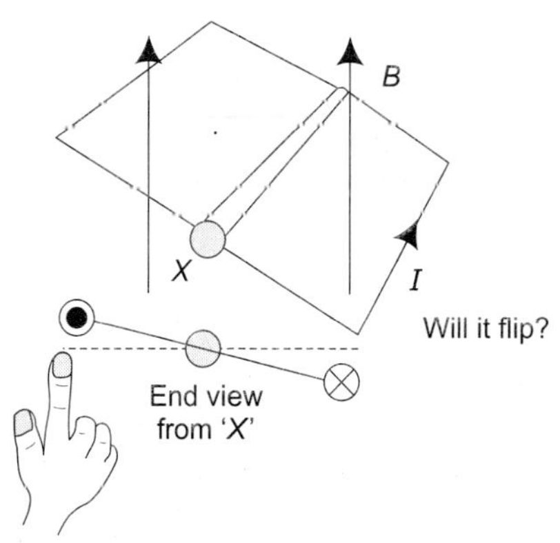

Fig. 8.11 Will the loop flip?

The loop is stable because if it is *rotated slightly* from the horizontal, the magnetic forces will cause it to turn *counterclockwise* and *return* to its original position as seen from Fig. 8.12.

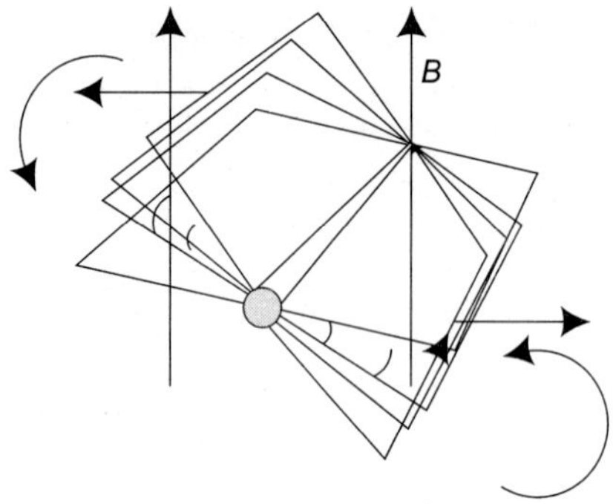

Fig. 8.12 The loop is stable.

If we *reverse* the direction of the current *I* as shown in Fig. 8.13 and if you *nudge* the loop slightly, will the loop flip?

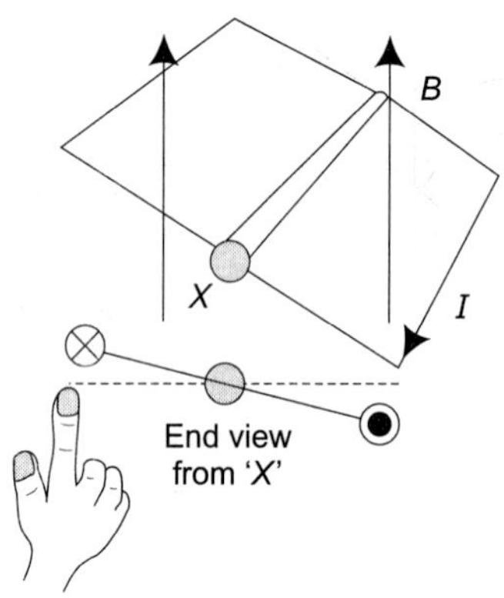

Fig. 8.13 Will the loop flip with the current reversed as shown?

This loop is unstable because if it is rotated slightly the magnetic forces as shown in Fig. 8.14 will cause it to *continue rotating* in the same direction and ***not*** return it to its original position.

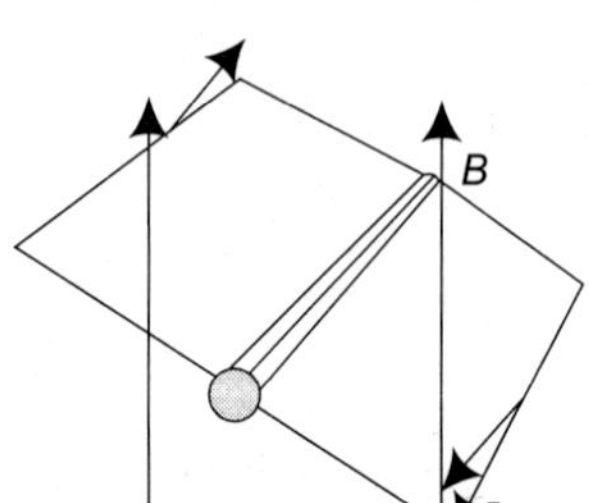

Fig. 8.14 The magnetic forces that act on the edges of the loop.

The loop will continue to rotate as shown in Fig. 8.15.

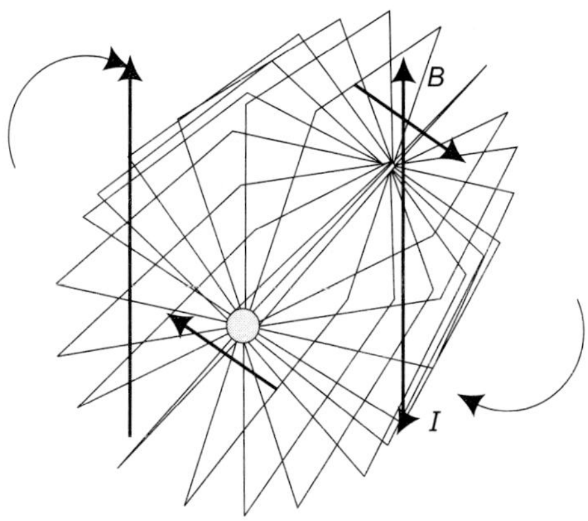

Fig. 8.15 The loop will continue to rotate.

The **magnetic moment (vector)** μ (see Sections 8.1 and 8.2) is ***not*** a piece of fiction that merely makes the formula for B look cleaner,

$$B_{axis} = \frac{\mu_o}{4\pi}\frac{2\mu}{x^3}$$

It is a vector that tells us what exactly a current carrying rectangular loop does when placed in a magnetic field and it is nudged.

One rectangular loop has an area $A = wh$, so its magnetic moment is $\mu = IA = Ihw$ as shown in Fig. 8.16. When we say a loop has a magnetic moment it implies that the loop is a current-carrying loop.

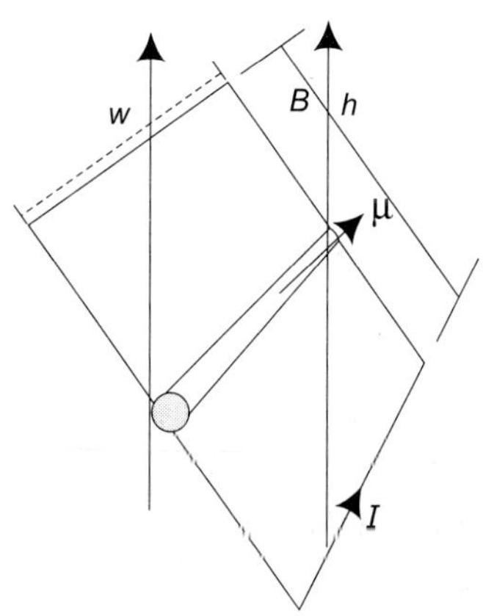

Fig. 8.16 The magnetic moment of a current carrying loop.

For the given current direction, the magnetic forces F_{mag} developed are shown in Fig. 8.17.

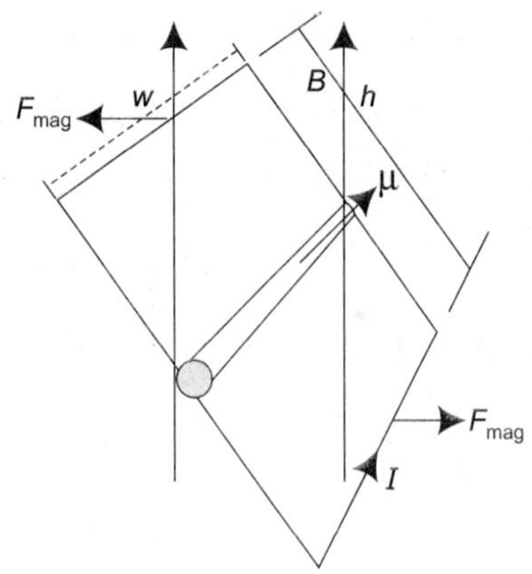

Fig. 8.17 The magnetic force F_{mag}.

And the coil *tends* to twist (counterclockwise here) to make μ line up with *B* as shown in Fig. 8.18.

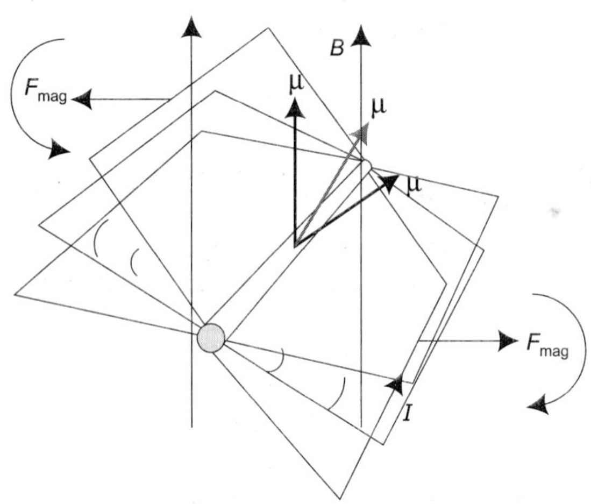

Fig. 8.18 The coil tends to twist to line up μ with *B*.

8.5 MAGNETIC MOMENTS OF ATOMS IN MAGNETS

Every atom consists of an electron occupying an orbital space around a positive atomic core: the neutron and the proton, as shown in Fig. 8.19.

The electron is a "moving charge" and some among these have a *non-spherically* symmetric probability distribution viz. the *'p,d, and f' orbital.*

Since the electron is a "moving charge", it produces a magnetic field and there is an associated non-zero magnetic moment with these orbitals.

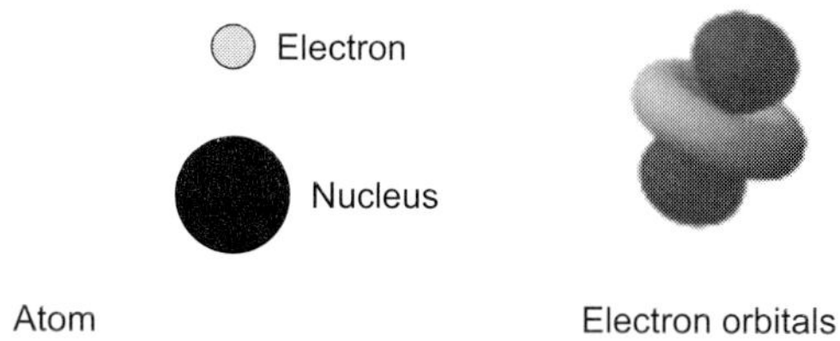

Fig. 8.19 An atom and the electron orbitals.

Figure 8.19 shows electron atomic orbitals also. The term *orbital* has become known as either the *mathematical function* or the *region* generated with the function. Specifically, atomic orbitals are the possible quantum states of an individual electron in the electron cloud around a single atom, as described by the mathematical function. Refer "Stable Atoms" in Appendix B.

In addition, some electrons also spin and these spinning balls produce a magnetic field.

In *many* materials, an atom has *no net orbital or spin* magnetization. But, in a few materials, notably *iron, nickel, cobalt and some alloys of these elements*, the orbital and spin motions in neighbouring atoms *line up with each other* and therefore produce a sizable magnetic field, though not significant enough for practical applications and these are called *ferromagnetic* materials.

8.6 A MAGNETIC COMPASS IN A MAGNETIC FIELD

Consider a magnetized light-weight compass needle (balanced on a nearly frictionless pivot point) as shown in Fig. 8.20(a). What is the *direction* of its atomic magnetic moments? The direction is shown by the arrow marked 'μ'. It points in the same direction as the needle.

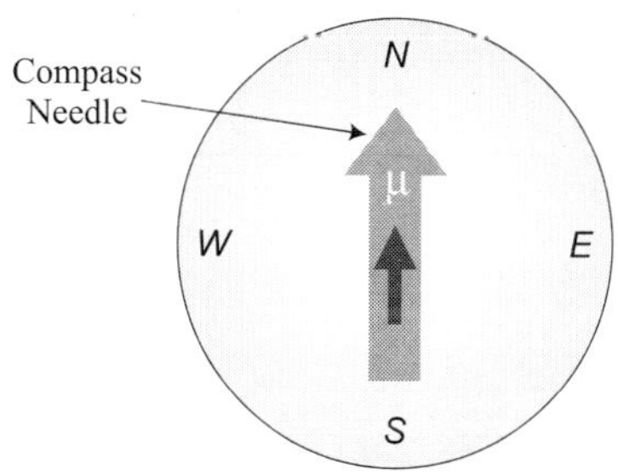

Fig. 8.20(a) A magnetized compass needle

If the needle is moved away from the North as shown in Fig. 8.20(b) and *then released* what will happen?

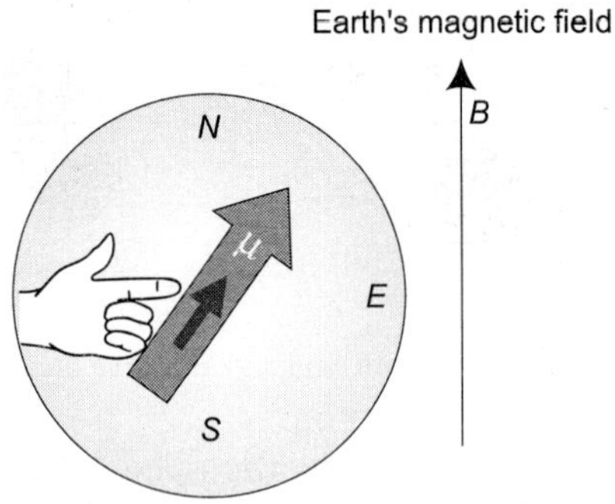

Fig. 8.20(b) The needle is moved away from the North and released.

The needle *swings back* to point North again as shown in Fig. 8.21.

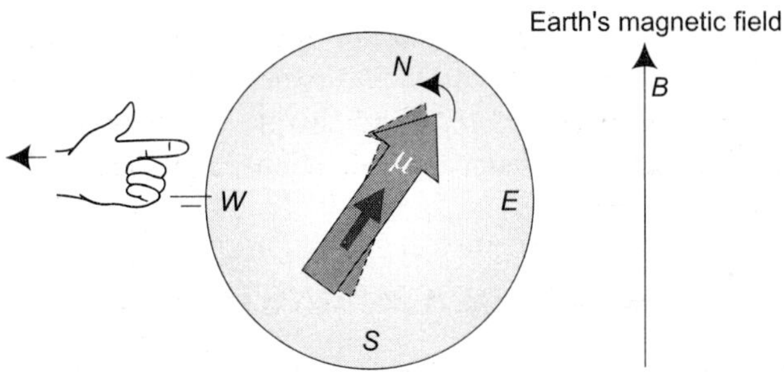

Fig. 8.21 The needle will swing back.

The needle is most *stable* when its magnetic moment is aligned with the earth's magnetic field, so the compass needle rotates until it is aligned as shown in Fig. 8.22.

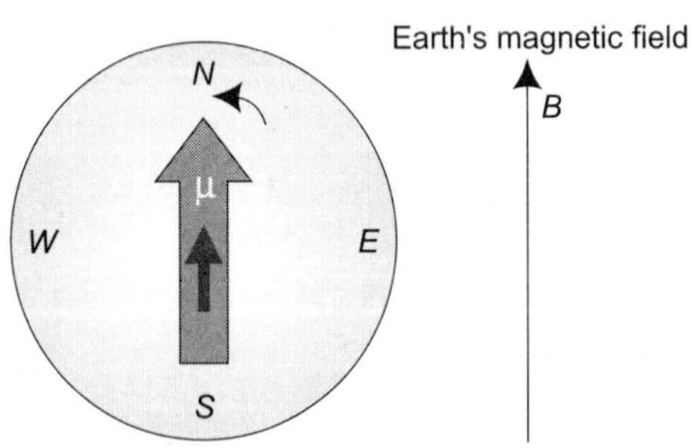

Fig. 8.22 The needle has swung back to the point North again

8.7 QUANTITATIVE TORQUE

The fact that a current-carrying coil *twists* in a magnetic field is very important. (Chabay & Sherwood, 1995, used by permission).[2]

The quantitative amount of *twist* is formally called the *torque* which is provided by each of the magnetic forces around the axle and is equal to the distance from the axle (the *lever arm*) times the component of the force *perpendicular* to the lever arm.

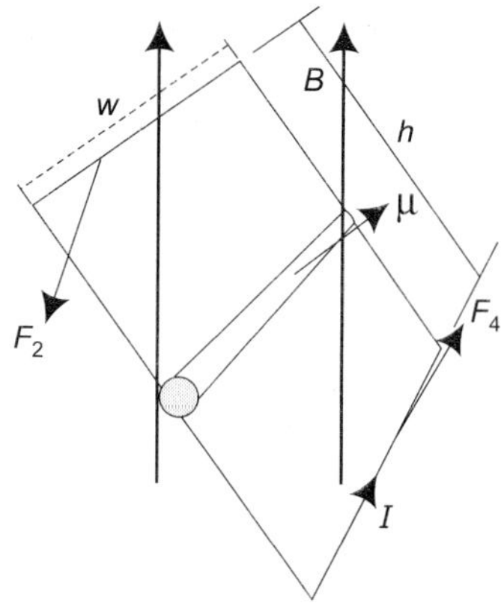

Fig. 8.23 The torque in a current carrying loop in a magnetic field.

Figure 8.23 shows a rectangular loop of length '*h*' and width '*w*' carrying a current *I* placed in a constant magnetic field *B* (Chabay & Sherwood, 1995 used by permission).

Using the relation for F_{mag} (Section 6.5, Fig. 6.15, $F = qv \times B$), the *perpendicular* component of each force is $F_2 = F_4 = F\perp = Iw\,B \sin\theta$, the lever arm is $h/2$ and 'θ' is the angle by which the loop is twisted from the horizontal as shown in the inset (side view) in Fig. 8.24.

The twist applied to this rectangular loop is due to the magnetic forces on the *w*-lengths of the loop.

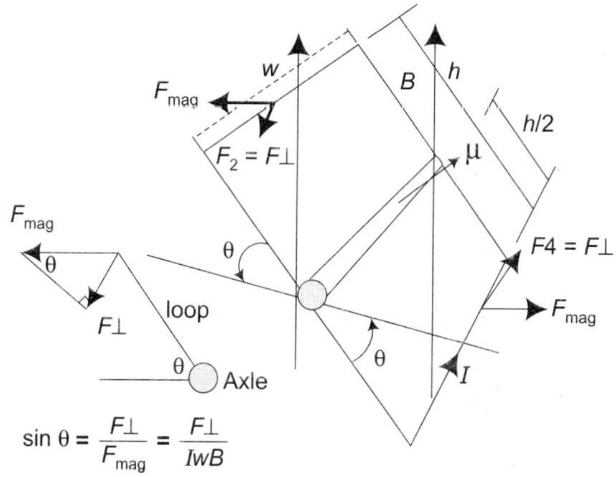

Fig. 8.24 Calculating the torque of a current carrying loop in a magnetic field.

The other **component** of *F*, and the forces on the *h*-length of the loop, just tends to *stretch* the loop without causing a twist as shown in Fig. 8.25.

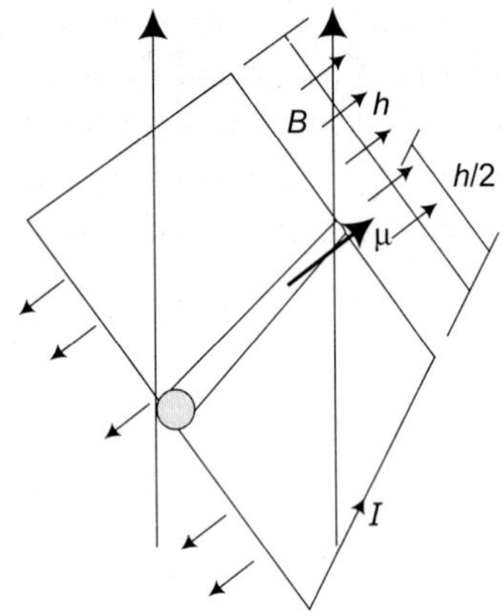

Fig. 8.25 Cancellation of forces on the *h*-length of the loop.

Each of the forces F_2 and F_4 exerts a "torque" of $(F_\perp)(h/2)$ as shown in Fig. 8.26.

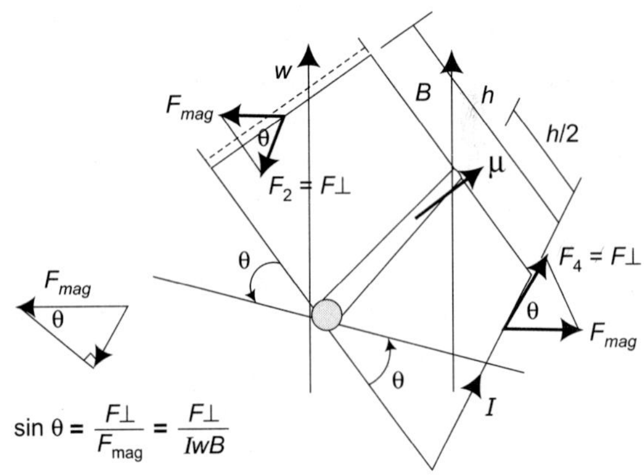

Fig. 8.26 Computing the total torque.

So, the total torque is: 2 IwB sinθ $(h/2)$ = (Iwh) $(B$ sinθ$)$.

8.8 UNIDIRECTIONAL TORQUE

If a rectangular current carrying loop is oriented with respect to a magnetic field as shown in Fig. 8.27 then, it will be subjected to a force F_{mag} on both sides of the loop.

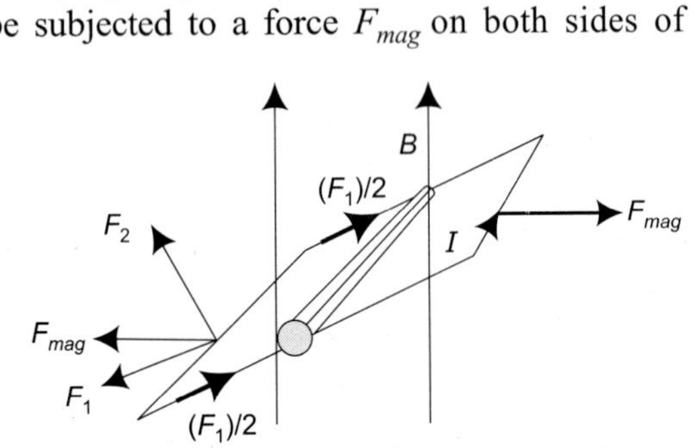

Fig. 8.27 The components of the (left) force F_{mag}.

The force F_{mag} acting on the left side loop can be resolved into 2 component force vectors F_1 and F_2 *perpendicular* to each other. Component F_1 is cancelled out by the reaction forces $2 \times (F_1)/2$ along the arms of the loop connected with the axle as shown in Fig. 8.27.

This leaves F_2 capable of imparting a movement to the left edge of the loop. A similar analysis of forces carried out on the right side of the loop shows component F_4 capable of imparting a movement to the right edge of the loop as shown in Fig. 8.28.

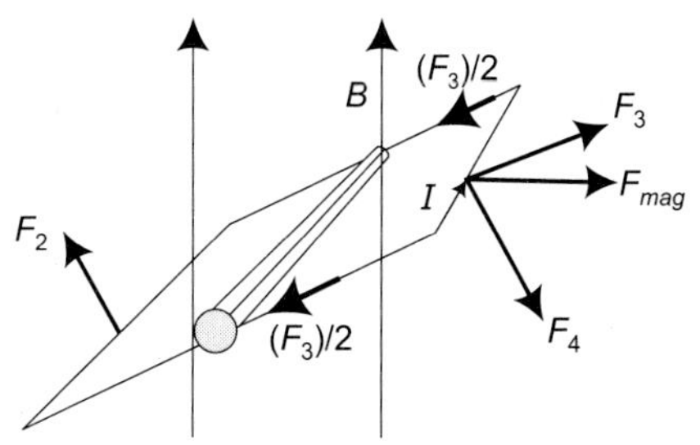

Fig. 8.28 The components of the (right) force F_{mag}.

Thus, the loop twists clockwise and attains the position as shown in Fig. 8.29 and will continue to rotate clockwise as shown in Fig. 8.30.

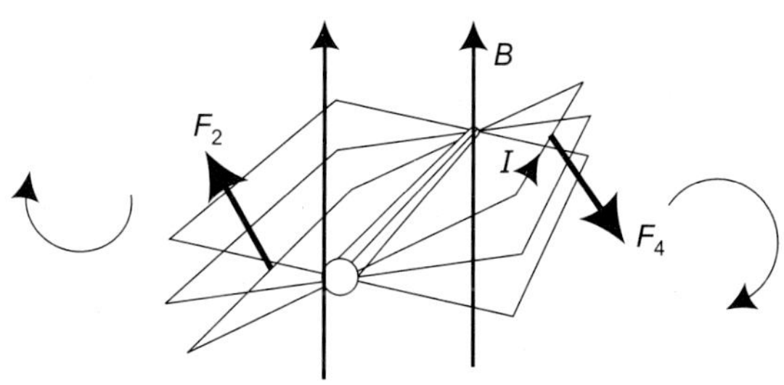

Fig. 8.29 The loop twists clockwise.

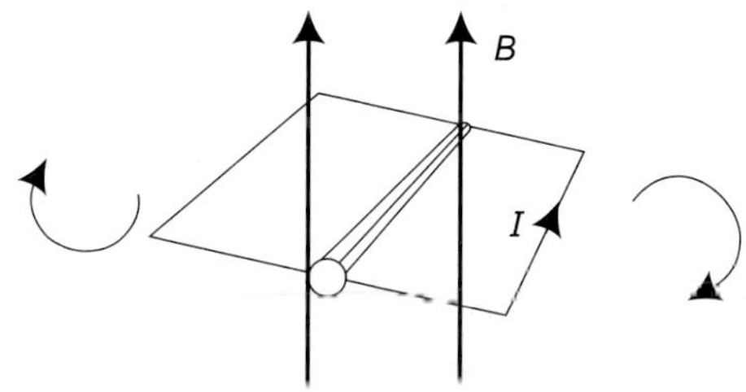

Fig. 8.30 The loop rotates and will continue to rotate.

Note that when the loop is oriented at position X with the current direction as in Fig. 8.31 the torque developed is *clockwise.*

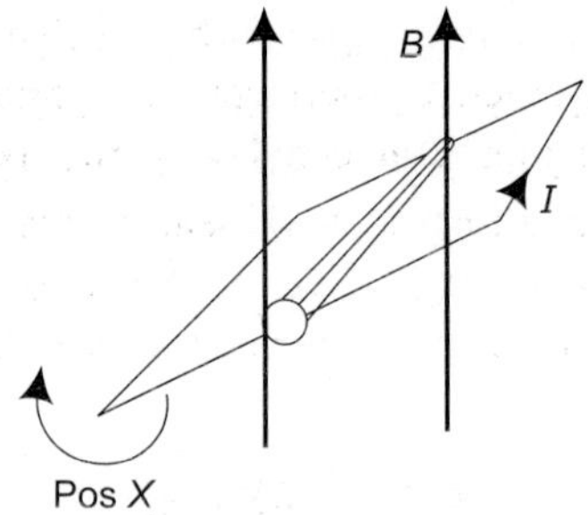

Fig. 8.31 Torque developed with the loop in position *X*.

This happens because a magnetic moment twists to its most *stable position* in a magnetic field or in other words, if allowed to pivot freely, a magnetic moment such as a current carrying loop *turns* towards alignment with the applied magnetic field.

Whereas, if the loop is oriented as in position *Y* as shown in Fig. 8.32, with the same current direction the torque developed is *counterclockwise.*

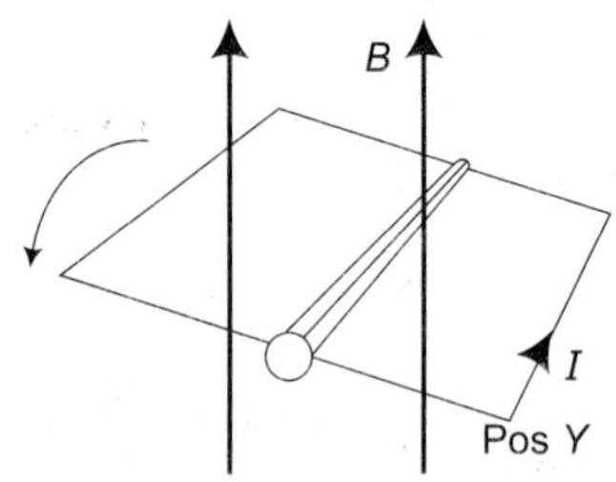

Fig. 8.32 Torque developed with the loop in position *Y*.

If we wish to build a rotating machine or a motor using such a loop and if the current direction *were not to be changed* as the coil rotates from say position *X* to *Y*, *uniform rotation will not be possible* and it will *not develop* unidirectional torque as shown with the torques developed in the positions *X* and *Y* superimposed in the Fig. 8.33.

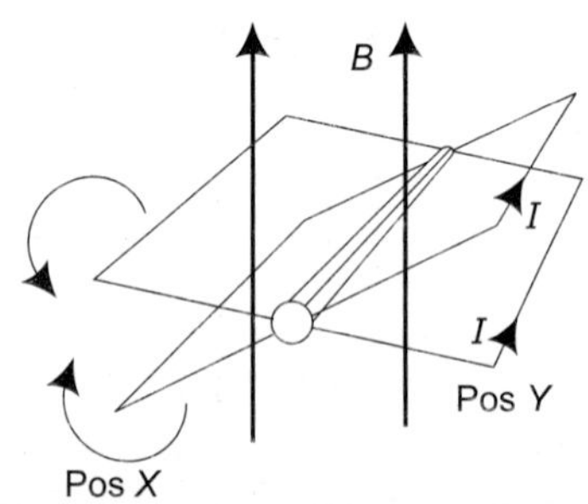

Fig. 8.33 Torques in positions *X* and *Y* (shown superimposed) will not develop unidirectional torque.

The key to develop unidirectional torque is to make electrical connections to the coil in such a way that *just as it is coming* to its stable position, we *reverse* the direction of the current.

Electric dc motors employ a *commutator* to change the direction of current as the shaft rotates similar to the commutator that was used to maintain a unidirectional emf in generators (Section 7.12). A description of how the commutator of a motor operates is given in Section 8.10.

Torque as a Vector

For a rectangular current (I) carrying loop ($w \times h$) in a magnetic field B, the torque, is given by torque $= (Iwh)\ (B \sin\theta)$ and, $\mu \times B = (Iwh) \times B = (Iwh)(B \sin\theta)$ ...(1)

where θ is the angle the loop makes with the horizontal, as we saw in Section 8.7.

The *torque* can also be expressed as a vector in the form $\mu \times B$ as seen in the equation 1 above and depicted in Fig. 8.34 (after Chabay & Sherwood, 1995).

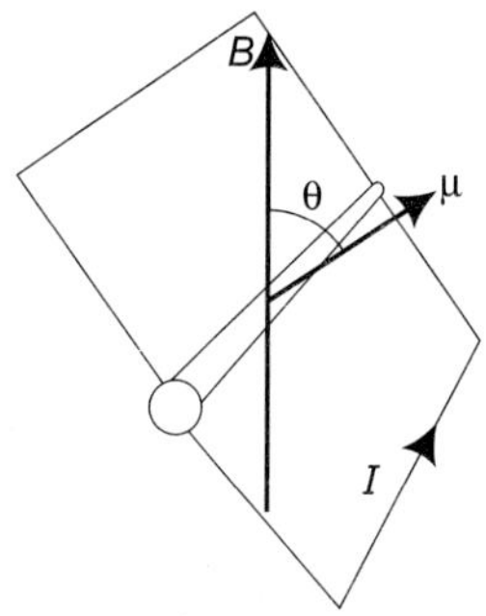

Fig. 8.34 Torque can be expressed as "$\mu \times B$".

The *torque* vector $\mu \times B$ has a magnitude $|\mu|\ |B| \sin\theta = \mu B \sin\theta$ and a direction along the axle around which the loop rotates as shown in Fig. 8.35.

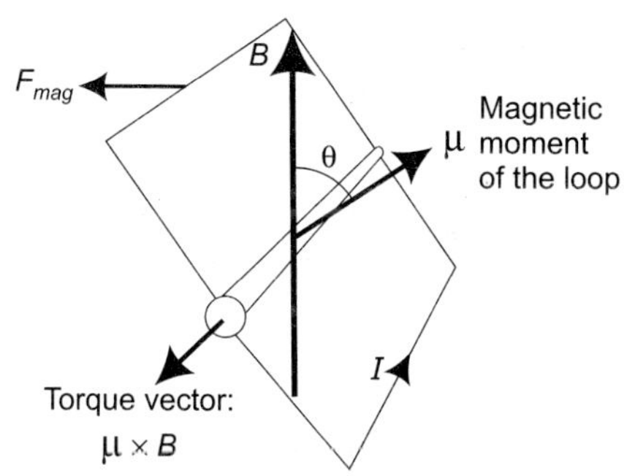

Fig. 8.35 The torque vector.

It will be observed that the force F_{mag} (see Fig. 8.35) is maximum when the loop is in line with the magnetic field B. This may be surprising when one compares with the movement of a compass needle which always aligns its length with the magnetic field (see Fig. 8.22). But, remember that it is the *magnetic moment* that aligns and while this is longitudinal for a compass needle it is perpendicular to a current-carrying loop (Section 8.2). That F_{mag} is a maximum can also be seen by substituting $\theta = 90°$ in the expression for the forces F_2 and F_4 (Refer Fig. 8.29): $IwB \sin\theta = IwB$.

The forces F_2 and F_4, as a matter of fact, reduce as the loop rotates and become zero when θ is 0° or perpendicular to the magnetic field.

If a real machine were to be built, the loop will be carried past this point as it rotates, by the momentum it had gained as it approached this point.

The non-uniformity of the force developed as it rotates creates torque pulsations and causes vibrations in the machine structure.

To remedy this, motor designers arrange to provide a radially directed magnetic field to the loop conductors so that the force that is developed is maximum at every point of their travel. The radial magnetic field is normally provided using suitably fashioned pole faces. See the pole head shapes of a real motor in Fig. 8.46.

8.9 TORQUE OF ANY SHAPED LOOP

We have seen that the torque for a rectangular loop is μB sinθ.

A loop of *any shape* can be *approximated* by a set of current-carrying rectangles along whose adjacent sides the current *cancels* as shown in Fig. 8.36.

The *torque* of *any* kind of loop is μB sin θ because *any kind* of loop can be *approximated* by a set of rectangles.

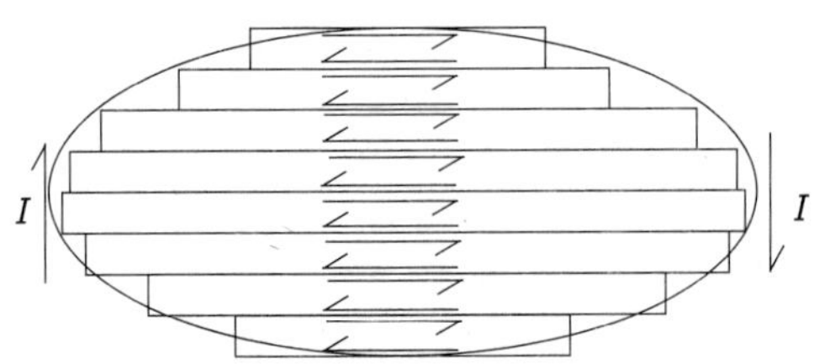

Fig. 8.36 The torque of any shaped loop.

8.10 ROTATION AND TORQUE PRODUCTION IN MOTORS

The *motor* is a device that is capable of producing a **torque** and is used to power *pumps, fans, blowers, compressors, conveyors, drilling machines, washing machines and many other mechanical loads.* Most of the time the motor rotates at a *constant number of revolutions per minute (rpm)* when, the load torque imposed upon its shaft by the mechanical load is constant.

However, the rpm ***may vary*** (maybe slightly or by large amounts) even during normal operation as in the case when the load torque ***varies,*** for example, by the *weights placed on a conveyor belt driven by a motor.*

In this section, we will see how a *motor* rotates and produces a torque.

The dc motor (see Fig. 8.46) consists of coils of conductors (windings) inserted into slots of an armature which is cylindrical in shape and made from metallic stampings. In the centre of the armature is the motor shaft to which the stampings are fixed. The armature shaft is mounted on the frame using bearings. The yoke that is fixed to the frame carries the magnetic poles which have windings that when energized produce a magnetic field in the air gap between the pole faces and the armature. Motors are designed for base or flange mounting on machines and the motor must be securely installed to a rigid foundation or mounting surface to minimize vibration and maintain alignment between the motor and shaft load. If not rigidly mounted, the motor *reaction* torque will cause the motor *frame to turn* in a direction *opposite* to the direction of rotation of the motor, and is called the *reaction torque* which can apply tremendous forces that could be dangerous; motors that are not mounted should never be energized.

The constructional details of a typical *dc motor* are not shown in Fig. 8.37 and focus on the *key mechanism of commutation* to generate unidirectional torque.

In Fig. 8.37 is shown a single loop of a small dc motor powered by a battery and the shaft of the armature fitted with a split-ring commutator. The commutator is made from a conductive ring split in two halves or segments and fixed on an insulated cylinder coupled to the shaft and which rotates along with the shaft of the motor. The voltage from the battery is applied through the commutator leaves to the two ends of the loop.

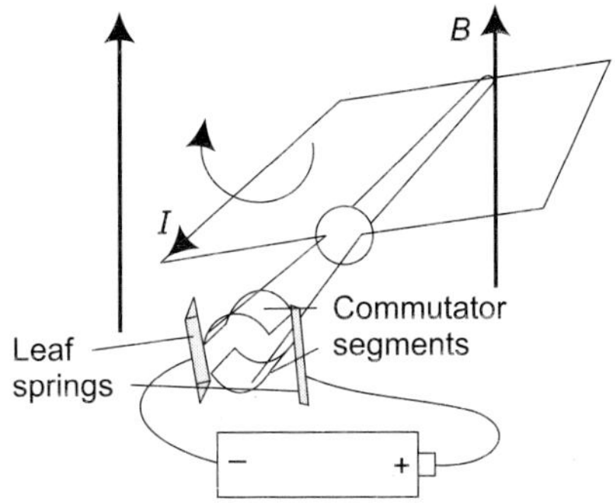

Fig. 8.37 The loop and commutator position before commutation.

The armature conductor carries a current as a result of application of the battery voltage and is shown in the position just before commutation occurs.

Figure 8.38 shows the loop and commutator position after commutation and also shown is the direction of the current that has reversed.

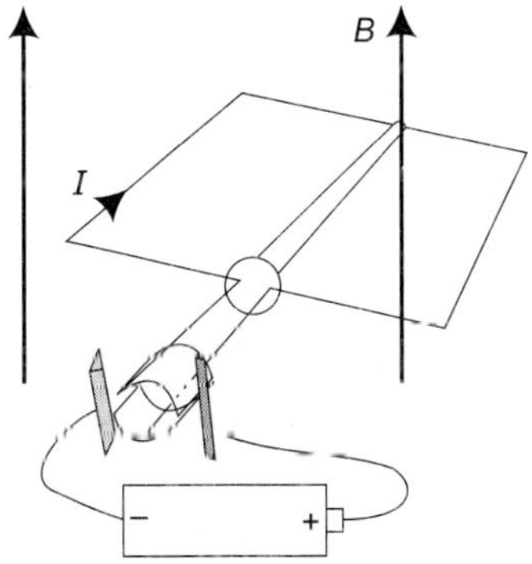

Fig. 8.38 The loop and commutator position after commutation.

The direction of the current having reversed in the conductors, the torque produced is unidirectional.

Remember, copper is *not* a magnetic material. Then, how does the conductor move?

Recollect from Section 6.2, how a metallic conductor that is carrying a current when subjected to a magnetic field will cause the electrons to be deflected downward as shown in Fig. 8.39 (apply $qv \times B$ to determine the direction taking care that the particles are electrons).

As time goes on, more and more electrons moved due to the *action of the magnetic field*, will pile up on the bottom of the wire with a corresponding deficiency of electrons on the top.

However, this piling-up won't continue forever.

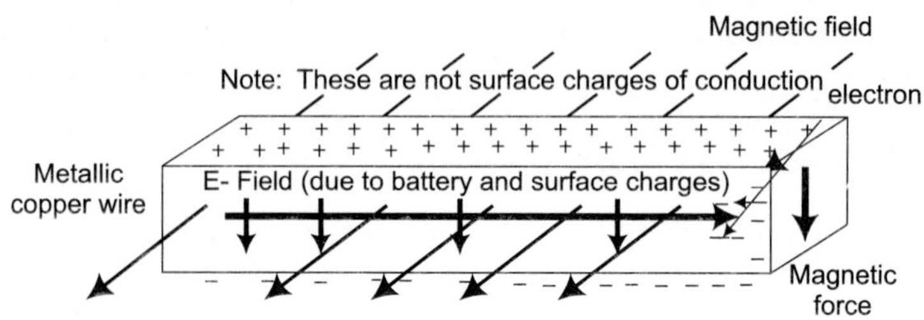

Fig. 8.39 Illustration to show how a wire moves.

The extra surface charges produce an *electric field* between the upper and lower surface of the metallic wire.

The extra surface charges exert an electric force qE (shown in grey in Fig. 8.40 and on electrons this will be "upward") on the *moving* electrons which in the steady state just balances the magnetic force qvB that caused their movement to the bottom, and so, $qE = qvB$.

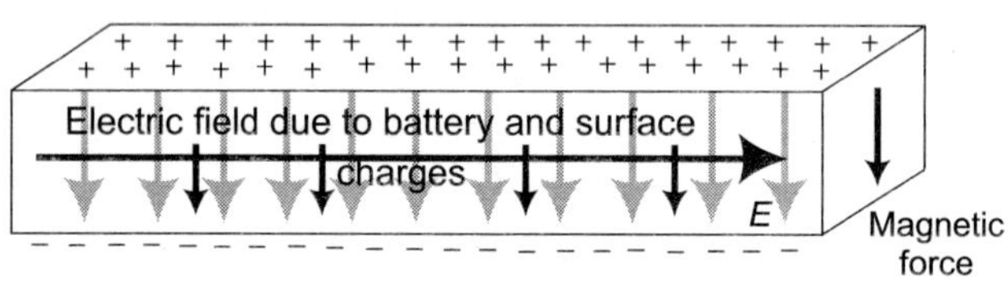

Fig. 8.40 Forces on the electrons are balanced.

Since qE becomes equal to qvB, the moving electrons now continue through the wire without being deflected as shown in Fig. 8.41.

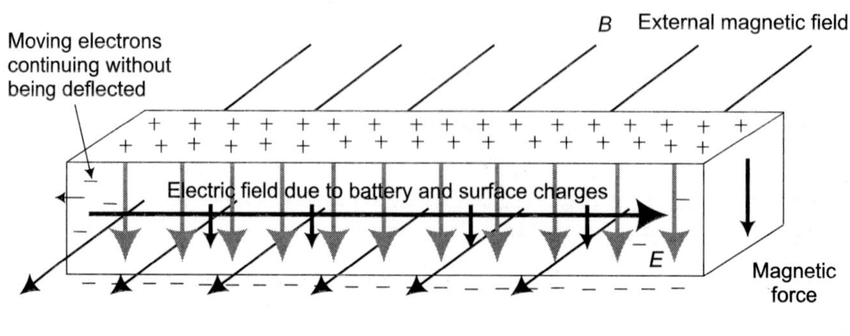

Fig. 8.41 Electrons continue to move without being deflected.

We have seen how the extra surface charges exert an electric force qE on the moving electrons which in the steady state just **...balances** the magnetic force qvB.

The extra surface charges also exert an electric force qE on the *stationary* positive atomic cores as shown in Fig 8.42.

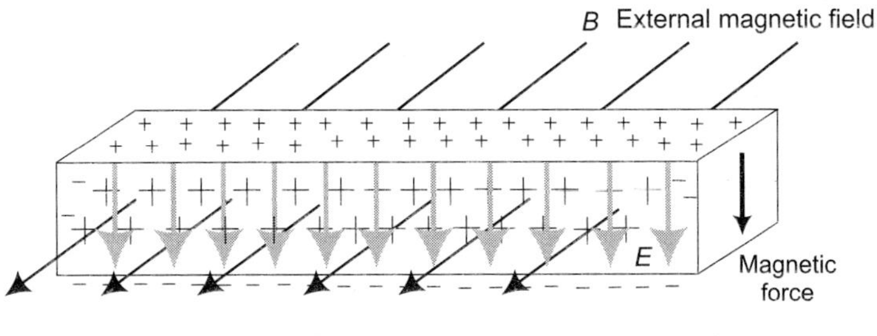

Fig. 8.42 The electric force acts on the positive atomic cores.

This electric force is an *unbalanced* force. There is no counter to this force because the atomic cores initially are not moving and hence are not subject to a magnetic force.

And that is the reason why *the wire moves*; an *electric force* acts on the positive atomic cores as shown in Fig. 8.43.

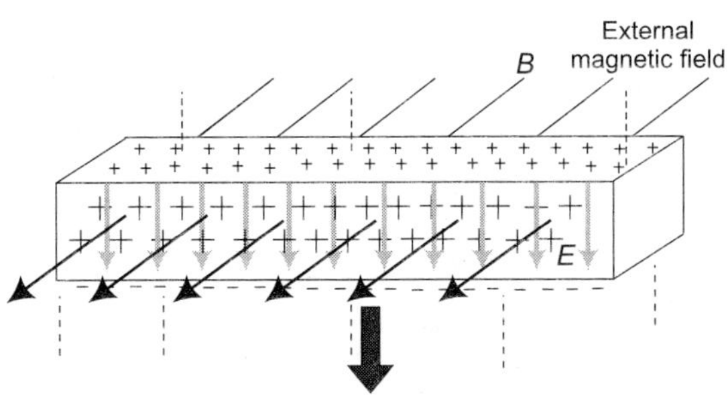

Fig. 8.43 The electric force acting on the positive atomic cores causing it to move.

The electric force due to the Hall effect surface charges forces the wire as a whole to move in the direction of $I\ (d\vec{L} \times \vec{B})$ where, dL is an elemental length of the conductor of length L carrying a current I, (also see Section 6.5) and the magnitude of the force is equal to this force.

Torque Production in Electric Motors

Fig. 8.44 shows the mechanism of torque production and rotation of motors.

It may be noted that the conductors of the rotating structure are made of copper and highlighted is the way the *Hall effect charges* exert a **force** on the *positive atomic cores* and is *responsible* for the production of torque.

This force is shown in Fig. 8.44 for one orientation of the armature and one set of conductors.

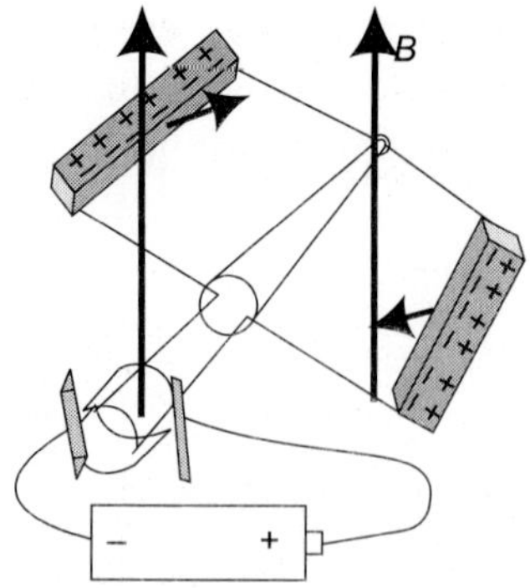

Fig. 8.44 The Hall effect charges in the armature conductors.

Figure 8.45 shows the motor (commutator not drawn for sake of clarity) driving an air blower load via a belt on a pulley fitted to the motor axle, with the forces in the conductors shown by arrows.

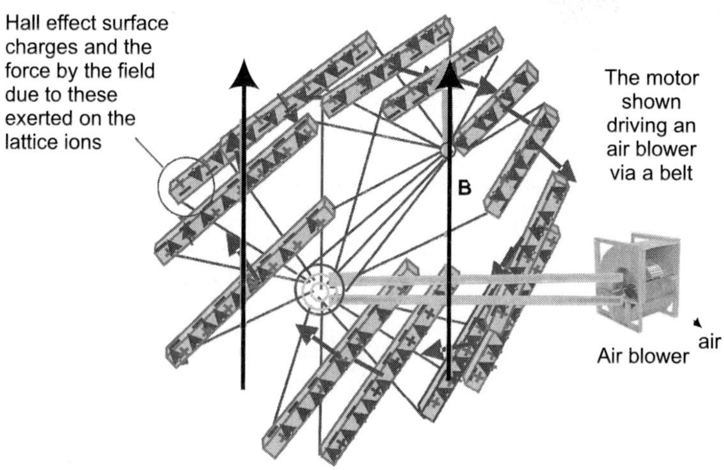

Fig. 8.45 The motor driving an air blower load.

A cut-away view of the principal components of a dc motor is shown in Fig. 8.46. The magnetic flux path is drawn in the figure and it will be noted that in dc generators and motors, the path of *least reluctance* lies in the frame or yoke of the motor and the field lines form *closed loops* taking the *following path:* the field poles, the air gap, inside the armature structure and through the air gap again and into the frame or yoke and return to the field poles.

The air gap is kept small to provide a low *reluctance* (Section 5.4) path for the magnetic field generated in the poles through to the armature conductors.

The motor conductors are placed in a structure commonly called the *armature* in a dc motor and the torque is produced that rotates the structure and the shaft.

The larger the number of conductors and these may be connected in series or parallel, then, for a given current in them, the larger the torque the motor is capable of developing.

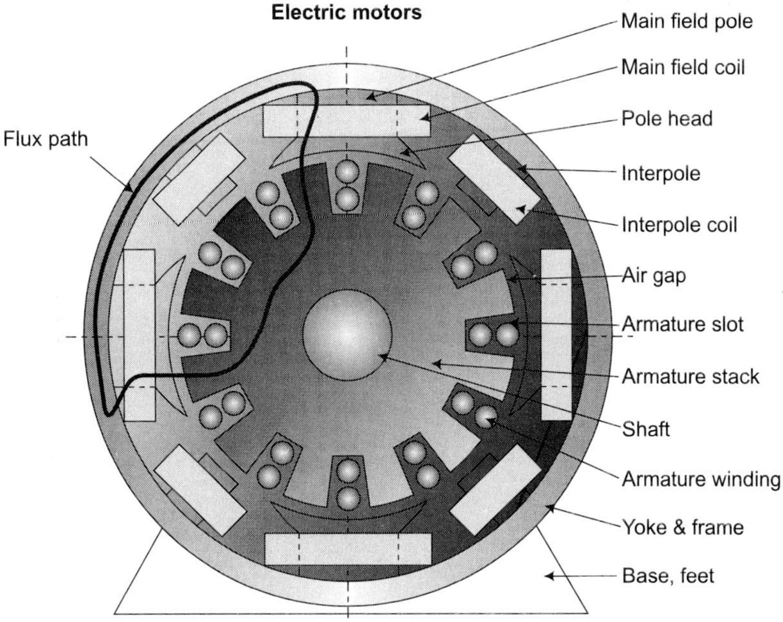

Fig. 8.46 Magnetic flux path in a dc motor.

The commutators of DC motors (not shown in Fig. 8.46) for industrial applications are multi-segment in construction as shown in Fig. 7.23. The two-segment commutator shown in Fig. 8.37 is found only in motors used to power small toys. As described in Section 7.12, hundreds of coil ends placed in the slots of the armature are soldered to the segments of the multi-segment commutator. The coils themselves are connected in series or in parallel, in a manner such that the motor is capable of delivering large power to mechanical loads.

The mechanical load such as a pump or fan or blower, compressor or conveyor belt or other mechanical device has a pulley mounted on its shaft and is driven by a belt running on a pulley mounted on the motor shaft. The load maybe directly coupled also to the motor as shown in the photograph of a motor directly coupled to a blower in the Fig. 8.47.

Fig. 8.47 Motor directly coupled to a blower.

8.11 LOAD TORQUE

In this section is described the effect of a load on a motor.

The action of the load torque is to impose a *braking effect* on the motor shaft as shown in the series of figures in Fig. 8.48 showing how the load torque imposed is increased as the tightness of the grip (of a hand) on the motor shaft is increased. This provides a symbolic interpretation only of the effect of the load torque on the motor shaft.

A motor is *normally used* by *coupling its shaft* to some mechanical device (the *load*) such as a pump, a compressor, a fan or a blower.

The devices extract energy from the motor and the motor *converts the electrical energy* (electromagnetic energy, voltage and current supplied by the battery or voltage source) from the power supply *into kinetic energy of rotation* of the motor shaft and the shaft of the coupled device.

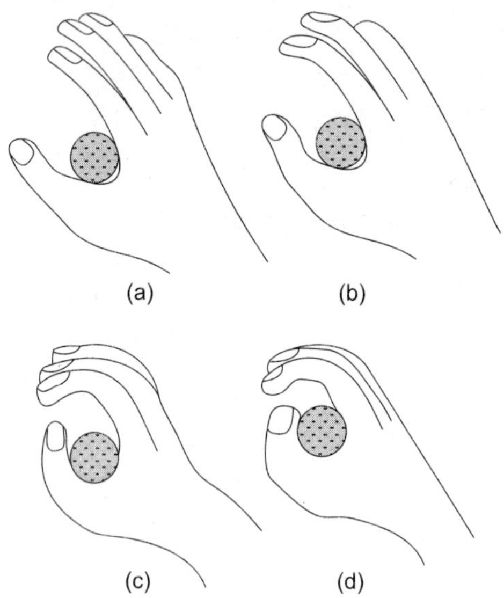

Fig. 8.48 Imposing a load torque on the motor shaft (increasing from *a* to *d*)

Warning! Do not try this anytime! You will lose your hand!

8.12 MOTOR TORQUE AND LOAD TORQUE

A motor develops a torque as it rotates which is small when there is no load, when no *pump or compressor or any other device* is coupled to it. The load torque is *small* when the motor shaft is *not coupled* to a pump or fan or compressor and it consists of only the frictional and other small losses of the motor bearings and heating of windings (also called a *no-load* condition). This is shown in Fig. 8.49a by different arrows representing the load and the motor torques.

The motor always develops a torque that *meets* the load imposed torque resulting in a *steady* rpm. If it were to be *more* than the load then the motor would *accelerate* and steady state cannot be achieved.

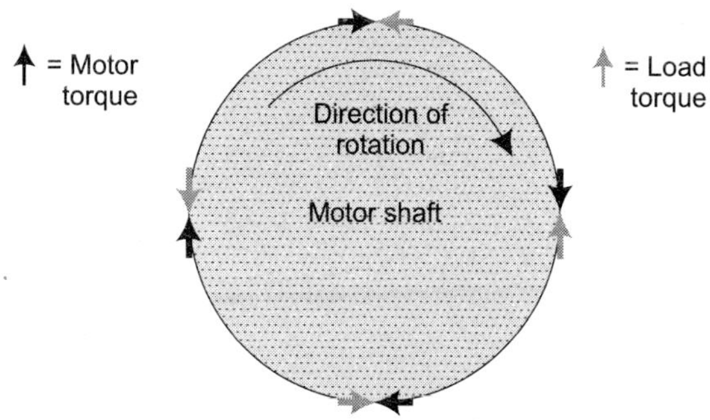

Fig. 8.49(a) A small load torque applied to the motor shaft.

The no-load rpm is large and greater than the *rated* load rpm where, the rated load is the torque the motor is capable of delivering when operating normally at its full capacity.

When a load (pump, fan or compressor) is coupled to the motor then, the load torque is large at the time of starting due to load standstill inertia and, the motor develops more than the running torque at the time of starting. This excess torque serves to accelerate the standstill motor as shown in Fig. 8.49b.

Conventional direct starting of motors always cause a large inrush current for a few seconds that provides the excess torque to accelerate the motor. Switchgear such as contactors, overload (overcurrent) and short-circuit protectors which carry the motor current, should be so selected as to allow for this short duration inrush current without tripping. If the devices prevent inrush currents, the motor will not accelerate and will surely stall or not start even. More on starting current and the back emf that determines its strength will be dealt with in Sections 8.17 to 8.22. The reader is urged to view the power point slides (Chapter 8) with animations on matching motor and load torques included in the accompanying CD.

When using resistive current limiters or electronic starters, a careful match of the consequent lower motor starting torque with the load torque is necessary to ensure that the starting motor current is yet strong enough to develop the torque required to accelerate the motor from standstill.

The standstill load torque will vary with the type of the load and also the construction of the rotating structure of motors. Long slender rotors (smaller diameter) will impose a smaller load torque as compared to rotors that are short and of large diameter. The load torque of machine tool type loads will be large and, will be small for fan and pump loads.

Usually, a few seconds after starting, and thereafter, the motor meets the load torque matching it exactly and the rpm remains steady. This is shown in Fig. 8.49c. When properly selected the load torque at steady rpm will be equal to the full load torque of the motor.

Should the motor not match the torque, it will accelerate or decelerate and, when the torques match, the rpm will remain steady.

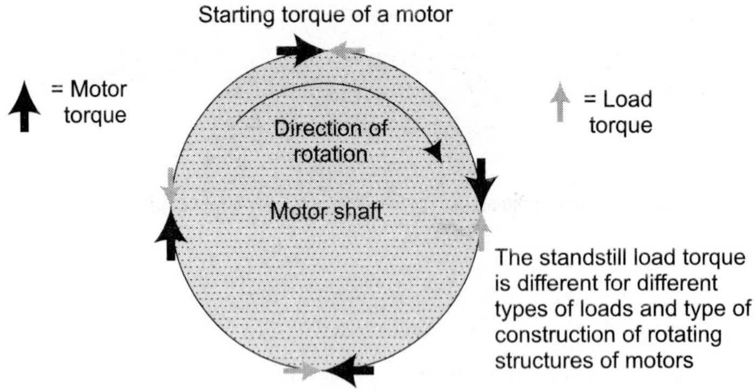

Fig. 8.49(b) Large motor torque that accelerates the load from standstill.

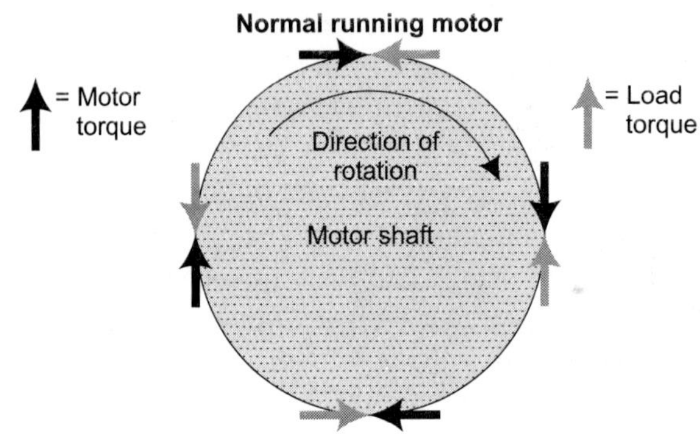

Fig. 8.49(c) Normal operation with full load torque.

If the load torque were to *decrease,* the motor will speed up or accelerate as shown in Fig. 8.50a. Such a situation will occur, for example, when parts are removed from a motor-driven moving conveyor belt or a load of parts that is rising on a hoist driven by a motor. The motor will then develop a reduced *torque* to meet the reduced load and will operate at a steady larger rpm. This is shown in Fig. 8.50b.

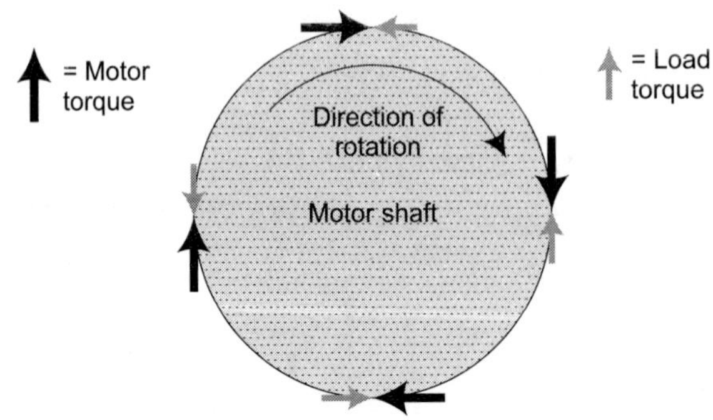

Fig. 8.50(a) In a normal operating motor, the load torque is decreased.

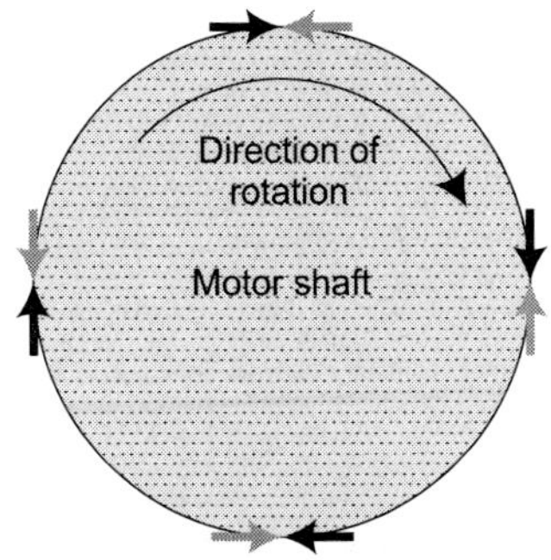

Fig. 8.50(b) The motor accelerates to a larger rpm and develops a smaller torque.

If the load torque were to *increase,* as shown in Fig. 8.51a, as will occur when more number of parts are placed on a conveyor belt driven by the motor, it will slow down or decelerate and in the process will develop an *increased torque* to meet the increase in the load torque and a steady *smaller* rpm will be reached. This is shown in Fig. 8.51b.

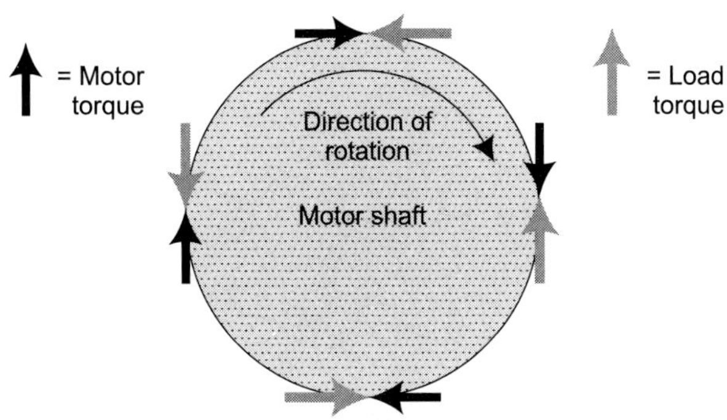

Fig. 8.51(a) The motor decelerates because of the increased load torque.

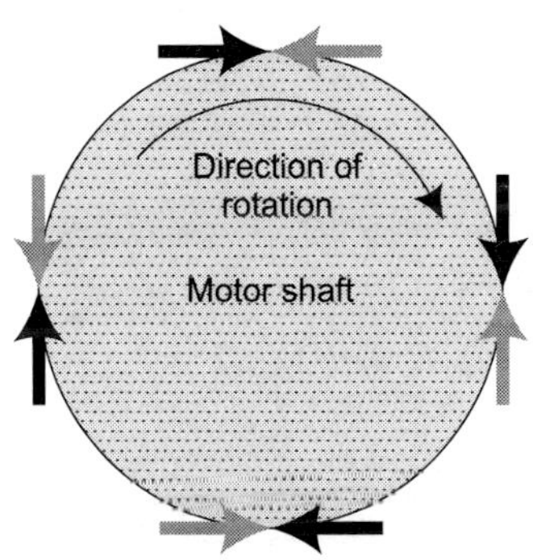

Fig. 8.51(b) The motor torque increases and the motor operates at a reduced rpm.

We shall study in Section 8.17 how the motor regulates its torque production to meet reducing or increasing load troques.

8.13 OVERLOADED MOTORS

What will happen if due to some reason a *large demand* is made on the torque developed by the motor ?

Such situations will occur if loads are *not* maintained properly especially due to *insufficient grease in large bearings* or *loss of oil in sintered bush bearings* in fan mountings and compressor housings. The motor will slow down as shown in Fig. 8.52a and eventually may even *stall* if the load were to *jam,* and because of the small resistance of the winding the motor will try to continue developing a *large torque.* The large current drawn by the motor to meet the increase in load torque, will flow in the windings which may be *insufficiently* thermally insulated and the motor will heat up and eventually burn out, as shown in Fig. 8.52b.

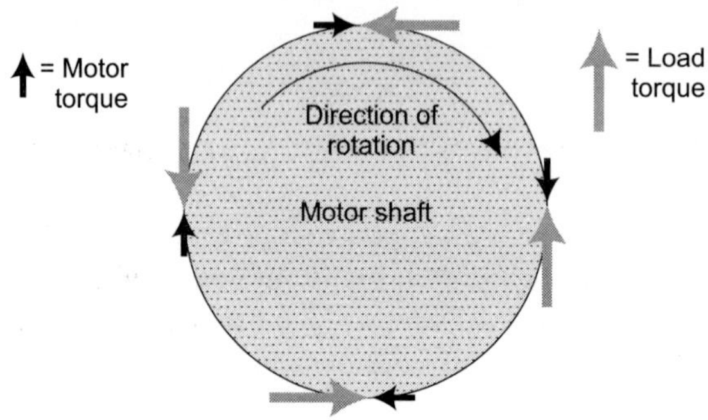

Fig. 8.52(a) The load torque is increased to a large value and the motor rpm is reduced.

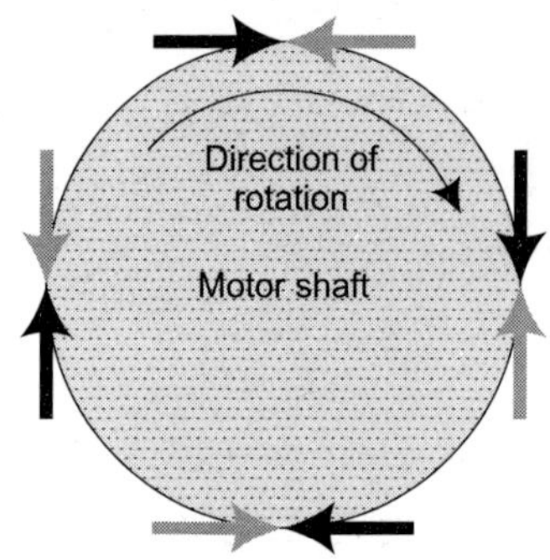

Fig. 8.52(b) The motor torque increases and the motor heats up.

8.14 MOTOR AND LOADS

The load torque that is *imposed* by the mechanical device such as a pump, a compressor, a fan or a blower has the effect of slowing down the motor shaft which would be otherwise rotating at a *greater no-load speed.*

The devices extract energy from the motor and the motor *converts the electrical energy* (electromagnetic energy, voltage and current supplied by the battery or voltage source) from the power supply *into kinetic energy of rotation* of the motor shaft and the shaft of the coupled device.

In short, the motor feels the *tug* of the mechanical device.

When we receive the *electricity bill* for operating motors the charge it contains is the price we pay for running the motor with a mechanical device. It is the price of the *tug* felt by the motor and the torque it develops to meet the *load.*

In Fig. 8.53 once again is shown a symbolic interpretation of the imposition of a load torque on a motor shaft. The load torque of motors generally is that of compressors, machine tools, fans, blowers and conveyor belts. Such loads are coupled to the motor using pulleys and belts, direct in-line flange coupling or using gear boxes that provide a reduction in rpm at the load shaft.

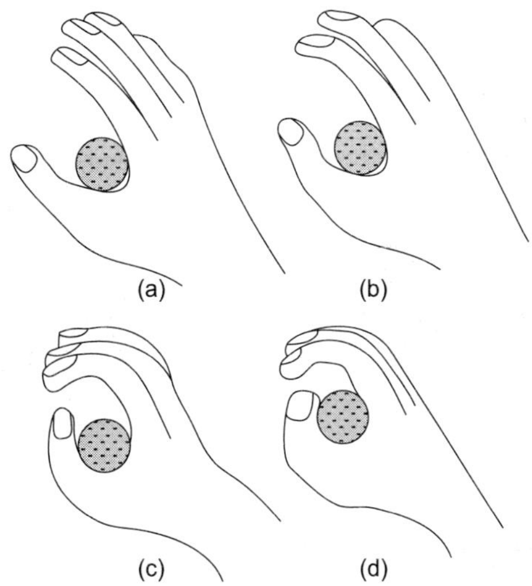

Fig. 8.53 Symbolic interpretation of imposing a load torque on the motor shaft (increasing from *a* to *d*).

8.15 TESTING MOTORS AT DIFFERENT LOAD TORQUES

Electrical equipment manufacturers commonly use a *brake dynamometer* to test and rate the various sizes of varying capacities of motors.

The brake dynamometer shown in Fig. 8.54 applies a torque (load) on the shaft of the motor that is under test using a belt that can be tightened or loosened via a nut and screw mechanism. Spring balances suspended to the ends of the belt record the amount of tension in the belt and are used to calculate the load torque being applied. For a given setting of the nut, the tightness of the belt is fixed and the brake applies a constant torque. The motor is energized and the current and speed recorded. In this manner the current and speed values are recorded at different load torque settings and the motor characteristics obtained.

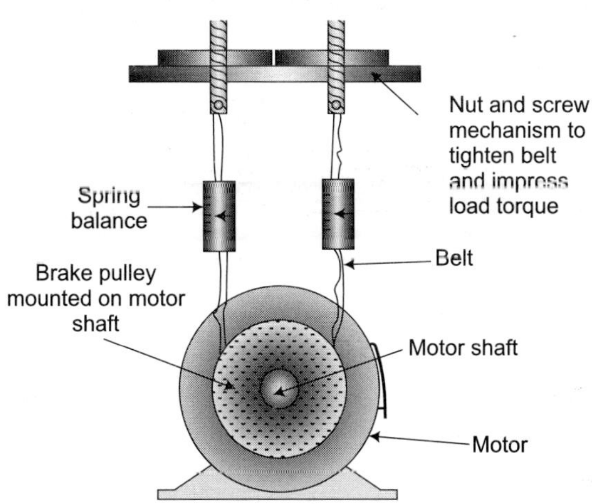

Fig. 8.54 Brake dynamometer.

8.16 DIFFERENT MOTOR SIZES AND RATINGS

Shown below are photographs of motors of various ratings and sizes, the more the power rating, the larger the motor.

Tiny 0.03 HP motor

0.75 HP motor

2 HP Motor

15 HP motor

Giant 200 HP moter

8.17 BACK EMF

The motor always attempts to develop the amount of torque that will meet the load torque.

How does it do it? Nature uses one trick which we shall see enables it to generate the required amount of torque.

To understand this, look at the torque equation: torque = $(Iwh)(B\sin\theta)$. Since B, w and h are constant the quantity that can be "varied" is the motor current I. How can the motor current I be varied? if, V the applied voltage is constant, as in say, dc operated motors.

For the most part in large motors, the motor current is *less* than volts applied /resistance of the winding conductor and is exactly (volts applied *minus* the *back emf*) /resistance of the winding.

The *back emf* is a trick nature uses that has a *regulating effect* in determining the amount of current the motor draws to develop a torque that meets the load torque and this "back emf" is a natural consequence of the induced emf process when a conductor *moves* in a magnetic field.

It must be borne in mind that in large motors the current given by volts applied / resistance of the conductor is quite large because the resistance of the conductors is small and this large current is drawn when the rotor does not rotate and is at standstill when there is no back emf. As the motor *picks up speed*, it is the *back emf* that *reduces* the effective voltage that a conductor is subject to and therefore, the current drawn is consequently *very less* than that it would draw when at standstill. The governing relation is $J = \sigma\ E_{NET}$.

The *back emf* is in reality the *motional emf* produced when a conductor moves in a magnetic field as we saw in Sections 7.7 and 7.8. This is shown in Fig. 8.55.

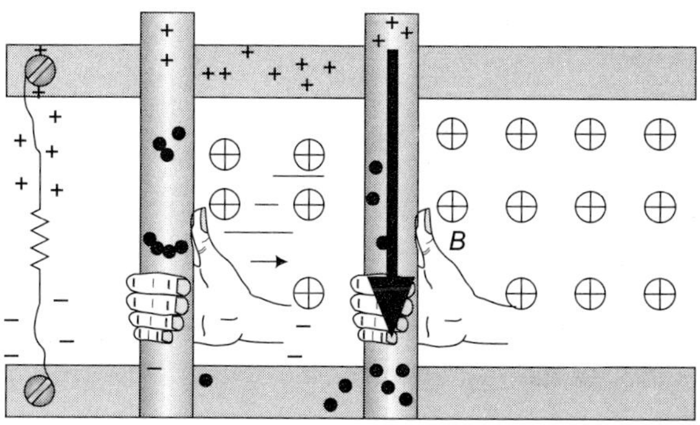

Fig. 8.55 Back emf is a type of motional emf.

As a result of the magnetic force (due to the presence of the uniform magnetic field), the free electron sea shifts and the bar is polarized producing the emf as shown in Fig. 8.55.

In a motor, the armature conductors carry the current due to the applied voltage and because of the constant magnetic field (produced by the field poles) in which these are immersed, produce an electric force on the stationary atomic cores in the conductors due to the Hall effect phenomenon (see Section 6.4). This action causes rotation of the armature and the process is called *motoring*. Now, as the conductors rotate in the magnetic field (like pulling the bar in Fig. 8.55), the electron sea in the conductors shifts laterally and because of the presence of the magnetic field is subjected to a magnetic force that polarizes the conductors to produce the motional emf. This emf is in a direction that *opposes* the applied voltage and therefore labelled *back emf*.

The motor current is governed by the relation $J = \sigma\ E_{NET}$ in which σ is the conductivity of the conductors (that does not change) and E_{NET} is determined by two electric fields, one, due to the applied voltage and the second, due to the motional emf. Because the back emf (or, motional emf) increases with increasing rpm of a motor, the motor current reduces in magnitude as it picks up speed.

If the load torque were to increase slowing the motor down, the back emf will be smaller and the motor draws more current causing an increase in torque to meet the load. Refer Section 8.12 and Section 6.5.

8.18 MOTOR CURRENT – START TO NORMAL OPERATING SPEED

When the motor starts from standstill, it draws a large current (there is no back emf which is a form of motional emf), to develop the torque necessary to overcome the inertia of the motor. As we studied in Section 7.7, motional emf is given by the expression vBL, where v = velocity of the conductor, B is the external magnetic field and L the length of the conductor. From this expression we deduce that there is no back emf developed in the conductors when the motor is at standstill and about to start because $v = 0$. This is also to be expected since the magnetic field deflects only moving charges. This is shown in Fig. 8.56.

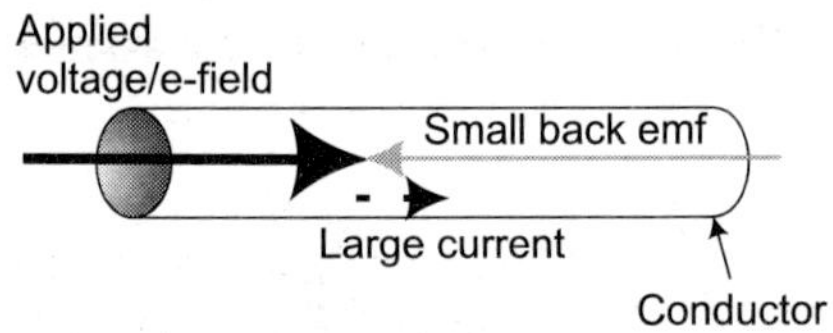

Fig. 8.56 Starting current.

After the motor accelerates from standstill, the initially drawn large current reduces because the back emf grows. This is because the conductors would now have attained some velocity and the electron sea in them is deflected by the magnetic field. The current is governed by $J = \sigma E_{NET}$. After the motor develops a large steady rpm, it draws a small current since the back emf would have grown in magnitude and nearly equals the applied voltage. This is shown in Fig. 8.57.

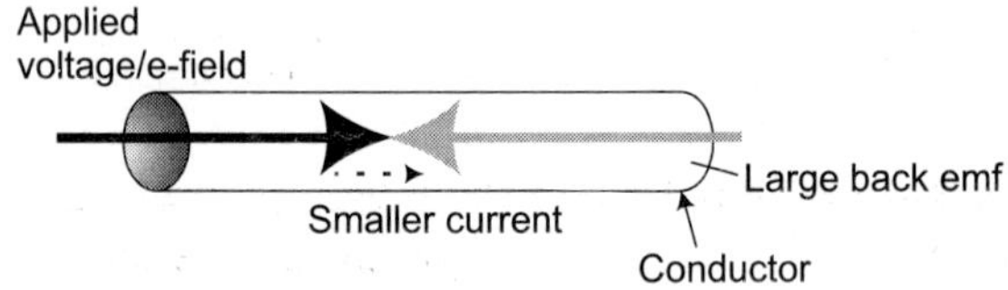

Fig. 8.57 Normal operating current.

Observe that the primary process when a motor is switched ON is the development of the motor driving torque which bears a relationship with the flux (total magnetic field in the air gap) and the armature current.

The torque is *independent* of the speed and this aspect is not apparent when the formulas are scanned for torque and speed.

The formulas mask the physical processes involved: of how the driving force applied to the load is the torque developed by the motor: the speed of the motor depends on this torque.

8.19 MOTOR SELECTION

In most applications, motors are selected largely based on motor rating or *capacity in hp and the rpm*.

Sometimes, when the load inertia is more, a *larger* motor may be *preferred* so that the motor accelerates quickly to the *operating speed* at which speed, the motor matches the load torque and does *not spend too much time* in drawing a large "starting current" and consequent heating.

Care should be exercised when making such large motor selections *because too rapid an acceleration* means tremendous forces are impacted (short of hitting the load with a sledge hammer! on thc load and may cause damage.

8.20 MOTOR TORQUE VS SPEED CHARACTERISTIC

Equipment designers select motors to match a particular type of load. This they achieve by using the torque vs. speed characteristics of motors and loads.

Shown below in Fig. 8.58 is a typical torque vs. speed characteristic of an ac motor. Observe the two regions of the motor characteristic: the starting region where, the motor spends little time after starting and, an operating region where, the motor normally operates, continuously.

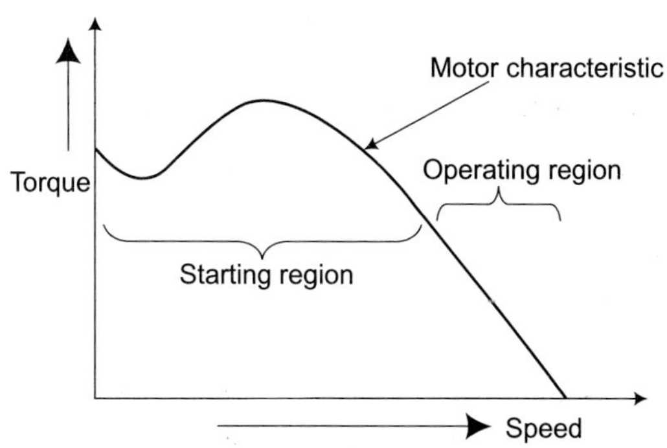

Fig. 8.58 Motor torque vs speed characteristic.

In Section 8.22 is given a description of the significance of the two regions and how motors are selected by matching the load torque with the motor torque.

A motor is useless unless it drives a load that could be a blower, fan, spindle of a lathe or a conveyor belt. It is necessary to understand the type and characteristics of such loads and the following section describes some of the more common loads found in industrial applications.

8.21 LOAD TORQUE VS SPEED CHARACTERISTIC

We have studied the characteristics of motors and since these are used to drive loads, it is important that we study loads and their characteristics in order to be able to match motors to suit loads.

Loads such as fans, pumps and compressors apply varying levels of *squeeze* or torque at different speeds at which they are made to rotate by the motors. They do not always apply a constant torque at different speeds when in operation.

The load torque vs speed characteristic helps equipment designers to select motors appropriately.

For most loads, the motor is made to operate *continuously* after starting, at the point where, its developed torque *matches* with the load torque therefore, it is useful to study the load torque vs speed characteristic.

Some common types of loads will be shown to illustrate.

Machine tools load motors during cutting, drilling and milling operations as shown in Fig. 8.59. Unlike fans, compressors and pumps, the torque applied by machine tool cutters, drills and grinders to the motor shaft *does not* vary with motor rpm or speed. The torque vs. speed characteristic is a straight line in the graph of Fig. 8.59.

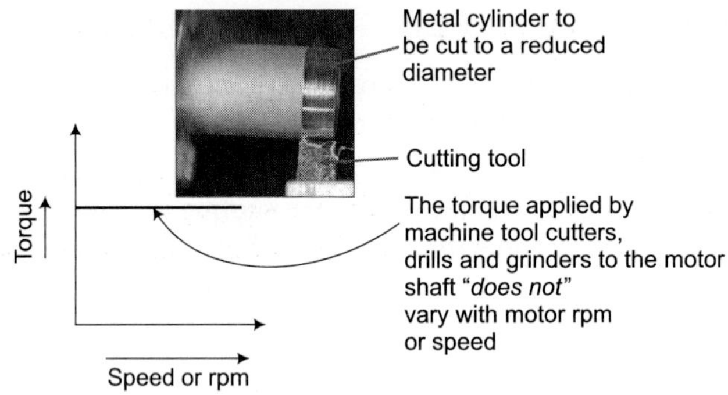

Fig. 8.59 Machine tool (load) vs. speed characteristic.

On the other hand, the torque vs. speed characteristic of axial fans, centrifugal pumps and blowers have a varying characteristic and follow a square law, as shown in Fig. 8.60. The torque is proportional to the square of the speed for such loads.

The torque applied by the fan and blower to the motor varies in *squared proportion* with motor rpm or speed.

The power consumed by such loads varies in cubic proportion with speed because the power consumed is proportional to torque *x* angular velocity.

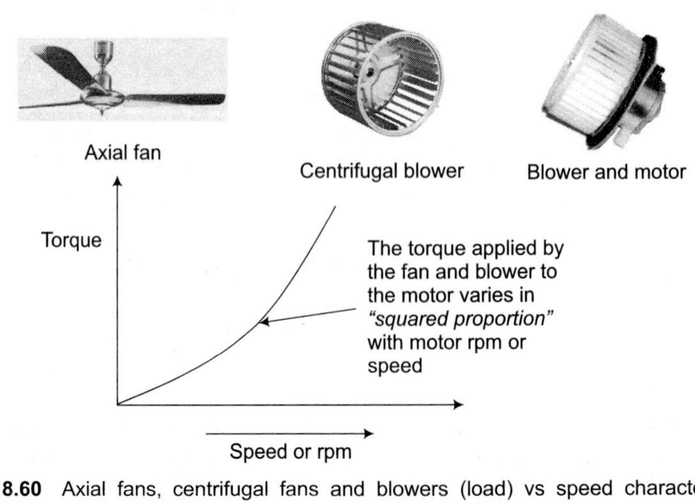

Fig. 8.60 Axial fans, centrifugal fans and blowers (load) vs speed characteristic.

These characteristics are obtained by coupling the devices such as fans, blowers to a *motor whose rpm can be varied* and from the corresponding power drawn at different speeds, the torque of these devices is computed and plotted.

In the case of machine tools, the *motor whose rpm can be varied* is coupled with a large cylinder and cutting forces increased by pressing the cutting tool at various forces, recording the power drawn at different speeds and the torque computed.

8.22 MATCHING MOTOR TORQUE TO LOAD TORQUE

Equipment designers can determine this speed or *the operating point* very simply by superimposing the load torque graph on to the motor torque speed characteristic.

The *point of intersection* of the two curves is the *operating point* for the given motor –load combination. This is shown in Fig. 8.61 for a cutting tool operation.

With the motor at standstill and energised, the torque it develops is T_A as shown in the Fig. 8.61. At the time of starting, the torque T_A it generates, enables it to accelerate the load with which it is coupled. The available torque to accelerate will be $T_A - T_{load}$. As the rpm increases, the operating point of the motor follows the motor torque curve.

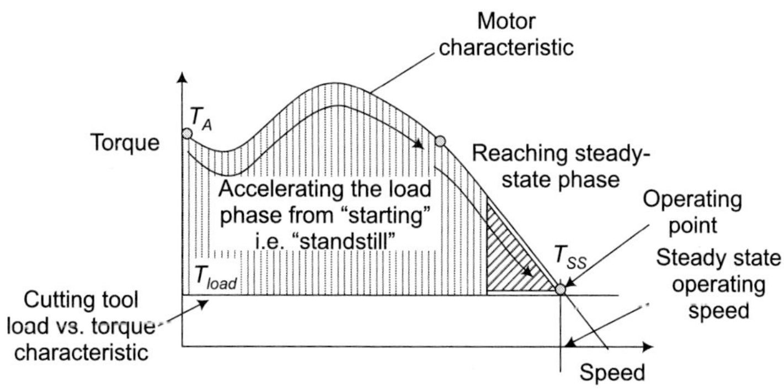

Fig. 8.61 Motor torque and speed variation from starting to steady-state operation.

Subsequently, after a few seconds as the back emf comes up, the motor torque gradually reduces to T_{SS}, the steady-state value, and the rpm becomes steady at the point of intersection of the load torque curve with the motor curve as shown in Fig. 8.61.

Let us assume that the designer selects a motor having *not enough* capacity for the given cutting tool torque; the cutting tool characteristic intersects the motor characteristic in the *starting region* (shown hashed) and the *operating point* is in this region. The situation is illustrated in Fig. 8.62.

Since, the motor is not designed to operate continuously in this region, it will heat up and eventually burn.

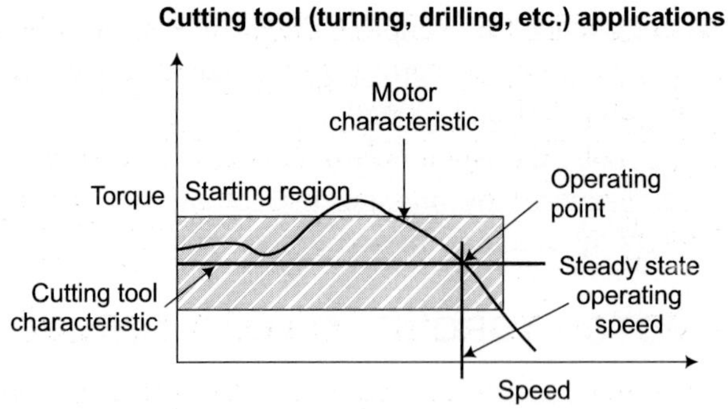

Fig. 8.62 Determining the operating point of a motor (wrong selection).

A more appropriate selection of the motor (larger capacity) is shown now in Fig. 8.63 where, the *point of intersection* of the two curves, the operating point lies in the operating region (shown hashed) of the motor.

The motor can now run *continuously* without heating up.

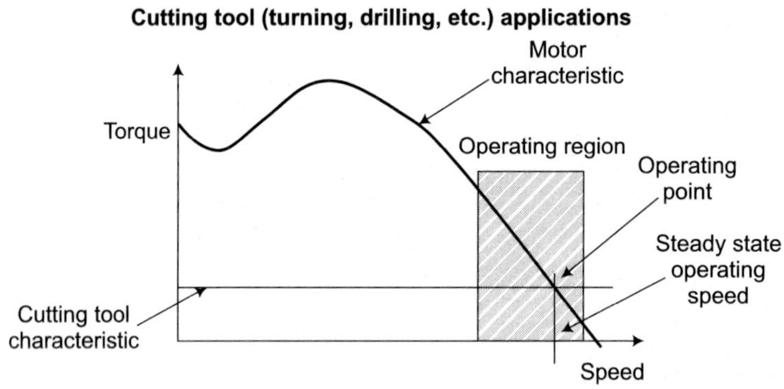

Fig. 8.63 Determining the operating point of a motor (correct selection).

In Fig. 8.64 is shown how we can determine the steady-state speed of the motor and load combination for fans, blowers and centrifugal pumps given the characteristics of the motor and the load. With such loads whose torque vs. speed characteristic follow a square-law, the starting regime is not demanding (because the load torque is zero at the time of starting) but, care must be exercised in not selecting too large a motor as too high an acceleration can damage the load.

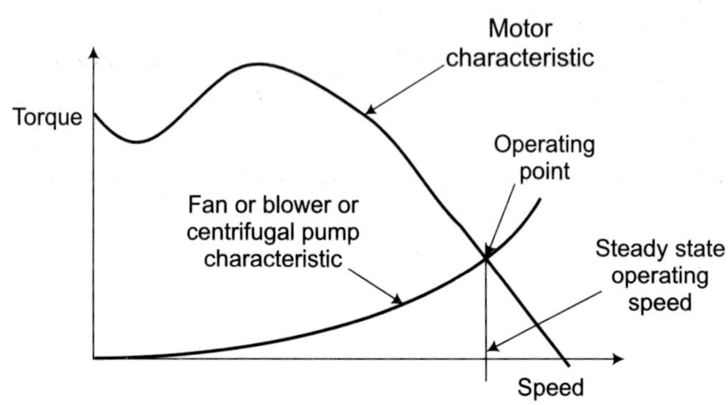

Fig. 8.64 Determining the operating point of a motor (fan or blower load).

8.23 ENERGY CONVERTED BY MOTORS

It is not right to assume that as the motor rotates the shaft and the coupling with the load, it does *not* deliver energy to the load because the *rpm is steady.* This is a common erroneous assumption made by students.

The reasoning that since there is *no acceleration* of the shaft *does not mean* that the motor is *not delivering energy* because energy = torque × angle and torque = force × lever arm = mass × acceleration × lever arm…and since there is no acceleration, it is "zero" and therefore *no energy is delivered !*

As long as the load is coupled and the supply is ON, every motor strains and develops the necessary forces continuously on its shaft to balance the opposing forces of the load *which, occurs when a steady rpm is reached.*

And, the *torque the motor develops multiplied with the angle of rotation or distance it travels along the circle of rotation constitutes energy.*

The energy delivery occurs all the time as long as the motor is ON and is rotating.

8.24 LOUDSPEAKERS

Loudspeakers operate on the principle of magnetic forces developed when a current carrying conductor is placed in a magnetic field.

Figure 8.65 shows the diagram of a loudspeaker: it consists of a single moving coil which is permanently and flexibly wired to the amplified electrical speech or music signal source.

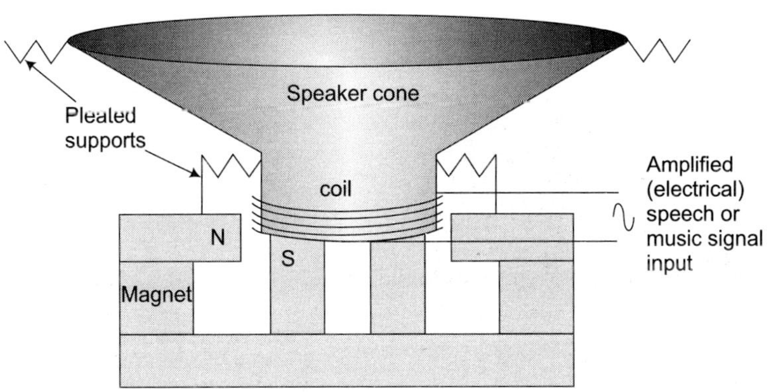

Fig. 8.65 Construction of a loudspeaker.

The coil moves in the field of a permanent magnet, which is usually shaped to produce maximum force on the coil. The coil is attached to a light weight paper cone, which is supported at its inner and outer edges by circular, pleated paper *springs*. This is shown in the cutaway view in Fig. 8.66.

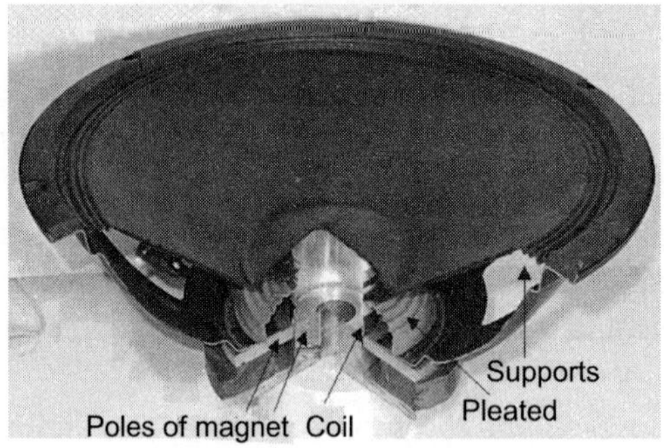

Fig. 8.66 Cutaway view of a loudspeaker.

The operation of a loudspeaker is shown in Fig. 8.67.

Figure 8.67 shows a view of how forces that make the cone vibrate are produced in a loudspeaker.

Shown in the figure are the coils (grossly exaggerated using rectangular cross-section wires) and the diaphragm that is caused to vibrate by the coil to which it is attached.

Shown in Fig. 8.67 is the permanent magnet that is cylindrically shaped and specially made to produce a magnetic field that is directed radially inward through the coil.

When the coil vibrates as a result of the Hall effect forces due to the applied audio signal emf and the radially directed magnetic field, it causes the diaphragm to vibrate along with it.

This causes sound waves to be propagated.

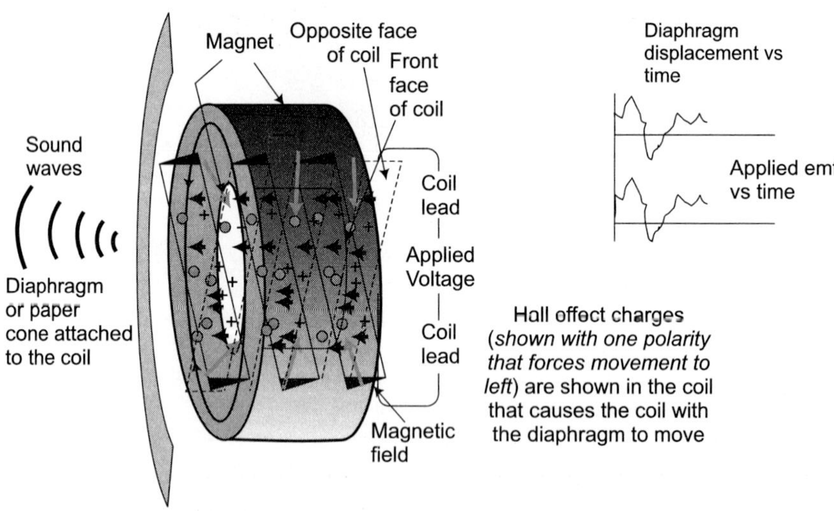

Fig. 8.67 Cutaway view of a loudspeaker.

8.25 POTENTIAL ENERGY OF A MAGNETIC MOMENT

In Chapter 1, we had studied different types of potential energy and to this we now add one more type, that of a magnetic moment.

We have seen how if allowed to pivot freely, a *magnetic moment* turns towards alignment with the applied magnetic field. See Section 8.3.

This means that an aligned magnetic moment has *lower potential energy*, and the system tries to *go to the lower potential energy configuration.*

Recollect from Chapter 1 that by separating the charges of a neutral atom (electrons and positive atomic cores), we could store the *work done* energy in the *electric field* of this *system* of charges—comprising the two charges.

....like stretching a spring....

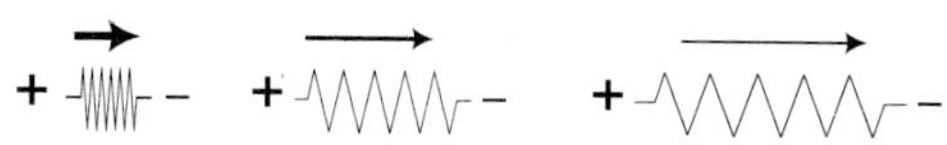

....and when you *released* the *negative charge*.....

.......the *electrostatic energy stored in the system of 2 charges was released* and converted into *kinetic energy* in the *negative charge*......

***.......like releasing a stretched* spring.....**

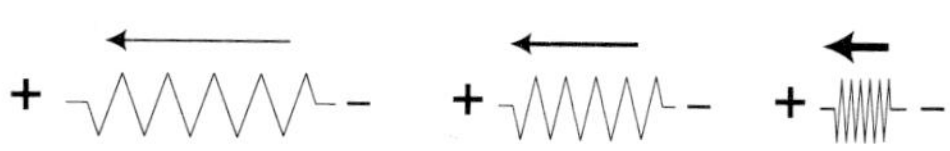

Analogously, observe carefully how when a current carrying loop in a magnetic field is displaced away from its equilibrium position, it "*attempts*" to return to the equilibrium position.

And, you have to hold your finger against the loop applying a *"force"* to return to a horizontal equilibrium position.

This is shown in Fig. 8.68a.

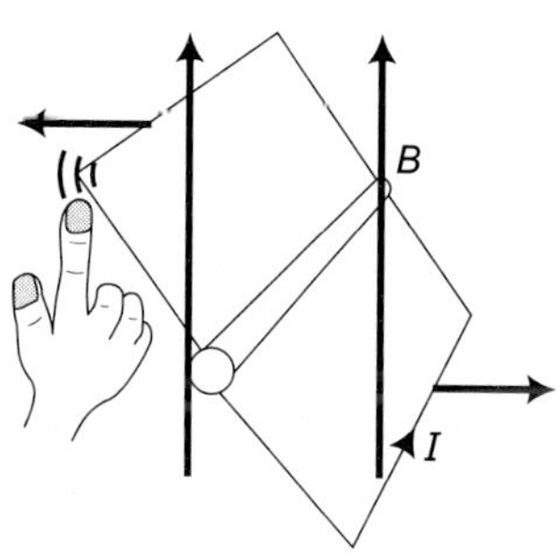

Fig. 8.68(a) Potential energy of a magnetic moment.

If you remove your finger, the magnetic forces will cause it to turn *counterclockwise* and *return* to a more stable horizontal position as seen in Fig. 8.68(b).

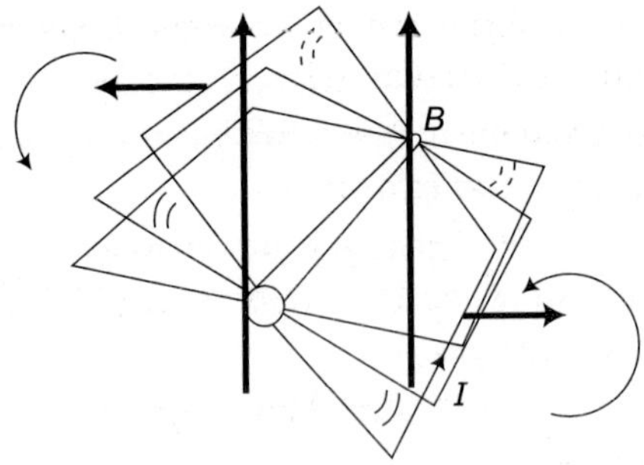

Fig. 8.68(b) Potential energy of a magnetic moment.

What we observe is that when a magnetic moment, a *current carrying loop* is placed at an angle in a magnetic field, it stores energy—potential energy—just like *separated opposite charges* and like a *stretched spring.*

This picture of the *potential energy* of a magnetic moment in a magnetic field is very important in atomic and nuclear physics.

In magnetic imaging equipment used in hospitals, the patient lies inside a large solenoid. In the presence of the magnetic field of the solenoid, nuclei in the patient's body have slightly different energies depending on whether the nuclear magnetic moment is *in the direction of* or *opposite to* the applied field.

Sensitive equipment detects transitions between these two slightly different levels.

Calculating the Potential Energy (PE) of a Magnetic Moment

We know that the potential energy stored by a mass m raised to a height h above the ground is given by mgh. Similarly, we can calculate the potential energy of a current-carrying loop in a magnetic field (Chabay & Sherwood, 1995 used by permission).[2]

Consider a rectangular loop (length h and width w) as shown in Fig. 8.69a with current I as in Fig. 8.23, and let us calculate how much **work** should be done to move it from an initial angle $\theta = \theta_i$ to a final angle $\theta = \theta_f$.

We move slowly, ***without*** changing the **kinetic energy** of the loop (that is by **not** applying a **force** that would cause the loop to accelerate away). (Chabay & Sherwood, 1995, used by permission).[2]

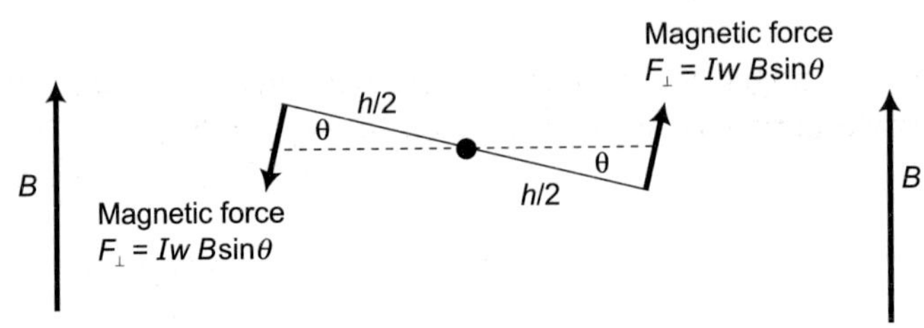

Fig. 8.69(a) A current carrying loop in a magnetic field.

We would have to exert a varying force $F_{us} = IwB\sin\theta$ (Section 8.7, Fig. 8.24) at both ends of the loop, *perpendicular* to the loop, to *balance* the perpendicular component $F\perp$ of the magnetic force on the loop as shown in Fig. 8.69b.

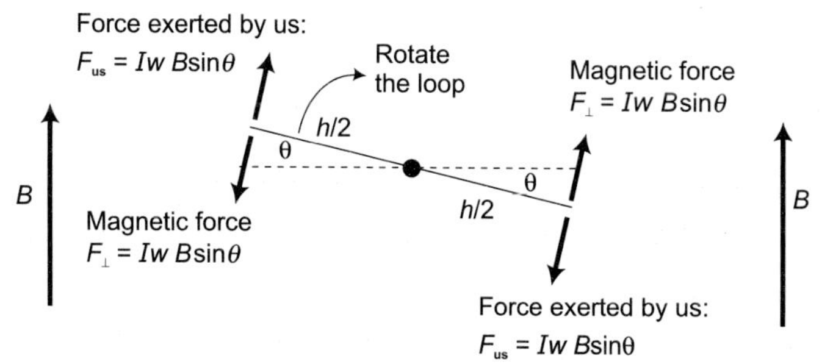

Fig. 8.69(b) Force exerted by us to balance the loop.

As we rotate the loop through a small angle $d\theta$, both of the forces that we exert act through a distance $\frac{h}{2}d\theta$ because arc length is radius ($h/2$) times angle ($d\theta$ measured in radians) as shown in Fig. 8.69(c).

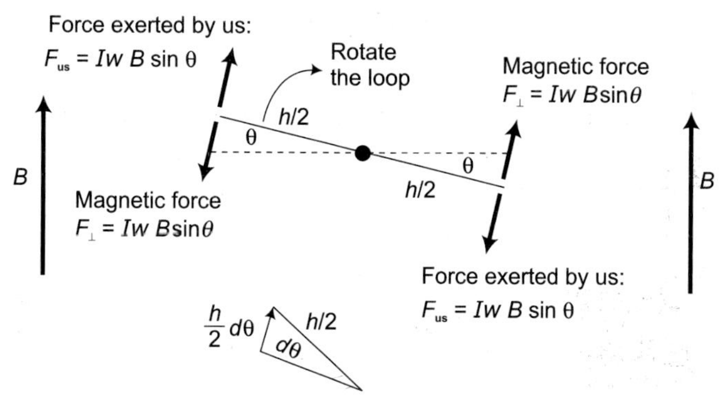

Fig. 8.69(c) Calculating the potential energy of a magnetic moment.

If we force the loop to rotate from the initial angle $\theta = \theta_i$ to a final angle θ_f, the **amount of work** we do by exerting our two forces must be calculated by **an integral**, because F is not a constant varying with the angle θ.

$$work = \Delta PE = \int_{\theta_i\, initial}^{\theta_f\, final} 2IwB\sin\theta\left(\frac{h}{2}d\theta\right) = IwhB\int_{\theta_i}^{\theta_f}\sin\theta\, d\theta$$

$$work = \mu B\left[-\cos\theta\right]_{\theta_i}^{\theta_f} = -\mu B\left[\cos\theta_f - \cos\theta_i\right]$$

$$= \Delta(-\mu B\ \cos\theta)$$

where, $\mu = I \times$ area of the loop $= Iwh$ (see Section 8.1).

Or,

$$work = \Delta\ (-\mu B\ \cos\theta)$$

It is customary to define the *potential energy* (PE) of a **magnetic moment** in a magnetic field **by** choosing the "zero" of potential energy in such a way that one writes:

$$PE = -\mu B \cos\theta$$

This can also be written as a vector dot product:

$$PE = -\mu.B$$

Therefore, the potential energy of a magnetic moment is $-\mu.B$.

8.26 THE POTENTIAL ENERGY OF MAGNETIC MOMENTS ALIGNED AT VARIOUS ANGLES WITH THE MAGNETIC FIELD

(Chabay & Sherwood, 1995, used by permission).[2]

Convention requires calling the potential energy ***zero*** when $\theta = 90$ degrees, since cos (90°) = 0 as shown in Fig. 8.70.

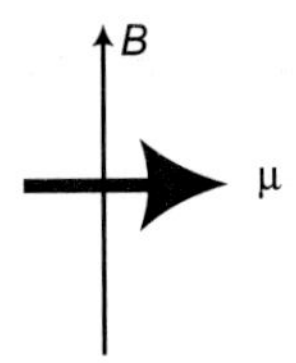

Fig. 8.70 PE 0 = when θ = 90°.

Applying this convention, in Fig. 8.71 are shown figures of the alignment between μ and *B* corresponding to the *lowest* potential energy (though negative) to the *highest* and this is *lowest* when the moment and the field are *aligned and in the same direction.*

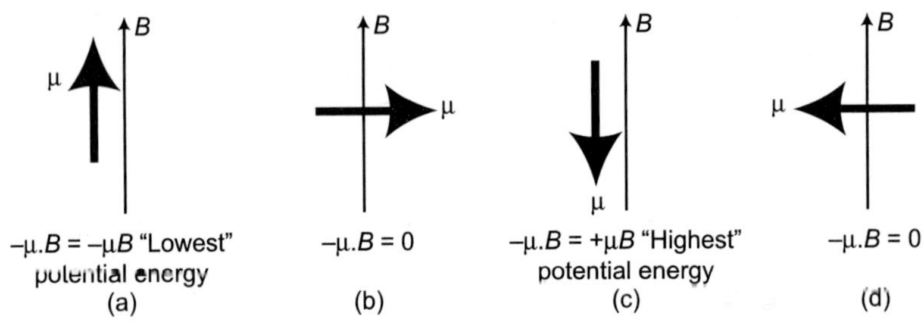

Fig. 8.71 PE at different angles for 'θ'.

The potential energy is the highest when the moment and the field are aligned in opposite directions as in Fig. 8.71(c).

8.27 MAGNETS AND MAGNETISM

We have seen the *twist* of a magnetic moment in a *uniform* magnetic field.

How can one explain the force on a magnetic moment in a magnetic field?

For example, how does a magnet *attract* a paper clip made of iron or steel?

It is not hard to show that the answer lies in placing a magnetic moment in a *non-uniform* magnetic field. Let us first see how magnetic fields are produced in magnets at the atomic level[28].

Cause of Magnetism

Magnetic fields are created by moving charges. Any time a charge moves (current) a magnetic field is created. Magnetic force is created by two magnetic fields occupying the same space. A magnetic field is defined as any region where another magnet would "feel" a force.

What are magnets? (atomic view)

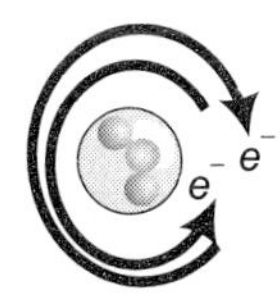

Most materials have paired up electrons spinning in opposite directions. The field created by one moving charge is cancelled out by the other, and so, no magnetic field is created.

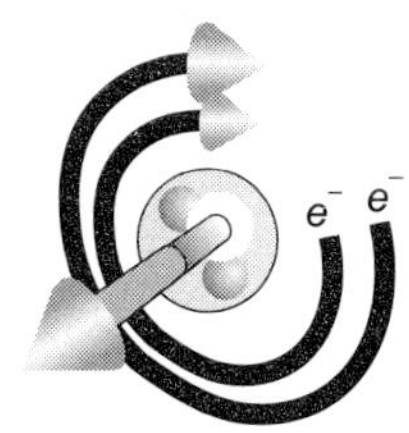

Some materials like iron, nickel, cobalt, or neodymium have a single electron or paired up electrons spinning in the same directions. The magnetic field created by one does not cancel the other. An atomic sized magnet is created.

Atomic to life size?

The atomic magnets line up to form domains within a sample. But, the lining up may not be uniform. Therefore, the metallic sample is not magnetized.

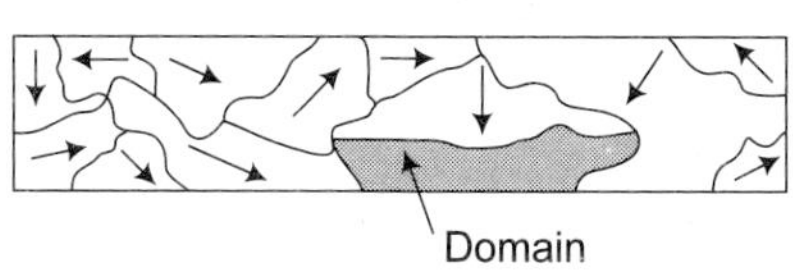

Magnetic domains can be encouraged to line up to form a magnet say, by the application of a strong magnetic field.

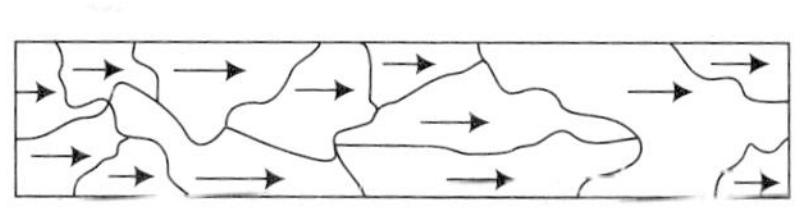

Conventions and Definitions

The direction of a magnetic field is defined as the direction that the north pole of a compass (magnet on a swivel) would point when placed in the field. The field lines drawn are imaginary lines of force-flux lines. The closer they are, the stronger the field.

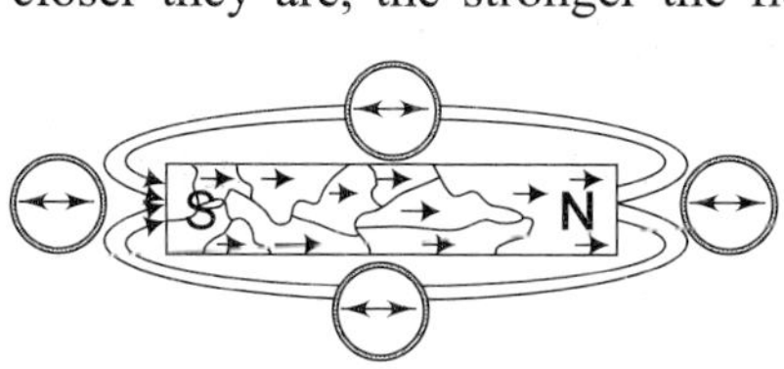

Permanent magnets

Always magnets have two poles-called "magnetic dipoles". There is no such thing as a magnetic monopole!

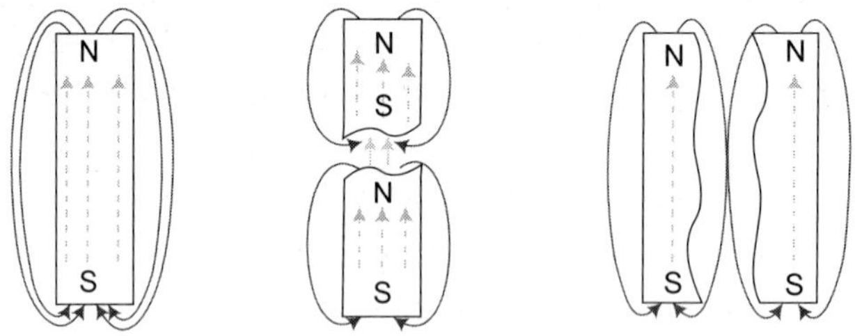

Would attract Would repel

Each atom consists of electrons orbiting about a central positive *nucleus* along with the electrons spinning about their own axes, as we saw earlier. Thus, an internal magnetic field is produced by electrons orbiting around the nucleus as in the left figure below or electrons spinnings as in the right figure below.

Electron in orbit around the nucleus Electron spin

Both of these electronic motions produce internal magnetic fields *B*, that are similar to the magnetic field produced by a current loop.

The equivalent current loop has a magnetic (dipole) moment of $\vec{m} = I_b S\vec{a}_n$, where *S* is the area of the loop and I_b is the bound current and $\vec{a}_n$ is a unit normal vector.

Since it is impossible ιo measure the magnetic moment of individual atoms, physicists use the *bound current* equivalent representation to analyse the effect of magnetic fields in matter.

8.28 MAGNETIC FIELDS IN MATTER

Like electric fields, magnetic fields penetrate through matter.

Unlike electric polarization, which is almost always in the same direction as *E*, the applied field, some materials acquire a magnetization parallel to the external applied magnetic field *B* (paramagnets) and some opposite to the external applied magnetic field *B* (diamagnets).

In the *presence* of the external magnetic field, matter becomes magnetized; that is, upon microscopic examination it will be found to contain many tiny dipoles, with a net alignment along some direction.[8] There are two mechanisms that account for this polarization: 1. paramagnetism in which the dipoles associated with the spins of unpaired electrons experience a torque tending to line them up parallel to the field, and 2. diamagnetism in which the orbital speed of the electrons is altered in such a way as to change the orbital dipole moment in a direction opposite to the field.

A few substances called ferromagnets (the most common example is iron) retain their magnetization even after the external field has been removed – for these the magnetization is not determined by the present field but by the whole magnetic "history" of the object.

Permanent magnets made of iron (see Section 8.27) are the most familiar examples of magnetism, though from a theoretical point of view they are the most complicated.

Diamagnetism and paramagnetism are extremely weak and it would take delicate experiments and a powerful magnet to detect them at all. The weakness of the effect is one of the reasons we are able to calculate the magnetic fields in copper wires, easily[8].

A magnetic dipole experiences a torque in a magnetic field and for a rectangular current loop of length a and width b carrying a current I, in a uniform field $\vec{B}$, the torque is given by

$$\vec{N} = \vec{m} \times \vec{B}$$

where, $m = Iab$ is the magnetic dipole moment of the loop.

The figure on the left below shows a piece of material of volume Δv before a magnetic field is applied and the right figure shows the magnetization M (the magnetic dipole moment per unit volume summing the magnetic dipole moments of all the atoms in the infinitesimal volume Δv/ Δv)after a magnetic field B is applied.

$$\vec{B} = 0; \quad \vec{M} = 0$$

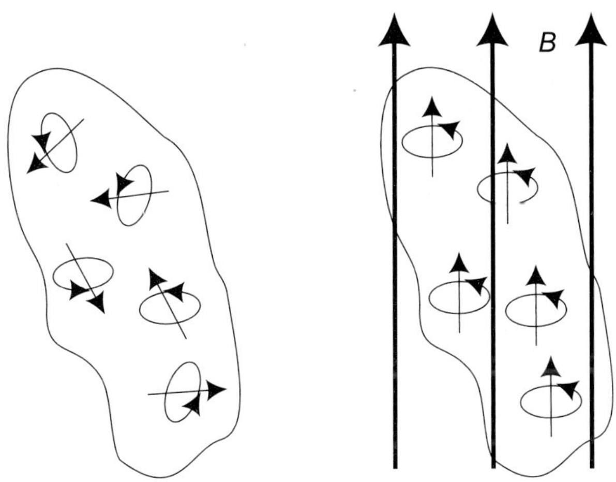

As we saw in Section 8.4 the torque is in such a direction as to line the dipole up parallel to the field. It is this torque that accounts for paramagnetism. While one could rightly expect paramagnetism to be a universal phenomenon, the laws of quantum mechanics dictate that the electrons within a given atom lock together in pairs with opposing spins, and this effectively neutralizes the torque on the combination. As a result, paramagnetism normally occurs in atoms or molecules with an odd number of electrons, where the "extra" unpaired member is subject to the magnetic torque. Even here the alignment is far from complete, since random thermal collisions tend to destroy the order.

The above descriptions of magnetic fields in matter are by no means complete and the reader is referred to the in-depth analysis provided in "Introduction to Electrodynamics" by David J Griffiths.[8]

In a uniform magnetic field B, the net force **F** on a current loop is zero:

$$\mathbf{F} = I \oint (d\vec{l} \times \vec{B}) = I\left(\oint d\vec{l}\right) \times \vec{B} = 0$$

the constant B comes outside the integral, and the net displacement $\oint d\vec{l}$ around a closed loop vanishes. In a non-uniform field this is no longer the case.

For example, suppose a circular wire is placed away from a bar magnet as in Fig. 8.72(a), the magnetic field fringes and B has a radial component, and there is a net force on the loop.

We will now show how when a circular current loop (assume that the loop is a coil and is a rigid body and R is the radius of the circular loop) is placed near the north pole of a bar magnet along the axis of the magnet, it experiences a force. Such a situation is shown in Fig. 8.72a (Chabay & Sherwood, 1995, used by permission).[2]

The magnetic field of the bar magnet diverges with an angle θ as shown and so is *non-uniform*.

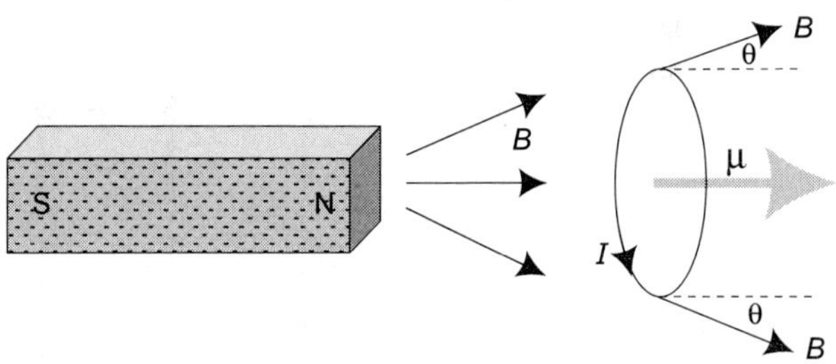

Fig. 8.72(a) Field of the magnet is non-uniform near the magnetic moment.

Each segment dl of the ring experiences a magnetic force $d\vec{F} = I\, d\vec{l} \times \vec{B}$ as we saw in Section 6.5 and is shown in Fig. 8.72b.

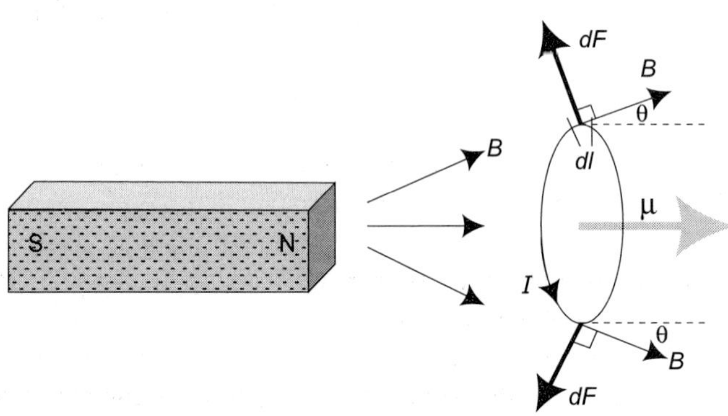

Fig. 8.72(b) Magnetic force on an elemental segment *dl*.

The horizontal component is $dF\cos(90° - \theta) = dF\sin\theta$ as shown in Fig. 8.72c.

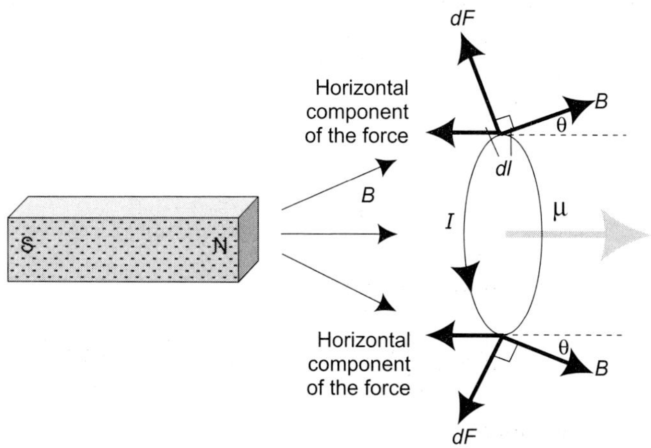

Fig. 8.72(c) Horizontal component of the force on a magnetic moment.

The vertical components (see Fig. 8.72d) being equal and opposite on the ring *cancel* because the coil is assumed to be a rigid body.

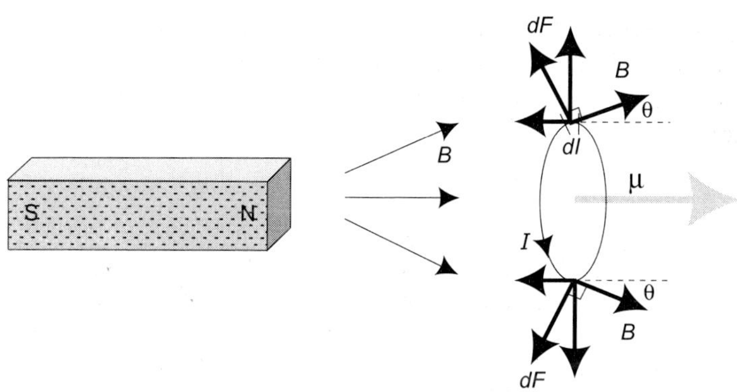

Fig. 8.72(d) The vertical components of the force cancel.

So, the net force is in the negative x direction, *towards the magnet.* This is shown in Fig. 8.72e.

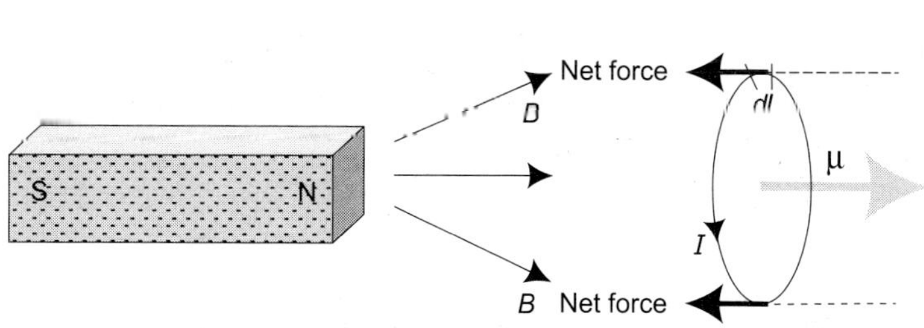

Fig. 8.72(e) Net force causes the ring to move towards the magnet.

The resultant force on the ring causes it to move towards the magnet. The net force on the magnetic moment, that is, the current carrying loop of radius R can be calculated as follows:

$$d\vec{F} = Id\vec{l} \times \vec{B}$$

$$F_{net} = F_x = \int_{loop} (IdlB\sin 90°)\sin\theta$$

$$= IB\ \sin\theta \int_{loop} dl = IB(2\pi R)\sin\theta$$

Since,

$$\mu = IA = I(\pi R^2) \qquad \text{(Refer section 8.2)}$$

$$F = \left(\frac{R}{R}\right) IB\ (2\pi R)\ \sin\theta = \frac{(I\pi R^2)2B\sin\theta}{R}$$

The net force F is, therefore

$$\frac{\mu(2B\sin\theta)}{R}$$

Force on a magnetic moment with the magnet reversed (Chabay & Sherwood, 1995, used by permission).[2]

If either the magnet is reversed or the *direction of the current* in the coil is reversed (but not both), as shown in Fig. 8.73a, then, by resolving the elemental force dF into components, we note that the components vertical to the axis of the coil cancel out because the coil is assumed to be a rigid body.

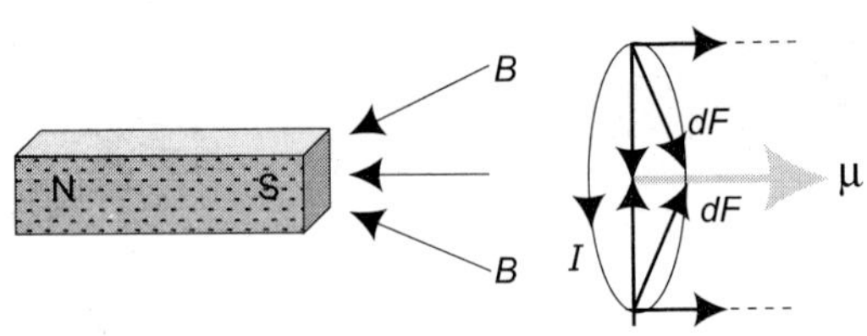

Fig. 8.73(a) The magnet is reversed.

The horizontal component of the magnetic force on the ring *is reversed* (it is now *away* from the magnet) as seen in Fig. 8.73b.

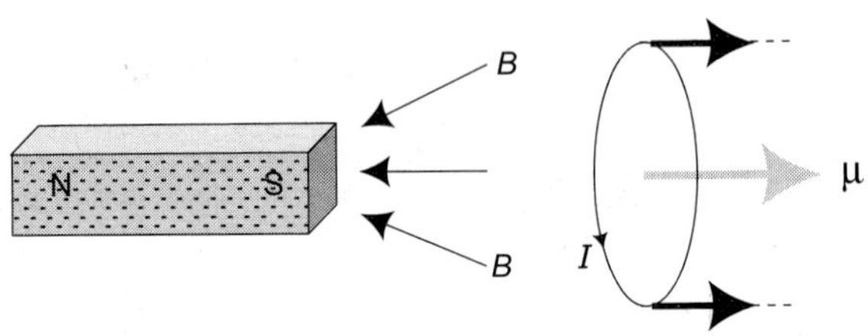

Fig. 8.73(b) Force on the magnetic moment is reversed.

We are now ready to understand how a magnet picks up a paper clip. Refer Fig. 8.74a.

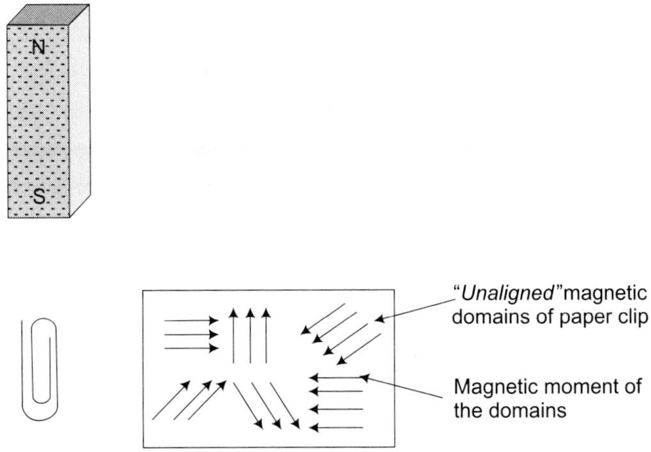

Fig. 8.74(a) The unaligned magnetic domains of a paper clip.

As shown in Fig. 8.74a, initially, the paper clip is *not* magnetized and all the magnetic domains inside it are pointing in random directions. The net force on the paper clip by the magnet is "*zero*".

However, the field of the magnet causes a torque on the magnetic moments of the iron atoms, and the domains tend to *align* with the magnet's field and are shown in Fig. 8.74b.

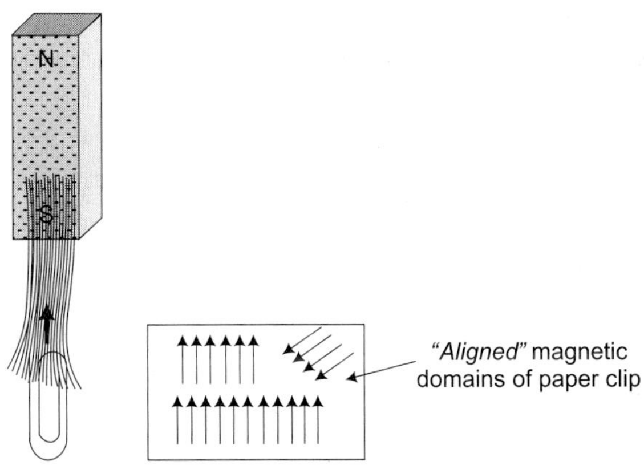

Fig. 8.74(b) A magnet picking up a paper clip.

Now there is a net force on the paper clip by the magnet. When the magnet is close enough for the magnetic force to equal the gravitational force, it can *pick up* the paper clip.

According to leading physicists, it is not known "who" does the work when magnets pick up metallic paper clips.

Questions

1. With reference to a loop of radius 'a' carrying a current I, define the magnetic moment of the loop and give its magnitude and direction as a vector.
2. Given the situation of a current carrying loop as shown in the figure below, state whether the loop will flip or not.

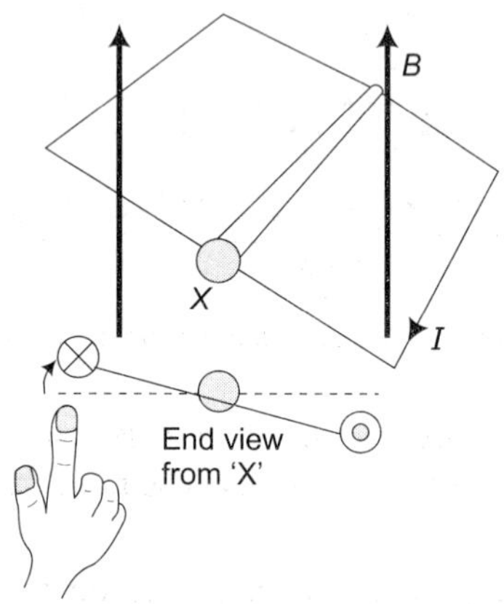

3. What is the magnetic moment of a magnet due to? Explain the action of magnetic needles placed in magnetic fields.
4. Derive an expression for the torque developed by a current-carrying loop placed in a magnetic field.
5. With the help of neat sketches showing a rectangular current-carrying loop in a magnetic field that develops a torque whose direction changes, explain how a unidirectional torque can be produced.
6. Describe the operation of a dc motor.
7. Explain what is meant by "load torque" and how this influences the back emf and the current consumed by a motor.
8. Describe some common industrial load torques.
9. Derive an expression for the potential energy of a magnetic moment for a rectangular loop (length h and width w) placed in a magnetic field of strength B. Determine the position the moment, if left free to rotate, will assume where the potential energy is a minimum.
10. How do magnets exert a force on a current carrying loop? Explain with the help of sketches. Who does the work when a magnet is used to pick up paper clips?

9

The Role of Surface Charges in Diodes; Formation of a p-n Junction; a Diode in Forward and Reverse Bias; Why V_o can Never be Made Zero; Rectification

While questioning students I found that there were several gaps in their understanding of the operation of pn-junctions and diodes and this is addressed in the sections of Chapter 9. Initial sections provide descriptions of charge movement processes after p-n junction formation followed by a description of the potential and kinetic energies in atoms. This is then followed by energy transfer processes at junctions, the appearance of the potential barrier and the behaviour of a junction to applied forward bias voltages. Two sections are devoted to explain how the potential barrier V_o of a junction cannot be made zero by a forward bias with a worked example.

The last sections are devoted to a description of the use of junction diodes for the rectification of ac sinusoidal voltages to dc.

9.1 SURFACE CHARGES IN A SIMPLE CIRCUIT WITH A DIODE

We review the development of surface charges in a simple circuit consisting of a battery, wires and a resistor. This is shown in Fig. 9.1.

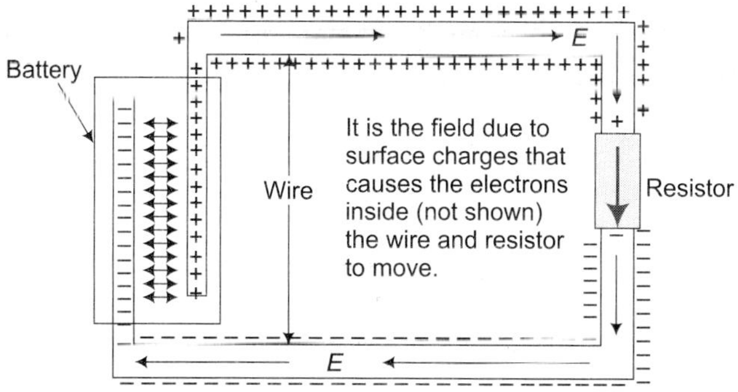

Fig. 9.1 Development of surface charges.

We now add a diode to the circuit and retain the resistor to restrict the overall current. The initial establishment of surface charges takes a few nanoseconds. This is shown in Fig. 9.2.

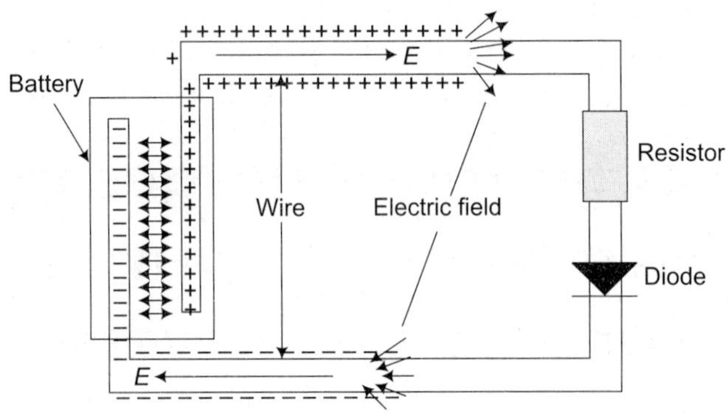

Fig. 9.2 Development of surface charges–intermediate stage.

In the final stage of development a strong electric field is set up in the resistor and the field across the junction in the diode is reduced (which we shall describe in a later section) as shown in Fig. 9.3 (not forgetting that an electric field is an invisible *alteration* of space).

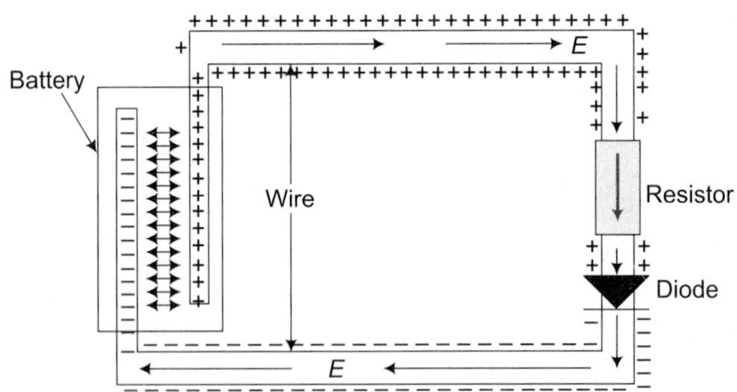

Fig. 9.3 Development of surface charges–final stage.

A current is then established and the circuit reaches steady state.

9.2 RAISING THE TEMPERATURE OF A SEMICONDUCTOR BAR

To understand the processes that enable a diode to conduct when it is forward biased and to block current when reverse-biased we will begin by studying how a semiconductor conducts. A silicon bar at zero kelvin is an insulator as shown in the visualization in Fig. 9.4.

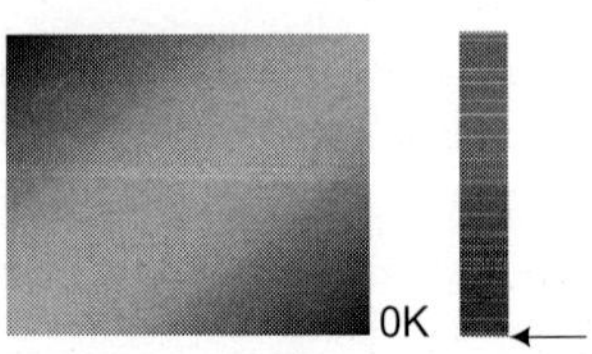

Fig. 9.4 A silicon bar at 0K.

However, when its temperature is **raised** above zero kelvin it turns into a **beehive of activity** as shown in the visualization in Fig. 9.5. To gain an intuitive understanding of temperature, please read "What is temperature" in Appendix B.

There is a **net** concentration of carriers at any instant of time above zero K.

The laws governing the distribution in energy of the electrons in the semiconductor are given in several textbooks and make use of the Fermi-Dirac probability function. For notes on the necessity to use probability theories and statistical methods that predict energy densities of electrons please see Appendix B "Quantum mechanics, probabilities and orbitals of electrons".

As the temperature is raised above 0K, the binding forces of the atoms—the covalent bonds break and make and Electron Hole Pairs (EHPs) are continuously generated and annihilated—and there is a net concentration of carriers.

There is always a net concentration at any instant of time—maybe not the same set of EHPs but there is *always* a net concentration as seen in the visualization in Fig. 9.6.

Fig. 9.5 A silicon bar when the temperature is raised.

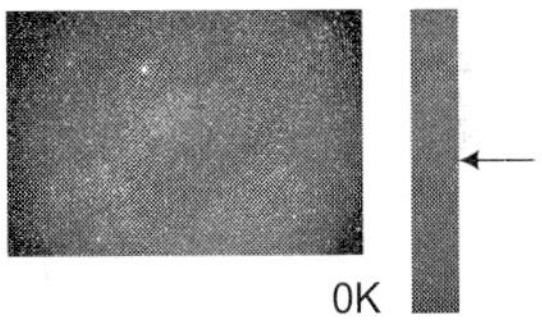

Fig. 9.6 Above 0K a net concentration of electron-hole pairs always exists.

9.3 CREATING A P-N JUNCTION

Since the 1880s, physicists were searching for means to control current direction in electrical circuits with a view to realize a device that could convert alternating voltages into direct voltages and block currents in circuits at will. The first such invention was the vacuum tube diode in 1904 by an American, John Ambrose Fleming. The vacuum diode relied on the priniciple of thermionic emission (see Appendix B "Thermionic Emission") for its operation.

A vacuum diode contains a filament sealed in a glass envelope which has been evacuated of all air (vacuum). When heated, the filament releases electrons into the vacuum creating a negatively charged cloud or space charge. These electrons will be drawn to a metal plate inside the envelope, if the plate (also called the anode), is positively charged relative to the filament (or cathode). The result is a flow of electrons from filament to plate.

This cannot work in the reverse direction because the plate is not heated and does not emit electrons. The device operates as a diode: a device that conducts current only in one direction.

The tubes were critical to the development of electronics and do still provide service in radio broadcasting stations but were very bulky and fragile and proved cumbersome especially in radio receivers that needed to be portable. That was soon to be changed after the Second World war with the discovery of semiconductors.

In their attempts to realize diodes using crystalline materials, physicists realized the impossibility of making the positive lattice ions of materials to move and that could have properties of controlling current direction!

The p-material and n-material

To produce a material of only positive mobile charges, one possibility was to produce a p-type material by cleverly doping a semiconductor bar with trivalent impurities. Such a material could not be totally devoid of conduction electrons but that was a small price to pay for realising vacant valence band states that could be filled by an electron supplied from an n-material touching it.

This electron that filled such a valence band state could, on application of a properly directed electric field "hop" along to a neighbouring say, silicon atom at the valence energy level, and not at the conduction energy level and in a similar fashion to another silicon atom and so on. The hole attracts an electron from a neighbouring sillicon atom's valence state leaving a hole exposed in the silicon atom which, in turn, attracts another electron from its neighbouring silicon atom thereby generating an apparent movement of positive charge in the opposite direction, or a movement of the hole. This action of an electron filling the vacant state is called *recombination*. Recombination is the merging of a free electron and a hole. Presumably, the term recombination was used for EHP annihilation in intrinsic material where the process occurs all the time.

The term *recombination* is used even when a neutral trivalent impurity atom creates a positivity (or empy valence band energy state) in the intrinsic semiconductor lattice that attracts an electron, which it never really lost in the beginning, for combination. *Recombination*, in a p-n junction diode is the important *link* between the free electron movement at the conduction band energy level and the *seemingly* opposite movement of holes (because positive charges or ions cannot move) at the valence level, that we shall see in Section 9.6. This maintains current conservation in a p-n junction diode and enables its rectifying properties. Remember only electrons are mobile and not lattice ions.

n-material is made by doping a semiconductor bar with pentavalent impurities that results in a profusion of conduction band electrons.

The movement of conduction band electrons that are attached to the lattice with a very weak coulombic electric force of an n-type material, cannot be construed as a positive charge movement in the opposite direction because, the conduction band electrons are everywhere, being *free*, and do not allow the creation of positive pockets. However, in the p-type zones wherein the electron attached to the lattice with a stronger coulombic electric force, hops at the valence level from one positive pocket to the next (nearer to the lattice ion), we can surely say there seems to be a positive charge or hole movement in the direction opposite to the valence electron movement.

The making of p-material and n-material created a milestone in the realization of a device, the p-n junction diode, that could block current in one direction and allow current in the opposite direction using semiconductor material that could operate at room temperatures and in compact dimensions. With its invention, equipment for telecommunication, radio, television, computers, medical electronics and industrial control were no longer the same. These became compact, smarter and more affordable.

To create a p-n junction, **neutral** atoms of dopants are added to a silicon or germanium bar as shown in Fig. 9.7: first, with a mask on the n-side when the p-dopants are added to the bar and then, with a mask on the p-side when the n-dopants are added to the bar. The fabrication of the p-n junction described is oversimplified and the methods adopted for manufacturing junctions are described in several textbooks[30].

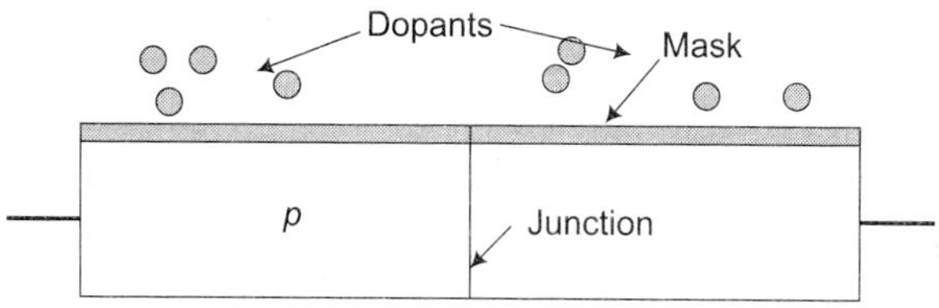

Fig. 9.7 Creating a p-n junction.

Immediately the mobile charges on either side of the p-n junction diffuse to the opposite side as shown in Fig. 9.8. "Diffusion" describes the spread of particles through random motion from regions of higher concentration (e.g. holes in p-region) to regions of lower concentration (n-region where very few holes are present).

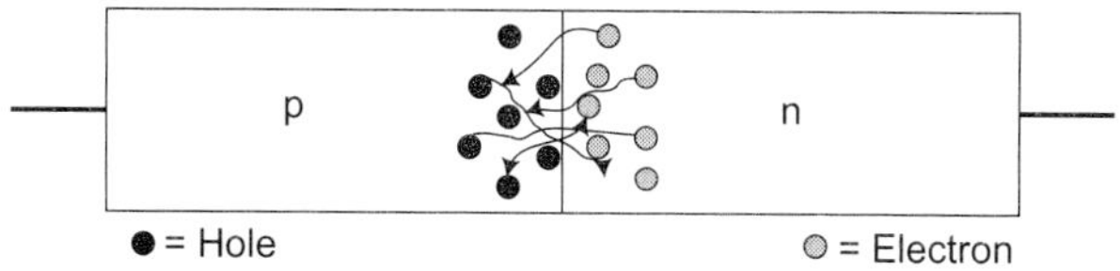

Fig. 9.8 Electrons move to the p-side.

In reality, it is the electrons that move to the p-side (the n-side electrons stretch far and break to recombine with P-side holes) exposing (immobile) lattice positive charges on the n-side. The electrons in excess there now, create a negative potential (at the p-side).

Why do the electrons move to the p-side and holes to the n-side?

Whenever a semiconductor is non-uniformly doped, or junctions are formed between different semiconductors which are crucial to the various types of electronic and optoelectronic devices made using semiconductors, an important concept should be established here regarding the demands of equilibrium, when there can be no net current in any direction.

That concept is that "*no discontinuity or gradient can arise in the equilibrium Fermi level* E_F".

Imagine having two steel cylindrical glasses of water at different temperatures, and because water is a liquid, the temperature in each glass can be the same everywhere at a molecular level; quickly, either by stirring the water, or, slowly, by the process of natural convection.

When you heat up a pan of water, the heat from the stove causes the molecules in the pan to vibrate faster causing the pan to heat up. The heat from the pan causes water molecules to move faster and heat up. So, when you heat something up, you are just making its molecules move faster. Heat is the total amount of energy possessed by the molecules in a piece of matter This energy is both kinetic energy and potential energy.

If a hot high-energy atom or molecule comes into contact with a cool low-energy atom or molecule, the excited atom or molecule will lose some of its energy to the cool atom or molecule. The two atoms or molecules will settle into an energy level that's between where they each started out. That level is called Thermal Equilibrium.

Now, when the glasses are brought into contact, their energies cxchange and then, the temperatures will equalize—the energy exchange stops when, there is no discontinuity in their energy level. The temperature is the same in both the glasses and thermal equilibrium is attained.

Within a semiconductor, which is really a crystal, there can be no evening out of the energies of individual electrons (potential energies, really); their energies are spread over the acceptable levels with a distribution, or in short, they possess distributed energies.

We cannot stir up the semiconductor material and make the energies of the electrons same everywhere as we could do with water molecules in the steel glasses. See Appendix B on how electrons occupy specific energy levels, with fixed potential and kinetic energies.

Therefore, we need a *characteristic* energy or a *representative* energy for these distributed energies and we call that the Fermi-level E_F. Consider two materials in intimate contact such that electrons can move between the two as shown in the figure below.

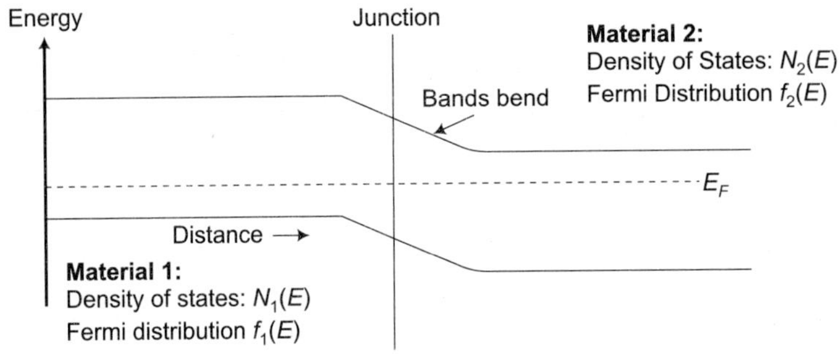

For the two materials in intimate contact forming a *junction*, then, at equilibrium, since the net motion of electrons should be zero, the equilibrium Fermi level must be constant throughout; else, it will result in a motion of the electrons, until the Fermi level becomes constant again.

Since nature changes gradually and continuously in space and time, there is a local change in the energy of electrons at a semiconductor junction, due to space charge effects. This exchange of charge *redistributes* the energies of those charge carriers which have been exchanged, giving a curvature to the energy vs. distance diagram near the junction and the bands appear to bend when the Fermi levels equilibriate as seen in the figure above. The general shape of the charge distribution results in the formation of a transition region at the junction which consists of immobile ions, negative in the p-side because of electrons *filling* the neutral acceptor or p-type atoms and, positive in the n-side because of electrons *leaving* the neutral donor or n-type atoms.

9.4 TRANSITION REGION

Electrons *away* from the junction on the n-side, *shift less* exposing lattice positive charges and, at the junction they fill the nearby holes in the p-side exposing a negativity there. These actions result in the production of an electric field that has maximum strength at the junction which decreases progressively away from the junction. The region where the immobile charges (these are positive ions on the n-side because neutral pentavalent atoms lost electrons and negative ions on the p-side because neutral trivalent atoms gained electrons) lie exposed is called the *transition region* as shown in Fig. 9.9, wherein a "dipole" is formed and there is an electric field pointing to the p-side from the n-side.

This field may be contrasted with the electric field between the plates of a charged capacitor that is *uniform* in the space between its plates caused by tight and closely spaced charges on its plates. In a p-n junction, the immobile positive and negative ions on either side of the junction are typically a hundred atoms in width and cause the electric field due to these to *vary* with distance as shown in Fig. 9.9[17 and 2 (2010 edition)]. Visit http://jas.eng.buffalo.edu/education/pn/biasedPN2/BiasedPN2.html or http://web.eng.fiu.edu/~npala/Applets/education/pn/biasedPN2/BiasedPN2.html for an animated interactive depiction of the electric field in the p-n junction.

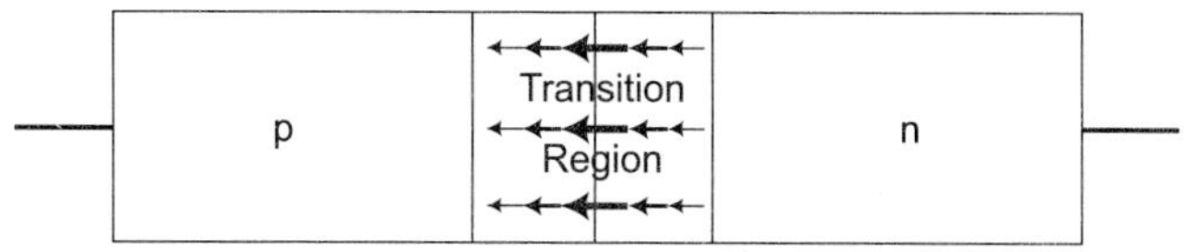

Fig. 9.9 The transition region of a p-n junction.

The voltage associated with this electric field is called the electrostatic potential barrier V_o because its direction is such that it opposes the further flow of mobile charges. This is shown in Fig. 9.10.

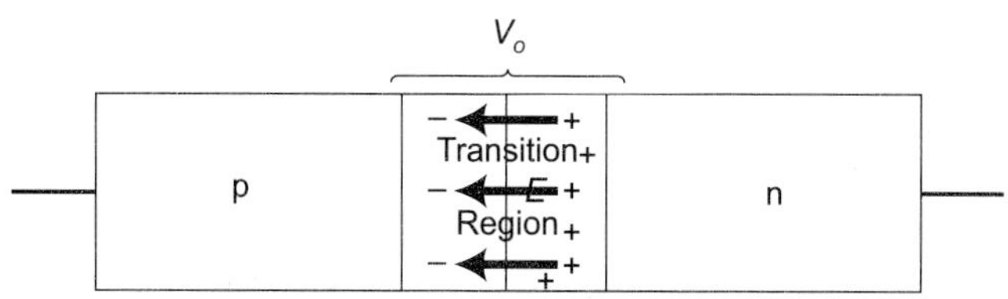

Fig. 9.10 The electrostatic potential barrier V_o.

The **electric field** E in turn causes a drift current of holes from the n-side (from thermally generated EHP pairs in the n-side) and electrons from the p-side (from thermally generated EHP pairs) to the opposite side as shown in Fig. 9.11.

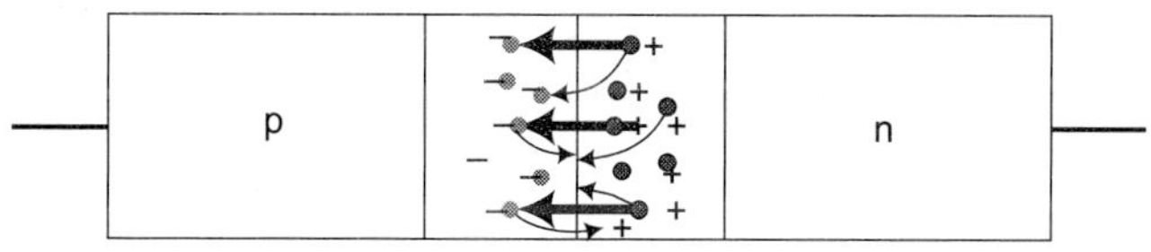

Fig. 9.11 The drift current in the transition region.

Notice that the drift current is opposite to that of the diffusion current and is of an equal magnitude so that the net current of the open-circuited diode is zero; otherwise, charge will continue to pile up and the diode will become a live wire!

9.5 DYNAMIC EQUILIBRIUM IN AN OPEN-CIRCUITED P-N JUNCTION

If we assume that the temperature is *not* anywhere near absolute zero and is above it, the diffusion current will always remain and so will the production of the electric field and the resultant drift current, all in balance, also called "dynamic equilibrium" as shown in Fig. 9.12.

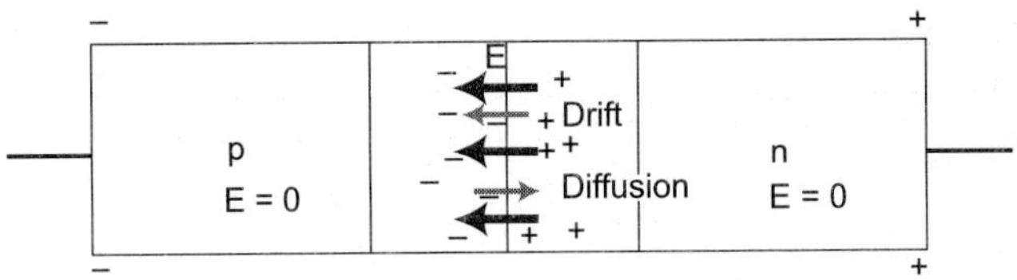

Fig. 9.12 Dynamic equilibrium in the open-circuited p-n junction.

The diffusion current which is a *forward* current (from p to n) is due to the *mobile* holes and electrons in the doped p- and doped n-regions respectively that would have gained sufficient kinetic energy from random thermal collisions (the only source of energy being the warm surroundings for the atoms in an open-circuited diode) to move through the transition region *against* the electric force. The drift current which is a *reverse* current (from n to p) as a result of an electric field is due to *thermally generated* holes and electrons in the n- and p-regions respectively that have survived long enough to wander near the edges of the transition region and to be swept across it by the electric field.

Note that we have not applied any voltage nor made any connections to the diode. It is open circuited yet the diffusion current will always remain and so does the production of the electric field and the resultant drift current "dynamic equilibrium". There are two reversible processes at work here; diffusion and, drift (Section 2.24).

There are no mobile charges in the transition region. This is why the diode has a high resistance end to end when left open-circuited. In equilibrium, the net electric field in the material outside the depletion layer or transition region of an open-circuited p-n junction is zero. This is so because as long as there is a nonzero electric field, the mobile charges (holes in the p-region and electrons in the n-region) will move in a direction to polarize the material in such a manner that will lead to a zero net electric field. This is another example of the fact that inside a conductor in equilibrium the net electric field must become zero through polarization shown by the tiny charges on the outer surfaces of the junction in Fig. 9.12. The tiny amount of charge on the outer surfaces plays no significant role in the behaviour of the p-n junction, so we will omit showing these charges in later diagrams.

We can create an imbalance in the two types of current, the forward and the reverse currents, to tilt in favour of the forward current by altering the potential difference across the diode with a battery. When a diode is forward biased, the thermal generation drift current (or *reverse leakage current*) is very tiny and still in the *reverse direction* to the forward diffusion current which will become large in a forward biased diode as we shall see in Section 9.6.

9.6 A DIODE IN FORWARD BIAS AND RECOMBINATION

When a voltage is applied to forward bias a diode, surface charges are arranged in a gradient and these produce a resultant electric field inside the diode developing a voltage across the diode. This is shown in Fig. 9.13.

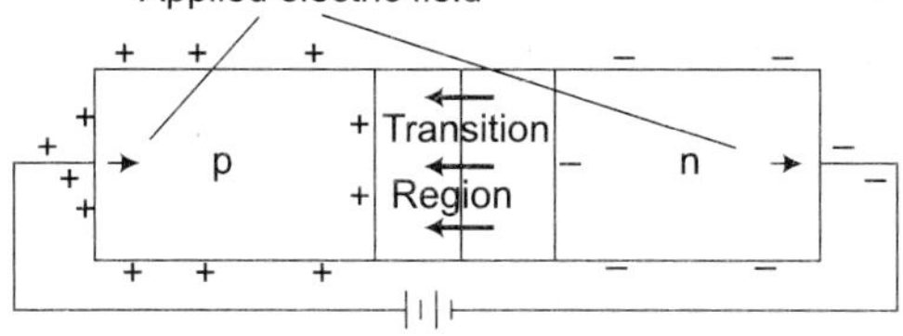

Fig. 9.13 A diode in forward bias.

The electric field in thc transition region is the net (resultant) field due to the applied field and the electrostatic potential barrier V_o (which is lowered because some of the immobile charges are neutralized or the stretch of the dipoles is reduced due to the voltage source) and is shown by the smaller length arrows in the transition region compared with their lengths in Fig. 9.12.

When a forward bias voltage V that is *less than* V_o (the potential barrier voltage), is applied to a diode, the *transition region width* is reduced and the holes in the p-side *gain* a tremendous amount of *kinetic energy*. The electric field in the neutral regions is very small (shown by the small arrows *near the diode contacts* in Fig. 9.13) and when deriving the diode equation assumed ideal, the electric field in the neutral regions is *considered to be zero* and the applied voltage is assumed to drop entirely across the junction. If a field and associated voltage drop were to be taken into account in the neutral regions, the diode equation becomes complex and introduces a complexity that is really needless when the relation derived without this field is fairly accurate.

With their large kinetic energy, the *holes overcome the electrostatic potential barrier and swiftly begin crossing* the junction as shown in Fig. 9.14.

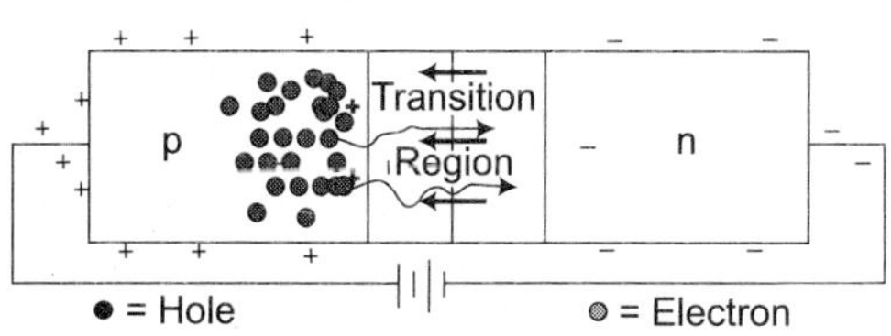

Fig. 9.14 Holes from p-side cross the junction.

It may be noticed that the holes seemingly are flowing against the electric field in the transition region. Their *large kinetic energy* enables them to *overcome* the potential barrier that was lowered by the applied forward bias voltage within the transition region. This is shown in Fig. 9.15. The process is one of drift within the p-region and within the transition region and, diffusion at the edge of the transition region near the n-region and a little into the n-region.

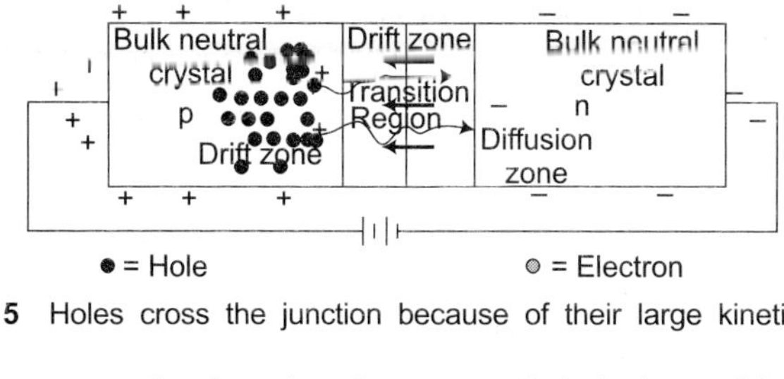

Fig. 9.15 Holes cross the junction because of their large kinetic energy.

The **electrons** also cross the junction because of their large kinetic energy due to the forward bias as shown in Fig. 9.16. It may seem here too that the electrons are flowing against the potential barrier that was lowered by the applied forward bias voltage in the transition region. They have sufficient kinetic energy to *overcome* the potential barrier *within the transition region*. The process is one of drift within the n-region and within the transition region and, diffusion at the edge of the transition region near the p-region and a little into the p-region.

The high resistance of the transition region which was earlier devoid of mobile charges is seemingly brought down by the presence of the carriers crossing over from either side due to the forward bias.

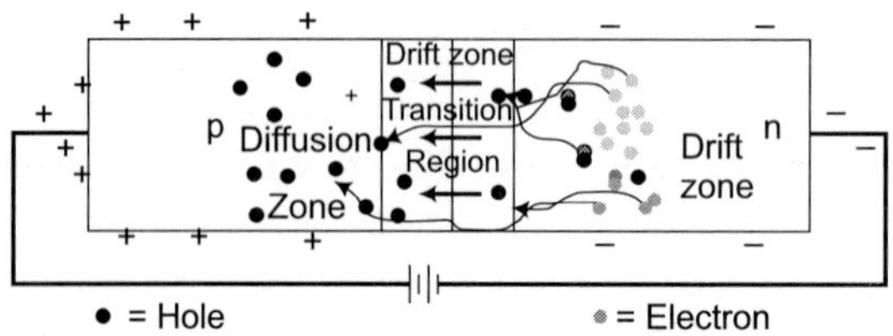

Fig. 9.16 Electrons also cross the junction because of their large kinetic energy.

The holes contribute to conduction when they are moving in the bulk p-regions and across the transition region and a little into the n-region, when they are not recombining with valence electrons.

For an animation of how the holes move in the lattice see http://www.ece.utep.edu/courses/ee3329/ee3329/Studyguide/Shockwave/Fundamentals/Demos/acceptor.html.

After a free electron leaves the negative terminal of the battery, another enters the right end of the diode. Another electron from the n-region reaches the junction. When the battery voltage is greater than a few tenths of a volt, the free (conduction band) electron has enough energy to crossover the depletion layer.

Recombination is the mechanism by which the conduction band electrons (n-side) can fill the valence band holes (p-side). This makes available valence electrons (transition region, p-side and a few in the n-side) that enable the continued movement of holes to the right in the p-side and a little into the n-side as shown in Fig. 9.16 and 9.17a.

In Fig. 9.17a, the *moving* free electron for the sake of clarity is shown with a '•'.

We represent each acceptor (trivalent) atom with a circled minus sign and the plus sign outside is the hole in its valence orbit. Similarly, we represent each donor (pentavalent) atom with a circled plus sign and the minus sign outside is the *free electron* it contributes to the semiconductor. Each piece (both p-and n-type) is electrically neutral because the number of pluses and minuses is equal.

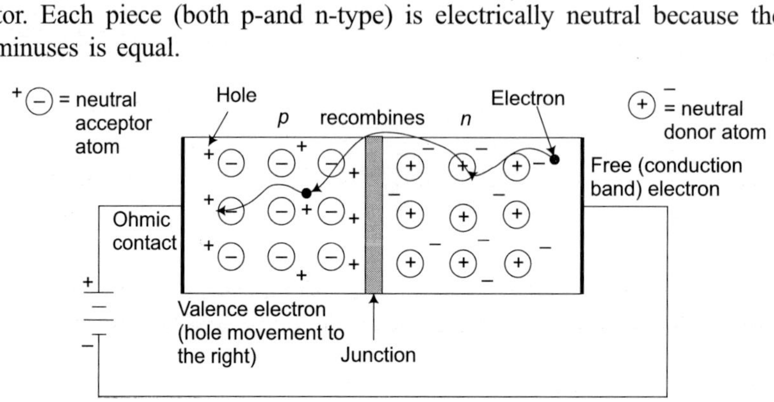

Fig. 9.17(a) Recombination in a forward biased diode.

Forward bias lowers the potential barrier at the junction and gives free electrons more energy equivalent to higher energy levels. Fig. 9.17(b) gives a comprehensive view of the movement of carriers in a forward biased diode.

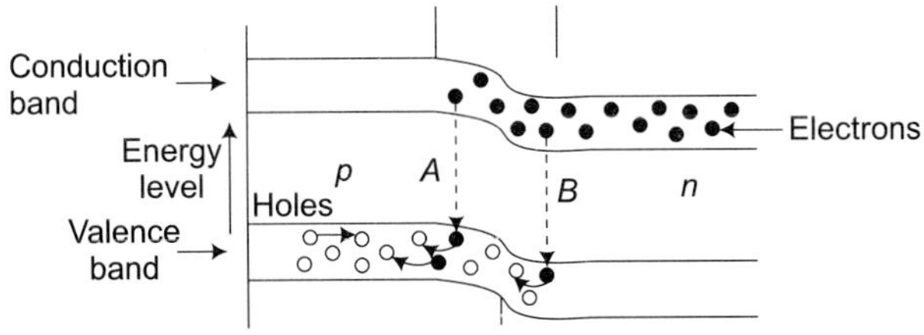

Fig. 9.17(b) Forward bias and the important role of recombination.

Soon after entering the p-region, the free electrons fall into holes (path *A*) due to mutual attraction between them. As valence electrons, they continue moving towards the left end of the crystal; this is equivalent to holes moving towards the junction. Some holes penetrate the n-region as well due to the lowered potential barrier as shown in Fig. 9.17(b). In this case, conduction-band electrons can follow recombination path B.

Regardless of where the recombination takes place, the result is the same. Free electrons move in a steady stream towards the junction and fall into holes to which they are attracted, near the junction. The captured electrons (now valence electrons) move left in a steady stream through the holes in the p-region. The movement of electrons in the n-region is similar to the movement of electrons described during conduction in metallic conductors in Section 1.18. There are millions and millions of electrons and holes in the n- and p-regions that participate in the process of conduction. The electron that leaves the p-terminal (having left a hole in the p-lattice) to drift towards the battery positive terminal is different from an electron that fills a hole at one point in the p-region and which is different from another electron that fills a hole at another point in the p-region and so on at millions of locations in the p-region. The overall snapshot at every instant during forward bias in the steady state is of millions of electrons filling holes throughout the p-region of the diode at various cross sections of the p-region resulting in an apparent movement towards the right in Fig. 9.17 of holes towards the junction and into the n-region. Hole movement in the p-region thus caused by the action of millions and millions of electrons filling holes and creating holes in the atoms they left is observed only within the p-region and extends slightly into the n-region at the junction.

In this way, a seemingly *continuous* flow of electrons (to the left in Fig. 9.17(a)) through the diode is observed.

In a forward biased diode, there is ever present a thermally generated *reverse leakage* current that was described in Section 9.5 and this is very small compared to the forward current due to the mobile holes and electrons that forms the forward current in a forward bias. The *reverse leakage* current will be manifest when the diode is reverse biased as will be described in Section 9.11.

Note that there are interface charges at the boundary of the transition region and the p and n layers to produce the necessary driving electric field that lowers the barrier as shown in Fig. 9.18.

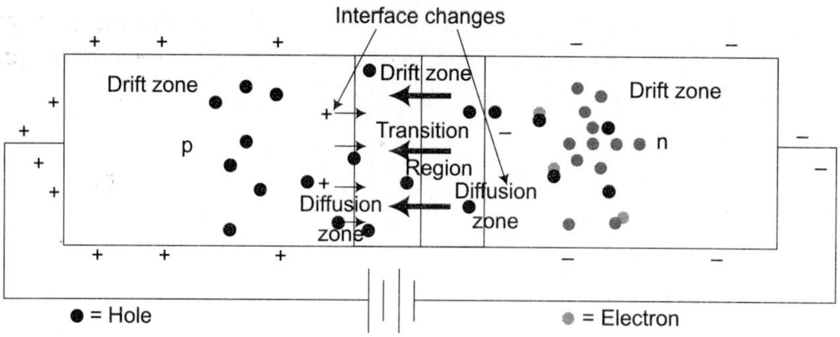

Fig. 9.18 Interface charges at the boundary of the transition region.

It may seem here too that the electrons are flowing *against* the potential barrier in the transition region; they cross-over due to their large kinetic energy. The holes then move by a *diffusion* process (that is statistical in nature) from a high concentration region (p-region and the transition region they find themselves in) to a low concentration region (n-region). See Fig. 9.15. The same holds true for the electrons in the n-region. They too move into the p-region by a *diffusion* process. See Fig. 9.16.

9.7 POTENTIAL AND KINETIC ENERGIES OF THE CHARGE CARRIERS OF A JUNCTION UNDER FORWARD BIAS

The distance between the conduction band edge, E_c, and the energy of a free electron outside the crystal (called the vacuum level labeled E_{vacuum}) is quantified by the electron affinity, χ multiplied with the electronic charge q as shown in Fig. 9.19.

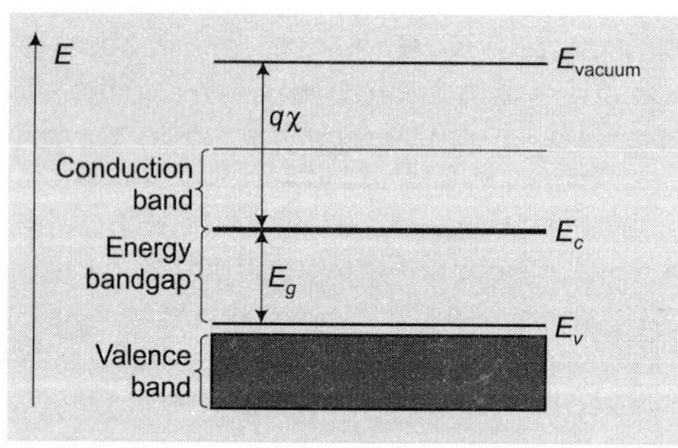

Fig. 9.19 Potential energy of an electron with reference to E_{vacuum}.

It is usually the practice to set the vacuum as the zero potential energy level. If the potential energy is set to zero at infinity, the usual convention, then bound (conduction and valence) electron states have negative potential energy (see Fig. 9.20) which means some amount of energy has to be supplied to the atom to shift an outermost electron from its orbital space.

Note that some of the electrons in the conduction band can gain energies that are positive. They possess energies *greater than* the vacuum considered zero potential energy (reference zero energy) and it is this *system* of the particles in the semiconductor and no particles in vacuum for which we compute the potential energies, as shown in Fig. 9.20.

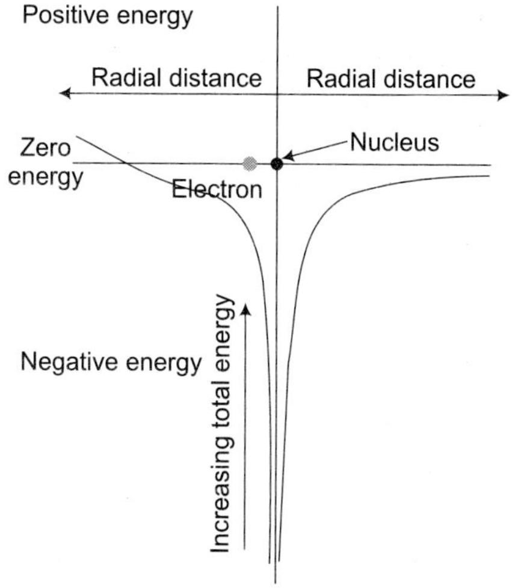

Fig. 9.20 Potential energy variation with distance of an electron from the nucleus.[19]

All potential energies have an arbitrary *zero point*: the point we designate a zero value for the potential energy.

Potential energy is relative to that datum or zero point. If work is done by external agents (heat, light and voltage) on the charge to move it against the forces of attraction from fields, its potential energy is increased with respect to that datum.

If, on the other hand potential energy is extracted or converted into kinetic energy from the charge by letting it fall with the electric force (of attraction say between electrons and lattice ions), then the potential energy gets lower and lower and possibly even below the datum (or reference) value, and hence negative.

Negative energy is a matter of convenience to represent confinement, like when the electron is confined within the atom due to its attraction by the nucleus (consisting of protons and neutrons and not including the electrons), else it should be at infinity or in vacuum, with zero potential energy! The potential energy profile for the simple case of a hydrogen atom as shown in Fig. 9.20 can be computed from e^2/r where e^2 is the product of charge of the nucleus and electron and r is the radial distance of the electron from the nucleus. The electric field that gives rise to the electrostatic force of attraction between the nucleus and electron is *not shown* in Fig. 9.20.

The potential energies (negative) of electrons are typically spread over a few tens of eV (electron volts) in conductors and semiconductors that are commonly used in the manufacture of electric and electronic components. Electron energy levels in atoms are separated by a few eV and the energy gap in semiconductors is about an eV or more and insulators have gaps of about 10 eV. As was mentioned in Section 1.20, physicists and engineers prefer the small energy unit eV (electron volt) to express electron energies over the joule. In electric and electronic circuits, millions and millions and millions of electrons in tiny energetic amounts (expressed in electron volts), work harmoniously as we learnt in Chapter 4, transporting and converting energy, which is measured in joules, which is a large unit of measure of energy.

Positive energy of electrons can result even when bound to the crystal and is *mostly kinetic* because Total E = potential E (small negative) + kinetic E (large positive) as shown in Fig. 9.21 where we see the electron gain kinetic energy and thus possesses a net positive energy.

For a description of the relationship between the energy levels and orbits of electrons please read "Energy levels of electrons" in Appendix B.

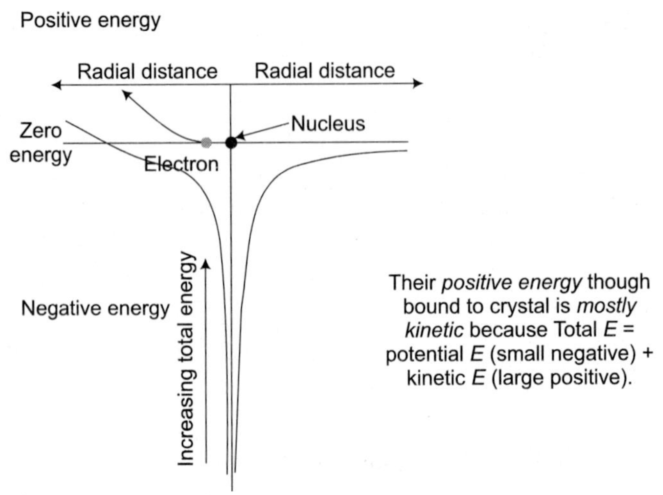

Fig. 9.21 Positive total energy of an electron.

That is how the electrons in the conduction band possess a total positive energy. This is depicted in the energy band diagram in Fig. 9.22.

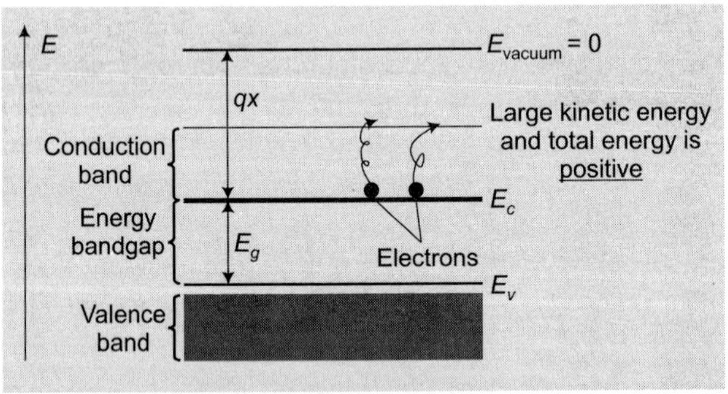

Fig. 9.22 Positive energy of an electron.

How Holes Surmount the Potential Barrier

We now see how holes in the p-region of a p-n junction surmount the potential barrier under the influence of a forward bias voltage.

When a p-n junction is left open-circuited, the **holes** in the *tail* band energies will be as shown in Fig. 9.23. There are several possible unfilled states.

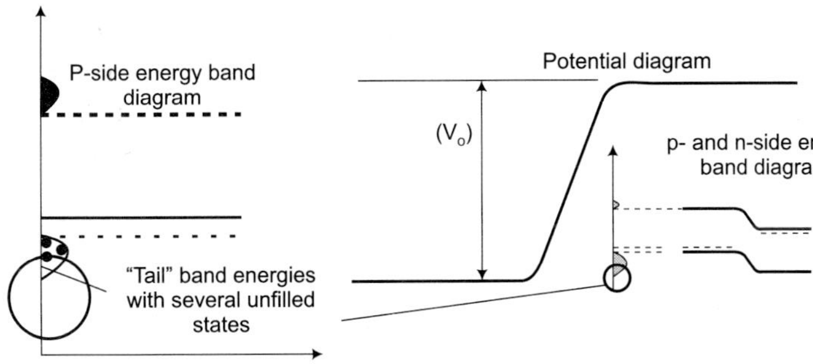

Fig. 9.23 Holes in the tail band.

When a forward bias voltage V_f with its accompanying e-field is applied then several of the unfilled states are filled and the filled energy band states are shown in Fig. 9.24. How do holes *fill* these states? When a valence (bound) electron gains energy, it moves to a higher level within the valence energy band. This is equivalent to a hole transiting *down* the vacant energy states to gain energy, thus *filling* a tail band energy and so giving the atom a stronger positive quality.

Note: $V_o - V_f$ refers to the potential *at* the transition region—the *true* junction and not in the bulk of the diode.

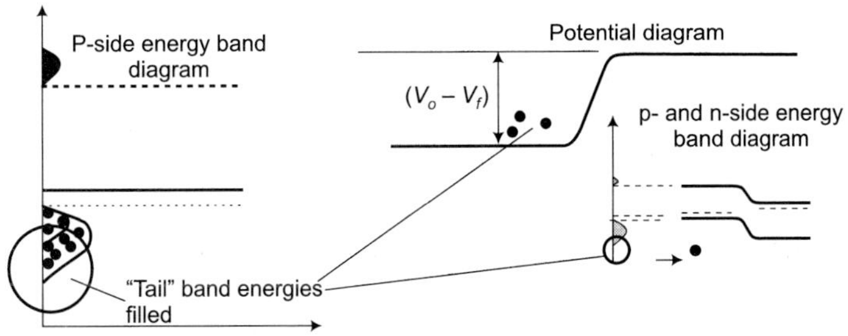

Fig. 9.24 Filled energy bands in the tail band after the forward bias is applied.

Due to their large kinetic energy the holes in the tail band energies can surmount the potential energy barrier as shown in Fig. 9.25.

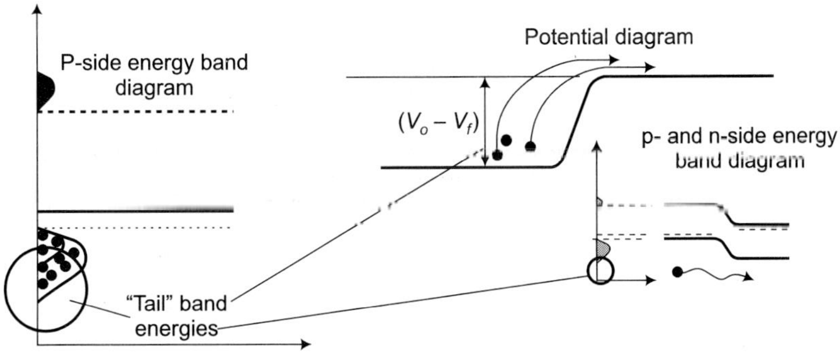

Fig. 9.25 Holes surmounting the potential energy barrier.

Intuitively, it is more instructive to visualize the holes overcoming the potential barrier in the potential diagram than in the energy band diagram and that is the reason both the diagrams are shown in the Figs. 9.23, 9.24 and 9.25. Electrons in the *tail* band energies for electrons are shown in Fig. 9.26. There are several possible unfilled energy states.

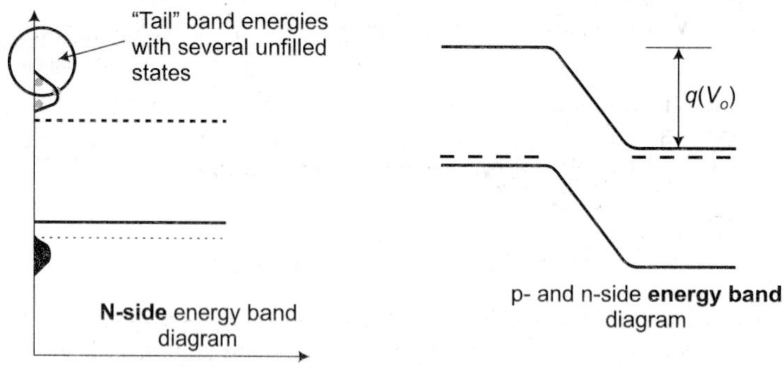

Fig. 9.26 Tail energy bands for electrons.

Similarly, electrons in the tail band energies for electrons which are filled by the effect of the applied forward bias voltage V_f and accompanying e-field, can surmount the potential/potential energy barrier due to their large *kinetic energy*. The filled energy bands are shown in Fig. 9.27.

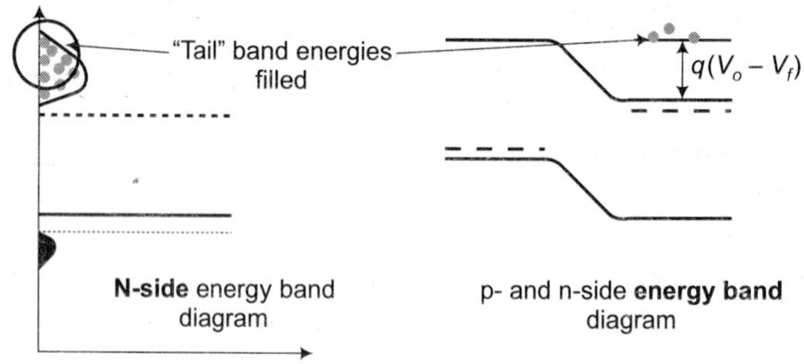

Fig. 9.27 Filled energy bands in the tail band after the forward bias is applied.

The electrons surmounting the lowered potential energy barrier with their large kinetic energies is shown in Fig. 9.28.

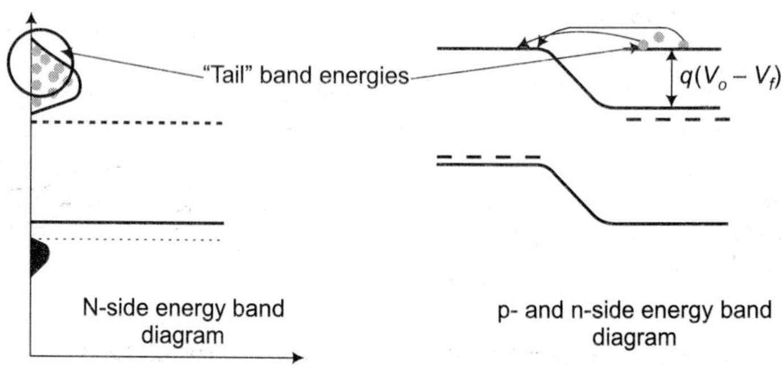

Fig. 9.28 Electrons surmount the potential energy barrier.

9.8 VOLTAGE DROPS IN A FORWARD BIASED DIODE

The potential barrier of a p-n junction can never be brought down to zero ….nor ever *reversed*, however strong the forward biasing voltage. The question then arises: *where does the applied voltage drop?*

Consider a silicon p-n junction with doping concentrations $N_A = 2.5 \times 10^{15}$ cm^{-3} and $N_D = 5 \times 10^{16}$ cm^{-3} at 300 K with the intrinsic concentration $n_i = 1.5 \times 10^{10}$. (Problem in "Electronic Devices and Circuits" by Milman, Halkias and Satyabrata Jit).[17]

Using $V_o = V_T l_n \frac{N_D N_A}{n_i^2}$, we obtain V_o = **0.703** volts at 300° K and $I_o = 9.17 \times 10^{-13}$ Amps.

Using the formula for current $I = I_o(e^{V/V_T} - 1)$, we obtain for a forward bias V of 0.6V, V_T = 0.026 V at 300 K or 27° C, current I = **10.45** milliamps. What we observe here is that a forward bias voltage of **0.6 V** which is *less than* V_o = 0.703 V has caused a current of 10mA in the diode. For a forward bias voltage **0.69 V < Vo = 0.703 V**, we obtain I = **340** milliamps.

An *increase* in forward bias voltage of 0.09 V has caused the forward current to *increase* by nearly 330 milliamps and the bias voltage is yet *less* than the barrier potential. With V = **0.7029 V < 0.703 V (V_0)**, I = **500 milliamps**. If a small diode, it will surely burn out with this current!

We assumed that the applied voltage appears across the junction entirely ignoring the voltage drops across the bulk of the p- and n-regions. It is reasonable to assume that, yet, as the applied voltage approaches V_o = 0.703 V, and the current increases, some voltage will begin to drop across the bulk p- and-n-regions and the resistance of the ohmic contacts to uphold the current conservation law while, the part of the applied voltage that lowers the barrier potential V_o, makes the voltage *across the junction close to zero* but, not equal to zero. And, the current will begin to level off before a current of **500 milliamps** flows.

Note: For an intuitive understanding of V_T please see Appendix B on the "Einstein relation".

9.9 A DIODE AND ITS OPERATION

One intuitive way to understand the operation of a diode is to visualize the operation of a **check valve** for water, which allows water to flow in one direction only. Such valves are used to prevent the downward flow of water back into a ground-level sump or reservoir when the pump that raised the water to an overhead tank is shut off. Such a pumping system is commonly found in multi-storeyed buildings. With very little or no fluid flow the valve is held closed by the ball pushed against the seat by the spring and water cannot flow to the other chamber. The spring displacement maybe likened to V_o, the potential barrier in the transition region of a diode as shown in Fig. 9.29a.

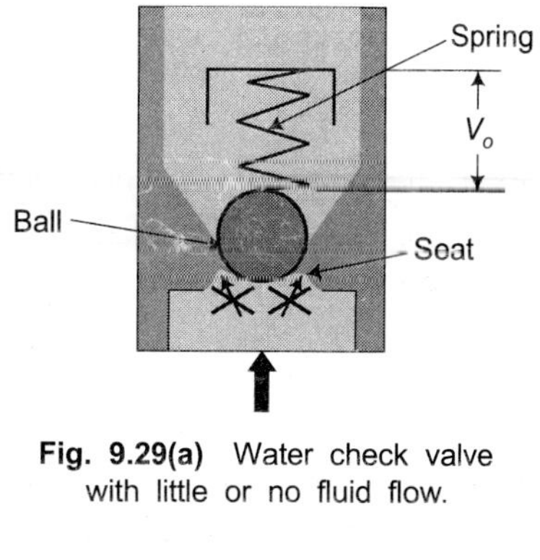

Fig. 9.29(a) Water check valve with little or no fluid flow.

With a large fluid flow, the **ball** is raised by the water pressure and this enables the water to flow into the other chamber.

The spring is compressed and the *reduced* spring displacement may be likened to $V_o–V_f$ the *lowered* potential barrier in the transition region of a diode as shown in Fig. 9.29b.

With large flows $V_o–V_f$ is small. With small flows $V_o–V_f$ is large.

For operation as a check valve, a **spring** is necessary. For operation as a diode, a **p-n junction** is necessary. As long as the valve is operating, the spring will **always** have a displacement, however small.

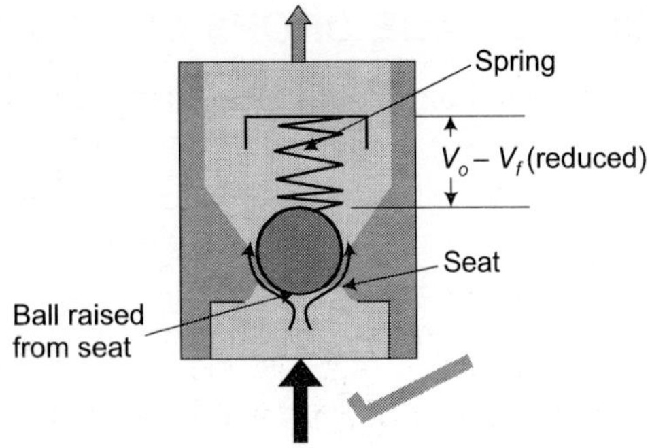

Fig. 9.29(b) Water check valve with large fluid flow.

Similarly, as long as the diode is operating the junction will always have a barrier potential, V_o, however small.

One of the *vexing* questions that students seek an answer to is, how much is $V_o–V_f$? For large applied forward voltages? Greater than V_o? say, greater than 1 V?

The answer : V_f *adjusts* itself to a value such that the conservation of current law is upheld. And this will not make $V_o = 0$. In other words, $| V_f | < | V_o |$. The bulk of the applied voltage in case of large forward biasing voltages, drops in the bulk of the diode material ....***away*** from the transition region. See the worked example in Section 9.8.

V_f *adjusts* itself to the value that results in the sum of the large diffusion current and the small drift current *in the transition region* equalling the drift current *in the bulk region where most of the applied voltage is dropped.*

Any deviation will cause a pile-up of charges and this will automatically cause a correction of V_f to the right value so that the current is uniform around the circuit.

That is the reason why we see the *V-I* characteristics of diodes plotted in data sheets and electronic manuals only upto about 0.6 V to 1V for silicon diodes and seldom beyond (See Fig. 9.41). And if applying voltages, say, 1V, a current limiting resistor is added in series with the diode.

When the forward bias voltage approaches V_o (typically about 0.6 to 0.7 volts for a silicon diode), the current will become arbitrarily large.

As a practical matter we can never reduce the barrier to zero, because as the current increases without limit, the bulk resistance of the crystal, as well as the resistance of the ohmic contacts, will limit the current and it is no longer possible to assume that all the voltage V appears as a change across the p-n junction.

One intuitive way of understanding this, is when the forward bias voltage is *much less* than V_o, the current is determined or rather restricted in a manner of speaking almost entirely by the diffusion components. The majority carrier concentrations are usually large in the bulk neutral regions, so that only a small electric field is needed to drive the drift currents.

However, with forward bias voltages that approach and exceed V_o, the drift component has to *catch up with and supply* the flood of diffused carriers to uphold the conservation of current law and this consumes larger portions of the applied voltage. A significant applied electric field therefore appears in the bulk neutral regions reflecting the increased voltage drops to sustain the large currents in these regions. This is shown in the figure below.

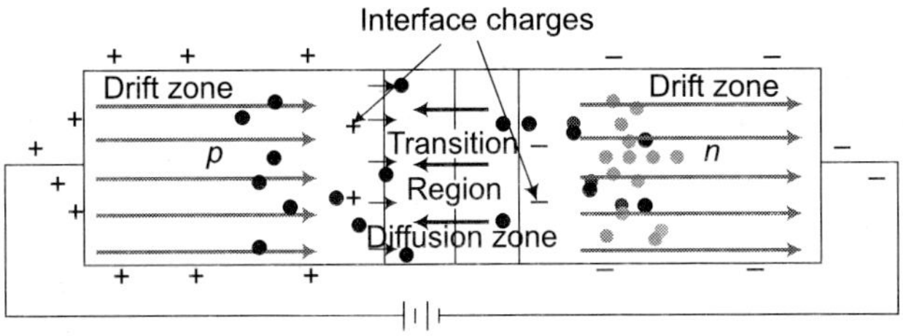

We conclude that, as the forward voltage V becomes comparable with V_o, the current through a real p-n diode will be governed by the ohmic-contact resistances and the crystal bulk resistance.

To better understand p-n junction operation, look at the check valve again. Take away the spring and ball, then, you will be left with a hollow piece of pipe.

Analogously, in the p-n junction, if there were no V_o, you would have a bar of semiconductor material of opposite polarity charge carriers across a dividing line.

A word of caution: It is possible to adjust the acceptor N_A and donor N_D levels of a p-n junction such that the potential barrier is **zero** or V_o = zero. Since, $V_o = V_T \, ln \frac{N_D N_A}{n_i^2}$ using $N_A = 1.5 \times 10^{10}$, $N_D = 1.5 \times 10^{10}$, $n_i = 1.5 \times 10^{10}$, we note that $V_o = 0$.

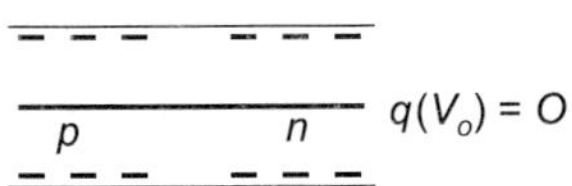

However, an **ever so slight forward bias** will trigger conduction as seen in the figure below.

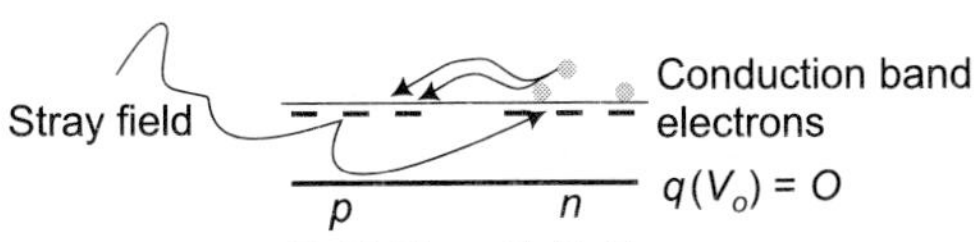

Visit http://jas.eng.buffalo.edu/education/pn/pnformation3/index.html or
http://web.eng.fiu.edu/~npala/Applets/education/pn/pnformation3/index.html
and enter the n_i, N_A and N_D values to check that V_o is indeed zero. This can achieve an ideal diode characteristic; when forward biased, the diode will begin conducting with a "zero" voltage drop and block reverse current with the application of a non-zero reverse bias voltage.

There is no change in the processes of the diode with ideal characteristics both in forward and reverse bias except that there is no barrier voltage for the holes to contend with and a mere voltage difference > 0 between the p- and n-sides is sufficient to trigger conduction in the forward bias and a voltage difference < 0 to block current in the reverse direction.

But there is little to be gained with such an ideal diode—even in the likelihood of its manufacture being successful—like the water valve with a ball *but without a spring*.

In noisy environments, circuits using diodes with such ideal characteristics will be so susceptible to interfering inputs (for example, conducting at unwanted times) that they would soon become a nuisance.

For most commercially available diodes, the diode (the bulk region and the junction) in forward bias behaves like a battery of voltage Vγ (also called the *cutin* voltage, which encapsulates the barrier potential and the geometric properties of the junction) connected in *series opposition* with a resistance $R_{forward}$ as shown in Fig. 9.30. In some textbooks, $R_{forward}$ is also called the average ac resistance, r_{av}.

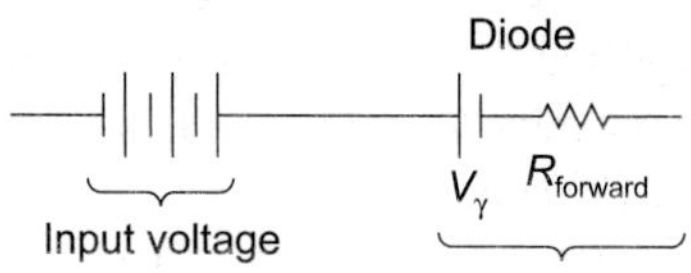

Fig. 9.30 Diode equivalent circuit.

Note that the voltage Vγ opposes the forward biasing applied voltage, whatever be the applied voltage strength. Typically, Vγ is 0.2 volts for germanium and is 0.6 volts for silicon diodes. For applied voltages < Vγ, the current is very small, and is less than 1 percent of the maximum rated value. Beyond Vγ, the current rises very rapidly as may be seen in the forward V-I characteristics of these diodes[17]. But remember, the barrier potential V_o of the depletion region is never reduced to zero when the diode is operated within rating. The *cutin voltage* is simply the potential where the current is drastically changed from a very low value (~0) to a high value (like a switch) and is *only* realized from the piecewise linear characteristics of a commercial diode.

When visualizing the current in forward bias, note that the movement of the holes from the p-side is a current, and, the movement of electrons from the n-side is a current, as well.

The *movement* of charges continues—moving and recombining *on and on and on*—and a *steady state* is established with the current *limited* by the *resistor* in the circuit.

Under forward bias, the diode current is due to recombination[18]. This recombination can occur within the quasi-neutral regions, within the depletion region or at the metal-semiconductor ohmic contacts. The thermal equilibrium EHP generation and recombination continues in the background but that is *swamped* by the dominant injected current components. The generation current becomes manifest when the diode is reverse biased and has swept away all excess mobile charges.

For recombination to occur, electrons must lose energy to fill the holes they occupy. How do they lose energy?

Recombination of electrons and holes is a process by which both carriers annihilate each other: the electrons fall in one or multiple steps into the empty state which is associated with the hole. Both carriers eventually disappear in the process. The energy difference between the initial and final state of the electron is given off.

In the case of radiative recombination this energy is emitted in the form of a photon or a packet of visible electromagnetic energy (light). This band-to-band transition is typically also a radiative transition in direct bandgap semiconductors as in light emitting diodes or LEDs. This usually does not occur in rectifier diodes.

In the case of non-radiative recombination, which is more commonly found in rectifying diodes and signal amplification transistors used in electronic circuit applications, the energy difference is passed on to one or more phonons (another name for the vibrating lattice ions) in the form of heat (more intense lattice vibrations) due to a collision with the momentum conserved.

In rectifying p-n junction diodes, the level does not change by a direct transition from a conduction energy level to a valence energy level but via a *trap* assisted process much like descending to a lower floor using a few steps. This is accomplished by allowable energy levels or traps created in the band gap by impurities within the semiconductor material.

In a p-n junction diode, there is **no generation** of **additional** electron-hole pairs due to the applied forward bias voltage under low-level injection conditions; the thermal equilibrium concentrations were established and that was over when doping was completed and the junction had formed.

For a fixed forward bias, the current components therefore, are due to the **movement** of the open-circuit thermal equilibrium number of the holes (acceptor) and the electrons (donor) and are decided by the number of injected minority carriers.

The *movement* of holes and electrons continues moving and **recombining** (n-side conduction band electrons recombining into p-side valence band energy levels) ***on and on and on*** and a ***steady state*** is established. The valence electrons in the bulk p-region hop along deep into the p-side and move on towards the p-side ohmic contact and deplete the positive charges on the voltage source positive terminal. The voltage source sensing the drop in voltage commences charge separation internally and deposits electrons on its negative terminal.

It should be borne in mind that a *high level* of hole current is present only in the p-region and slightly into the n-region.

There is *no* hole current at the diode contacts or within the wires connecting it to the source of emf. This way, the circuit action is continued and a steady state is observed.

Given a forward bias, the currents will change only if the thermal equilibrium conditions were to change. For example, if the diode was allowed to heat up without provision of a heat removal mechanism, the elevated temperatures could result in additional electron-hole pair generation and the forward current for a given forward bias voltage will increase.

The currents in forward bias under constant thermal equilibrium conditions are therefore due to the *movement* of the holes and the electrons moving and **recombining** ***on and on and on***.

9.10 A DIODE AS A RECTIFIER

A diode when used as a half wave rectifier enables the production of a positive voltage when a *sine wave* is applied to it. The process of rectification is shown in Fig. 9.31.

The diode blocks out the **negative swinging** voltage of the input as seen in Fig. 9.31. A half wave rectifying circuit can be used to achieve rectification.

Such a circuit is shown in Fig. 9.32.

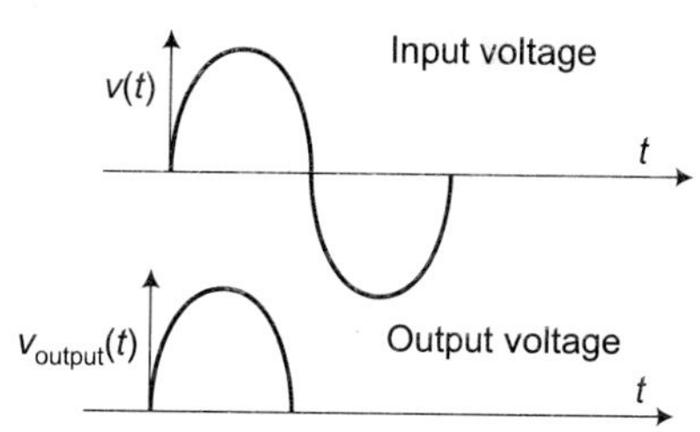

Fig. 9.31 Rectification process.

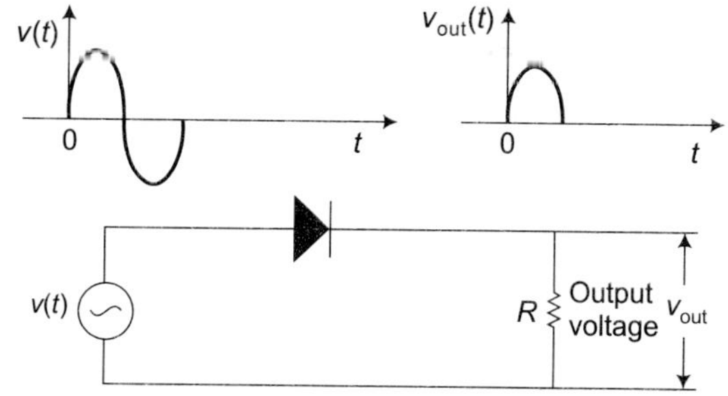

Fig. 9.32 Half-wave rectifying circuit using a diode.

During the *positive half* cycle the surface charges develop as shown in the Fig. 9.33, resulting in the output rectified voltage v_{out} and is equal to $v(t) - V\gamma$ (the cutin voltage and neglecting the drop across $R_{forward}$ of the diode) to a good approximation for all currents within rating.

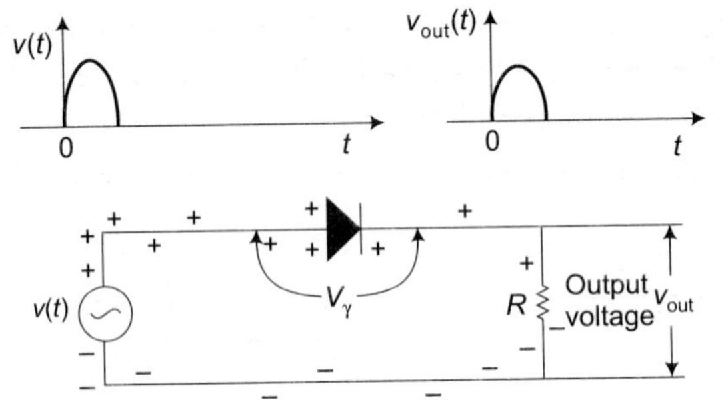

Fig. 9.33 The positive cycle applied to the rectifying circuit.

But what happens during the negative cycle?

9.11 REVERSE BIASING A DIODE

During the *negative* half cycle there is a rearrangement of surface charges.

In the negative half cycle, the surface charges cause the appearance of a reverse bias across the diode as shown in Fig. 9.34.

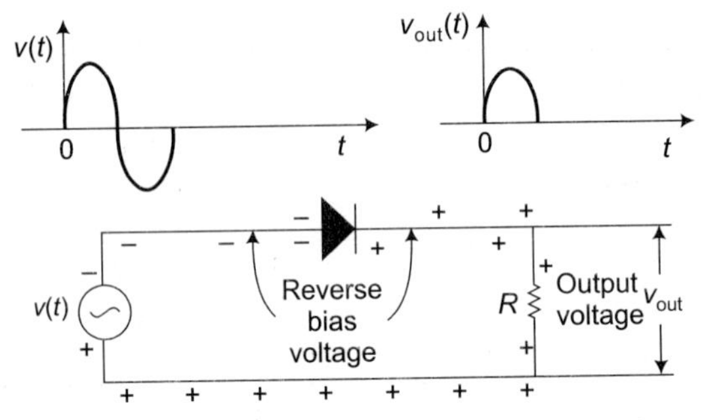

Fig. 9.34 Reverse biasing voltage during the negative cycle.

It should be borne in mind that the *maximum magnitude* of the reverse voltage across the diode is the peak negative cycle voltage of the input. This is shown in Fig. 9.35.

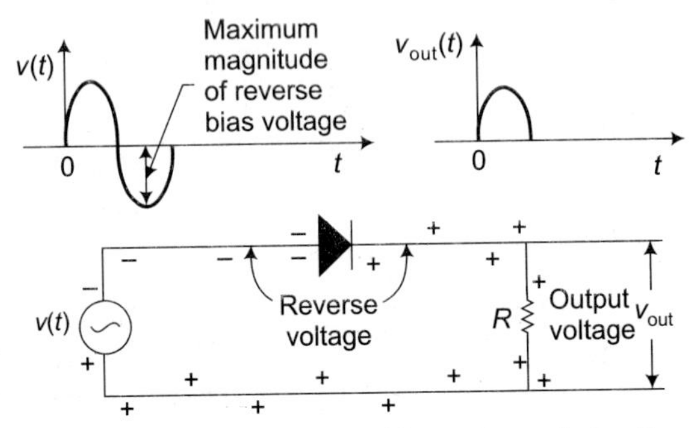

Fig. 9.35 Maximum magnitude of reverse bias voltage during the negative cycle.

Under reverse bias, the current is due to generation of electron-hole (EHP) pairs. Several mobile charges of the extrinsic p- and n-materials would have been absorbed by the reverse bias as they move away from the junction and nominally a zero current results.

Actually, a small current does flow because a small number of electron-hole pairs are generated throughout the crystal as a result of thermal energy (even room temperature is sufficient as a source of thermal energy) contributing to what we call *generation* or *leakage* current as was described in Sections 9.5 and 9.6.

We have observed the output voltage of the half-wave rectifier with the load resistor connected. See Appendix C "Dielectric in a uniform electric field", for a discussion of why it is difficult to characterize the output voltage with the load resistor *not* connected.

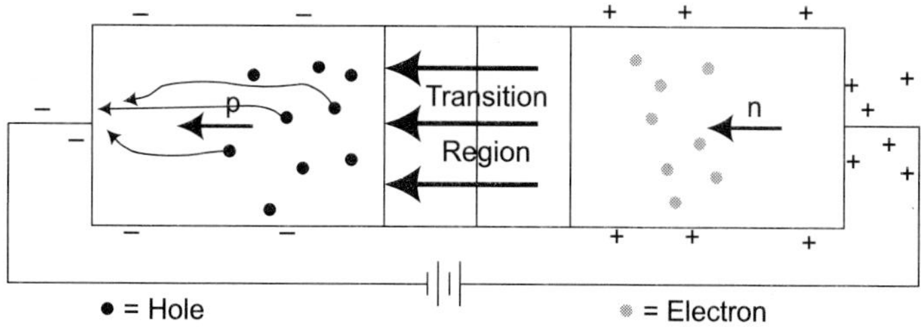

Fig. 9.36 Removal of a few mobile charges by the reverse bias.

When a diode is reverse biased, the electric field acts to move the *mobile* charges away from the junction and some remain in the diode the voltage source *"absorbs"* the ones that are moved as shown in Fig. 9.36. More immobile charges are exposed at the junction in comparison with the number when the diode was open circuited.

The battery pulls some electrons out of the n-region, leaving a thicker positive portion of the depletion layer. Similary, the battery pulls some holes out of the p-region leaving a thicker negative region of the depletion layer. The battery does not pull out *all* the mobile holes and electrons from the diode. Only a mere fraction of these are pulled out to cause the depletion layer to "drop" the applied reverse bias voltage across the diode. A thicker depletion layer produces a larger average electric field in the layer, which creates a larger potential difference across the layer.

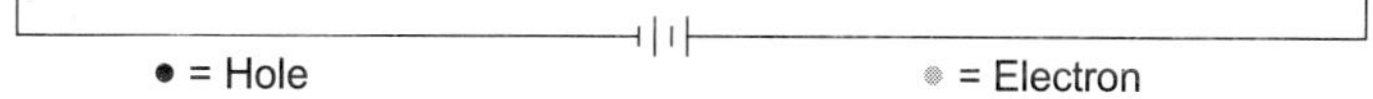

Fig. 9.37 The voltage across the diode is $V_m + V_o$.

The voltage *across* the diode is the *full* reverse voltage V_m applied plus V_o (the electrostatic potential barrier voltage) as shown in Fig. 9.37.

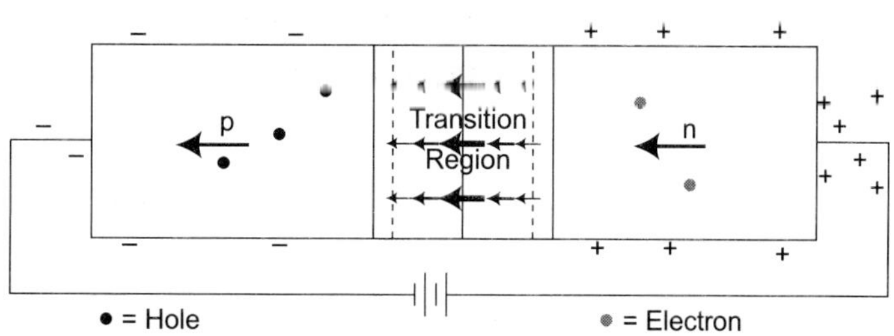

Fig. 9.38 Increase in the width of transition region.

The reverse biased diode causes an increase in the width of the transition region as shown in Fig. 9.38.

The electric field in the transition region due to the exposed immobile charges, is not uniform and is a maximum at the junction progressively decreasing towards the bulk of the neutral p- and n-regions and becomes very small within the bulk of the p- and n-regions (with large numbers 'n' of mobile carriers) as seen in Fig. 9.38 and is just sufficient to ensure that the current given by neAμE (electron current of Section 1.20 is multiplied by charge 'e' to obtain current in amperes) to be same throughout the diode.

To intuitively understand the operation of a diode during the negative half cycle, we consider the operation of a check valve that we described in Section 9.9, with the water flow reversed.

Such a situation will result, after a pump raises water to an overhead tank from a ground level reservoir and the pump is shut off. The operation of the valve is shown in the figure below.

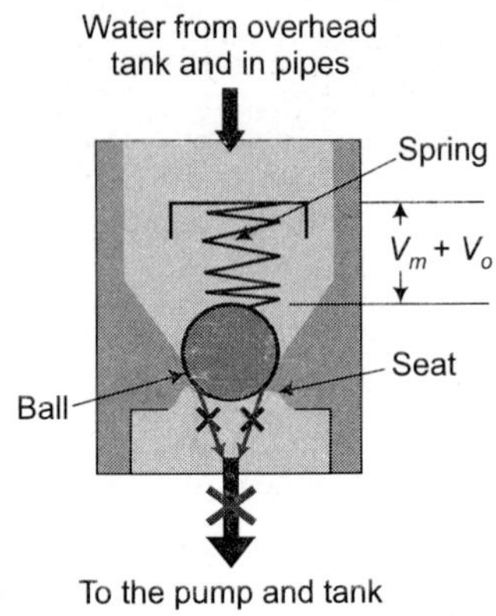

The water in the pipes from the overhead tank will exert a downward pressure on the check valve after the pump is shut off. This causes the ball to exert a pressure and the spring to stretch, aided by the head due to the column of water in the pipes running upto the overhead tank located a few tens of metres from the ground, thereby closing the orifice in the seat completely.

This prevents the water, all that was pumped into the overhead tank (!), from draining back into the ground level reservoir. It is this mode that makes the check valve most useful—preventing water flow in the reverse direction.

If the water was pumped to a great height, greater will be the pressure exerted by the water present in the pipes (after the pump is shut off) on the ball and greater will be the sealing effect on the seat.

This is similar to the reverse bias voltage V_m (that is applied to reverse bias the diode) + V_o (the barrier voltage, that came into existence when the junction was formed). And it is this current blocking mode of operation that makes the diode a useful device in circuits.

An interactive animation that depicts the current components of a p-n junction under forward and reverse bias is available at http://jas.eng.buffalo.edu/education/pn/biasedPN/index.html or http://web.eng.fiu.edu/~npala/Applets/education/pn/biasedPN/index.html.

I urge the reader to visit http://jas.eng.buffalo.edu/education/pn/biasedPN2/BiasedPN2.html or http://web.eng.fiu.edu/~npala/Applets/education/pn/biasedPN2/BiasedPN2.html and view the variation of the electric field with distance of a p-n junction under reverse bias.

After the reverse bias is applied, the diode is in thermal equilibrium after a small duration.

Remember that there is continuous production and annihilation of electron hole pairs when a semiconductor is in thermal equilibrium (dynamic) and net p_{no} and n_{po}, thermal equilibrium concentrations of holes in the n-side and electrons in the p-side regions respectively, exist.

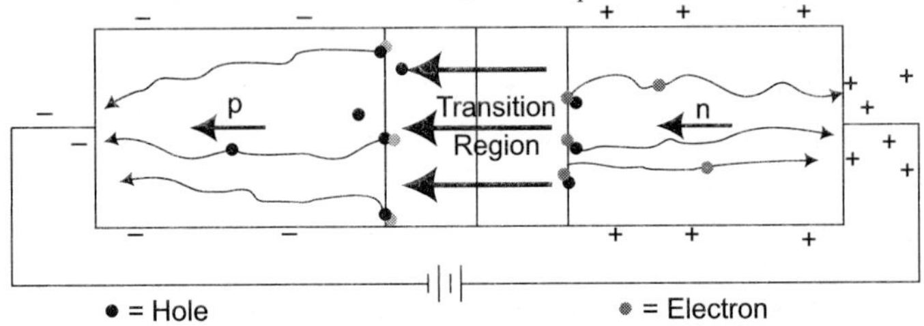

Fig. 9.39 EHPs migrate towards the diode terminals

The electron hole pairs generated in thermal equilibrium at the *edges of the transition region* move towards the diode terminals urged by the electric fields to recombine there…with charges from the voltage source as shown in Fig. 9.39.

This is what we call the reverse saturation current I_o.

How is the supply of holes replenished at the metal wire to p-material contact?

If a depletion region must be avoided at the metal wire to p-material interface, an *ohmic* contact should be formed (review how the Fermi levels equilibriate at the junction of two materials in Section 9.3 and of the transition region in Section 9.4), which would have a linear I-V characteristic with minimal resistance and no tendency to rectify signals. When a metal of work function φ_m is brought into contact with a p-type semiconductor whose work function φ_s is less than φ_m, holes transfer from metal to semiconductor until their Fermi levels are aligned. However, note that a depletion layer is formed in the metal though not in the semiconductor since the transferred holes are majority carriers in the semiconductor[30].

A practical method for forming ohmic contacts is by doping the semiconductor heavily in the contact region. Thus, if a barrier exists at the interface, the depletion width is small enough to allow carriers to *tunnel* through the barrier as shown in the figure below.

Ohmic contact

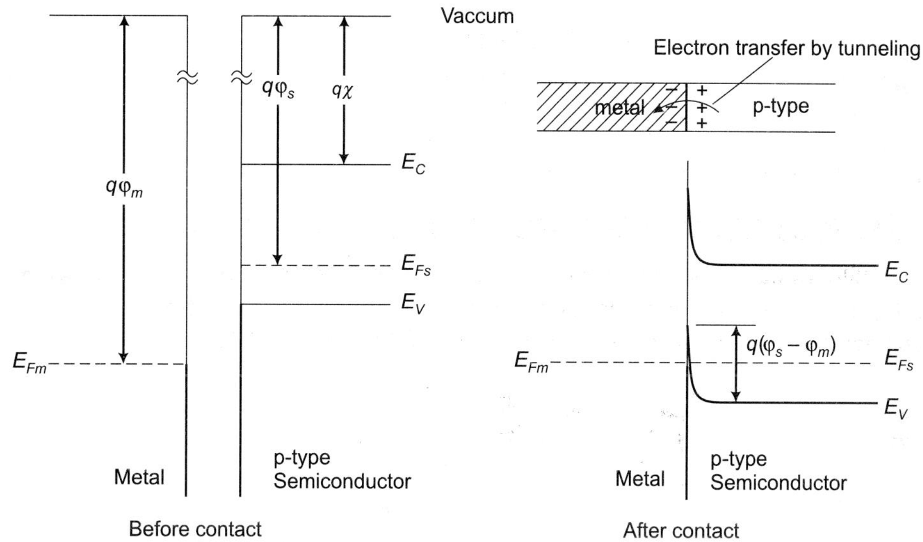

Legend: φ_m, φ_s = work function of metal, semiconductor
χ = electron affinity
E_{Fm}, E_{Fs} = Fermi level metal, semiconductor

In the case of an aluminium wire on p-type silicon, the metal contact, i.e., aluminium also provides the acceptor dopant. Thus, the required p+ layer is formed during a brief heat treatment of the contact after the aluminium is deposited. The additional holes exert a relentless pull of attraction, thereby increasing the strength of the confining force on nearby electrons of silicon atoms. The system electron energies tend to more negative and the electrostatic potential is raised[30] (made less negative). **Note:** The electrostatic potential cannot become positive, because if it did, electrons will no longer be confined and would be emitted from the p-type semiconductor! The reader is urged to view the power point slides in presentation Chapter 9 on electrostatic potential included in the accompanying CD for an electric field viewpoint on the ohmic contact development for both metal to p-type and n-type semiconductors and is also referred to the article at http://academic.reed.edu/chemistry/roco/Potential electrostatic_potential.html.

Now, when a forward voltage is applied to the diode, the electric field drives holes to the right (Fig. 9.15), mobile electrons are created (at the top of the *valence* band of the p-region being in near alignment with the top of the *conduction band* of the metal due to alignment of the Fermi levels of the metal and the p-type semiconductor) in the p-region near the junction and these "tunnel" into the metal wire, the depletion width being extremely small, thereby replenishing the supply of holes near the junction [30]. The contact is very nearly ohmic.

9.12 DIODE – PEAK INVERSE VOLTAGE (PIV)

The diode chosen for half wave rectification duty should be capable of withstanding the *peak negative voltage* of the sinewave voltage input as shown in Fig. 9.40.

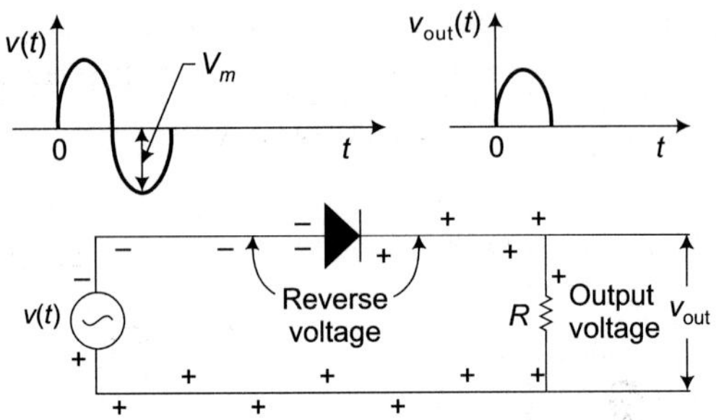

Fig. 9.40 In the reverse cycle, the maximum voltage across the diode is V_m.

9.13 PEAK INVERSE VOLTAGE (PIV) RATINGS OF DIODES

Diodes are chosen such that their Peak Inverse Voltage (PIV) is ***greater*** than the peak negative voltage V_m of the applied sinewave, else the diode will breakdown and may short-circuit. The diode characteristic is shown indicating the PIV voltage in Fig. 9.41.

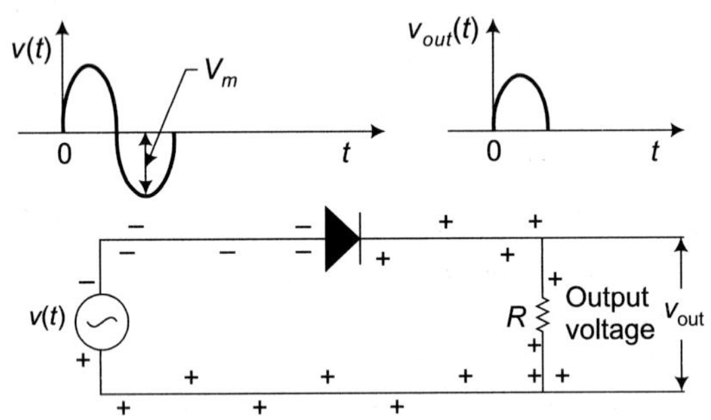

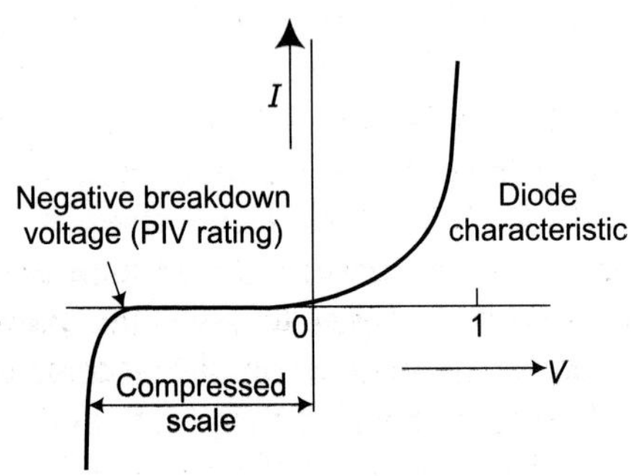

Fig. 9.41 The diode PIV rating should be greater than V_m.

The table below shows some commonly available diodes, their part numbers, PIV ratings, reverse saturation currents and applications.

Diode Part No.	*Peak Inverse Voltage (PIV)*	I_o *(reverse saturation current)*	*Application*
1N4002	400 V (can be used with 230 VAC)	30 μA	D.C. Power Supply
1N4007	1000 V (can be used with 230 VAC)	30 μA	D.C. Power Supply
1N3605	40 V	50 nA	Small signal rectifying
1N914	110 V	20 nA	Small signal rectifying

The Diode as a Protection Device for a Transistor that Switches a Relay

The relay as shown in the sketch below consists of a coil which forms a part of an electromagnet that attracts a metal piece called the armature, that is pivoted to move. The armature carries one contact, the moving contact called the common or COM. The other contacts are stationary on the relay structure and, these are the normally open (NO) contact and the normally closed (NC) contact.

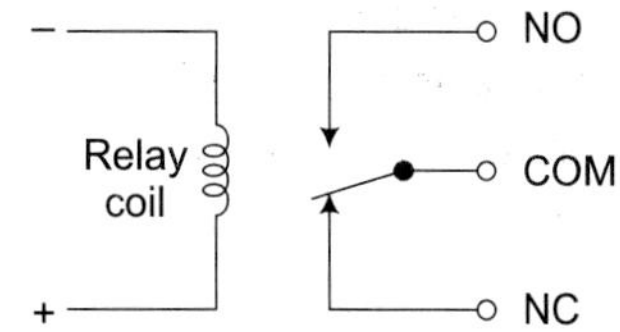

When the relay coil is energized, the moving contact closes with the stationary NO contact and, can be used in circuits to switch ON large loads like fan motors and industrial valve actuators. Line voltage from a power supply is connected to the COM contact and the load connected to the NO contact. The second terminal at the load is returned to the neutral wire.

The contacts open when the coil is de-energized. In electronic control circuits, the relay is usually switched on and off using an electronic device such as a transistor. See the figure below.

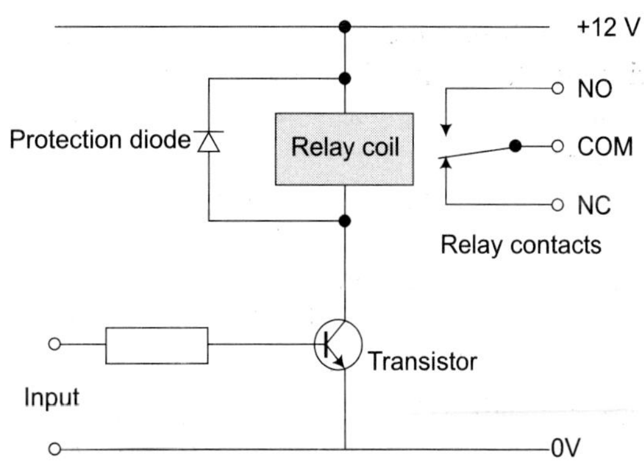

The transistor can be turned on by applying a suitable positive voltage at its input terminals and can be turned off by applying zero volt at the input.

At the moment when the relay coil is de-energized by applying a zero volt input to the transistor, the attempt being made to stop current flow through the coil which in essence is an inductor, $d\phi/dt$ or the rate of change of flux in the relay coil is high (see Section 5.12). This results in the production of a large induced emf across it that could destroy the transistor with which it is connected. Transistors are essentially not capable of handling voltages beyond a few hundred volts while the induced emf could exceed these values.

To protect sensitive transistors operating in this manner, a protection diode is connected as shown in the figure. The diode does not conduct when the transistor is ON because it is reverse biased, but conducts (forward-biased) whenever the relay coil is de-energized when the transistor is switched OFF. It *free-wheels* the current through the coil sustained by the induced emf allowing the magnetic field to die away quickly rather than instantaneously. So, the magnetic energy in the coil is dissipated in its resistance and the diode whose resistance and voltage when conducting is small. This prevents a sudden rise in the voltage that could damage the transistor.

9.14 TRANSITION AND DIFFUSION CAPACITANCES IN DIODES

It will not be proper to conclude this chapter without a reference to the capacitive effects in diodes. Capacitance effects are not solely associated with the classic lumped capacitor elements such as the disk type parallel-plate capacitor as was described in Chapter 2. These capacitors when charged cause separated point charges (shifting electrons in the electron sea) to be tightly packed and remain stationary on the surfaces of its conductive plates, electrons on one plate surface and lattice ions on the other. In p-n junctions, different spatial arrangements of unlike charges such as positive and negative lattice ions, holes and electrons can also give rise to a capacitive effect.

Reverse biasing a p-n junction causes majority carriers to move away from the junction, thereby uncovering more immobile charges or lattice ions. This uncovering of unlike charges on either side as the thickness of the space-charge (or transition) region increases with reverse voltage is a capacitive effect since, unlike stationary charges accumulate over a distance with voltage. The capacitance is known as *transition* capacitance. The distribution pattern of separated charges is different from the parallel-plate capacitor (see Section 9.4). The difference is that the depletion layer *width* and hence the capacitance is *voltage dependent*. The parallel plate expression still applies since charge is only *added* at the *edge* of the depletion regions[18]. The distance between the added negative and positive charge equals the depletion layer width. Derivations of expressions of the non-uniform net field, its pattern and transition capacitance are described in several textbooks[17].

The diffusion capacitance effect is qualitatively different from the transition capacitance in that the charges involved are not immobile. As a p-n diode is forward biased, the minority carrier distribution in the quasi-neutral region increases dramatically. In addition, to preserve quasi-neutrality, the majority carrier density increases by the same amount due to carriers supplied from the contact terminals. This effect leads to an additional capacitance called the diffusion capacitance. Unlike a parallel plate capacitor, the positive and negative charge is not spatially separated. Instead, the electrons and holes are separated by the energy bandgap *everywhere in* the crystal upto a certain distance from the edges of the transition region, away from the junction. Nevertheless, these voltage dependent charges yield a capacitance just as the one associated with a parallel plate capacitor.

It should suffice here to state that the electrons and holes that wait to recombine give rise to a resultant electric field around them giving rise to the diffusion capacitance effect. There is ever present a voltage dependent store of charges waiting to recombine, and in short, all the elements of a capacitance; holes and electrons separated by a distance.

When a junction is forward-biased, the injected minority carriers (holes and electrons) on either side of the must recombine with their opposite types to maintain current conservation. Recombination is a very orderly event governed by energy and momentum conservation principles. It requires several conditions to be fulfilled by both the carrier types before the energy give and take takes place, when it occurs, and there is no margin for error. A description of the conditions is beyond the scope of this textbook. For details the reader is referred to "Solid State Electronic Devices" by Ben G. Streetman and Sanjay Kumar Banerjee.

The derivation of expressions for diffusion capacitance is available in several textbooks[17]. The derivation of an expression for the electric field, however, is difficult because, the unlike charges involved are neither on the surface as in parallel-plate capacitors nor arranged spatially like the immobile charges in the space-charge region, but are *within* the lattice, the bandgap separating them.

When a diode is driven from the reversed condition to a forward state or in the opposite direction, the transition and diffusion capacitances cause the diode response to be accompanied by transients and an interval of time elapses before the diode recovers its steady state[17].

Questions

1. What happens to a silicon bar when its temperature is raised from 0 K.
2. How do the energies of the electrons distribute when a p-n junction is formed and explain how the equilibrium Fermi level is made constant throughout?
3. Describe the nature and strength of the electric field in the transition region and the electrostatic potential barrier that is formed when there is dynamic equilibrium.
4. Describe in detail the processes when a diode is forward biased with special emphasis on how the potential barrier is lowered and the process of recombination to enable conduction.
5. With reference to the potential and kinetic energies of electrons during forward bias explain in detail how the holes and electrons surmount the potential barrier.
6. Is the potential barrier reversed when a diode is forward biased? With the help of a suitable analogy explain how the potential barrier can never be made 0 volt.
7. Is the potential barrier reversed when a diode is forward biased? With the help of the diode equation $I = I_o(e^{V/V_T} - 1)$ and assuming values for the doping concentrations explain how the potential barrier can never be made 0 volt.
8. Explain where the voltage drops when a diode is sufficiently forward-biased into conduction and thereon for large forward bias voltages exceeding V_o.
9. Explain the processes that occur when a diode is reverse biased.
10. Describe the operation of a half-wave rectifier and how it can be used to convert a sinusoidal voltage into a steady DC voltage.
11. Explain how the current and voltage ratings, in particular, the peak inverse voltage ratings are important in selecting diodes for rectifier operation.

Appendix A

From "Electric and Magnetic Interactions" by Chabay and Sherwood, 1995, used by permission.

Electric Field

Define the electric field by the following relation

$$\vec{F}_{\text{electric on } q} = q\vec{E}$$

where $\vec{F}$ is the force acting on charge q and $\vec{E}$ is the value of the electric field at the location of charge q as shown in Fig 1. The electric field therefore has units of newtons per coulomb or *N/C*.

$$\vec{E} = \frac{\vec{F}}{q} \qquad \text{...(1)}$$

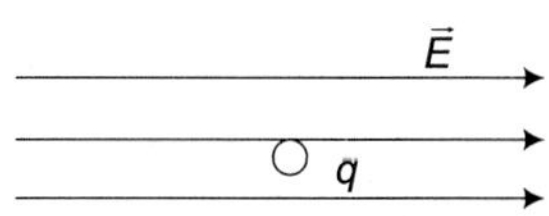

Fig. 1 Force $\vec{F}$ acting on a charge q due to the electric field $\vec{E}$ at the location of q.

Example

If $a + 2\ nC$ (nano coulomb $= 1 \times 10^{-9}$ C) is placed at a location and it experiences a force 3×10^{-7} Newtons to the right, what is the magnitude and direction of the electric field at that point?

$$\vec{E} = \frac{\vec{F}}{q} = \frac{3 \times 10^{-7}\, N}{2 \times 10^{-9}\, C} = 150 \text{ N/C}$$

The electric field points to the right because the charge is positive.

The Electric Field Due to a Point Charge

If there is a charge q_1 at some location in space, we know from Coulomb's law that if we place another charge q_2 at a location a distance '*d*' from q_1, the other charge q_2 will experience a force whose magnitude is as shown in Fig. 2.

$$\vec{F}_{\text{on } q2} = \frac{1}{4\pi\varepsilon_o}\frac{q_2 q_1}{d^2}$$

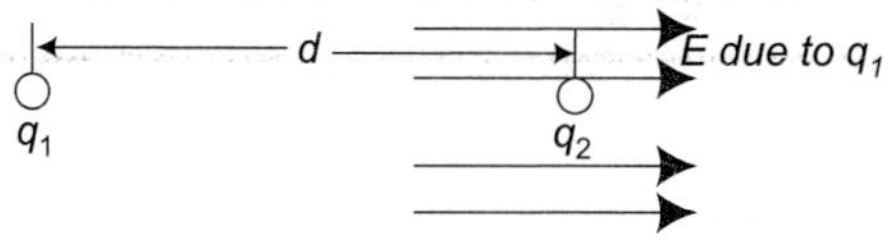

Fig. 2 Electric field at location q_2 due to a charge q_1.

This can be rewritten as

$$\vec{F}_{\text{on } q2} = q_2\left[\frac{1}{4\pi\varepsilon_o}\frac{q_1}{d^2}\right] = q_2\vec{E}$$

Note: The electric field of a charge cannot exert a force on itself. It is the electric field due to q_1 that exerts a force on q_2. IT IS AN EXTERNAL FIELD THAT EXERTS THE FORCE ON q_2.

So the electric field of point charge q_1 has magnitude

$$\vec{E} = \left[\frac{1}{4\pi\varepsilon_o}\frac{q_1}{d^2}\right] \quad \text{...(2)}$$

What is the electric field $\vec{E}$, 20 cm away from a positive point charge of one nano-Coulomb ($+1 \times 10^{-9}$ Coulomb).

You may use $\frac{1}{4\pi\varepsilon_o} = 9 \times 10^9\ N\text{-}m^2/C^2$.

Magnitude of $\vec{E} = \frac{1}{4\pi\varepsilon_o}\frac{q_1}{d^2} = \left[\frac{9\times10^9\ Nm^2}{C^2}\right]\left[\frac{10^{-9}C}{(0.2m)^2}\right]$

$= 225\ N/C$

The direction of $\vec{E}$ is radially outward away from the positive charge as shown in Fig. 3.

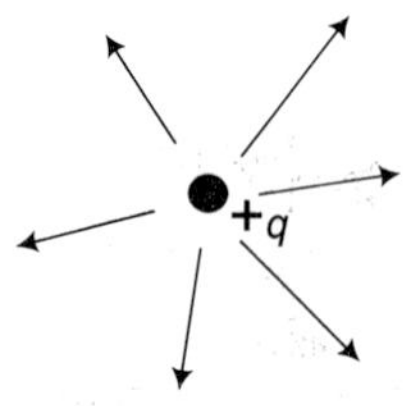

Fig. 3 Field due to a positive charge.

The Electric Field of a Dipole

Fig. 4 An electric dipole.

A dipole is composed of equal and unlike charges separated by a distance.

What is a dipole?

Matter contains both positive and negative charges (protons and electrons). Dipoles occur frequently in nature. Certain molecules occur in nature as dipoles with inherent polarization and are called permanent dipoles, for example, water molecules and HCl molecules. For example, a single HCl molecule is an electric dipole, the H end is somewhat positive and the Cl end is somewhat negative. A molecular dipole could be either an induced dipole created by external charges or a permanent dipole such as the water molecule.

Is an atom a dipole? See Appendix B for an answer to this.

The analysis of a single dipole provides a basis for analyzing more complicated forms of matter.

Superposition of Electric Fields

A very important property of electric fields is a consequence of the superposition principle which states: "The net electric force on a charged object is the vector sum of the individual electric forces on the object due to all other charged objects. Each individual interaction is unaffected by the presence of other charges." What this means is that the force between one particle and another is unaffected by the presence of a third particle. As a consequence of this principle, the net electric field due to two or more charges is unaffected by the presence of other charges. Superposition as a vector sum holds true for forces in general, so it also holds true for fields, because field is a force per unit charge.

Force on a point charge by a dipole (Chabay & Sherwood, 1995, used by permission)[2]

The *net charge* of a dipole is zero, so we might expect that the force exerted by a dipole on a charged object having charge Q would be zero (see Fig. 5a), using Coulomb's law:

Fig. 5(a) Illustrating the force exerted by a dipole on a charge Q.

$$F_{\text{on } Q \text{ by dipole}} = \frac{1}{4\pi\varepsilon_o}\frac{(dipole\ charge)Q}{d^2}$$

$$= \frac{1}{4\pi\varepsilon_o}\frac{(0)(Q)}{d^2} = 0\ !$$

However, if we draw a diagram of the situation, we realize that this may not be the case, since one of the ends of the dipole is closer to Q than the other.

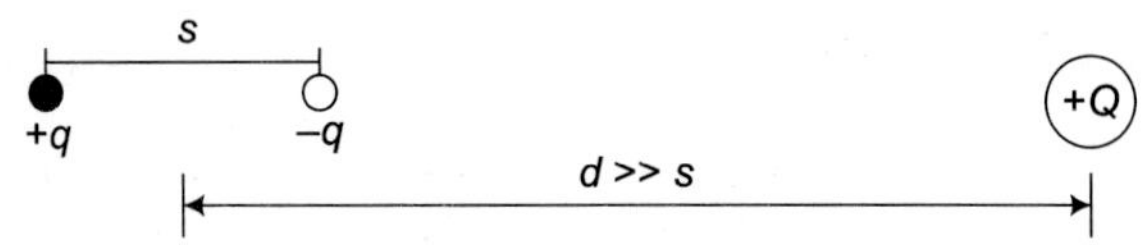

If the distance '*d*' is much larger than the dipole separation '*s*', perhaps this difference won't matter, and the net force will be zero, then. Just to be sure, let us first draw two vectors showing the force on *Q* by the negative end of the dipole, and the force on *Q* by the positive end of the dipole as shown in Fig. 5(b).

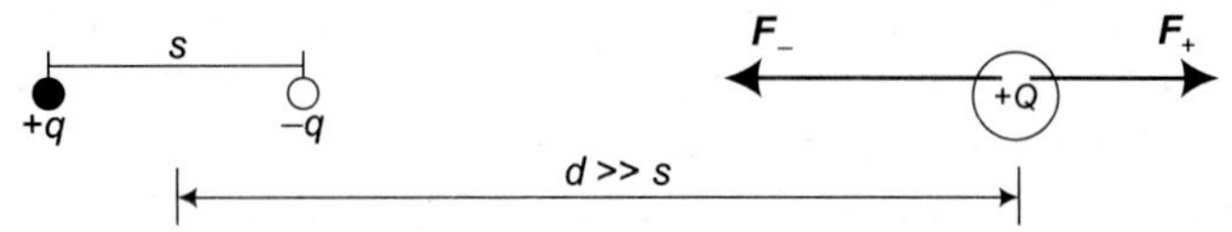

Fig. 5(b) The forces F_- and F_+ due to the two ends of the dipole on Q.

Since the two force vectors point in opposite directions, the magnitude of the net force is the difference of the magnitude of the two individual forces:

$$F_{\text{on}Q \text{ by dipole}} = F_{\text{on}Q \text{ by } -\text{ve end}} - F_{\text{onQ by +ve end}}$$

Do the two forces have exactly the same magnitudes?

No ! The magnitude of F_+ is less than the magnitude of F_- because the positive end is farther away from charge *Q* by the separation '*s*'. The net force on *Q* due to the dipole is small, and points to the left.

Therefore, the net force will be one of attraction as shown in Fig. 6.

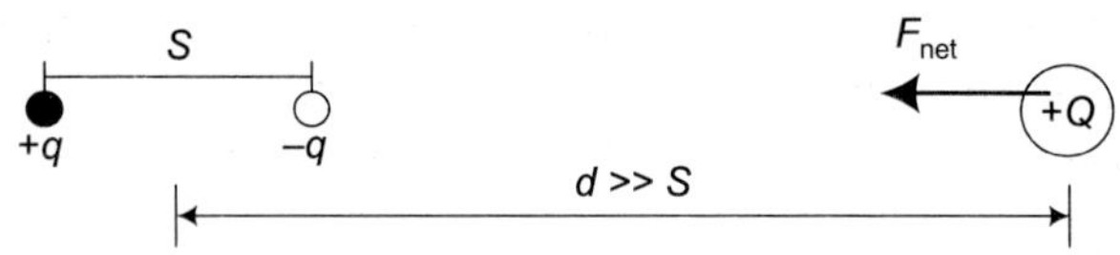

Fig. 6 Showing the resultant force on Q due to the dipole.

Let us calculate the resultant force.

$$F_{\text{on}Q \text{ by } -\text{end}} = \frac{1}{4\pi\varepsilon_o} \frac{qQ}{\left(d - \frac{s}{2}\right)^2}$$

$$F_{\text{on}Q \text{ by } +\text{end}} = \frac{1}{4\pi\varepsilon_o} \frac{qQ}{\left(d + \frac{s}{2}\right)^2}$$

$$F_{\text{net on } Q} = \frac{1}{4\pi\varepsilon_o} \frac{qQ}{\left(d - \frac{s}{2}\right)^2} - \frac{1}{4\pi\varepsilon_o} \frac{qQ}{\left(d + \frac{s}{2}\right)^2}$$

from the principle of superposition

$$= \frac{1}{4\pi\varepsilon_o} qQ \left[\frac{1}{\left(d - \frac{s}{2}\right)^2} - \frac{1}{\left(d + \frac{s}{2}\right)^2} \right]$$

...common factor

$$= \frac{1}{4\pi\varepsilon_o} qQ \left[\frac{\left(d + \frac{s}{2}\right)^2 - \left(d - \frac{s}{2}\right)^2}{\left(d - \frac{s}{2}\right)^2 \left(d + \frac{s}{2}\right)^2} \right]$$

...common denominator

$$= \frac{1}{4\pi\varepsilon_o} qQ \left[\frac{\left(d^2 + 2d\frac{s}{2} + \left(\frac{s}{2}\right)^2\right) - \left(d^2 - 2d\frac{s}{2} + \left(\frac{s}{2}\right)^2\right)}{\left(d - \frac{s}{2}\right)^2 \left(d + \frac{s}{2}\right)^2} \right]$$

...expand numerator

$$= \frac{1}{4\pi\varepsilon_o} qQ \left[\frac{2ds}{\left(d - \frac{s}{2}\right)^2 \left(d + \frac{s}{2}\right)^2} \right]$$

...simplify numerator

Upto now we have done straightforward algebra, but in order to gain some physical insight into the nature of this force, we will make an important approximation.

We have already stated that $d >> s$, so

$\left(d - \frac{s}{2}\right)^2 \approx d^2$ and $\left(d + \frac{s}{2}\right)^2 \approx d^2$, and we can replace the denominator by d^4.

$$F_{\text{net on } Q} = \frac{1}{4\pi\varepsilon_o} qQ \left[\frac{2ds}{\left(d - \frac{s}{2}\right)^2 \left(d + \frac{s}{2}\right)^2} \right] = \frac{1}{4\pi\varepsilon_o} qQ \left[\frac{2ds}{d^4} \right].$$

How good is the approximation?

Let us take an example.

Let $d = 1000$ and $s = 1$. Calculate the value of $\left(d-\frac{s}{2}\right)^2\left(d+\frac{s}{2}\right)^2$ exactly using your calculator.

$(1000 + 0.5)^2 (1000 - 0.5)^2 = [9.999995 \times 10^{11}]$. (Some calculators can't handle enough significant figures to do this calculation accurately.)

What value did you get using d^4?

$(1000)^4 = 1.0 \times 10^{12}$.

What is the per cent error? $\left[\dfrac{1.0\times 10^{12} - 9.999995\times 10^{11}}{9.999995\times 10^{11}}\right] = 0.0000005 = 0.00005\%$.

Note that even if d is only 100 instead of 1000 the error is only 0.005%. So if $d >> s$, we can be confident that $\left(d+\frac{s}{2}\right)^2 \approx d^2$ is a very good approximation.

Why cannot we use $s = 0$ in the numerator? How much is ds? You cannot put zero here! Coulomb's law is applicable to point charges only, not dipoles!

So, using the approximation that $\left(d-\frac{s}{2}\right)^2 = \left(d+\frac{s}{2}\right)^2 = d^2$ when $d >> s$, and cancelling d in the numerator and denominator, we get:

$$F_{\text{on } Q \text{ by dipole}} = \frac{1}{4\pi\varepsilon_o} qQ \frac{2s}{d^3}$$

Electric Field Due to a Dipole

(Chabay & Sherwood, 1995, used by permission)[2]

In the previous section, we calculated the force exerted by a permanent dipole on a charge placed along the x-axis. We will now calculate the electric field E_1, due to the dipole at a distance x from the centre of the dipole.

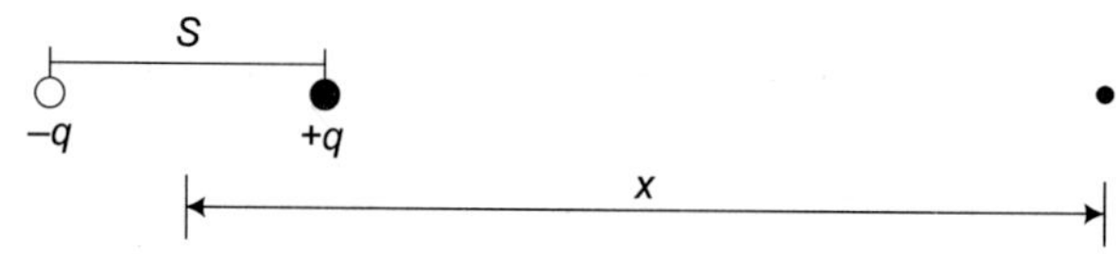

Fig. 7 Electric field due to a dipole at a distance x from the dipole.

The net field E_1 is horizontal to the right, because at this location, the field E_+ of the nearer $+q$ charge is larger than the field E_- of the more distant $-q$ charge. The magnitude is $E_1 = E_+ - E_-$.

Let us find a simplified expression for E_1.

Using the formula for $\vec{E} = \left[\frac{1}{4\pi\varepsilon_o}\frac{q_1}{d^2}\right]$ due to a point charge q_1 at an observation location distance '*d*' away from the point charge applied to our above system of a dipole:

$$E_1 = E_+ - E_- = \frac{1}{4\pi\varepsilon_o}\frac{q}{\left(x-\frac{s}{2}\right)^2} - \frac{1}{4\pi\varepsilon_o}\frac{q}{\left(x+\frac{s}{2}\right)^2}$$

...superposition

$$= \frac{q}{4\pi\varepsilon_o}\left[\frac{1}{\left(x-\frac{s}{2}\right)^2} - \frac{1}{\left(x+\frac{s}{2}\right)^2}\right]$$

...common factor

$$= \frac{q}{4\pi\varepsilon_o}\left[\frac{\left(x+\frac{s}{2}\right)^2 - \left(x-\frac{s}{2}\right)^2}{\left(x-\frac{s}{2}\right)^2\left(x+\frac{s}{2}\right)^2}\right]$$

...common denominator

$$= \frac{q}{4\pi\varepsilon_o}\left[\frac{\left(x^2+sx+\frac{s^2}{4}\right) - \left(x^2-sx+\frac{s^2}{4}\right)}{\left(x-\frac{s}{2}\right)^2\left(x+\frac{s}{2}\right)^2}\right]$$

...expand numerator

$$= \frac{1}{4\pi\varepsilon_o}\left[\frac{2qsx}{\left(x-\frac{s}{2}\right)^2\left(x+\frac{s}{2}\right)^2}\right]$$

...simplify numerator ...(3)

If we are far away from the dipole, as is normally the case when we interact with molecular dipoles, the magnitude of the field can be further simplified.

This situation is depicted as $x >> s$, then

with
$$E_1 = \frac{q}{4\pi\varepsilon_o}\left[\frac{2sx}{\left(x-\frac{s}{2}\right)^2\left(x+\frac{s}{2}\right)^2}\right]$$

from eqn. (3) above

$x - \frac{s}{2} \approx x$ and $x + \frac{s}{2} \approx x$ as well, so

$$E_1 = \frac{q}{4\pi\varepsilon_o}\left[\frac{2sx}{(x)^2(x)^2}\right] = \frac{q}{4\pi\varepsilon_o}\frac{2sx}{x^4}$$

$$= \frac{1}{4\pi\varepsilon_o}\frac{2qs}{x^3} \qquad ...(4)$$

Notice how the electric field drops away rapidly, proportional to $1/x^3$ at distances along the x-axis that are far away from the dipole.

Next, consider the electric field $\vec{E}_2$ at a location y along the y-axis as shown in Fig. 8. Note that the distance from either charge to the location y is, by Pythagorean theorem,

$$\left[y^2 + \left(\frac{s}{2}\right)^2\right]^{1/2}$$

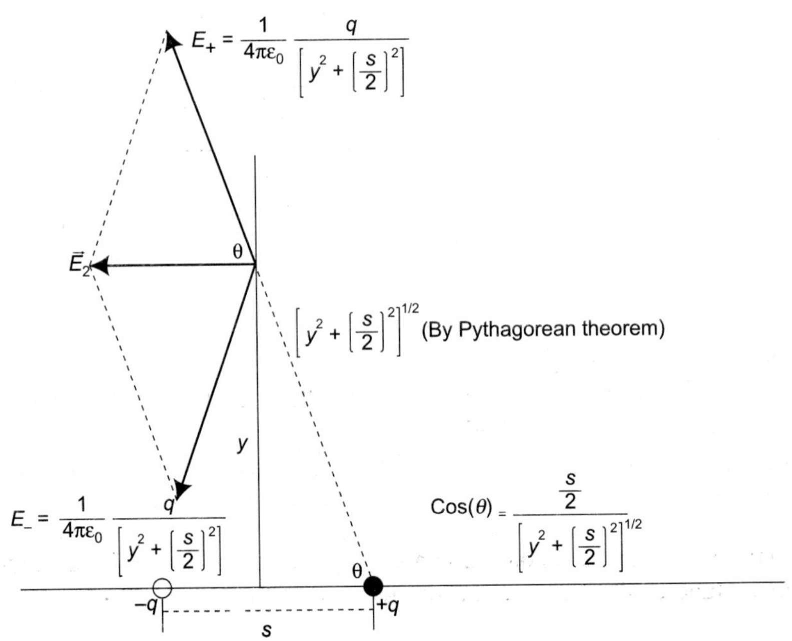

Fig. 8 Vector diagram for calculating the field at a location y along the y-axis.

The net field $\vec{E}_2$ is the vector sum of $\vec{E}_+$ and $\vec{E}_-$.The vertical components of the two fields $\vec{E}_+$ and $\vec{E}_-$ cancel, and the net field $\vec{E}_2$ is therefore horizontal to the left.

Note that both $\vec{E}_+$ and $\vec{E}_-$ have an x-component that involves $\cos(\theta) = \dfrac{\frac{s}{2}}{\left[y^2 + \left(\frac{s}{2}\right)^2\right]^{1/2}}$.

The magnitude is $E_2 = E_+ \cos\theta + E_- \cos\theta = 2E_+\text{co}\theta$.
From Fig. 8 above we compute E_2 as:

$$E_2 = 2\frac{1}{4\pi\varepsilon_o}\frac{q}{\left[y^2 + \left(\frac{s}{2}\right)^2\right]}\left[\frac{\left(\frac{s}{2}\right)}{\left[y^2 + \left(\frac{s}{2}\right)^2\right]^{1/2}}\right]$$

$$= \frac{1}{4\pi\varepsilon_o}\frac{qs}{\left[y^2 + \left(\frac{s}{2}\right)^2\right]^{3/2}}$$

If we consider points far away along the y-axis where $y >> s$, then

$$E_2 = \frac{1}{4\pi\varepsilon_o}\frac{qs}{\left[\left(y + \frac{s}{2}\right)\right]^{3/2}} \approx \frac{1}{4\pi\varepsilon_o}\frac{qs}{\left[y^2\right]^{3/2}} \quad (\text{since } y >> s)$$

$$= \frac{1}{4\pi\varepsilon_o}\frac{qs}{y^3}$$

Since $\vec{E}_2$ points to the left (in a direction that is opposite to the normal positive x-direction), the x component of the vector E_2 is negative. See Fig. 9.

Notice that the electric field far away from the dipole along the y-axis drops off rapidly proportional to $1/y^3$.

The reason we consider the cases when $y >> s$ is because while the stretch of the atomic dipoles are tiny in orders of magnitude of 10^{-12} m in most circuit applications, the points where the field is of significance in conduction processes may be at distances several molecular distances away say, at 10^{-9} m or more.
Note that 10^{-9}m >> 10^{-12}m.

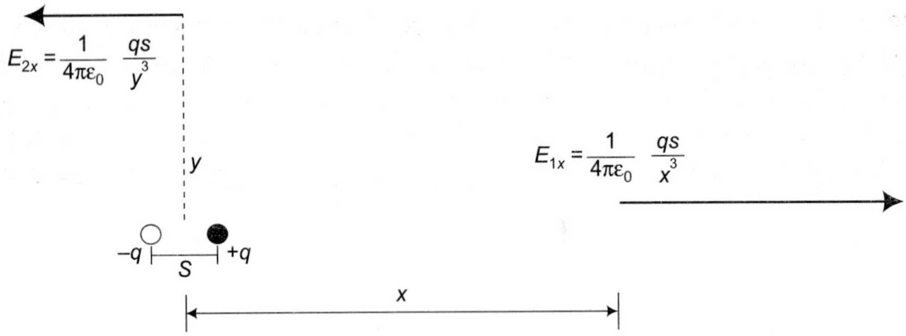

Fig. 9 The field strengths along the *x*–axis and *y*-axis of a dipole.

The electric field at both locations is proportional to the product qs, called the "dipole moment" and denoted by $p = qs$. Far away from a dipole, the same electric field could be due to a small q and a large s, or a large q and a small s. The only thing that matters is the "dipole moment" qs.

Moreover, this is the quantity that is measurable for molecules such as HCl and H_2O that are permanent dipoles, not the individual values of q (the amount of charge on an end) and s (the stretch of the molecule).

Force on a point Charge Due to a Permanent Dipole

(Chabay & Sherwood, 1995, used by permission)[2]

We can easily calculate from the electric field calculations above the force acting on a point charge at distances away from a dipole of separation s as depicted in Fig. 10.

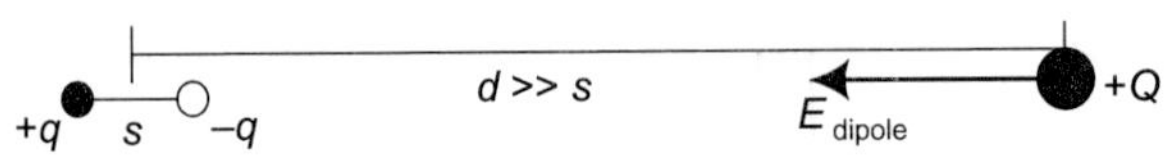

Fig. 10 Force acting on a point charge due to a dipole.

Assume that there is a point charge $+Q$ located a distance $d >> s$, then

Direction of $F_{\text{on } +Q \text{ by dipole}}$ is to the left;

From eqns (1) and (4)

$$F_{\text{on } +q \text{ by dipole}} = QE_{\text{dipole}} = Q\frac{1}{4\pi\varepsilon_o}\frac{2qs}{d^3}$$

Force on a Point Charge Due to an Induced Dipole

(Chabay & Sherwood, 1995, used by permission)[2]

When a small electric field E is applied to an atom or molecule that is not a permanent dipole, it is found experimentally and theoretically that the charge separation s (and the induced dipole moment $p = qs$) is proportional to E. Double the electric field E, double the stretch s. This proportionality does not hold for very large electric fields such as the electric fields in the electromagnetic radiation in a tightly focused beam of a powerful laser. But under most conditions it is an excellent approximation to say that the dipole moment $p = qs = \alpha E$, where α is a constant that expresses the polarizability of the particular kind of atom or molecule.

Consider an induced dipole, a neutral atom or molecule polarized by a point charge $+Q$ that is a distance $d >> s$ from the centre of the dipole. The dipole moment of the induced dipole is $p = qs = \alpha E_Q$, where E_Q is the electric field at the location of the dipole, due to the point charge $+Q$.

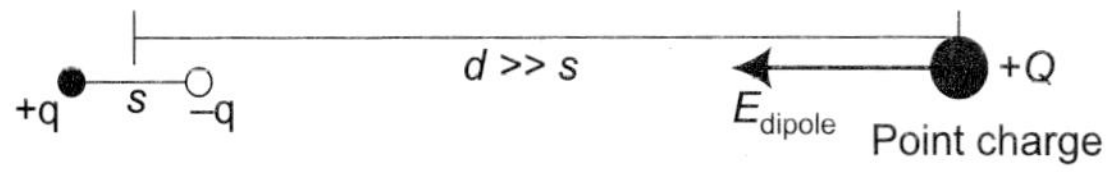

Separation s, dipole moment $p = qs = \alpha E_Q$

Fig. 11 Force on a point charge due to an induced dipole.

The direction of $F_{\text{on } +Q \text{ by the induced dipole}}$ is to the left.

The electric field E_Q due to the point charge at the location of the dipole is given by eqn. (2)

$$E_Q = \left[\frac{1}{4\pi\varepsilon_o}\frac{Q}{d^2}\right] \quad \text{...(5)}$$

Therefore,

$$F_{\text{on } +Q \text{ by dipole}} = QE_{\text{dipole}} = Q\left[\frac{1}{4\pi\varepsilon_o}\frac{2qs}{d^3}\right] \text{ using expression (4)}$$

Since $$p = qs = \alpha E_Q;$$

Therefore, $$F_{\text{on } +Q \text{ by induced dipole}} = QE_{\text{dipole}} = Q\left[\frac{1}{4\pi\varepsilon_o}\frac{2qs}{d^3}\right]$$

$$= Q\frac{1}{4\pi\varepsilon_o}\frac{2}{d^3}\left[\alpha\frac{1}{4\pi\varepsilon_o}\frac{Q}{d^2}\right]$$

$$= \left[\frac{1}{4\pi\varepsilon_o}\right]^2 \frac{2aQ^2}{d^5}$$

Notice that the force F on $+Q$ by an induced dipole falls away at the rate of $1/d^5$.

By Newton's third law, *the magnitude of the force exerted by Q on the dipole is the same.*

Amount of the Electron Cloud Shift in the Carbon Atom of Small Bits of Paper that is Attracted to a Charged Plastic Pen

The electric field of the charged pen shifts the electron cloud of the carbon atoms in the small bits of paper a small amount, creating an induced dipole (ignoring the effects that the other induced dipoles in the paper have on this carbon atom).

When the pen is a height h above the paper, the carbon atom's outer shell of four electrons (charge –4e) shifts an unknown distance s relative to the nucleus. The inner shell of 2 electrons shifts a negligible amount relative to the nucleus, because the inner electrons on the average are much closer to the nucleus and much more strongly attracted to the nucleus.

Assume that the pen carries a charge of about $-Q = -10^{-7}$ coulomb and height $h = 1$ cm, we will calculate how far the carbon atom's outer shell of four electrons must be shifted, s, in order that the upward electric force of attraction between the pen and the carbon atom equals the downward gravitational force on the carbon atom.

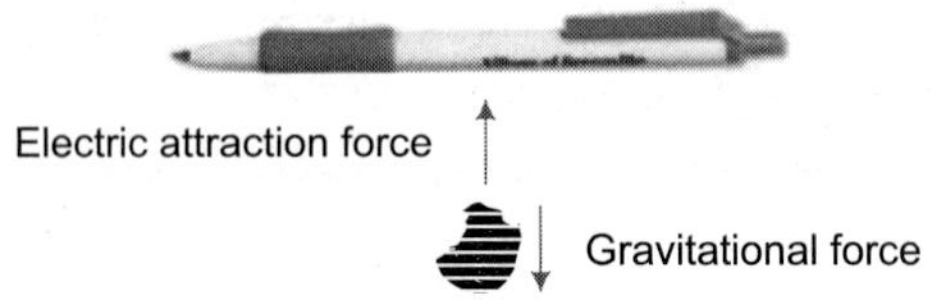

The mass of the carbon atom is $(12 \times 10^{-3}$ kg/mole)/(6×10^{23} atoms/mole).

Force on the carbon atom by the pen will be given by $F_{\text{on Q by dipole}} = \frac{Q}{4\pi\varepsilon_o}\frac{2qs}{h^3}$ where Q is on the plastic pen and the dipole is in the bit of paper, using expression (4).

The downward force due to gravity on the carbon atom will be

$$\text{mg} = \frac{12\times10^{-3}}{6\times10^{23}} \times 9.81 \text{ newtons}$$

When the pen is $h = 1$ cm away, the force of attraction of the pen should be slightly greater than the downward force of gravity in order that the pen is able to lift the bit of paper. Therefore,

$$\frac{12\times10^{-3}}{6\times10^{23}} \times 9.81 = \frac{Q}{4\pi\varepsilon_o}\frac{2qs}{h^3} = \frac{10^{-7}}{4\pi\varepsilon_o} \times \frac{2\times4\times1.6\times10^{-19}\times s}{(0.01)^3}$$

[**Note:** $\frac{1}{4\pi\varepsilon_o} = 9 \times 10^9$ Nm^2/C^2]

Therefore,

$$s = \frac{12\times10^{-3}\times9.81}{6\times10^{23}} \times \frac{1}{9\times10^9} \times \frac{(0.01)^3}{10^{-7}\times2\times4\times1.6\times10^{-19}}$$

$$= \frac{19.62}{72} \times 10^{-15} \text{ metre}$$

$$= 0.72 \times 10^{-15} \text{ metre.}$$

The distance the electron cloud shifts turns out less than 10^{-15} metres.

This is roughly the radius of a single proton. The radius of the electron cloud of the carbon atom is about 10^{-10} metres, or about 100,000 times larger than a proton.

The shift is remarkably tiny to have such a marked effect. This is another example of the fact that electric interactions are much stronger than gravitational interactions.

Electric Field Due to Distributed Charges

The electric field due to two individual charged particles can be calculated as the superposition of the fields due to each particle. In a similar way, we could calculate the net electric field due to three or four charged particles.

But how can we calculate the electric field due to the superposition of the fields of millions of individual charged particles?

For this, we apply the superposition principle, and add the contributions of all the point charges and follow the general steps as listed below in the calculation of the field due to a charged ring.

Uniformly charged Ring (Chabay & Sherwood, 1995, used by permission)[2]

Field of a single ring at a point on the axis

Step 1: Cut up the charge distribution into pieces (Δl) and draw ΔE.

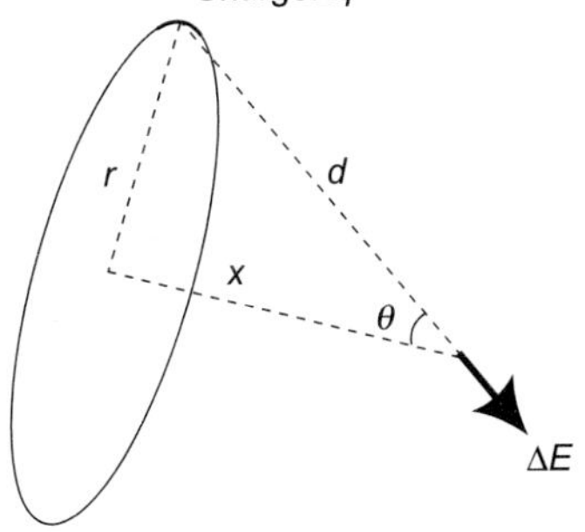

Fig. 12 Electric field along the axis of a charged ring.

We cut up the ring into very short sections each of length Δl with positive charge Δq. Here Δl is a small portion of the total circumference 2πr, and Δq is a small portion of the total charge q of the ring.

Step 2: We invent an integration variable to refer to each piece. We use the variable "l" measured circularly around the ring with Δl a small increment of this integration variable. There is a small amount of charge Δq on a piece of length Δl. We show the components of ΔE parallel to the axis (ΔE_x) and perpendicular to the axis ($\Delta E\perp$):

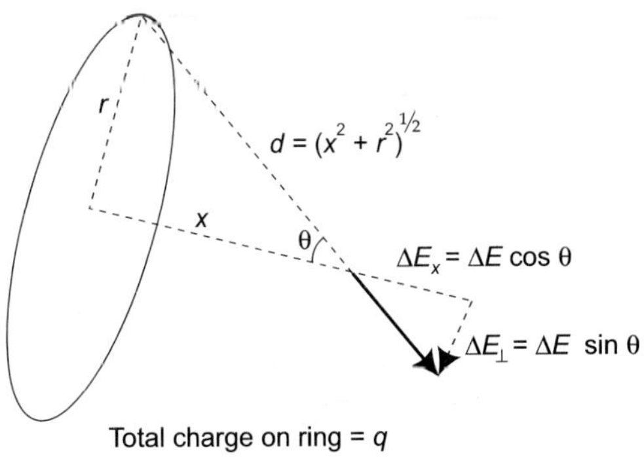

At a distance along the axis from the centre of the ring, the little piece of charge Δq is a distance $d = (x^2 + r^2)^{1/2}$ away, and it makes a field $\Delta\vec{E}$ of magnitude

$$\Delta E = \frac{1}{4\pi\varepsilon_o}\frac{\Delta q}{d^2} = \frac{1}{4\pi\varepsilon_o}\frac{\Delta q}{(x^2 + r^2)}$$

Since the ring is uniformly charged with a total charge q, and $\frac{\Delta l}{2\pi r}$ is the fraction of the whole ring represented by Δl, so the amount of charge on a section of length Δl is

$$\Delta q = \left(\frac{\Delta l}{2\pi r}\right)q = \left(\frac{q}{2\pi r}\right)\Delta l$$

When we add up the contributions of all the little pieces that make up the ring, the components perpendicular to the axis add up to zero.

For each ring segment with charge Δq_1 that makes a field $\vec{E}_1$ whose y-component is $\vec{E}_{1y}$, there is a piece Δq_2 (located 180 degrees around the ring from Δq_1) that makes a field $\vec{E}_2$ whose y-component $\vec{E}_{2y}$ is equal in magnitude and opposite in direction to $\vec{E}_1 y$. So the components all add up to zero, and only the x-component is non-zero.

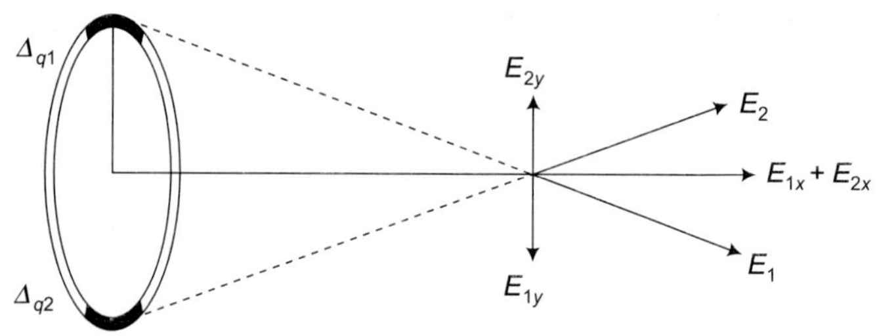

And, the net electric field of the ring is along the x-axis. Each Δq contributes $\Delta E_x = \Delta E\cos(\theta)$ to the net field

$$\Delta E_x = \Delta E \cos\theta$$

Writing $\cos\theta$ in terms of x and r

$$\cos\theta = \frac{x}{d} = \frac{x}{\left(x^2 + r^2\right)^{1/2}}$$

Now put it all together.

$$\Delta E_x = \frac{1}{4\pi\varepsilon_o}\frac{\Delta q}{\left(x^2 + r^2\right)}\cos\theta$$

$$\cos\theta = \frac{x}{d} = \frac{x}{\left(x^2 + r^2\right)^{1/2}}$$

and $$\Delta q = \left(\frac{q}{2\pi r}\right)\Delta l$$

So, $$\Delta E_x = \left[\frac{1}{4\pi\varepsilon_o}\frac{q}{2\pi r}\frac{\Delta l}{(x^2+r^2)}\frac{x}{(x^2+r^2)^{1/2}}\right]$$

$$= \frac{1}{4\pi\varepsilon_o}\frac{q}{2\pi r}\frac{x}{(x^2+r^2)^{3/2}}\Delta l$$

Step 3: Add up all the contributions of all the pieces. The net field lies along the axis and is the summation of ΔE_x contributed by each little piece of the ring:

$$E_x = \frac{1}{4\pi\varepsilon_o}\frac{q}{2\pi r}\frac{x}{(x^2+r^2)^{3/2}}\sum \Delta l$$

The integration variable "*l*" runs around the circumference of the ring, and Δl is an increment of the integration variable, corresponding to a short step along the circumference of the ring. If we go all the way around the ring, we will have accounted for every small piece of the ring, and adding up all the steps, we get the circumference of the ring:

$$\sum \Delta l = 2\pi r.$$

More formally, if we let $\Delta l \rightarrow dl$ (an infinitesimally small increment with no bounds), we can integrate

$$\int dl = 2\pi r.$$

Adding up all the *dl*'s gives us the total length (the circumference). So, the result which does not contain the integration variable *l*, is:

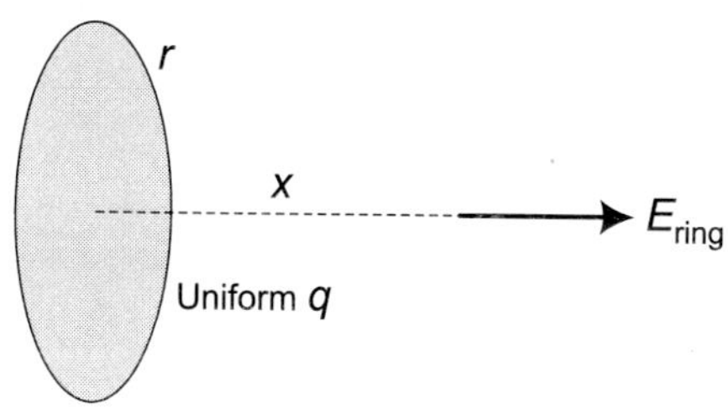

$$E_{\text{ring}} \quad \frac{1}{4\pi\varepsilon_o}\frac{qx}{(x^2+r^2)^{3/2}} \qquad \text{...(6)}$$

for a ring of radius *r* and charge *q*, a distance *x* from the centre, along the axis.

Special case

If $x = 0$, were at the centre of the ring, the electric field should be zero since, then the contributions of all portions of the ring point inward and cancel each other. Setting $x = 0$ in the expression given by eqn. (6), we indeed find $E_{\text{ring}} = 0$.

If *x* is very much larger than *r*, the distant ring looks almost like a point charge, so the net field ought to look like the field of a point charge.

If $x >> r$, then $x^2 >> r^2$

$$\frac{x}{(x^2+r^2)^{3/2}} = \frac{x}{(x^2)^{3/2}} = \frac{x}{x^3} = \frac{1}{x^2}, \text{ so}$$

$$E_{\text{ring}} = \frac{1}{4\pi\varepsilon_o}\frac{q}{x^2}.$$

The above expression does indeed look like the electric field of a point charge when we compare this with expression (2).

Field of two rings (Chabay & Sherwood, 1995, used by permission)[2]

Two rings of radius of 5 cm are 20 cm apart and concentric with a common axis. The ring on the left carries a uniformly distributed charge of +35 nC and the ring on the right carries a uniformly distributed charge of –35 nC.

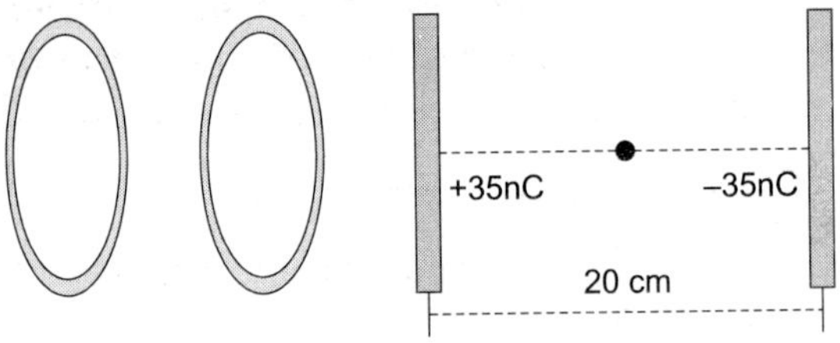

What is the magnitude and direction of the electric field on the axis, halfway between the two rings?

The net field is the vector sum of the field due to ring 1 and the field due to ring 2.

The vectors $\vec{E}_1$ and the $\vec{E}_2$ are shown displaced for clarity, they are equal and collinear.

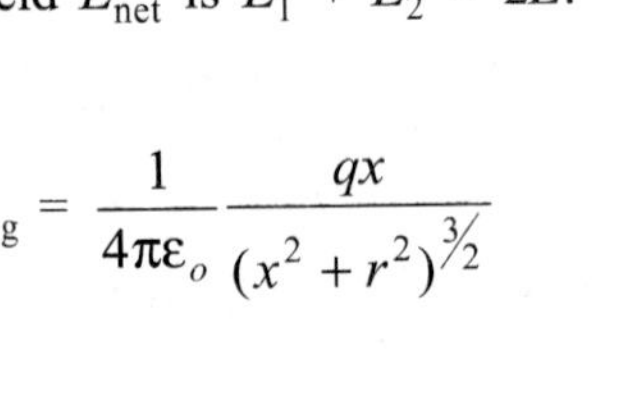

The magnitude of the net field E_{net} is $E_1 + E_2 = 2E$.

Using the expression

$$E_{\text{ring}} = \frac{1}{4\pi\varepsilon_o}\frac{qx}{(x^2+r^2)^{3/2}}$$

We can compute

$$E_{\text{net}} = 2.\frac{1}{4\pi\varepsilon_o}\frac{(35\times10^{-9}\,C)(0.1m)}{\left((0.1m)^2+(0.05m)^2\right)^{3/2}} = 4.5\times10^4 \text{ N/C}$$

If the rings both carry a charge of +35 nC, i.e., like charges are present on both the rings, then the magnitude and direction of the electric field midway between the rings will be zero.

Field of Uniformly Charged Disks

(Chabay & Sherwood, 1995, used by permission)[2]

We now consider the important case of a uniformly charged disk made from insulating material say, glass that has been rubbed with silk. We will see in the next section that two charged metal disks can form a "capacitor", a device that is important in electric circuits.

Field of a Uniformly Charged Disk, at a Point on the Axis

Step 1: Cut up the charge distribution into pieces and draw ΔE. We will calculate the electric field of a uniformly charged disk of radius R.

We cut up the uniform disk into rings, each contributing some known electric field ΔE.

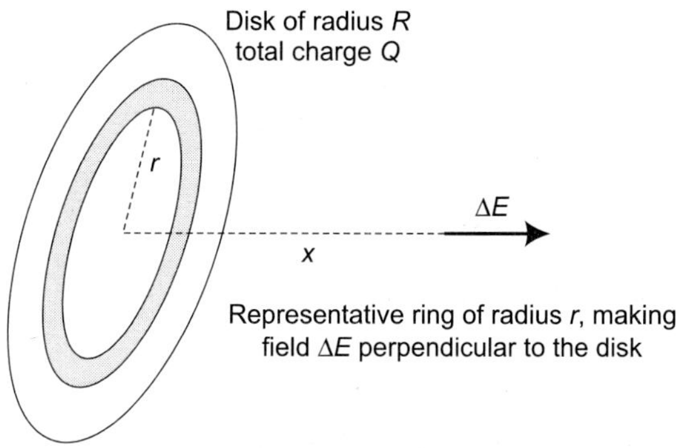

Let us divide the disk of radius R into concentric rings, and approximate each ring as having some average radius r to the middle part of the ring (the smallest ring is actually a small disk).

Each ring has some amount of positive charge ΔQ (not necessarily the same on each ring). Each ring contributes an amount ΔE to the electric field perpendicular to the disk.

Step 2: Write an expression for the electric field due to one piece.

We have already invented an integration variable: the radius r of a ring. (The distance x from the centre of the disk to the observation point is a given constant, but r is a variable that we invented to refer to pieces into which we cut the disk). Using our previous result for the electric field of a single ring of charge, at a location on the axis, we can write: (See expression (6))

$$\Delta E = \frac{1}{4\pi\varepsilon_o}\frac{\Delta q x}{(x^2 + r^2)^{3/2}}$$

where Δq is the amount of charge on the ring, and r is the radius of the ring. Both Δq and r will be different for different rings.

What is Δq for a particular ring in terms of the integration variable r?

Imagine dividing the disk into a very large number of rings, and consider one ring with radius r and very narrow width Δr.

If you cut such a ring and roll it straight, you get an approximately rectangular strip of length of $2\pi r$ (the circumference of a circle) and width Δr, so this area is

$$\Delta A = 2\pi r \, \Delta r$$

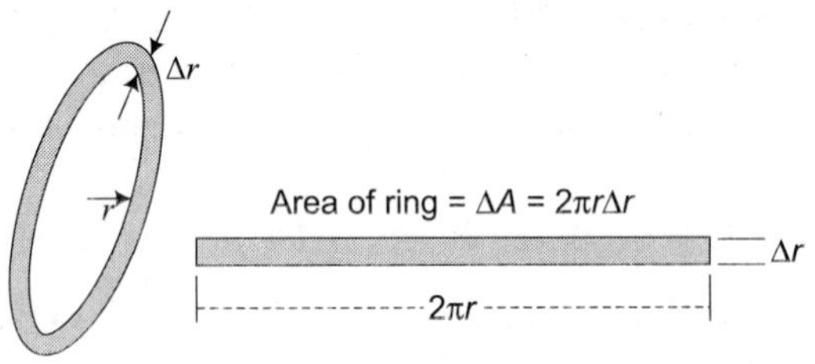

The total charge Q on the disk is spread over a total area A $(= \pi R^2)$, so the uniform charge density is Q/A (coulombs per square metre). Therefore, $\Delta q = (Q/A)(2\pi r \Delta r)$, since the ring area is $2\pi r \Delta r$.

Now that we know the amount of charge on a ring, we can write an algebraic expression for the contribution of one ring, in terms of integration variable r:

$$\Delta E = \frac{1}{4\pi\varepsilon_o}\left(\frac{Q}{A}\right)\frac{x}{(x^2 + r^2)^{3/2}}(2\pi r \Delta r)$$

Step 3: Add up the contributions of all the pieces

Numerical summation

If we consider a finite number of rings—say five—we can add up the electric fields of all five rings to get the approximate field of the disk.

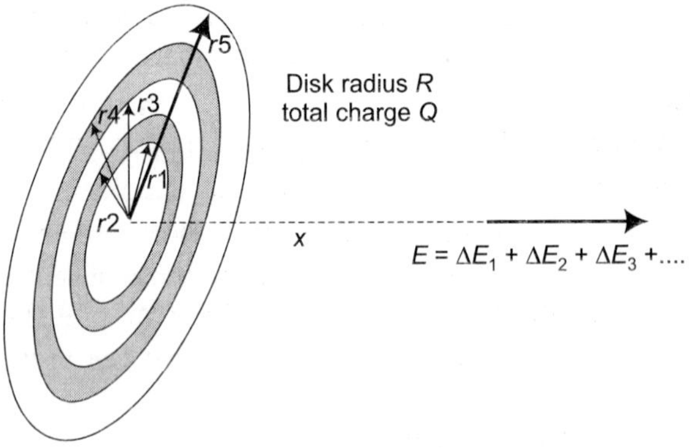

The total field E is the summation of the field contributions of the 5 rings:

$$E = \sum_{n=1}^{n=5}\frac{1}{4\pi\varepsilon_o}\left(Q/A\right)\frac{x}{(x^2 + r_n^2)^{3/2}}(2\pi r_n \Delta r)$$

This result will not be exact, because we need an average radius r_n for the nth ring. We could evaluate the summation numerically to get an approximate result for a particular disk. We would get a more accurate result if we divided into ten rings, or a hundred, or a thousand rings. If we divide the disk into an infinite number of infinitesimally narrow rings, we get an exact result, and we consider that procedure next.

Integration

Let the number of rings become very large, so Δr becomes dr, an infinitesimally small thickness. Remember, that the integral sign $\int$ is just a distorted S standing for "Summation".

Since $$dA = 2\pi r dr,$$

$$E = \int \frac{1}{4\pi\varepsilon_o} \frac{x}{(x^2 + r_n^2)^{3/2}} \left(\frac{Q}{A}\right)(2\pi dr)$$

Since r ranges from 0 to R,

$$E = \int_0^R \frac{\left(\frac{Q}{A}\right)}{2\varepsilon_o} \frac{x}{(x^2 + r^2)^{3/2}} r dr = \frac{\left(\frac{Q}{A}\right)}{2\varepsilon_o} x \int_0^R \frac{r}{(x^2 + r^2)^{3/2}} dr$$

Let $u = x^2 + r^2$, so that $du = 2rdr$ and $dr = du/2r$ and rewrite this integral

$$E = \frac{\left(\frac{Q}{A}\right)}{2\varepsilon_o} x \int_{r=0}^{r=R} \frac{1}{2} \frac{1}{u^{3/2}} du = \frac{\left(\frac{Q}{A}\right)}{2\varepsilon_o} x \left[-\frac{1}{u^{1/2}}\right]_{r=0}^{r=R}$$

$$E = \frac{\left(\frac{Q}{A}\right)}{2\varepsilon_o} x \left[-\frac{1}{(x^2 + R^2)^{1/2}} - \frac{-1}{(x^2)^{1/2}}\right]$$

$$= \frac{\left(\frac{Q}{A}\right)}{2\varepsilon_o} \left[1 - \frac{x}{(x^2 + R^2)^{1/2}}\right] \quad ...(7)$$

We have calculated the electric field due to a uniformly charged disk of radius R, along its axis. Note that the final result does not contain the integration variable r:

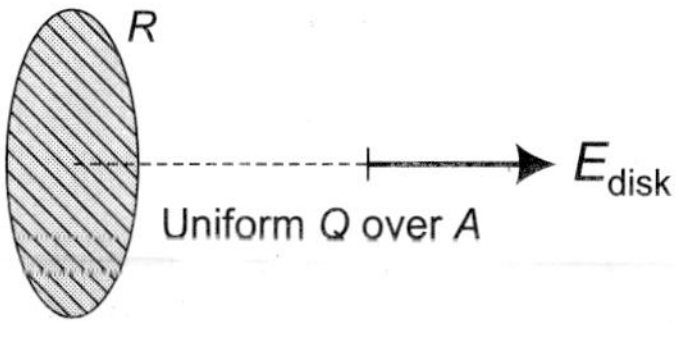

$$E_{\text{disk}} = \frac{\left(\frac{Q}{A}\right)}{2\varepsilon_o} \left[1 - \frac{x}{(x^2 + R^2)^{1/2}}\right]$$ perpendicular to the centre of the disk ($A = \pi R^2$).

Electric field near the centre of a uniformly charged disk

A special case turns out to be of considerable interest. Let us see what the electric field is like *near* the centre of a uniformly charged disk (x small, but not zero). Suppose that we are so close to the disk that x is very small ($x << R$), or equivalently that we are some distance away but the disk is very large ($R >> x$).

Since $$E_{disk} = \frac{\left(Q/A\right)}{2\varepsilon_o}\left[1 - \frac{x}{(x^2 + R^2)^{1/2}}\right]$$ then, with $R >> x$, $(x^2 + R^2)^{1/2} \approx R$ so

$$E_{disk} = \frac{\left(Q/A\right)}{2\varepsilon_o}\left[1 - \frac{x}{R}\right] \qquad ...(8)$$

Further, if x/R is very very small, $\left[1 - \frac{x}{R}\right]$ reduces to 1, so we have approximately

$$E_{disk} = \frac{\left(Q/A\right)}{2\varepsilon_o}\left[1 - \frac{x}{R}\right] = E_{disk} = \frac{\left(Q/A\right)}{2\varepsilon_o}$$ near the centre of a disk.

The result is approximately nearly independent of the distance x and is true near any large uniformly charged plate, not just a circular one.

What does this result mean?

Assuming a disk of radius $R = 20$ cm and the total charge is distributed uniformly all over the disk is $Q = 6 \times 10^{-6}$ C and that $\varepsilon_o = 9 \times 10^{-12} C^2/N - m^2$, we find the electric field 1 mm from the centre of the disk using expression (7) to be 2.639×10^6 N/C and that 3 mm from the centre of the disk it is only very slightly smaller: 2.613×10^6 N/C.

This electric field is nearly strong enough to cause breakdown in air – the breakdown field strength is roughly 3×10^6 N/C.

You find that the field at a distance of 1 mm from the centre of the disk and the field at a distance of 3 mm are nearly the same.

Here is a way to make a "nearly uniform field" in a small region; an electric field that doesn't vary much in direction or magnitude in that region.

Does this contradict Coulomb's law that the electric field varies as $1/d^2$?

Coulomb's law describes the electric field of a point charge. A uniformly charged disk is not a point charge, so its field does not necessarily look like $1/d^2$.

The field of the disk is calculated by integrating the contributions of point-like pieces of the disk over the whole surface of the disk. The contribution of each point-like piece does follow Coulomb's law, *but the sum does not.*

Two uniformly charged disks: A capacitor (Chabay & Sherwood, 1995, used by permission)[2]

Consider two uniformly (oppositely) charged metal disks placed very near each other (separation or gap distance s), carrying charges of $-Q$ and $+Q$.

This arrangement is called a "capacitor" and such devices are used in electric and electronic circuits.

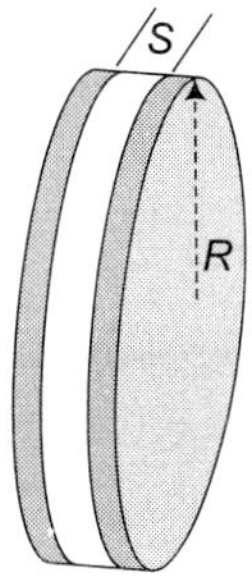

Two oppositely charged metal disks very near other

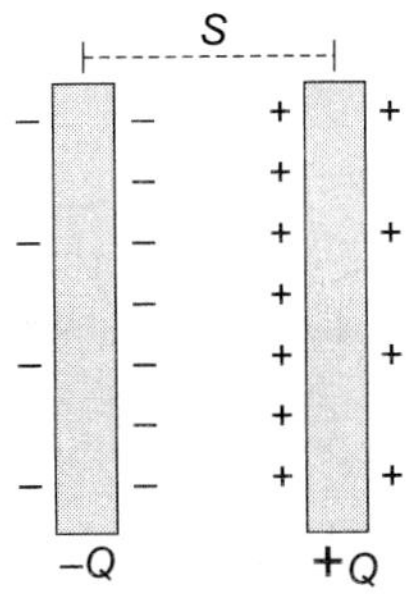

Side view of the 2 disks

A single charged metal disk cannot have a truly uniform charge density, because the mobile charges tend to push each other to the edge of the disk. Nevertheless, our results for disks made of insulating material are approximately correct for metal disks, especially in this two disk configuration if the disks are very close together ($s << R$). Due to the attraction by the neighbouring disk, almost all of the charge is uniformly distributed on the inner surfaces of the disks, with very little charge on the outer surfaces of the disks.

We'll calculate the strength of the electric field at locations near the centre of the disks, both inside and outside the capacitor.

Step 1: Cut up the charge distribution into pieces and draw $\Delta\vec{E}$.

We cut up the distribution into just two pieces because we know the electric field made by a single uniformly charged disk.

The diagram below shows a side view of the region near the centre of the disks, blown up. The disks are very close together ($s << R$), and extend up and down beyond the boundaries of the diagram. Drawn in the diagram at each of the locations labelled 1, 2 and 3 are arrows labelled E_+ and E_- showing the contributions of each disk to the net electric field at that location. **Note:** The tails of the arrows are shown at the location points' different vertical levels only for the sake of clarity.

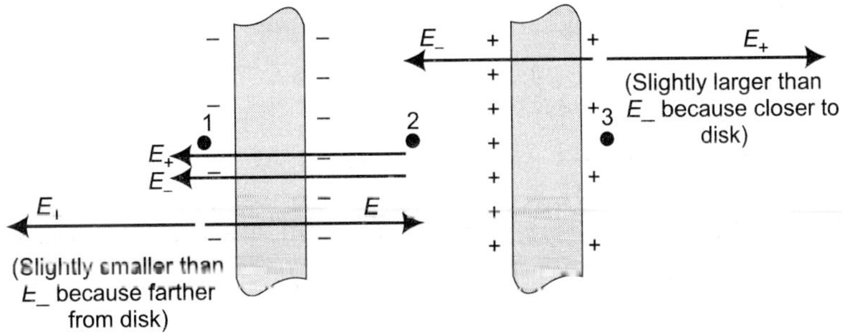

Notice that at locations 1 and 3, $E_+ \neq E_-$ because the distances to the two disks are slightly different. On the diagram below are drawn the "net" electric field at points 1, 2 and 3. The vector lengths reflect the relative magnitudes of the electric fields.

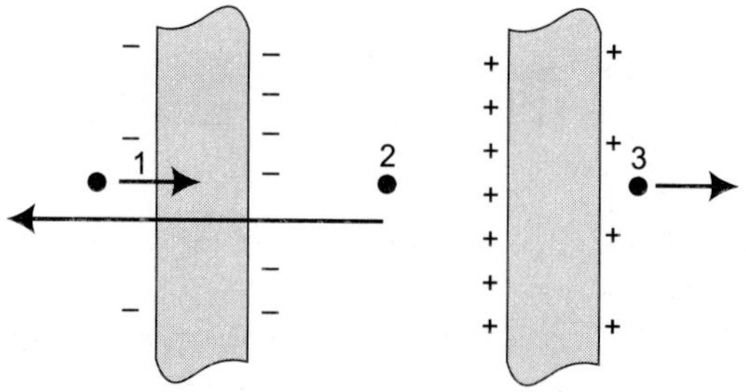

The above figure is important to understand the behaviour of capacitors.

Step 2: Write an expression for the electric field due to one piece.

Assuming a uniform charge density on each disk (and ignoring the small amount of charge on the outer surfaces of the disks) we know that the field *Edisk* near the centre of a single disk is approximately using eqn. (8)

$$E_{\text{disk}} = \frac{\left(Q/A\right)}{2\varepsilon_o}\left[1 - \frac{x}{R}\right]$$

We're assuming that the plate separation s is very small ($s << R$), and that locations 1 and 3 are very near the disks ($x << R$).

Step 3: Add up the contributions of all the pieces

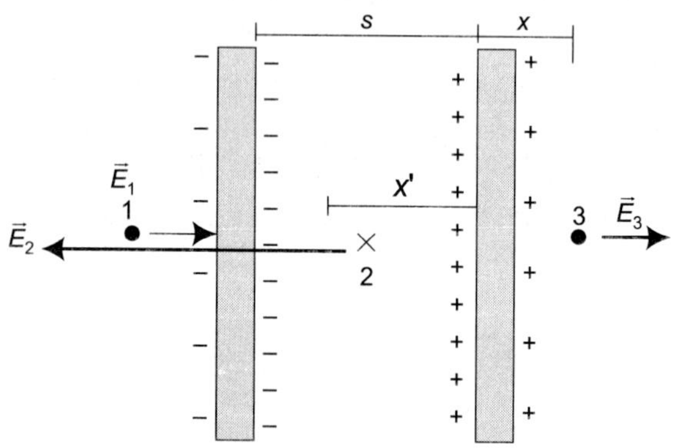

We calculate the magnitude of $\vec{E}_2$ inside the gap as the superposition of the field of the negative disk and the field of the positive disk, both of which point to the left inside the gap. Consider a point a distance x' from the positive plate (and a distance $s - x'$ from the negative plate).

$$E_2 = \frac{\left(Q/A\right)}{2\varepsilon_o}\left[1 - \frac{x'}{R}\right] + \frac{\left(Q/A\right)}{2\varepsilon_o}\left[1 - \frac{s - x'}{R}\right]$$

$$= \frac{\left(Q/A\right)}{2\varepsilon_o}\left[1 - \frac{x'}{R} + 1 - \frac{s - x'}{R}\right] = \frac{\left(Q/A\right)}{2\varepsilon_o}\left[2 - \frac{s}{R}\right]$$

$$= \frac{\left(\frac{Q}{A}\right)}{\varepsilon_o}\left[1-\frac{s/2}{R}\right] \qquad = \frac{\left(\frac{Q}{A}\right)}{\varepsilon_o} \quad \text{to the left}$$

The electric field between the plates is essentially twice the field due to one plate. The magnitude hardly depends on x', so the field is remarkably "uniform". As you move away from the positive plate, the very slightly smaller contribution of the positive plate is nearly compensated by the very slightly larger contribution of the negative plate.

Capacitance

The potential difference between the plates will be $V = Es = \frac{Qs}{\varepsilon_o A}$. The straight multiplication Es was possible because the field E between the plates is uniform from one plate to the other. Therefore, the capacitance which is the amount of charge held in the plates divided by the voltage $\left(\frac{Q}{V}\right)$ is given by $\frac{\varepsilon_o A}{s}$. The reader may appreciate that capacitance depends entirely on the geometry of the distribution of charges on the conductors. Since all conductors have an enormous store of mobile charges, their number and distribution on capacitor plates is what determines its capacitance and not its resistivity. It is to be regarded that the conductive material of the plates enabled the capacitor to guide and hold the distribution (tight packing) of the few mobile charges (out of the millions available in the conductor) on its plates. This is true for all conductive materials used for making capacitors.

Fringe Field (Chabay & Sherwood, 1995, used by permission)[2]

Next, let us calculate the magnitude $\vec{E}_3$ just outside the capacitor.

Location 3 is a distance x from the main positive charges on the inner surface which makes a field to the "right", and a distance $x + s$ from the main negative charges, which make a field to the "left". Therefore, the magnitude of $\vec{E}_3$ is the "difference" of the fields of the two disks:

$$E_3 = \frac{\left(\frac{Q}{A}\right)}{2\varepsilon_o}\left[1-\frac{x}{R}\right] - \frac{\left(\frac{Q}{A}\right)}{2\varepsilon_o}\left[1-\frac{x+s}{R}\right]$$

$$- \frac{\left(\frac{Q}{A}\right)}{2\varepsilon_o}\left(\frac{s}{R}\right) \quad \text{to the right}$$

Similarly, we find in the same way at location 1 just to the left of the capacitor, we have

$$E_1 = \frac{\left(\frac{Q}{A}\right)}{2\varepsilon_o}\left[1-\frac{x}{R}\right] - \frac{\left(\frac{Q}{A}\right)}{2\varepsilon_o}\left[1-\frac{x-s}{R}\right]$$

$$= \frac{\left(\frac{Q}{A}\right)}{2\varepsilon_o}\left(\frac{s}{R}\right) \text{ to the right}$$

The most important thing to note is that the field outside the capacitor is very small compared to the field inside the capacitor. Inside the gap, the fields of the two disks add, but outside the gap the fields of the two disks, subtract.

By calculating the ratio of the outside field to the inside field, we can show that the ratio is very small if $s << R$.

$$\frac{E_{outside}}{E_{inside}} = \frac{E_3}{E_2} = \frac{\frac{Q/A}{2\varepsilon_o}\left(\frac{s}{R}\right)}{\frac{Q/A}{\varepsilon_o}} = \frac{s}{2R} \rightarrow 0 \text{ if } s << R. \qquad ...(9)$$

The small field outside the capacitor is often called the "fringe field". We have calculated the fringe field at a location outside the capacitor but very close to it, where the fringe field hardly changes with distance x. (If however, you are very far away, so $d >> R$, the capacitor looks like an electric dipole, and the fringe field falls off like $1/d^3$.)

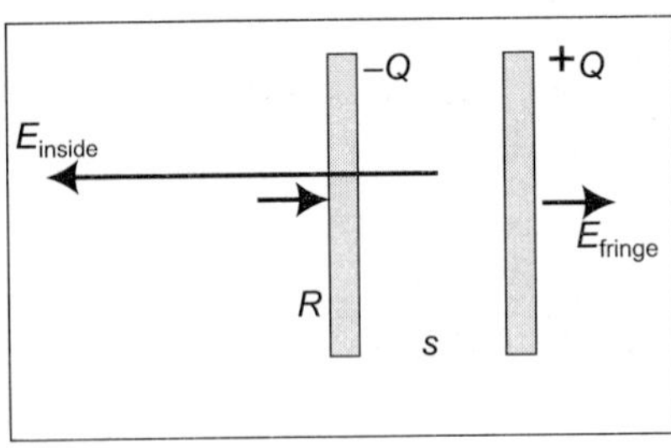

Two-disk capacitor: plate area $A = \pi R^2$, near centre

$$E_{inside} = \frac{Q/A}{\varepsilon_o}$$

$$E_{fringe} = \frac{Q/A}{\varepsilon_o}\left(\frac{s}{2R}\right)$$

The results we obtained apply not only to circular capacitors but also to capacitors with rectangular or other shapes or plates, where A is the area of one of the plates. The same restriction applies as with circular capacitors; the plate separation s must be much smaller than the width or height of a plate. As long as the gap is very small, E_{inside} is nearly equal to $\frac{Q/A}{\varepsilon_o}$ near the centre of the plates. $E_{outside}$ near the centre of the plates depends on the shape of the plates, but is always much smaller than E_{inside} by a factor of about (s/L) where L is the width or height of a plate (or diameter $2R$ for circular plates).

We saw that the electric field inside the capacitor near the centre of the plates is remarkably uniform across the gap. It is also true that the magnitude and direction of the electric field are practically the same everywhere in the gap; not just near the centre of the plates. Only if you get very near the outer edges of the plates does the electric field vary much from its centre value Here is a close-up view of the gap in a parallel plate capacitor, not necessarily circular, far from the outer edges of the plates (the small fringe field is not shown).

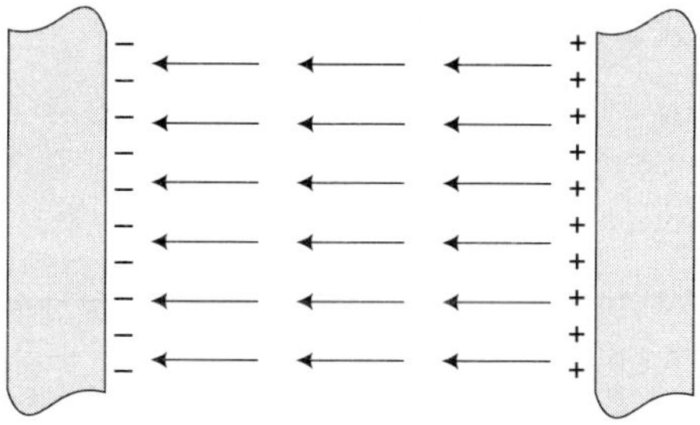

Notice again that despite the $1/d^2$ character of electric interactions among point particles, the electric field due to a distribution of a large number of charges can have a very different distant dependence. Here the electric field is uniform in magnitude and direction throughout a region.

The small fringe field plays an important role in the discharge of capacitors and in the current pattern that results whenever a capacitor is charged when a voltage is applied to a capacitor.

Example: If the electric field in air exceeds roughly 3×10^6 newtons/coulomb, the air breaks down and a spark forms. For a two-disk capacitor of radius 50 cm with a gap of 1 mm, what is the maximum charge (positive and negative) that can be placed on the disks without a spark forming (which would permit charge to flow from one disk to the other?)

We can calculate how much charge it would take on the capacitor ($+Q$ and $-Q$) to produce a field sufficiently strong to breakdown air:

$$E = \frac{Q/A}{\varepsilon_o} = \frac{Q}{\pi R^2 \varepsilon_o} \quad \text{because } A = \pi R^2$$

So,

$$Q = \pi \varepsilon_o R^2 E$$

$$= 9 \times 10^{-12} \frac{C^2}{Nm^2} \pi (0.5m)^2 \left[3 \times 10^6 \frac{N}{C}\right]$$

$$= 2.08 \times 10^{-5} C$$

Under these conditions, what is the strength of the fringe field just outside the centre of the capacitor?

The fringe field just outside the capacitor is approximately (using expression (9) with E_{fringe} being the field E_{outside}:

$$E_{\text{fringe}} = \left[\frac{s}{2R}\right] E_{\text{inside}}$$

$$= 0.5 \left[\frac{0.001m}{0.5m}\right] 3 \times 10^6 \ N/C = 3 \times 10^3 \ N/C$$

We also see that E_{fringe} is a fraction ($s/2R$) times the electric field inside E_{inside} and in this example is about 1/thousand times E_{inside}.

It is interesting to note that if the radii of the disks are infinitely large, then, the fringe field using the ideal approximation (see Section 2.11) is zero.

A Note on the Units of the Electric Field

The electric field may also be expressed in volts/metre. If we write the units of volts in joules (J) per coulomb (C) then, since joules (J) are expressed in Newton (N)-meter (m), this becomes evident:

$$\frac{V}{m} = \frac{J}{C-m} = \frac{N-m}{C-m} = \frac{N}{C}$$

Energy Stored in Stretched Springs and Unlike Charges that are Separated

In case of a real spring anchored at one end to a wall with a mass attached to the other end, you have to apply a force that is non-constant to slowly pull the mass to a distance x_2 away from x_1 where x_1 is the initial rest distance of the mass from the wall. This is because the spring exerts a force that is k times x (where k is the spring stiffness) that is opposite to the force you exert to pull the mass. The work you did will be stored in the spring as "elastic potential energy" and will be $\int_{x_1}^{x_2} kxdx = \frac{1}{2}kx_2^2 - \frac{1}{2}kx_1^2$. Similarly, if you pull a positive charge q_2 that is a distance r_1 away from a (fixed) negative charge $-q_1$ you have to apply a non-constant force of magnitude $\frac{1}{4\pi\varepsilon_o}\frac{q_1q_2}{r^2}$ that is opposite to the force exerted by the charge $-q_1$. If you move it to a new distance that is r_2 away, the work done will be stored as electric potential energy and will be $\int_{r_1}^{r_2}\frac{1}{4\pi\varepsilon_o}\frac{q_1q_2}{r^2}dr = \frac{q_1q_2}{4\pi\varepsilon_o}\left[\frac{1}{r_1}-\frac{1}{r_2}\right]$.

The observant reader would have noted that the force required to cause separation of an object of mass m from the earth (to raise it) to a few metres above the earth will be constant (mg) and the force required to pull the mass attached to the spring increases with distance (kx) of separation from the wall (within reasonable distance limits before the spring snaps !). The force required to pull the positive charge away from the negative charge decreases (proportional to $1/r^2$) with increasing separation distance r.

Energy Stored in Capacitors

Suppose that at some point in the process of charging a capacitor, the charge on the positive plate is q, so that the potential difference is q/C. Then, the work you must do to transport the next piece of charge, dq, is $dW = \left(\frac{q}{C}\right)dq$. The total work necessary, then, to go from $q = 0$ to $q = Q$, is $W = \int_0^Q \left(\frac{q}{C}\right)dq = \frac{1}{2}\frac{Q^2}{C}$ or, since $Q = CV$, $W = \frac{1}{2}CV^2$, where V is the final potential of the capacitor.

Appendix B

Comparison of electric and gravitational forces (Chabay & Sherwood, 1995, used by permission)[2]

To get a feel for the strength of electric interactions, let us compute the magnitude of the electric force between two particles, and compare it to the magnitude of gravitational forces.

The radius of an atom is of the order of 10^{-10} metres. The electric force of attraction between a proton and an electron that are this far apart (as they might be in a hydrogen atom consisting of 1 proton and 1 electron) is

$$F = \frac{1}{4\pi\varepsilon_o}\frac{e^2}{d^2} = \left(9\times10^9\,\frac{Nm^2}{C^2}\right)\frac{(1.6\times10^{-19}\,C)(1.6\times10^{-19}\,C)}{(10^{-10}\,m)^2} = 2.3\times10^{-8}\ N$$

The force does not seem to make much sense unless a meaningful comparison is made with some other force such as a gravitational force.

We can calculate the gravitational force on a hydrogen atom due to the earth:

$$F_{\text{grav}} = mg\ (1.7\times10^{-27}\ kg)(9.8N/kg) = 1.7\times10^{-26}N$$

How does the magnitude of the electric force binding the hydrogen atom together compare to the gravitational force exerted by the entire earth on the hydrogen atom?

$$\frac{F_{electric\ on\ electron\ by\ proton}}{F_{gravitational\ on\ atom\ by\ earth}} = \frac{2.3\times10^{-8}\,N}{1.7\times10^{-26}\,N} = 1.3\times10^{18}$$

So, the electric force exerted by the proton on the electron is about 10^{18} times larger than the gravitational force exerted by the earth on the hydrogen atom.

The electric force of the outermost electron on the proton of an atom in a copper bar will be small, say, 2.3×10^{-8} newtons by a similar calculation as above and, a hammer and sharp chisel that exerts a force of 100 newtons could prise electrons from the bar. It should be noted that the force actually transmitted to individual atoms of the bar would be quite small, however sharp the chisel point is. The chisel point would be several atomic diameters thick resulting in a stress or force per unit area transmitted to be of the order of 10^{-8} *N*.

Observing static electric phenomenon of this kind between metals require careful and controlled experimentation and have not been done, but it has a fair chance to be proven given the fact that if two optically flat metallic pieces were brought in contact, it would be diffcult to prise them apart on account of electron transfer.[24]

The force that holds inner electrons to the nucleus is far greater than the force of 2.3×10^{-8} newtons and these cannot be affected much by the forces exerted by the hammer and chisel.

The force exerted by a voltmeter probe will still be tinier and will not get "inside" a surface charge on a conducting wire !

See Appendix D for a very simple description with images of the forces at work at the atomic level when an object is touched.

The gravitational attraction between the proton and electron is (Chabay & Sherwood, 1995, used by permission)^2

$$F = G\frac{m_e m_p}{d^2} = \left(6.7\times10^{-11}\frac{\text{Nm}^2}{\text{kg}^2}\right)\frac{(9\times10^{-31}\text{kg})(1.7\times10^{-27}\text{kg})}{(10^{-10}\text{m})^2} = 1.0 \times 10^{-47}N$$

How does the magnitude of the electric attraction between the proton and the electron compare with the magnitude of the gravitational attraction between the proton and the electron?

$$\frac{F_{\textit{electric on electron by proton}}}{F_{\textit{gravitational on electron by proton}}} = \frac{2.3\times10^{-8}\,N}{1.0\times10^{-47}\,N} = 2.3 \times 10^{39}$$

So, the electric force exerted by the proton on the electron is about 10^{39} times larger than the gravitational force exerted by the proton on the electron.

Stable Atoms (Made to Nature's specifications)

Why does an electron not fall into the nucleus?

If one takes a positive nucleus and a stationary negative electron, the nucleus will attract the electron, and the electron will start moving towards the nucleus until the electron falls into the nucleus. This cannot be a model for an atom. The electrons cannot be hanging stationary above the nucleus. They must be moving. We can think of circular motion, just as the planets move around the sun. The planets (uncharged objects) are attracted towards the centre of the sun, by gravity, in a manner which constantly changes their *direction, but not magnitude* of tangential velocity given to them since the time that they were created. If halted, the planets would fall in towards the sun. As a matter of fact, they are going too fast to actually do this. This circular motion protects the planets against falling into the sun. And, also note there is no friction where the planets orbit the sun, because there is no atmosphere, and there is no friction where the electron orbits the nucleus as well ! Therefore, there is no loss of energy of the frictional type. So one might think, this would work for electrons as well. But......

The Model Works for Planets but does not work for electrons! Why?

Classical mechanics teaches us that if a charged object (such as the electron) moves with an acceleration (in other words, the magnitude or the direction of its velocity is changing in time), this object RADIATES ELECTROMAGNETIC WAVES ![25]

If an electron (which is a charged object) would be doing a circular motion around a nucleus, the direction of its velocity would be changing, so it would radiate out energy; therefore it would lose energy, which means it would spiral down to the nucleus.

Classical mechanics **does not** give a satisfactory answer, as to why atoms are stable.

The answer is found in Quantum Mechanics (QM).

1. If an electron is in the electric field of a nucleus, the electron can occupy only certain energy levels. When it is sitting on one of these energy levels, it does not radiate, it does not lose energy. QM shows the way how to calculate these energy levels. Electrons are allowed to move around the nucleus ONLY IN CERTAIN SPECIFIC orbits, called Sommerfeld orbits and the enhancements of Schrodinger and Pauli. In these orbits, the electrons DO NOT RADIATE!!!! So, they are stable. And again, the orbits are not definite.
2. It is possible for electrons to change energy levels, but they have to either absorb, or emit a quantum of energy. This energy is an INTEGER number multiplied by a Planck's constant.
3. It is now known that the electrons are not localized in certain, well defined and precise positions or orbits around the nucleus, but occupy *orbital spaces* as seen in the images below of simulations on a computer of these: the *s*-orbital (left image) and *p*-orbital (right image).

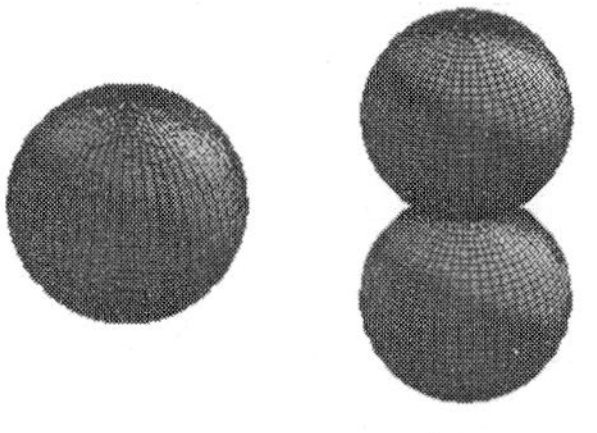

In fact, there is a finite "nonzero" probability of finding an electron anywhere in our universe.

Quantum Mechanics, Probabilities and Orbitals of Electrons

Quantum mechanics gives a prediction and shows how to calculate the probability of finding an electron at a given space point of our universe. Universe? Yes, solutions to the electron wave equation are available at several points upto the ends of the universe! And probability?

We have learnt that the electron is very close to the nucleus (say, within a 10^{-10} metre radius of the nucleus), and this is so because an electron prefers to be so near the nucleus where it will be in stable equilibrium. There is a *high probability* of it being near the nucleus than elsewhere in the universe, just as at the end of the day after your energies are spent in college or in a playground or a movie theatre or in a shopping mall, there is a high probability of finding you in the comfort of your room! High probability? Not sure? Yes, it is entirely probable that you may go away to a relative's house and spend the night!

In the interpretation of large numbers of events or data gathered about physical phenomena, as your movements in the course of a day, mathematicians use statistical methods. These phenomenon cannot be described by equations that predict the future unambiguously and deterministically, as you would with equations of motion, that you use to determine the position and velocity of an object after a certain time dropped from a height with an initial velocity.

When studying physical phenomenon that involve a large number of parameters, the more the number of observations of events or experimental inferences, better are the statistical models and their predictions.

With the development of the atomic structure of matter and electricity, physicists found it necessary to formulate statistical laws to describe the energies of electrons.

Since it was not possible to experimentally *determine* the movement of electrons, physicists used a process of elimination to predict the likelihood of electron positions around the nucleus and in the entire universe; for example, for an object lying on a table it is not entirely possible for one of the electrons of its atoms to be in the corner of the room, so a very tiny probability is assigned to that position, and one metre away the probability is a tiny fraction more, and so on.

That is why the orbitals of electrons are expressed using probabilities.

The electron cloud model represents a sort of history of where the electron has probably been and where it is likely to be going. Imagine, as the electron moves it leaves a trace of where it was. This collection of traces quickly begins to resemble a cloud.

For an animation of the quantum mechanical model of the hydrogen atom please visit http://dwb4.unl.edu/chemAnime/BOHRQD/BOHRQD.html and click on "Quantum Mechanical Model". Visit http://www.av8n.com/physics/orbitals.htm#htoc6 for an animation of electron position from positional probability data extracted from orbitals.

In semiconductor physics, how is it possible to obtain an expression for the number of electrons and holes in an intrinsic semiconductor using such probabilistic models?

In semiconductor physics, the origin of the Fermi-Dirac probability function can be traced to probabilistic considerations. We first associate a density of possible states or energy levels with the probabilistic occupancies of these states by electrons and holes to determine the number of holes in the valence band and the number of electrons in the conduction band, at a given thermal equilibrium temperature.

An integration is performed over a band of energies to "capture" those electrons and holes that have *probable* occupancies into the *verifiable* number of conduction electrons and valence holes that fill these energy bands.

This is similar to the summation of the number of students who have obtained equal grades over the entire range of possible grades to obtain the total number of students who appeared in an examination.

In semiconductors, the number of electrons in the conduction band and the number of holes in the valence band are essential to determine the conductivity of the material which is used in the derivation of the volt-ampere relationship and the current components of p-n junctions.

Energy Levels of Electrons

To a good approximation, we can identify the total energy of an electron with the size of its orbit. That is, we can think of each radius on the left of the figure below as equivalent to an energy level of the figure on the right.

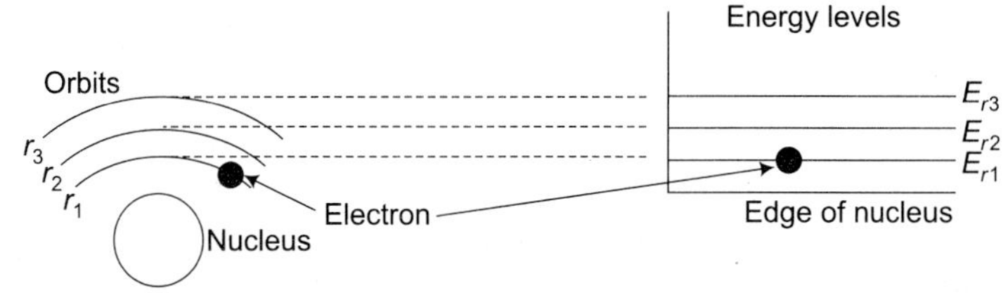

Energy level is proportional to orbit size (radius)

Electrons in the smaller orbit r_1 are on the first energy level E_{r1}; electrons in the second orbit r_2 are on the second energy level E_{r2}, and so on.

Is an atom a dipole?

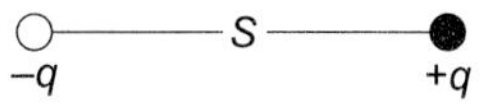

The electric dipole moment (product of the charge in coulombs 'q' and the separation distance 's' = qs) is caused from an electric dipole (a separation between positive and negative charge).

Picture an atom with its positively charged nucleus and negatively charged electrons.

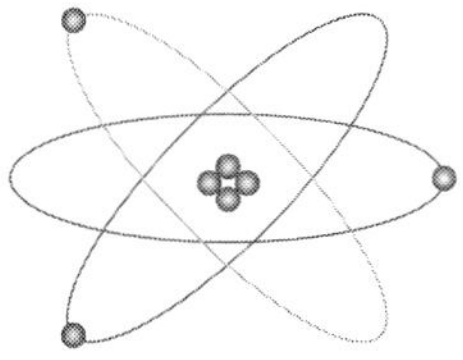

The separation between the positive and negative charge is constantly changing direction (i.e., electric dipole is a vector), and the average (if you wish) is just zero.

A second way is to imagine the electron as a "cloud" that surrounds the atom, then it appears that the dipole moment is zero. This is essentially a classical model.

With an asymmetric distribution of charge, one can expect a dipole to appear. But even with a perfectly symmetric charge distribution (assuming point particle and a stable atom in the ground state) there can result an electric dipole moment (qs), quantum mechanically.

However, the electric dipole moment (qs) of an atom is yet vanishingly small due to symmetries. This is the cause for near perfect field cancellation around the atom and the resultant negligibly tiny electric field outside that it qualifies to treat the atom as being "neutral" and not as a dipole.

In metals, where millions of atoms are packed closely together, there is a resultant electric field around the atoms that is not zero, yet vanishingly small—so negligibly small, that it qualifies to treat the interior of an isolated metal as being neutral, not forming significant dipoles.

These tiny insignificant dipoles, however, produce "noise" emfs that superimpose on small signals transmitted in sensitive electronic circuits !

How is a molecule a dipole?

Consider the water molecule. A single H_2O molecule is an electric dipole, the H end is somewhat positive and the O end is somewhat negative.

A molecular dipole could be either an induced dipole created by external charges or a permanent dipole such as the water molecule.

In a molecule there is an electric field *around the component atoms* in so far as treating the atoms independently.

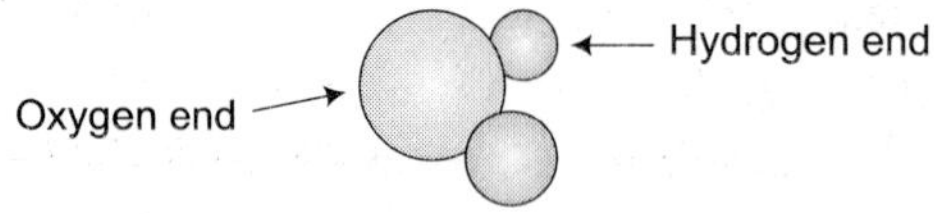

Molecules bond when their electron "clouds" overlap substantially enough that there are in fact electrical fields within each atom induced by the radial fields between the electron "cloud" and nucleus of the other.

To view the electric field (resultant) surrounding a pair of unlike charges...that grows (uncancelled fields of each charge) around the pair when pulled apart to form a dipole and collapses when brought near each other, please visit http://ocw.mit.edu/ans7870/18/18.013a/textbook/HTML/chapter28/section04.html. On the interactive exercise in the webpage there, please make all charges = 0, except q1 and q2, to view the field due to these two charges.

You can even drag to join them together...and the field collapses totally around the pair.....however, note that this is not possible to do *practically* with a real nucleus and its valence electron (a pair of unlike charges)....nature prevents you from doing this ! There will always be a tiny...ever so tiny resultant electric field around the atom!

Will it so happen that if the atoms were in a place where the temperature is extremely cold (say, in the depths of outer space) the electrons will move so slowly and will "fall" into the nucleus?

Quantum mechanics states that among all the possible energy levels an electron can sit in the presence of a nucleus, there is one, which has THE MINIMAL energy. This energy level is called the "ground state". So, even if atoms are in a *very very cold* environment, QM prohibits electrons from falling to the nucleus. The worst that can happen is that they will sit in their ground states.

Metallic bonding and Metallic structures (a special kind of great importance to all engineers and scientists)

Unlike the ionic compounds (sodium chloride) and atoms joined together by the classic covalent bond (diamond and graphite) that gives rise to their structures, the manner in which the atoms are arranged in a lattice of the crystal, for example, in a grain of salt and in a diamond in your ring, a bond *may not always determine* a structure. We consider a metallic structure, which is of special importance to mechanical and electrical and electronic engineers, where the bond does not determine its structure.

This is due to the manner in which the *atoms were packed* when the metal was created, whether occurring naturally or man-made as in a furnace.

The structure of the metal is determined by the fact that each atom tries to be as close to as many other atoms as possible. Metals are well known to be solids (except for Mercury!). The bonds between atoms in metals can loosely be described as *covalent bonds* (due to sharing electrons), except that the metal atoms do not just share electrons with 1, 2, 3 or 4 neighbours, as in covalent bonding, but *with many atoms.*

Take the metal Sodium that has the electronic structure $1s^2 2s^2 2p^6 3s^1$. Sodium has an electrical conductivity that is about a quarter of that of aluminium.

When sodium atoms come together, the electron in the $3s$ atomic orbital of one sodium atom shares space with the corresponding electron on a neighbouring atom to form a molecular orbital — in much the same sort of way that a covalent bond is formed.

The difference, however, is that each sodium atom is being touched by eight other sodium atoms — and the sharing occurs between the central atom and the $3s$ orbitals on all of the eight other atoms. And each of these eight is in turn being touched by eight sodium atoms, which in turn are touched by eight atoms—and so on and on, until you have taken in all the atoms in that lump of sodium.

Therefore, this kind of bonding is given a special name "Metallic Bonding".

All of the $3s$ orbitals on all of the atoms overlap to give a vast number of molecular orbitals which extend over the whole piece of metal. There have to be huge numbers of molecular orbitals, of course, because any orbital can only hold two electrons.

The electrons can move freely within these molecular orbitals, and so each electron becomes detached from its parent atom.

The electrons are said to be delocalised. The metal is held together by the strong forces of attraction between the positive nuclei and the delocalised electrons. They must be holding the metal together....else, there would not have been pieces of metal...there would be only metal ions in the universe ! No metal pieces would have been possible ! It could be that nature decided in some inexplicable way not to make an ideal (noble gas style) bonding in favour of a structure, that we could hammer, bend and shape to any object we desire !

Metals tend to have high melting points and boiling points suggesting strong bonds between the atoms.

Metals are therefore, giant structures of atoms held together by metallic bonds. "Giant" implies that large but variable numbers of atoms are involved — depending on the size of the bit of metal.

The delocalized electrons are the conduction band electrons or the outermost electrons that gained just enough energy to become free and roam the lattice freely.

Energy Stored in a Two Electron Configuration

The rest energy of the electron 0.511 Mega electronvolts or 0.511×10^6 eV was calculated using Einstein's famous mass-energy relation: $E = mc^2$ where, m = rest mass of the electron = 9.109×10^{-31} kilogram and c = the speed of light = 3×10^8 m/sec. We assume that the electron is at rest for the calculation, an impossible achievement! We have idealized the electron as a point charge.

Using $E = mc^2$, we get, $E = 9.109 \times 10^{-31} \times (3 \times 10^8)^2 = 82 \times 10^{-15}$ Joules. Since 1 Joule $= 6.2 \times 10^{18}$ electronvolts, therefore, 82×10^{-15} Joules $= 82 \times 10^{-15} \times 6.2 \times 10^{18}$ electronvolts $= 0.508 \times 10^6$ electronvolts or, approximately, 0.51 Mega electronvolts(eV).

If we could compress two electrons together to a distance of say, 10^{-10} metres, then, the potential energy (PE) stored in this configuration is given by $PE = kqQ/r$

where, $q = Q = 1.6 \times 10^{-19}$ Coulombs, $r = 10^{-10}$ metres and, $k = \dfrac{1}{4\pi\varepsilon_o} = \left(9 \times 10^9 \dfrac{Nm^2}{C^2}\right)$.

Substituting these values in the relation for the PE, we obtain, $PE = 23 \times 10^{-19}$ Joule.

Using, 1 Joule = 6.2×10^{18} electronvolts, we get the PE = 14.2 electronvolts(eV).

This is the interaction energy of the two electrons separated by a distance of 10^{-10} metres or, in other words, the energy required to assemble this configuration.

While it does not make sense to compare the energy due to the field interaction of the two electrons, or pontential energy, with their rest energy, the compasison puts into perspective the tininess of the amount of the potential energy; the two electrons have a total rest energy of approximately 10^6 electronvolts while the energy of their interaction a distance 10^{-10} metres apart, is a mere 14.2 electronvolts.

Resistivity of Metals

If the metal crystal lattice were to be perfect, geometrically perfect and flawless (free from lattice imperfections) an electron will move freely with a mean time between collisions of *infinity*.

But it has been observed experimentally that the mean free time in a piece of sodium metal between collisions is 3×10^{-14} sec and the metal begins to exhibit resistance to the flow of electrons. This finite mean time is the reason for the 10^{14} collisions per second an electron experiences as was described in Section 1.18 "Neutrality in current carrying conductors".

The progress of the electron-wave like movement (of conduction band electrons) that interacts with large regions in the crystal is *interrupted* by irregularities in the array or so-called lattice imperfections and that is one of the reasons for the observed mean time and the resistivity of metals[22]. While atomic arrangements are characterized by translational periodicity in three dimensions in ideal crystals, real crystalline materials are rarely perfect.

Several types of lattice imperfections are present in metallic crystals. Lattice imperfections are classified into: (i) point defects (ii) line defects (iii) planar defects and (iv) bulk or volume defects. Among these the most significant lattice imperfections that contribute to the resistivity of metals are the point defects.

Atoms in a material are arranged in a periodic manner. A pure copper crystal structure is face-centred cubic (FCC) and copper atoms occupy all the sites of the FCC lattice. Occasionally, a few lattice positions remain vacant. A missing atom from a normal lattice site in a crystal is called a "vacancy" as shown in the figure below.

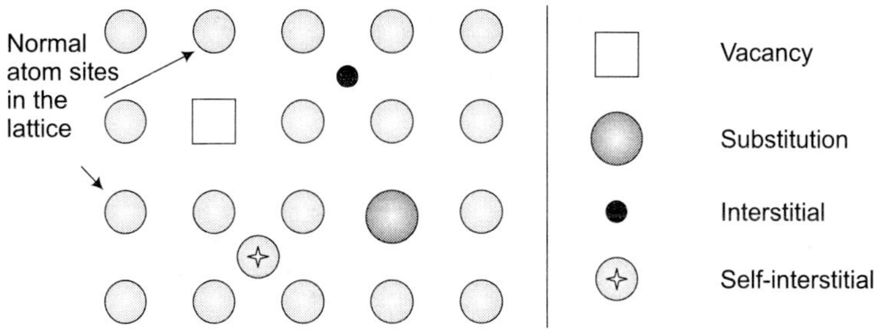

Vacancies influence the property of the conductivity of materials. Sometimes, instead of a vacancy an impurity atom may occupy a regular lattice site. The impurity atoms can be accommodated either at normal lattice sites or in the interstices between the matrix (lattice site) atoms.

If the impurity atom of nearly the same size as the lattice's original matrix atom occupies a site as in the figure above, it is called "substitution". Sometimes if the impurity atoms are very tiny they may occupy interstitial sites between the atoms. In steel (iron-carbon alloys), for example, carbon atoms occupy interstitial positions as shown in the figure above. Another defect occurs when the lattice's original atom occupies an interstitial site. Such a defect is called self-interstitial defect as seen in the figure above.

Point defects such as vacancies and interstitials influence the electrical conductivities of materials. The defects of one of the above types or several in combination may be present in materials.

The other significant contributor to resistivity is the vibrations of the lattice ions at elevated temperatures. The vibrations are random in nature which makes the lattice at any moment slightly irregular geometrically, and the more so, the higher the temperature.

It is this effect which further impedes the progress of the electrons inside the lattice structure and makes the resistivity of a pure metal to increase as the temperature is raised.

If a real piece of copper is cooled below about 20 kelvin, we may expect to attain very low resistivity. But this does not happen; as a matter of fact, there is some residual resistivity due to the lattice defects.

For conductors, $v(drift) = \mu E$ where μ is the electron mobility and E the applied electric field strength. With increasing number of lattice imperfections as in resistors, the mobility decreases and consequently a smaller current "i" is established in them for a given applied electric field since $i = nAv(drift)$, where n is the number density of mobile elecrons and A the area of cross section of the conductor.

In electronic circuits, carbon composition resistors are the most commonly used components. The carbon atom has six electrons, 4 of the electrons are in its valence shell (outershell). There are three types of carbon, fullerenes, diamonds, and graphite. The difference between fullerenes, diamonds, and graphite is in their molecular structure.

Graphite has a layered molecular structure and there are very weak bonds between the layers, which gives graphite its slippery feel as opposed to diamonds which have rigid bonds between their atoms which allows very little or no movement.

Graphite has many forms and uses in our everyday life. Natural graphite can be found in China, Mexico, Canada, Ukraine, and Brazil, and it is used as a lubricant, filler for flexible graphite, artistry tools, KEVLAR, and of course, pencils. It has a very low resistivity ranging from 9 to 40 μΩm.

The different forms of graphite accounts for the range in resistivities. Some graphite is manufactured while some is mined from the above locations. Because of its layered molecular structure, graphite has parallel and perpendicular resistivities, depending on how the graphite is oriented in a circuit.

Graphite is an interesting material since it displays properties of both metals, and nonmetals. However, like a metal, graphite is a very good conductor of electricity due to the mobility of the electrons in its outer valence shells.

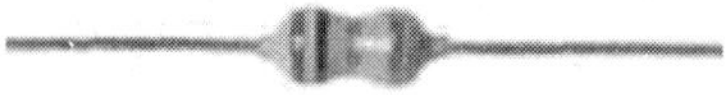

The resistive element consists of finely ground carbon (graphite) particles with refractory filling. These are bonded together using a synthetic resin binder and the mixture is then either compressed into shape and fired in a kiln or coated onto a substrate such as glass and then fired. The composition of the mixture determines the resistance value.

Carbon Film Resistors have many of the same characteristics as carbon (graphite) composition resistors and are made by coating ceramic rods with a mixture of carbon materials. This material is applied to these rods in a variety of means, the ones most familiar are dipping, rolling, printing, or spraying the rods in the appropriate solution.

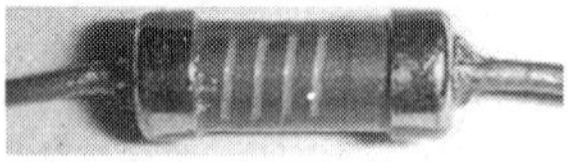

These rods are cut to the length required for a specific size of resistor. These rods can then be spiral cut to a wide range of values. The original method of spiralling these was done with grinding wheels on a machine similar to a lathe. Later processes use lasers that are programmed to cut to specific values.

How is heat produced in conductors and resistors?

Heat is a form of energy and an increase in the random kinetic or vibrational energy of lattice ions manifests itself as heat and a rise in the temperature of the conductor. The heat energy of a substance is determined by how active its atoms and molecules are. A hot object is one whose atoms and molecules are excited and show rapid movement. A cooler object's molecules and atoms will be less excited and show less movement. Please also refer Section 2.23 "The process of capacitors sharing charges" for a description of heat, its production and dissipation as e-m waves in conductors (wires).

Now that we have understood heat, let us understand what types of collisions occur in nature. A collision can be inelastic which is what most collisions in nature are and fortunately (else, there would not be the electric light bulb !) for us, occurs when an electron collides with a lattice ion.

So what is different between an inelastic collision and an elastic collision? An inelastic collision can result in the production of heat while an elastic collision (if done to perfection!) cannot.

The nearest thing to an elastic collision we can produce is the collision of two billiard balls. The smoothness of their surfaces, the smoothness of the billiard table surface (very little friction) and the point-like area of contact when they collide ensures that kinetic energy is conserved (almost) along with momentum. Therefore, after the balls collide their kinetic energies macroscopically change but are conserved and most importantly they move at the same temperature they were before collision.

Another example is the gravity assisted slingshot that spacecraft use to alter the spacecraft's path and movement using their relative movement and the gravity of a planet. The linear momentum gained by the spaceship is equal in magnitude to that lost by the planet, though the planet's enormous mass compared to the spacecraft makes the resulting change in its speed negligibly small. Typically gravity assist is used to save propellant, time, and expense. Note that there is no physical contact between the spacecraft and the planet!

With an electron and lattice ion collision, which is inelastic, momentum will be conserved but the kinetic energies will not. So, where is the missing energy?

The electron in an inelastic collision with a lattice ion acquires some extra kinetic energy, as well as momentum, between collisions. A collision, or at most a few collisions, redirects its momentum at random (conserving it) but does not necessarily restore the kinetic energy to normal. Some of the missing energy is radiated away due to acceleration and lost movement of the neutrons and protons of the lattice ions. For kinetic energy to be conserved, the electron has to transfer kinetic energy to the lattice ion that deflects it. Again, note that the collision between the electron and the lattice ion is not a physical contact due to strong electron repulsions ! And since it has a smaller mass than the ion it collides with, like a tiny marble colliding say, with a cricket ball, the electron will continue to accumulate extra energy until its average kinetic energy is so high that its average loss of energy equals the amount gained between collisions. In this way, by first "heating itself up", the work done by the electrical force driving the electrons is eventually passed on to the rest of the medium (to the lattice ions of the conductors and then to the air molecules with which the conductors are in contact) as random kinetic energy (vibrations), or heat[22].

What is temperature?

Most of us would have done several calculations in secondary school and pre-University Physics level involving temperature and heat. Seldom do we pause to understand what temperature is ....apart from some vague notion that temperature is somehow related to energy !

So what really is temperature? Temperature *partitions* energy. If you put the magnitude of energy of a single atom on a scale, then temperature can be used to "scribe" the energy (kinetic, the incessant movement of atoms and molecules) scale into partitions of constant energy !

The **Boltzmann constant** (k or k_B) is the physical constant relating temperature to energy. It is named after the Austrian physicist Ludwig Boltzmann, who made important contributions to the theory of statistical mechanics (because millions of atoms and molecules are involved), in which this constant plays a crucial role. Boltzmann's constant k is a bridge between macroscopic and microscopic physics; a proportionality factor that scales empirical temperature to the microscopic mean kinetic energy.

The microscopic properties of molecules introduces Boltzmann's constant

$$pV = NkT$$

where, p = pressure, V = Volume, N is the number of molecules of gas, and k is Boltzmann's constant 1.380×10^{-23} J/K, T = absolute temperature in K.

Thermionic Emission

Thermionic emission is the heat-induced flow of charge carriers from a surface or over a potential-energy barrier. This occurs because the thermal energy given to the carrier overcomes the forces restraining it. The charge carriers can be electrons or ions, and in older literature are sometimes referred to as "thermions".

In vacuum tubes, a **hot cathode** is a cathode electrode which emits electrons due to thermionic emission. Hot cathodes are the main source of electrons in electron guns in cathode ray tubes, electron microscopes, vacuum tubes, and fluorescent lamps.

Heating the metal vibrates the lattice structure vigorously increasing its kinetic energy which is passed on to the electrons by collisions. This results in a copious emission of electrons. In the analogy of the cricket ball (lattice ion) and marble (electron) collision (see Section1.18), heating a metal causes the heavier cricket balls to transfer their energy to the lighter electrons. (Applying an electric field to cause emission of electrons is very difficult because of the tremendous field strength and uniformity in strength with distances that will be required to cause the electrons to be ripped off.)

After emission, a charge will initially be *left behind in the emitting region* that is equal in magnitude and opposite in sign to the total charge emitted. But if the emitter is connected to a battery, then this charge left behind will be neutralized by charge supplied by the battery, as the emitted charge carriers move away from the emitter, and finally the emitter will be in the same state as it was before emission.

The thermionic emission of electrons is also known as thermal electron emission. Hot cathodes may be either **directly heated**, where the filament itself is the source of electrons, or **indirectly heated**, where the filament is electrically insulated from the cathode; this configuration minimizes the introduction of hum when the filament is energized with alternating current. The filament is most often made of tungsten.

With indirectly heated cathodes, the filament is usually called the *heater* instead. The cathode for indirectly heating is usually realized as a nickel tube which surrounds the heater.

Einstein Relation

To gain an intuitive insight into the Einstein relation let us perform a thought experiment. Supposing we remove a bunch of electrons from a bunch of positive charges in a neutral body B1 a small distance x, such that the potential energy of the system is a value P. Mark the energy P and the voltage V of the separated charges on a scale.

Change the temperature of a body $B2$ to a temperature T from 0°K and mark the energy of the body kT on a scale.

Fix the temperature T of the body $B2$ and separate the charges of the body $B1$ to a distance until their potential energy $P = kT$ the energy of the body $B2$.

Proceed to do this for several values of the temperature T of the body $B2$, plotting T and kT against corresponding P and voltage V of the separated charge centers always maintaining $P = kT$ for the two bodies $B1$ and $B2$.

We will observe that the relationship between the voltage of the charge pairs in body $B1$ and the temperature of the body $B2$ is linear and, when the temperature of the body is 25deg Centigrade then the corresponding voltage (V_T) of the separated charge centers is 25.7 millivolts. The relationship between the voltage and the energy will be governed by $V_T = kT/e$ and V_T is called the *volt-equivalent of temperature*.

In the above processes of charge separation and the increase of the kinetic energy of the particles, the values we would have recorded are essentially *statistical* averages because of the large number (millions and millions!) of particles involved in the above thought experiment.

For example, consider a long semiconductor bar that is at equilibrium in which no net current flows. The role of thermal energy is to knock charges around (with an average energy kT, but having a typical Gaussian-like distribution). Some with higher energy (the mobile charges have a mobility μ) can fly out farther against the electric field created from separated charge centres.

Thus, there is a tug-of-war between the thermally knocked charge diffusing out, with drift induced by the electric field that results from the charge centers separating from each other.

Since, both diffusion D and mobility μ are *statistical* thermodynamic phenomenon, D and μ are not independent. The relation between them is given by the Einstein relation:

$$D_p/\mu_p = V_T \text{ (for holes)}$$

and,

$$D_n/\mu_n = V_T \text{ (for electrons)}$$

that relates the diffusion constant D and mobility μ to the volt-equivalent of temperature, V_T.

Appendix C

SURFACE VOLTAGE AND FIELD STRENGTH OF INSULATORS

By definition, insulators do not have a voltage. Before we understand how the charge distribution inside an insulator prevents us from analyzing an insulator's voltage profile, it is instructive to see what really makes a conductor (made of copper, aluminium or steel or any metal) an equipotential object. The description following is reprinted with permission from Compliance Engineering, "Surface Voltage and Field Strength: Part I, Insulators", September/October 2001. Copyright © 2001 Canon Communications LLC. The original article is by Niels Jonassen, DSc of the Technical University of Denmark and is available for viewing at http://www.ce-mag.com/archive/01/09/mrstatic.html.

Figure 1 shows an insulated conductor *A*, which is a conductor surrounded by insulation to prevent current leakage or short circuits, with a charge *q*.

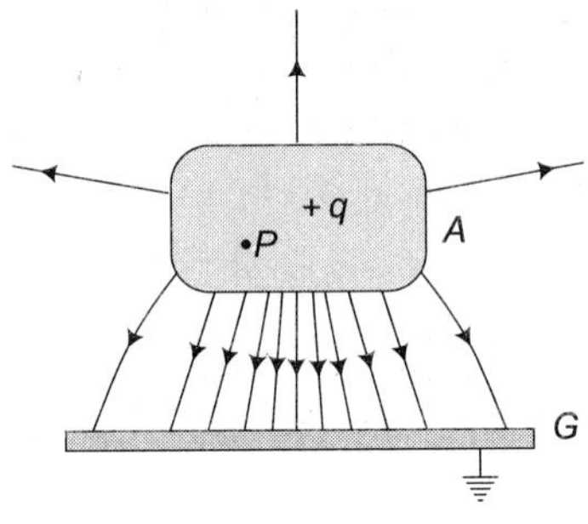

Fig. 1 Charged conductor.

The charge will automatically distribute itself on the surface of the conductor in such a way that the field in the interior of the conductor will be zero, the field will be perpendicular to the surface, and the integral of the field strength *E* from any point *P* in or on the conductor to a ground point *G* is constant and given by

$$V = \int_P^G E.da \qquad ...(1)$$

where *V* is the voltage or potential of the conductor.

The potential will progressively decrease at points farther and farther away in space from the surface of the conductor.

Interpret the equation 1 like this: it would take more work to push a test unit positive charge a distance Δa from some point in space towards the conductor….against the repulsive force of the positive charge (encapsulated in the field E….also recollect from Chapter 1 the visualization using springs that push like charges apart) on the conductor….as the test unit charge gets nearer …you would need to do more work to push a distance Δa towards the conductor against the repulsive force …..and nearer…..more work…..until you reach the conductor.

V will remain the same whichever way you approach because the field strength around the conductor has been adjusted by a trick of nature….thin from far away, like infinity and, thick from the grounded surface which is nearer!

The voltage V and the charge q are proportional, and q is usually written as

$$q = C.V \qquad ...(2)$$

where C is the capacitance of the insulated conductor and is determined by the conductor's size and shape, and its placement relative to other conductors and ground.

The charged system stores an electrostatic energy of

$$W = \frac{1}{2}\ C.V^2 \qquad ...(3)$$

which can be dissipated in a single discharge or current pulse. Figure 2 shows a charged insulator B.

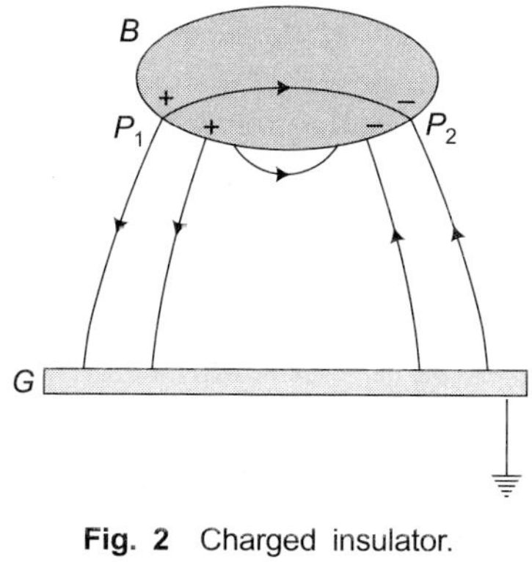

Fig. 2 Charged insulator.

The field conditions here are very different from those at a charged conductor.

The polarity of the charge may be different from point to point, the field in the interior may be different from zero, the field is not necessarily perpendicular to the surface, and the integral of the field strength from a point on or in the insulator to ground is usually different from point to point.

In Figure 2, the integrals of the field strength for P_1 and P_2 are

$V_{P1} = \int_{P1}^{G} E.da > 0$ and $V_{P2} = \int_{P2}^{G} E.da < 0$ respectively and which can be vastly different. V_{P1} and

V_{P2} are the surface voltages (or surface potentials) of the two points.

In general, the surface voltage of an insulator will vary from point to point, as will the potential of any point in the interior.

It is therefore not possible to characterize a charged insulator with a potential of single value.

In other words, an insulator *does not have a single value of potential throughout* its volume and surface.

There are cases in which the surface of an insulator has a constant surface voltage.

But apart from such instances, rare as they are, there is only one situation in which all points in and on an insulator can be ascribed as well-defined (but unmeasurable) voltage. This is shown in Fig. 3.

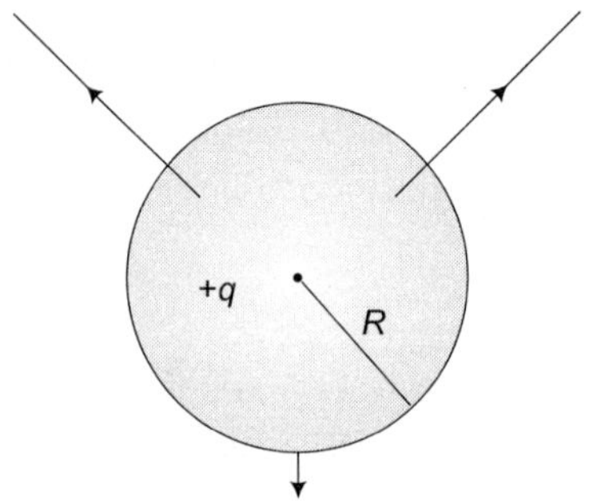

Fig. 3 Uniformly charged, spherical insulator.

If a spherical insulator with radius R and uniform charge q as in Figure 3, is placed infinitely far (a distance much greater than R) from any conductors, the sphere would have a voltage of

$$V_{\text{sphere}} = \frac{q}{4\pi\varepsilon_o R} \qquad \text{...(4)}$$

However, this very theoretical situation is the only case in which it makes sense to talk about the voltage of an insulator. Similarly, the concept of an insulator's capacitance is meaningless.

Although it is possible to get a discharge from a charged insulator, the discharge will always be partial, and the energy dissipated can neither be related to the total charge nor be related to any kind of voltage.

In other words, voltage and capacitance are quantities of a conductor, not an insulator.

Dielectric in a Uniform Electric Field

When a sphere of a homogeneous linear dielectric material is placed in an otherwise uniform field E_o, the electric field flux lines will be as shown in Fig. 4.

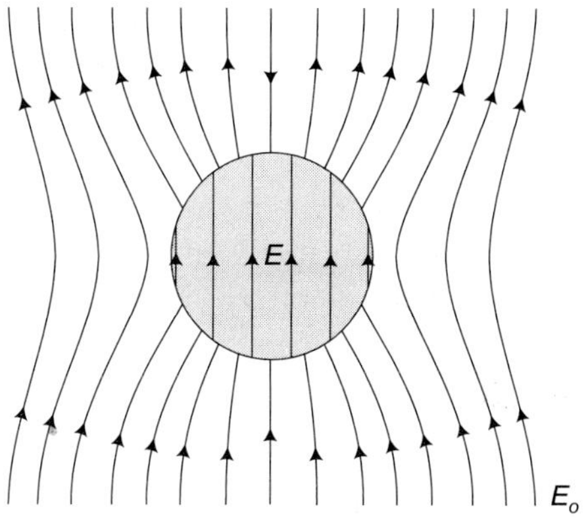

Fig. 4 Dielectric sphere in an otherwise uniform electric field E_o.

Recollect how polarization can result in tiny dipoles (atomic or molecular) in the dielectric material as well as on the surface as was described in Section 1.14.

In dielectrics, all charges are attached to specific atoms or molecules – they're on a tight leash, and all they can do is move a bit within the atom or molecule.

Such microscopic displacements are not as dramatic as the wholesale rearrangement of charge in a conductor, but their cumulative effects account for the characteristic behaviour of dielectric materials.

The reason there is a field inside the sphere is due to the fact that the cancellation of the external field E is only partial.

Physicists study polarized material that is, an object containing a lot of microscopic diploes lined up using the concept of *bound volume charges* and surface charges.

The field of a polarized object is identical to the field that would be produced by a certain distribution of *bound* charges and this equivalence makes it easier to compute the fields both inside and outside objects that are regular and symmetrical in shape such as rectangular boxes and spheres.

You may have noticed that the field in Fig. 4 is uniform in the interior of the sphere. The reader is referred to "Introduction to Electrodynamics" by David J. Griffiths for a calculation of the internal field

$$E = \frac{3}{\varepsilon_r + 2} E_o \text{ where } \varepsilon_r \text{ is the relative permittivity of the dielectric.}$$

The effect of "cancellation" of the electric field within the dielectric is what makes manufacturers build capacitors of small dimensions with a dielectric between its plates that can store larger amounts of energy.

When the space between the plates of a capacitor is filled with a dielectric, the dielectric will reduce E and hence also the potential difference V.

Accordingly, the capacitance $C = Q/V$ is increased by a factor of the dielectric constant $C = \epsilon_r C_{vacuum}$ where ϵ_r is the relative permittivity of the dielectric and C_{vacuum} is the capacitance of the capacitor without the dielectric.

The formation of the dipoles being not so well-defined in the bulk in dielectrics, is one among the many reasons why it is difficult to characterize the no-load output voltage (without a filter capacitor) of a half-wave rectifier circuit made using a diode when subjected to an input sinusoidal voltage.

There is a transition region in the diode that is devoid of mobile charges which, when treated as a dielectric that is already polarized (that caused the barrier potential as described in Section 9.4) makes the computation of the electric field and the potential quite difficult.

Conductor in a Uniform Electric Field

In sharp contrast to the effect of a dielectric placed in a uniform electric field, when an uncharged conducting sphere is placed in a uniform electric field, the field due to the induced charge (only surface, no volume charge) completely cancels E_o within the sphere.

This effect is a result of the movement of mobile charges in a conductor. This is shown in Fig. 5.

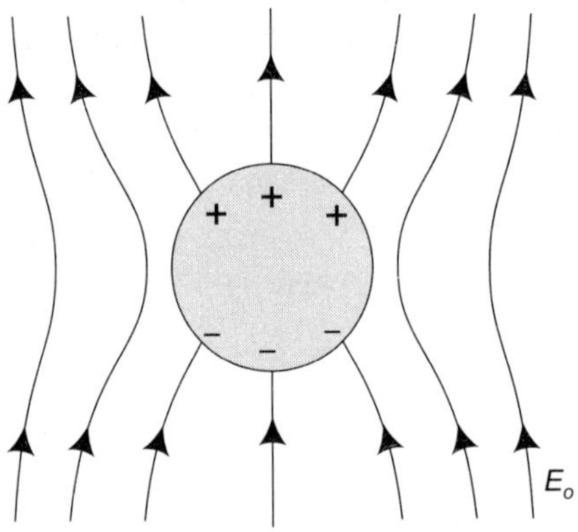

Fig. 5 Uncharged conducting sphere is placed in an otherwise uniform field E_o.

What are insulators and how do they prevent electric shocks?

An insulator, also called a *dielectric*, is a material that resists the flow of electric current. An insulating material has atoms with *tightly* bonded valence electrons. These materials are used in parts of electrical equipment, also called *insulators* or insulation, intended to support or separate electrical conductors without passing current through themselves. The term insulator is also used more specifically to refer to insulating supports that attach electric power transmission wires to utility poles or pylons.

Some materials such as glass, paper or Teflon are very good electrical insulators. A much larger class of materials, for example rubber-like polymers and most plastics are still "good enough" to insulate electrical wiring and cables even though they may have lower bulk resistivity. These materials can serve as practical and safe insulators for low to moderate voltages (hundreds, or even thousands, of volts).

It is not true that an electrical insulator such as the one that is used on an electrician's tester, "blocks" the voltage from being felt by its user; in reality, the voltage on the insulated handle is difficult to calculate and measure and is rarely blocked entirely. It is an insulator's high resistivity to *current* (to less than a few thousandth of a milliamp), that enables it to prevent electric shocks to personnel, because it is the currents within the range of a fraction of a milliamp to a few milliamps that is the cause of electric shocks leading to fibrillation and fatalities.

A piece of wood, which is an insulator, in contact with a live wire, will have an internal and surface electric field profile (and voltages at various points in its interior and on its surface) that is *same*, whether *dry or wet.* A person who touches the surface of the wood will not receive a shock when the wood is dry, but may feel a shock when the wood is wet because of the availability of mobile charge carriers that cause currents in the wood and through his body.

Why does a balloon charged by rubbing in one's hair not give a shock?

The voltage of the balloon may be quite large, perhaps several thousands of volts and would be several thousand joules of energy if the charge were 1 coulomb. But, the amount of charge that is transferred on to the balloon is typically less than a millionth of a coulomb. A high voltage means a lot of energy only if a lot of charge were involved. We also see from this an important difference between electric potential energy and electric potential.

Appendix D

DO WE REALLY TOUCH AN OBJECT

Before we discuss what really happens when we touch an object, let us understand the four fundamental forces in nature.

There are four known fundamental forces and these are listed below in the order in which these occur: beginning from the innermost regions of an atom (inside the nucleus) and moving outwards to encompass the entire universe.

1. **Weak Force:** This causes Beta decay (the conversion of a neutron to a proton, an electron and an antineutrino) and various particles (the "strange" ones) are formed by strong interactions but decay via weak interactions. The weak force **is a short range force** whose range is of the order of 10^{-16} centimetres. ***Not of much concern to electrical and electronics engineers**.

2. **The Strong Force:** This force binds neutrons and protons together in the cores of atoms. The strong interaction is very strong, and like the weak force, is **short-ranged**. It acts only over ranges of order 10^{-13} centimetres and is responsible for holding the *nuclei of atoms* together. It is basically attractive, but can be effectively repulsive in some circumstances. ***Not of much concern to electrical and electronic engineers**.

3. **Electromagnetic:** This acts between electrically charged particles or pieces of matter carrying an electrical charge. Electricity, magnetism, and light are all produced by this force and it has **infinite range. It can be attractive or repulsive,** and acts only between pieces of matter carrying electrical charge. ***of concern to scientists, mechanical, civil, chemical, metallurgical, electrical and electronic engineers**.

   This force can be separately studied in electrostatics and magnetostatics, but is always treated together as a *fundamental force*.

   Taken singly, the fields store energy: electric energy between the plates of a charged capacitor and magnetic energy in the field of a bar magnet or magnetic energy in the field surrounding a steady current carrying shorted *pure* inductor.

   Together, they transport energy (electromagnetic) as in circuits and as in radiation (a vibration of pure electromagnetic energy).

4. **Gravity:** This force acts between all mass in the universe and like the electromagnetic force it also has **infinite range**. ***of concern to all engineers and scientists**. The *gravitational force* is weak, but very long ranged. Furthermore, it is always attractive.

The weak and electromagnetic interactions have been unified under electroweak theory (Glashow, Weinberg, and Salaam were awarded the Nobel Prize for this in 1979).

Grand unification theories attempt to treat both strong and electroweak interactions under the same mathematical structure; attempts to include gravitation in this picture have not yet been successful.

Important Note: All the other forces are derived from these four fundamental forces; for example, frictional forces and reaction forces that prevent you from falling from your chair when you sit to read these notes are electromagnetic in origin !

You may use one type of force to overcome another ! **Example:** An electromagnet uses electromagnetic forces to lift a car which is resting on the ground due to the force of gravity. Thank nature for permitting us to do that !

What happens when an object is touched?

At the atomic level? What's in a Touch?

When two physical surfaces make contact, we say they are 'touching' each other. This contact may make it appear that two different surfaces are actually touching, but in fact, no atoms of any *one object can ever touch the atoms of another*![23]

*The man on **the left*** is leaning on a wall, which his jacket is 'touching'. At least, it seems that way. But when you look very closely at the two surfaces 'in contact', and understand a little about what atoms are like, you discover that

no jacket atoms are touching any wall atoms at all!

Here's a simplified explanation of why this is so: An atom could be described as a positively charged nucleus, surrounded by a cloud of negatively charged electrons. The tangential velocity keeps the negatively charged electrons in stable orbits, or actually orbitals, around the positively charged nucleus and prevents them from falling into the nucleus.

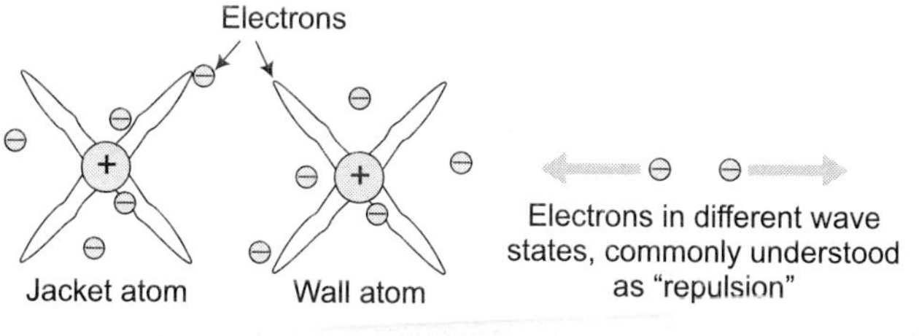

When two atoms are forced into 'contact' or interact closely (here, the jacket touching the wall), a force due to *electromagnetism*, keeps negatively charged electrons from actually making contact with each other. The only interaction is electromagnetic and weak force. There is NO repulsive force keeping them apart. It is the fact that the electron must be in a particular wave state that keeps them apart. The separation is a quantum mechanics effect. Electrons are point particles and cannot touch because they do not have surfaces. An infinitely small object has a surface area = 0. In reality the electron clouds penetrate each other and overlap, so in a sense electrons occupy the same space, but that is not the same thing as touching.

Push as hard as you want; you cannot make two electrons come into contact. The force you apply is about 20 newtons resulting in a stress (force/the area of contact) not even close to 2.3×10^{-8} newtons that is the force of attraction between the electron and the proton as we saw in Appendix B. The electron bound to one atom in "contact" with another atom has a negligibly small probability of being found in the location of the nucleus of the near atom. So the only thing actually 'in contact' is *the electromagnetic repulsive force between* the electrons, and there is always space between them! When the letter blocks are placed one above the other, electrons of the contact surface atoms of the upper and lower blocks overlap such that the 'normal' force of the lower block counters (by repulsive interactions of overlapping electrons) the force (say, gravitational or hand pressure on the upper block) exerted by the upper block.

It's as if you are pushing with similar poles of two very strong bar magnets; you can push one around with the other, but can never force them together.

Even standing on the floor, you are really supported by a very thin layer of electromagnetic force; no atoms of your foot actually touch atoms of the floor! The electron clouds of your shoe actually interpenetrate the electron clouds of the floor atoms, and the electron clouds of the blocks interpenetrate each other, so one might say they do touch. But in reality since the atom has no definite surface the idea of touch could be misleading. Again, forces do not contact each other! Forces are not physical objects. Forces are interactions between physical objects and so can be considered pushes or pulls on physical objects. Forces can be considered to add together and are independent of each other.

But wait a minute ... what about when you cut something!?!!

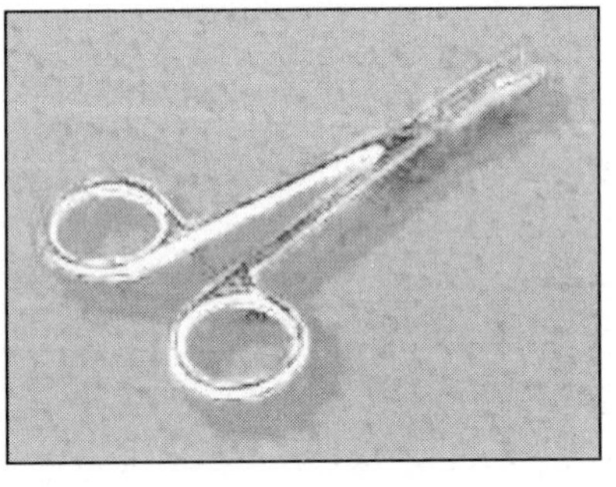

Don't the scissors blades have to touch what they're cutting, in order to cut?

No, they don't. Atoms can't touch, remember? They just push against each other with an electromagnetic force. That's what happens with the scissors:

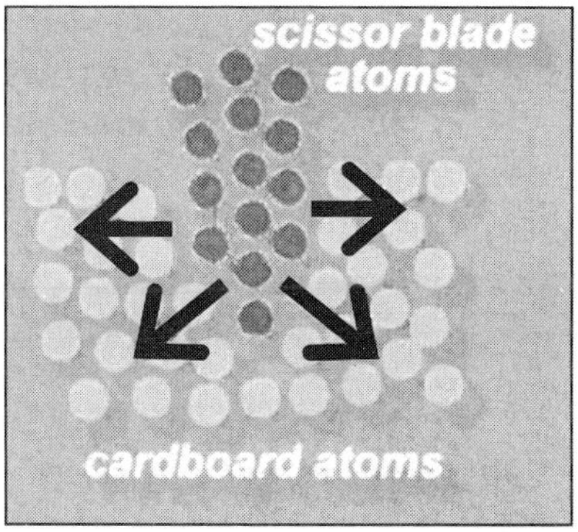

Even the sharpest knife doesn't 'cut' ... it pushes apart the atoms of the material it's being used on. The atoms of the knife or scissors blade push electromagnetically against the atoms of c ardboard (see the figure above), spreading them apart. Even when the blade is withdrawn, and the material rejoins (as with flesh cut by a scalpel), the sealing is because of the positive and negative attraction between atoms and molecules with net charges (chemistry!) ... but this is not strong enough to force electrons together, just as it wasn't in the original uncut material.

How do we feel things if we are not touching them at the atomic level?

You may be wondering how you can feel things, if you don't actually make contact with anything. For example, when you pick something up in your fingers, you can certainly 'feel' the contact !

What's happening? The caveman certainly felt the club used by the cavewoman when it made contact with his head ! Let's examine what actually took place when that happened.

Atoms of the club pushed against atoms of his head, and probably moved them around a lot (electromagnetically) !

Buried in the skin are nerve cells or (sensory) neurons that gather and transmit electrochemical signals. Neurons have axons which are long, thin, cable-like projections of the cell that carry electrochemical messages (nerve impulses or action potentials of the order of a few tens of millivolts) along the length of the cell.

The nerve endings are dendrites which by a method of changing action potentials carry messages between the billions of (receptor) neuron dendrites in our bodies, eventually reaching the brain, where other cells interpret this electrical signal as 'pain'.

At no time did any atoms or electrons actually touch each other; the only 'contact' was the electromagnetic force (positive and negative) between particles !

Friction is the effect of atoms pushing against each other. On a smooth surface, with few bumps, there is little friction. Most of the atoms can slide right by each other without actually pushing *against* each other. But, no surface is totally free of friction. Watch the man skiing on snow.

Man skiing on snow

If you examine any surface under a powerful enough microscope, no matter how smooth it appears to the naked eye, there will be bumps. This will cause atoms to come face-to-face.

If the forces between one set of atoms (snow, in this case), are weak, those atoms get pushed out of the way (by the ski atoms, in this case). But this uses up energy. It's called 'friction'.

If the atoms are joined by strong forces, as in rock, then they can't easily be pushed away. The forces between these atoms and your ski atoms, supposing that you were skiing on rock, might be enough to bring you to a stop!

In either case, it's atoms pushing against each other, at a distance, (electromagnetically), that causes friction. Generally, the energy "dissipated" by friction remains as molecular motion or what is called heat.

When we hit an object, the physical force felt would be the normal force, this force is an electromagnetic force, the electrons in the object repelling the electrons in the hand.

When a batsman hits a cricket ball, the force involved is electromagnetic; the electrons in the bat repelling the electrons in the ball.

The idea of touch appears to be so simple using classical mechanics concepts but is the quantum mechanical concept consistent with the classical descriptions?

While it is true that theoretically electrons cannot touch because of the repulsion, and of course the nuclei do not touch each other, the surfaces of the atoms (defined in some mathematical fashion) can touch and actually penetrate each other. Forces are interactions between two objects.

All atoms are not solid and have no definite surfaces. One can make a mathematical definition of where the surface is located, but that is subject to interpretation. Actually when you touch something, bonds form between the two surfaces so in a sense sub-microscopically the two surfaces penetrate each other slightly. So sub-microscopically there is no gap.[24]

Remember that outside an atom there is no static electrical field because the forces due to the protons and electrons are balanced on any nearby object. But if you think about it classically there is only a repulsive force if the "edges" of the atoms overlap each other.

The model that you use for physics depends on the scale; at a macroscopic level when objects touch, use classical mechanics but sub-microscopically use quantum mechanics and here classical concepts such as touch can no longer be applied.

Note:

The electrons in an atom have a probability distribution which is non-zero at the nucleus, that is, there is a non-zero (though small) probability of finding an atomic electron inside the nucleus of the atom.

It is possible to make electrons interact with the protons in a nucleus. This is accomplished artificially by making them interact due to their high momentum and energy at relativistic speeds in a cyclotron.

One of the things that can happen when the electron hits a proton is that they can combine to create a neutron. This process also occurs in a neutron star, where the intense gravitational pressure collapses all the atoms of the star, so that all the electrons combine with all the nuclear protons, forming a solid soup of neutrons.

References

1. Ruth W. Chabay and Bruce A. Sherwood, *"Unified treatment of electrostatics and circuits",* Center for Innovation in Learning and Dept. of Physics, Carnegie Mellon University.
2. Ruth W. Chabay and Bruce A. Sherwood, *"Electric and Magnetic Interactions",* John Wiley, USA, 1995.
3. Sherwood Parker, "*Electrostatics and Current flow*", American Journal of Physics, Vol. 38, June 1970.
4. Paul G. Hewitt, "*Conceptual Physics*", Little, Brown and Company, Boston, USA.
5. Paul Falstad, "*Electrostatics Applets*", http://www.falstad.com/vector2de/
6. Hermann Häertel, *"Constant electric current and the distribution of surface charges1",* Institute for Science Education, University Kiel.
7. MIT Open Course Ware, Massachusetts Institute of Technology, Massachusetts, USA.
8. David J. Griffiths, "*Introduction to Electrodynamics*", Prentice-Hall, India, 1999.
9. Norris W. Preyer, *"Surface charges and fields of simple circuits",* American Journal of Physics, Volume 68, No. 11, November 2000.
10. J.D. Jackson, *"Surface charges on circuit wires and resistors play three roles",* American. Journal of Physics, Volume 64, pp. 855-870, 1996.
11. William J. Beaty, http://amasci.com/miscon/eleca.html.
12. Lecture presentations from the Massachusetts Institute of Technology, Boston, USA
13. David Haliday, Robert Resnick, Jearl Walker, *"Physics"*, Wiley India, 2001.
14. The Physics Classroom Tutorial – Glenbrook South Physics (http://www.glenbrook.k12.il.us/gbssci/phys/phys.html)
15. Edward Hughes, "*Electrical Technology*", Longman, 7th edition, 1995.
16. Edward Jordan and Keith Balmain, "*Electromagnetic Waves and Radiating Systems*", Prentice-Hall of India, 2nd edition, 2003.
17. Jacob Milman, Christos Halkias, Satyabrata Jit, "*Electronic Devices and Circuits*", Tata Mc-Graw-Hill, India.
18. Bart Van Zeghbroeck, http://ecee.colorado.edu/~bart/book, "*Principles of Semiconductor Devices*", 2007.
19. Richard F.W. Bader, "*An introduction to the electronic structure of atoms and molecules*", http://www.chemistry.mcmaster.ca/esam/Chapter_1/intro.html, Department of Chemistry, McMaster University, Hamilton, Ontario, Canada.

20. Paul Malvino, "Electronic Principles", Tata McGraw-Hill, India.
21. George Kennedy and Bernard Davis, "Electronic Communication Systems", Tata McGraw-Hill, India.
22. Edward Purcell, "Electricity and Magnetism", Tata McGraw-Hill, India.
23. Worsley School Online, Alberta, Canada, "What's in a touch". http://www.worsleyschool.net/science/files/touch/touch.html.
24. John M. Clement and John Denker, "[Phys-l] Fields and touch", Archives of forum replies https://carnot.physics.buffalo.edu/archives/2010/2_2010/threads.html.
25. Fermilab, U.S. Dept. of Energy, "Inquiring minds" http://www.fnal.gov/pub/inquiring/questions/bob.html.
26. Elementary Physics II, Boston University, "Electric energy and potential" http://physics.bu.edu/~duffy/PY106/Potential.html.
27. Physics and Astronomy, University of St. Andrews, "Energy and moving electrons" http://www.st-andrews.ac.uk/~www_pa/Scots_Guide/audio/part6/page2.html.
28. Steve Wirt, "Introduction to Magnetism" http://www.regentsprep.org/Regents/physics/phys03/cintromag/
29. Lee, Y.C. and Mei, W.N. (1996), Self-inductance of a long wire: infinite multiple loop approach, International Journal of Mathematical Education in Science and Technology, 27:1,79-86.
30. Ben G. Streetman and Sanjay Banerjee, "Solid State Electronic Devices", Pearson Prentice Hall, 2006.

Images

1. **Chapter 1**

   Structure of metals and conductors

   mysite.verizon.net/kdrews47/solids/metal.gif

   Electric field of a point charge

   http://www.falstad.com/vector3de/ (reproduced with permission)

2. **Chapter 2**

   Potential and equipotential lines of a dipole

   http://www.falstad.com/vector3de/ (reproduced with permission)

   Electric field lines of a capacitor

   http://www.falstad.com/emstatic/ (reproduced with permission)

Index

F

G